Nutraceuticals

The global nutraceutical and health supplement market is experiencing unprecedented growth, driven by increasing consumer awareness and evolving healthcare needs. However, navigating the diverse and fragmented regulatory landscape across regions poses a significant challenge.

Nutraceuticals: Global Regulatory Status and Challenges provides a comprehensive overview of global regulatory frameworks, empowering professionals to understand and comply with varying standards. From ingredient approval to labeling requirements, this book offers critical insights into key markets, as well as emerging markets.

Features

- Detailed analyses of regulatory frameworks, terminology, and safety aspects.
- Expert perspectives on global trends and country-specific guidelines.
- Strategic insights for product development and market forecasting.

Nutraceuticals: Global Regulatory Status and Challenges is an indispensable guide for regulatory professionals, R&D teams, marketers, and business leaders seeking to capitalize on the global nutraceutical market.

Nutraceuticals: Basic Research and Clinical Applications

Series Editor
Yashwant Pathak, PhD

Nutraceuticals for Prenatal, Maternal and Offspring's Nutritional Health,
Priyanka Bhatt, Maryam Sadat Miraghajani, Sarvadaman Pathak, Yashwant Pathak

Bioactive Peptides: Production, Bioavailability, Health Potential and Regulatory Issues,
edited by John Oloche Onuh, M. Selvamuthukumaran, Yashwant Pathak

Nutraceuticals for Aging and Anti-Aging: Basic Understanding and Clinical Evidence,
edited by Jayant Lokhande, Yashwant Pathak

Marine-Based Bioactive Compounds: Applications in Nutraceuticals,
edited by Stephen T. Grabacki, Yashwant Pathak, Nilesh H. Joshi

Applications of Functional Foods and Nutraceuticals for Chronic Diseases: Volume I,
edited by Syam Mohan, Shima Abdollahi, and Yashwant Pathak

Flavonoids and Anti-Aging: The Role of Transcription Factor Nuclear Erythroid 2-Related Factor2,
edited by Karam F.A. Soliman and Yashwant Pathak

Molecular Mechanisms of Action of Functional Foods and Nutraceuticals for Chronic Diseases: Volume II,
edited by Shima Abdollahi, Syam Mohan and Yashwant Pathak

Plant Food Phytochemicals and Bioactive Compounds in Nutrition and Health,
edited by John O. Onuh, Yashwant Pathak

Anxiety, Gut Microbiome, and Nutraceuticals: Recent Trends and Clinical Evidence,
edited by Yashwant Pathak, Sarvadaman Pathak, and Con Stough

Wealth out of Food Processing Waste: Ingredient Recovery and Valorization,
edited by M. Selvamuthukumaran

Nutraceuticals Inspiring the Contemporary Therapy for Lifestyle Diseases,
edited by Mala Trivedi, Sachidanand Singh, Parul Johri, Pedro Lopez-Sanchez

Medicinal Fruits of the Middle East,
edited by M. Selvamuthukumaran

Nutraceuticals: Global Regulatory Status and Challenges,
edited by Yashwant Pathak and Niyati Acharya

For more information about this series, please visit:
https://www.crcpress.com/Nutraceuticals/book-series/CRCNUTBASRES

Nutraceuticals
Global Regulatory Status and Challenges

Edited by
Yashwant Pathak and Niyati Acharya

CRC Press
Taylor & Francis Group
Boca Raton London New York

CRC Press is an imprint of the
Taylor & Francis Group, an **informa** business

Designed cover image: Shutterstock

First edition published 2026
by CRC Press
2385 NW Executive Center Drive, Suite 320, Boca Raton FL 33431

and by CRC Press
4 Park Square, Milton Park, Abingdon, Oxon, OX14 4RN

CRC Press is an imprint of Taylor & Francis Group, LLC

ISBN: 978-1-032-72984-8 (hbk)
ISBN: 978-1-032-73602-0 (pbk)
ISBN: 978-1-003-46504-1 (ebk)

DOI: 10.1201/9781003465041

Typeset in Times
by Apex CoVantage, LLC

Contents

Foreword vii
About the Editors ix
List of Contributors xi

Chapter 1 Nutraceuticals: Historical Perspectives in Different Countries 1

Mandar Kulkarni, Khushi Pandya, Niyati Acharya, and Yashwant Pathak

Chapter 2 Understanding the Different Systems of Diverse Terminologies for Dietary Supplements and Related Products 24

Khushi Pandya, Niyati Acharya, and Yashwant Pathak

Chapter 3 Regulatory Framework for Nutraceutical Registration in CARICOM 43

Sandeep B. Maharaj, Darren Dookeeram, and Nila-Marie Ramdeo

Chapter 4 Safety and Toxicity Aspects for Nutraceuticals and Related Products: Regulatory Provisions across the Globe 54

Jiya Patel, Aabir Pramanik, Sonali Maheshwari, Niyati Acharya, and Yashwant Pathak

Chapter 5 Regulatory Requirements and Challenges for Nutraceuticals in the United States... 82

Anita Jacob and Yashwant Pathak

Chapter 6 Nutraceuticals and Regulatory Compliance in Latin America 96

Juan Montilla, Niyati Acharya, and Yashwant Pathak

Chapter 7 The UK and the Medicines and Healthcare Products Regulatory Agency (MHRA): Regulatory Requirements and Challenges for Nutraceuticals 109

Ramesh Pennamareddy, Shivkumar Somani, and Kandi Purushothama Reddy

Chapter 8 Regulatory Frameworks and Guidance for Food Products Registration, Safety, Labeling, and Consumer Protection Guidelines: A Comprehensive Review on German Standards 120

Nikunjana Patel, Nishit Parekh, and Sanjeev Acharya

Chapter 9 Regulatory Landscape of Nutraceuticals in the European Union: Compliance and Standards 147

Arindam Chatterjee, Mayank Bansal, Rakesh Kumar Gupta, Devid Chutia, Deepak Sharma, and Madan Mohan Gupta

Chapter 10 Regulatory Requirements for South Africa for Marketing and Importing Nutraceuticals, Herbal Products, and Functional Foods 163

Pravin Semnarayan, Deon du Rand, and Megan Saffer

Chapter 11 Regulatory Provisions for Food Supplements in India 178

Maitreyi Zaveri and Bhavita Dhru

Chapter 12 Navigating Regulatory Requirements: A Comprehensive Guide to Nutraceuticals and Traditional Chinese Medicine 192

Risha Fillah Fithria and Jinfu Peng

Chapter 13 Nutraceutical Regulation in Indonesia 215

Ayu Shabrina

Chapter 14 Complementary Medicines: Australian Regulatory Requirements, Market Trends, and Challenges in the Australian Market 232

Aditi Kirtane

Chapter 15 A Comprehensive Guide to South Korea's Food Regulation 243

Sahil Sorathiya and Hardik Bhatt

Chapter 16 Navigating the Maze of Regulatory Guidelines for Nutraceuticals in the Gulf Countries 271

Nikunjana Patel, Tasneem Rangwala, Niyati Acharya, and Sanjeev Acharya

Chapter 17 Global Future Trends Related to Nutraceuticals 305

Aabir Pramanik, Sonali Maheshwari, Niyati Acharya, and Yashwant Pathak

Chapter 18 Nutraceuticals in Czech Republic 330

Ales Franc and Vladimír Bíba

Index 341

Foreword

It is with immense pleasure and a deep sense of appreciation that I write this preface for the book *Global Regulatory Requirements and Challenges for Nutraceuticals*, an expertly curated and scholarly compilation edited by Dr. Niyati Acharya and Dr. Yashwant Pathak. My association with Dr. Pathak dates back to 1975, when we were both pursuing our master's degrees at the Department of Pharmaceutical Sciences, Nagpur University. Since those formative years, our professional paths have intertwined, and our friendship has remained steadfast. I am especially proud to see how our shared passion for pharmaceutical sciences has led to significant contributions, like this one, which address a topic of growing global importance.

The global nutraceutical market has witnessed phenomenal growth in recent years. Valued at USD 383.69 billion in 2022, it is projected to grow to USD 868.38 billion by 2031, at an impressive compound annual growth rate (CAGR) of 9.50% from 2024 to 2031. This burgeoning market comprises diverse categories, such as dietary supplements, functional foods, medicinal foods, and pharmaceuticals. Dietary supplements, for example, are concentrated sources of nutrients derived from food products, delivered in convenient forms, such as capsules, powders, or pills.

India, with its rich heritage in traditional medicine and natural healing, has emerged as a significant player in the nutraceuticals sector. The country's age-old knowledge systems, such as Ayurveda and Siddha, have laid a strong foundation for the development of nutraceutical products that blend ancient wisdom with modern scientific advancements. As a result, Indian nutraceuticals have gained widespread acceptance globally, especially in markets that value natural and holistic health solutions. The Indian nutraceutical market is expected to grow substantially, driven by increased health awareness, rising disposable incomes, and a robust demand for preventive health-care solutions.

Despite the significant opportunities in the nutraceutical industry, companies face stringent regulatory challenges worldwide, including in India. The regulatory landscape in India is overseen by the Food Safety and Standards Authority of India (FSSAI), which enforces comprehensive guidelines to ensure product safety and efficacy. These regulations are evolving rapidly to keep pace with global standards while accommodating the unique aspects of traditional Indian nutraceuticals. Compliance with FSSAI guidelines requires rigorous quality control measures, precise labeling, and adherence to safety protocols, reflecting India's commitment to maintaining high standards in this growing industry.

In the United States, nutraceuticals must adhere to several regulatory requirements, which are equally rigorous:

- **Safety.** Manufacturers are responsible for ensuring the safety of their products before they reach consumers. The US Food and Drug Administration (FDA) has the authority to act against any unsafe products that pose a risk to public health.
- **Labeling.** Labels on nutraceutical products must be accurate and non-misleading. They should include a clear statement of identity, net quantity, and necessary product information. Importantly, claims related to the treatment or prevention of specific diseases are strictly prohibited.
- **Notification.** Manufacturers are required to notify the FDA at least 75 days before distributing products containing new dietary ingredients. This notification should detail safety information and describe the ingredients used.
- **Production and quality control.** Consistent batch quality must be ensured through comprehensive production and quality control systems. Detailed records of all processes must be maintained, and products must ideally be manufactured under Good Manufacturing Practices (GMP).
- **Adherence to Good Manufacturing Practices (GMP).** Compliance with GMP is mandatory to ensure product safety and quality.

- **Adverse event monitoring.** Manufacturers must monitor and report adverse events associated with their products, ensuring ongoing safety oversight.

The primary legislation regulating dietary supplements in the United States includes the Federal Food, Drug, and Cosmetic Act and the Dietary Supplement Health and Education Act (DSHEA). Globally, other countries have developed their own regulatory frameworks, often imposing strict compliance requirements for nutraceutical products.

This comprehensive book addresses these critical regulatory aspects, consisting of 18 meticulously researched chapters. It covers the historical evolution of nutraceuticals, clarifies diverse terminologies, and examines safety and toxicity concerns. Furthermore, it provides an in-depth exploration of regulatory requirements across various regions, including the United States, United Kingdom, European Union, CARICOM countries, Latin America, India, China, Australia, South Korea, and other Asian nations.

The chapters devoted to India's regulatory framework offer a detailed analysis of FSSAI guidelines and highlight how Indian manufacturers can navigate both domestic and international regulatory landscapes. Given India's growing prominence in the nutraceutical market, this book serves as an invaluable resource for Indian nutraceutical companies seeking to expand their presence on the global stage. It offers essential insights for navigating the complex regulatory landscape, enabling these companies to effectively position their products in different jurisdictions while capitalizing on India's rich tradition of natural health products.

I extend my heartfelt congratulations to the editors, Dr. Niyati Acharya and Dr. Yashwant Pathak, for bringing this important work to fruition. I also wish to acknowledge and commend the chapter authors for their exceptional contributions, which provide well-researched, authoritative information enriched by their expert commentary.

With warmest regards and best wishes to the editors for the success of this book.

Shirish Belapure

Shirish Belapure holds a post-graduate degree in pharmacy from Nagpur University and a management certification from the Indian Institute of Management (IIM), Ahmedabad. He has experience in managing operations in pharmaceuticals in a wide range of areas, including formulation development, manufacturing, quality assurance, regulatory affairs of API, and formulations, for about 46 years. Currently, he is the senior technical advisor of the Indian Pharmaceutical Alliance and manages the quality forum of the IPA. He is also a venture partner with Eight Roads, a private equity firm specializing in investing funds in health-care-related companies and technologies. Prior to that, he was the managing director of Zydus Hospira Oncology Pvt. Ltd., a joint venture company between Zydus and Pfizer. Earlier experience includes working as the president (global manufacturing), responsible for the entire manufacturing network of formulations and projects globally, of Zydus Cadila. In addition to these, he has also worked with Sun Pharma, Cipla, and Lederle, to name a few, in various capacities. He has several initiatives to his credit, notably the "Strategic Lean Integrated Manufacturing" (SLIM) operational excellence at Zydus Automation to reduce human intervention in the manufacturing processes. He was the president of the PDA India chapter for two years, from December 2019 to November 2021. During this pandemic period, PDA India continued to organize training programs online. He is also associated with Nirma University through the academic council of the Pharmacy Institute, and with Lovely University as a member of its corporate advisory board. His current focus area is quality forum in IPA, where various quality initiatives (like best practice sharing, organizing advanced GMP workshop, a global quality summit, industry–academia collaboration, as well as training for the pharma industry in collaboration with WHO) are undertaken. He is also on the board of several companies as an independent director—to name a few, Jubilant Pharma, Albert David, etc.

About the Editors

Yashwant Pathak, MPharm, Executive MBA, MSCM, PhD, FAAAS Yashwant Pathak received his PhD from Nagpur University, India, and his executive MBA and master in conflict management from Sullivan University, Louisville, USA. He has over 45 years' experience in academia and the industry. Presently, he is a professor and an associate dean for faculty affairs at Taneja College of Pharmacy, University of South Florida, Tampa, USA. Yashwant Pathak is also an adjunct professor at the Faculty of Pharmacy, Airlangga University, Surabaya, Indonesia. Yashwant has published over 400 research papers, reviews, and chapters, with close to 5,000 citations in Google Scholar. He has edited over 70 books in nanotechnology, nutraceuticals, drug delivery systems, artificial intelligence, and artificial neural network. He has keen interest in conflict management and has edited several books, namely, *Organizational Conflict Management*, *Handbook of Organizational Conflict Management*, *Eastern and Indigenous Perspectives in Conflict Management*, *RSS Way of Resolving Conflicts*, and these two volumes on nonprofit organizations conflict management. He is the recipient of several international and national awards, including as Fulbright senior scholar, 2016–2017, at Indonesia, for six months; Fulbright specialist scholar, 2019, at Witwatersrand University, South Africa; adjunct professor, Witwatersrand University, South Africa; adjunct professor, Faculty of Pharmacy, Airlangga University, Surabaya, Indonesia; fellow of the American Association for the Advancement of Science (FAAAS); Fulbright specialist, Universidad Distrital Francisco José de Caldas, Bogota, Colombia, 2023; and Fulbright senior scholar, Masaryk University, Czech Republic, 2024. He is actively involved in many nonprofit organizations, including HSS.org, ICHRRF.org, IACDSC.org, Uberoi Foundation for Religious Studies, and so on. Dr. Pathak recently received the Nelson Mandela Washington Reciprocal Exchange Fellowship for Nigeria, West Africa.

Dr. Niyati Acharya, PhD, MPharm (Pharmacognosy) Dr. Niyati Acharya, currently working as an assistant professor and the head, Department of Pharmacognosy, Institute of Pharmacy, Nirma University, Ahmedabad, has over 22 years of experience in teaching and natural product research, has published 47 research papers, and has 15 book chapters to her credit (1,191 citations, 17 h-index, 22 index). She has completed research projects for the Department of Biotechnology, New Delhi, and GUJCOST, Gandhinagar, of more than 50 lacs and is doing many minor research and consultancy projects for the herbal drug industry and research centers. Recognized as a PG/PhD guide, she has guided more than 63 MPharm and 12 MSc cosmetic technology students; 6 PhD students have been awarded, and 3 are pursuing PhD at Nirma University under her able guidance. She has presented more than 40 oral and 80 poster presentations in various international and national-level programs and delivered expert talks. She is a recipient of the Best Oral Presentation award by the Society of Ethnopharmacology at SFEC 2022, the **P. K. Debnath Memorial Award: SFE Young Ethnopharmacologist Award 2020** for Best Oral Presentation in the Field of Ethnopharmacology, the **GABTP National Woman Scientist Award 2019** at the 29th APSI Scientist Meet and International Conference on Drug Discovery and Development in Agrobiotechnology and Pharmaceutical Sciences, the **Best Research Paper Award** at the International Symposium on Cancer Research at GCRI in 2014, and the **Dr. P. D. Sethi Research Paper Award 2006**. She is a recipient of the Best Innovation Idea under the Rising Startup Idea Award, Biothon 2025, in Gujarat Biotechnology University. The research area of her interest includes phytopharmacology-based investigations and characterization of biomarkers from medicinal plants for the management of wounds, obesity, liver disorders, and neurodegenerative disorders, like Alzheimer's disease, and the development of novel and target-based herbal formulations. She is a life member of various professional bodies, like the APTI, ISP, IPA, ISTE, Society of Ethnopharmacology, Indian Academy of Neurosciences, Indian Society of Chemists and Biologists, Indian Pharmacological Society, and IPGA, and has been working as an editorial member and reviewer for many journals of Elsevier, Taylor & Francis, and Willey.

Contributors

Niyati Acharya
Institute of Pharmacy
Nirma University
Ahmedabad, Gujarat, India

Sanjeev Acharya
Institute of Pharmacy
Ganpat University
Mehsana, India

Mayank Bansal
Jaipur College of Pharmacy
Jaipur, Rajasthan, India

Hardik Bhatt
Institute of Pharmacy
Nirma University
Ahmedabad, Gujarat, India

Vladimír Bíba
Law Office of JUDr. Pharm
Dr. Vladimír Bíba
Prague, Czech Republic

Arindam Chatterjee
Institute of Pharmacy
Assam Don Bosco University Tapesia,
Assam, India

Devid Chutia
Institute of Pharmacy
Assam Don Bosco University
Tapesia, Assam, India

Bhavita Dhru
K. B. Institute of Pharmaceutical Education and Research
Gandhinagar, Gujarat, India

Darren Dookeeram
Sangre Grande Hospital
Eastern Regional Health Authority
Sangre Grande, Trinidad and Tobago

Risha Fillah Fithria
Faculty of Pharmacy
Wahid Hasyim University
Indonesia

Ales Franc
Department of Pharmaceutical Technology
Masaryk University
Czech Republic

Madan Mohan Gupta
School of Pharmacy
The University of the West Indies
St. Augustine, West Indies

Rakesh Kumar Gupta
Jaipur College of Pharmacy
Jaipur, Rajasthan, India

Anita Jacob
Judy Genshaft Honors College
University of South Florida
Tampa, Florida, USA

Aditi Kirtane
Private Consultant
Inspirez Pty Ltd
Heidelberg, Australia

Mandar Kulkarni
Institute of Pharmacy
Nirma University
Ahmedabad, Gujarat, India

Sandeep B. Maharaj
School of Pharmacy
The University of the West Indies, Stg. Augustine Campus
Trinidad and Tobago

Sonali Maheshwari
Institute of Pharmacy
Nirma University
Ahmedabad, Gujarat, India

Juan Montilla
Taneja College of Pharmacy
University of South Florida
Tampa, Florida, USA

Khushi Pandya
Institute of Pharmacy
Nirma University
Ahmedabad, Gujarat, India

Nishit Parekh
SSR College of Pharmacy
Sayli, Union Territory of Daman and Diu and Dadra and Nagar Haveli, India

Jiya Patel
Institute of Pharmacy
Nirma University
Ahmedabad, Gujarat, India

Nikunjana Patel
Institute of Pharmacy
Ganpat University
Mehsana, India

Yashwant Pathak
Department of Pharmaceutical Sciences, Taneja College of Pharmacy
University of South Florida
Tampa, Florida, USA
Faculty of Pharmacy
Airlangga University
Surabaya, Indonesia

Jinfu Peng
School of Pharmaceutical Science
Central South University
China

Dr. Ramesh Pennamareddy
Masuu Global Solutions
Hyderabad, Telangana, India

Aabir Pramanik
Institute of Pharmacy
Nirma University
Ahmedabad, Gujarat, India

Nila-Marie Ramdeo
School of Pharmacy
The University of the West Indies, St. Augustine Campus
Trinidad and Tobago

Deon du Rand
Vecteng SciTech
Sandton, South Africa

Tasneem Rangwala
Meteoric Biopharmaceuticals Pvt. Ltd.
Ahmedabad, Gujarat, India

Kandi Purushothama Reddy
Masuu Global Solutions
Hyderabad, Telangana, India

Megan Saffer
Vecteng SciTech
Sandton, South Africa

Pravin Semnarayan
Vecteng SciTech
Sandton, South Africa

Ayu Shabrina
Faculty of Pharmacy
Universitas Wahid Hasyim
Indonesia

Deepak Sharma
Institute of Pharmacy
Assam Don Bosco University
Tapesia, Assam, India

Shivkumar Somani
Masuu Global Solutions
Hyderabad, Telangana, India

Sahil Sorathiya
Institute of Pharmacy
Nirma University
Ahmedabad, Gujarat, India

Maitreyi Zaveri
K. B. Institute of Pharmaceutical Education and Research
Gandhinagar, Gujarat, India

1 Nutraceuticals
Historical Perspectives in Different Countries

Mandar Kulkarni, Khushi Pandya, Niyati Acharya, and Yashwant Pathak

ABBREVIATIONS

ARGCM Australian Regulatory Guidelines for Complementary Medicines
BSE bovine spongiform encephalopathy
DIN drug identification number
EC European Commission
EFSA European Food Safety Authority
EMA European Medicines Agency
EU European Union
FBO food business operators
FDA Food and Drug Administration
FSKAN Food Safety Knowledge Assimilation Network
FSSAI Food Safety and Standards Authority India
GMO genetically modified organisms
GMP Good Manufacturing Practices
HPRG Health Products Regulation Group
NHPD Natural Health Products Directorate
NHPR Natural Health Products Regulations
NZFSA New Zealand Food Safety Authority
SAMR State Administration for Market Regulation
SFS safety food standards
TGA Therapeutic Goods Administration
TSE transmissible spongiform encephalopathy

1.1 INTRODUCTION

The term *nutraceutical* blends the concepts of nutrition and pharmaceuticals. Hippocrates gave a statement that clarifies the connection of food with medicine: "Let food be your medicine and medicine be your food." The word *nutraceutical* was derived for the first time in 1989 by Stephen Defelice, who was the chairman and founding member of the Foundation for Innovation in Medicine, Cranford, New Jersey. The term *nutraceutical* focused on food containing additional health benefits, along with basic nutrition value. The emergence of modern science in the 19th and 20th centuries led to the discovery of active compounds in foods and herbs, which set the stage for the development of modern nutraceuticals. The finding of vitamins in the early 20th century revolutionized our understanding of nutrition and made it possible to enrich foods with essential nutrients. [1]

DOI: 10.1201/9781003465041-1

In the second half of the 20th century, there was a big increase in industry and business in the food and pharmaceutical sectors, and this resulted in the large-scale production and promotion of fortified foods, supplements, and functional foods designed for certain health issues.

1.2 HISTORICAL PERSPECTIVE OF NUTRACEUTICALS IN INDIA

In India, nutraceutical products and health supplements are regulated under the FSSAI, that is, the Food Safety Standard Authority of India. The FSSAI provides the regulatory framework for the registration and approval of nutraceutical products in India. The Food Safety and Standards Act, established in 2006, aimed to govern the production, distribution, marketing, and importation of food items. Its main goal is to guarantee the safety and quality of food products for human consumption, making sure that safe and high-quality food is readily available. FSSAI gives nine regulations and various orders regarding food safety, which are governed by different ministries under one umbrella.

Before the implementation of FSSAI 2006, the applicable laws followed to manage food quality and safety in India were as follows:

- Prevention of Food Adulteration Act 1954
- Food Products Order 1955
- Meat Food Products Order 1973
- Vegetable Oil Products (Control) Order 1947
- Edible Oils Packaging (Regulation) Order 1998
- Milk and Milk Products Order 1992
- Solvent-Extracted Oil, De-Oiled Meal, and Edible Flour (Control) Order 1967
- Insecticide Act 1968
- Bureau of Indian Standards Act 1986
- Environmental Protection Act 1986

1.2.1 The Prevention of Food Adulteration Act of 1954 (Amended in 1964, 1976, and 1986)

The primary aim of implementing this regulation is to guarantee that there is no adulteration in food and its related products. This act covers regulations for chemicals, pesticides, colorants, and flavorants used in the preparation of food and related products. This Act also covers the procedures and analysis of food products, to ensure whether they complied with safety and prescribed standards or not. The Act also provides the provisions regarding penalties for various offenses for food adulteration. However, this Act was replaced by the Food Safety Standards Act 2006.

1.2.2 The Food Product Order of 1955

The Food Product Order (FPO) was implemented under the Essential Commodities Act 1955. This Act provides the availability and adequate distribution of food products. It ensures the regulation of quality, packaging, distribution, sale, import, and export of essential food products categorized as essential commodities. FPO also ensures the prevention of black marketing and food adulteration, as well as ensuring enough food products are available in the market. This Act was also replaced by the Food Safety Standards Act 2006.

1.2.3 The Meat Food Products Order of 1973

The Meat Food Products Order (MFPO) was issued under the Essential Commodities Act 1955. This Act was established for the regulation of slaughterhouses for meat processing and distribution,

TABLE 1.1
Types of Application for Manufacturers of Meat and Related Products

Category A	This applies to manufacturers who produce meat food products solely from animals that are slaughtered and processed in their own factory.
Category B	Includes a manufacturer that makes meat food products exclusively from animals slaughtered and processed in a recognized slaughterhouse, which may include facilities operated by a Category A licensee.
Category C	A producer of meat products made exclusively from poultry and/or pig meat in regions lacking authorized slaughter facilities.

Note: This Act was replaced by the Food Safety Standards Act 2006.

as well as the sale of meat and related products, with proper hygienic conditions. As per the MFPO, there as three categories of application for manufacturers of meat and related products.

1.2.4 The Vegetable Oil Products (Control) Order of 1947

This Act was framed to regulate the production of vegetable oil, its quality, and its distribution standards. The implementation of standards for vegetable oil production ultimately leads to consumer protection and the prevention of adulteration and black marketing. This law provides information regarding laboratory facilities for testing the quality of vegetable oil, which is mandatory to obtain registration of products. After inspection, the Commissioner has the right to issue or reject the application of registration. (The Commissioner or Controller is appointed by the central government of India.) The applicant has 30 days to appeal the rejection of their application to the central government, but the decision given by the central government is considered final. The applicant needs to comply with standards given under Schedules III (for vanaspati), IV (for bakery shortening), V (for margarine), VI (for blended edible vegetable oil), VII (for mixed fat spread and vegetable fat spread), and VIII (for refined oils). Under this provision, the Commissioner can inspect the premises of manufacturing facilities and related vehicles and seize the stocks of vegetable oil products and collect samples for examination. Now, the Food Safety Authority of India, in collaboration with state food authorities, is responsible for the regulation of vegetable oil and related products.

1.2.5 The Edible Oils Packaging (Regulation) Order of 1998

As per the Act, vegetable oils and fats exclude margarine, vanaspati, bakery shortening, and fat spreads, according to the Prevention of Food Adulteration Act of 1954 and its regulations for human consumption. The person who obtained the registration certificate, as per subclause 4 of clause 4, for business purposes is known as the registered packer. The rules also include safety and hygiene standards to avoid contamination and adulteration of edible oil, as well as guidelines for the manufacture, storage, and transportation of quality edible oils. It also covers the sizes, packaging, and testing requirements for edible oil products.

1.2.6 The Milk and Milk Products Order of 1992 (MPPO-92)

This legislation provides the guidelines, under the provisions of the Essential Commodities Act 1955, to regulate the dairy sector of India. As per the Act, any dairy plant that processes more than 10,000 L of milk or 500 mt of milk solids per year must register with the registering authority, which is under the control of the central government of India. The main objective of the Act is to maintain the superior quality of milk, and related products, and its supply. For better understanding of milk products manufacturers, the MPPO Act provides schedules, like Schedule I, which stands for the application for registration, which consists of information like the details and content of the application for registration and the detailed description about the collection of milk. After successful application, a registration certificate is issued in Schedule II. Schedules III and VI have been omitted. Schedule IV contains information on the half-yearly returns and the specific marketing strategies for each product, which can differ from one product to another. Schedule V provides all the necessary details about the requirements for registering dairy establishments. It covers everything from general sanitary conditions and hygiene standards for the establishments to the hygiene guidelines for the staff. Additionally, it also outlines the sanitary and hygienic requirements for storage, as well as wrapping and packaging specifications.

This Act was amended in 2002, including the registration process, which includes sanitary/hygienic conditions, while quality measures of milk products are specified in Schedule V of MPPO-92. Also, the ability to examine dairy products has been improved. Additionally, products will now be approved within 45 days instead of 60 days, if the application is submitted correctly. After this amendment, applicants will enjoy the benefit of the raised state registry limit from 1 lakh liters to 2 lakh liters per day. Earlier, MPPO-92 was governed under the Ministry of Dairying and Fisheries, but now it is implemented by the Food Safety Standards Act 2006 under the Ministry of Health and Family Welfare.

1.2.7 The Solvent-Extracted Oil, De-Oiled Meal, and Edible Flour (Control) Order of 1967

This particular provision falls under section 3 of the Essential Commodities Act 1955. In order to grasp the essence of these terms, namely, *extracted oil*, *de-oiled meal*, and *edible flour*, it is necessary to explore their meanings. **De-oiled meal** means the residual material left over when oil is extracted by a solvent from any oil-bearing material. **Edible flour** is defined as the edible ground material prepared from de-oiled meal derived from oil cakes or oilseeds or oil-bearing materials as a result of solvent extraction of oil from such materials. Vegetable oils in any form obtained using

TABLE 1.2
Types of Licenses under the Solvent-Extracted Oil, De-Oiled Meal, and Edible Flour (Control) Order 1967

Category A	Producer with a factory that has not only a solvent extraction plant but also the capability to pre-clean oilseeds, unless the oilseeds are directly processed.
Category B	Producer's factory is not only equipped with a solvent-extraction plant but also has the necessary equipment for refining.
Category C	Producers whose factory is equipped with a solvent-extraction plant only.
Category D	Producer falls under category C and is involved in refining solvent-extracted oil produced by them in the factory of another producer in category B or a registered user engaged in the same business.
Category E	In respect of the producer: • Even without a solvent-extraction plant, the factory continues to produce solvent-extracted oil or de-oiled meal using oil cake from their own mill or imported oilseeds. • If someone does not have their own oil mill, they can still process oil-bearing materials in another producer's factory, but only if the Controller allows it and the central government agrees. They need to get permission in writing and show that there are valid reasons for doing so. • Whoever owns a factory that has a solvent-extraction plant can process oil-bearing material from another producer's factory if the Controller and the central government approve it in writing.

Note: This Act was substituted by the Food Safety Standards Act in 2006.

solvents are known as solvent-extracted oil. As per the provisions of this Act, the licenses for these products are divided into five categories.

At the start of the decade, the National Nonprofit Association suggested that it should be obligatory for all food manufacturing companies to provide evidence supporting the safety of their products, as this directly contributes to safeguarding consumers. Furthermore, the Association highlighted the need to establish a regulatory environment specifically for nutraceuticals.

In the year 2003, to strengthen current food laws, the expert group from the Ministry of Health recommended the inclusion of two specific categories, namely, functional food and dietary supplements. This proposal also emphasized the need for safety testing to be conducted for these categories. Additionally, the Ministry made concerted efforts to ensure compliance with the standards established by the Bureau of Indian Standards (BIS). These standards encompassed various aspects, such as product certification, quality system certification, testing, and consumer affairs.

The year 2005 witnessed the enactment of measures by several committees of the Parliament of Agriculture which aimed at establishing a unified regulatory body and implementing comprehensive regulations and laws for functional foods and dietary supplements in India.

1.2.8 The Food Safety and Standards Act of 2006 (FSSA) [2]

It is a standard regulatory authority that consolidates the various laws and regulations related to food safety and standards in India. It mainly regulates the manufacturing, storage, distribution, sale, import, and export of food supplements and nutraceutical products. The FSSAI is majorly involved in creating the scientific-based regulatory framework for food supplements having additional health benefits. It also provides the platform for the licensing and registration of food businesses, with essential standards to be followed. To ensure compliance with standards, the FSSAI conducts regular on-site inspections. Another function within the ambit of the FSSAI is the surveillance of nutraceutical products to ensure safety, quality, and efficacy. The FSSA covers a wide range of categories, like food safety standards for excipients, additives, and contaminants and the labeling and packaging of nutraceutical products.

The FSS Act 2006 serves as a catalyst in achieving global standards, thereby simplifying the process of exporting nutraceutical products. Under the umbrella of the FSSA, there are 12 chapters, with Chapter IV specifically delving into the realm of nutraceuticals, functional food, and dietary supplements. Article 22 within this chapter is dedicated to the regulation and approval process for manufacturing these products in India. Additionally, Article 23 and 24 provide comprehensive guidelines for labeling, packaging, and advertisement of the mentioned categories. Chapter IX includes the provision for offenses and their penalties linked to adulterated, misbranded, and substandard food products; false advertisement; and failure to achieve food safety standards.

In 2011, significant revisions were introduced to the FSSA 2006 amendment for strengthening the regulatory structure for nutraceutical products in India. It outlines the guidelines for overseeing food additives, contaminants, toxins, labeling, packaging, and hygiene criteria. Furthermore, the mandate for the licensing and registering of food businesses in India was launched in 2011, necessitating a valid license issued by the FSSAI.

The year 2015 marked a significant milestone for FSSAI, as it was appointed as Asia's regional coordinator for the Codex Alimentarius Commission. This esteemed international body, governed by the Food and Agricultural Organization (FAO) and the WHO, is committed to establishing regulatory standards, guidelines, and codes to protect the quality and safety of food products.

1.2.9 The FSSAI 2016 Amendment

The latest amendment in FSSA 2006 was done in 2016, regulating the eight categories under the umbrella of nutraceuticals as follows: food for special dietary use (FSDU), food for special medicinal purpose (FSMP), specialty food containing plants or botanicals, food containing probiotics and

FIGURE 1.1 FSKAN logo, India.

prebiotics, and novel food products. This regulation contains vital details, such as guidelines for claims, labeling requirements, approved additives usage, and acceptable levels of contaminants, toxins, and residues. Additionally, it explicitly bans the utilization of hormones and steroids in food products.

1.2.10 The Food Safety Knowledge Assimilation Network (FSKAN) 2017

The objective is to facilitate the integration of scientific evidence into food safety and nutrition research, to identify gaps, and to fund research projects in those gaps.

1.2.11 Novel Food Regulations in India

Novel foods fall under the category of non-specified food in India. As per FSSAI regulations 2016, the term *novel food* is defined as:

- Food with no previous history of consumption for at least 30 years in the origin country or 15 years in India.
- Ingredients used in preparation with no previous history of consumption for at least 30 years in the origin country or 15 years in India.
- Use of newer technologies with an innovative engineering process, altering the composition or structure or size of food significantly. This procedure may change the size, composition, or structure of the food or its ingredients, which may, in turn, change its nutritional value, metabolism, or level of undesirable substances.

In 2017, FSSAI issued a guidance, called the Food Safety and Standards (Approval of Non-Specified Food and Food Ingredients), which include novel food or novel ingredients processed with the use of novel technology; new additives; new processing aids, including enzymes and articles of food; and food ingredients isolated from microorganisms, bacteria, yeast, fungi, or algae.

The latest amendment in the FSSAI 2016 regulations was made in 2022, called the FSS (Health Supplements, Nutraceuticals, FSDU, FSMP, Functional Food, and Novel Food) Regulation 2022. Furthermore, the FSSAI also allowed food business operators (FBO) to use some additives and enzymes in health supplements, namely, ammonium acetate, ammonium lactate, choline salts and esters, croscarmellose sodium, potassium hydrogen malate, and sodium starch glycolate.

1.3 HISTORICAL PERSPECTIVE OF NUTRACEUTICALS IN THE USA

The term *nutraceutical products* is used for dietary supplements, as per the DSHEA 1994. Before the implementation or the establishment of the DSHEA, there were no stringent and harmonized regulations for dietary supplements in the USA. Prior to the establishment of this Act, there were no rigorous and uniform regulations governing the production and sale of dietary supplements. Initially, these products were regulated under the Food, Drug, and Cosmetics Act of 1940 and its accompanying rules from 1945. However, the absence of regulations and the lack of standardized guidelines posed difficulties in maintaining the quality, safety, and effectiveness of these products. The FD&C Act served as the foundation for the establishment of several acts and guidance documents that regulated the utilization of food products, food additives, vitamins, and minerals before the DSHEA came into effect.

1.3.1 The Pure Food and Drug Act of 1906

The Pure Food and Drug Act was signed by US president Theodore Roosvelt in 1906. The main objective of the Act was the prevention of the manufacture, distribution, sale, and transportation of adulterated or misbranded, poisonous, or deleterious food, drugs, and medication.

The Pure Food and Drug Act of 1906 owed a great deal to the influential efforts of Harvey Wiley. Serving then as the chief chemist at the United States Department of Agriculture, Wiley played a pivotal role in its passage. One of his notable contributions was the establishment of the Poison Squad, a group of individuals who willingly participated in experiments to assess the effects of food preservatives and additives on the human body. These experiments involved the consumption of meals by healthy volunteers containing various chemicals and preservatives, allowing Wiley and his team to observe their impact. The outcomes of these experiments not only raised awareness on but also underscored the significance of food regulations and the adoption of safer food practices.

1.3.2 The Food, Drugs, and Cosmetics Act of 1938

The Pure Food and Drug Act of 1906 was replaced by the Food, Drugs, and Cosmetic Act of 1938. This new Act granted the FDA greater authority to regulate the therapeutic claims found on product labels and drug advertisements. It introduced the term *misbranded* to identify food products with false or misleading labels. Additionally, the Act mandated pre-market approval for new drugs. The driving force behind this legislation was the elixir sulfanilamide tragedy of 1937, a public health crisis that resulted in numerous deaths due to the inclusion of a toxic solvent called diethyl glycol in drug products.

The **Delaney clause**, designed by the Delany Committee in 1950, which is a crucial part of the D&C Act, stipulates that any food additive cannot be deemed safe if it is found to induce cancer in both humans and animals. This clause highlights the criticality of evaluating the carcinogenicity factor when determining the safety of food additives.

1.3.3 The Food Additive Amendment of 1958

The FDA has officially disclosed a list of substances "generally recognized as safe" (GRAS), encompassing close to 200 substances.

In the year 1969, to address concerns regarding public health and safety, the Food and Drug Administration (FDA) introduced sanitation programs targeting a range of products, including milk, shellfish, and food service. These programs were designed to prevent instances of poisoning and to ensure the overall quality and safety of these products. By implementing stringent regulations and guidelines, the FDA aimed to mitigate the risk of contamination and maintain optimal sanitation practices across the production and distribution channels.

1.3.4 The Vitamin and Mineral Amendments of 1976 (Proxmire Amendments)

The Proxmire Amendments prohibit the FDA from imposing regulations that limit the potency of vitamins and minerals in food supplements or to classify them as drugs based solely on their potency.

1.3.5 The Nutrition Labeling and Education Act of 1990

The primary objective of this legislation was to ensure that consumers have access to precise and reliable information regarding the contents of the food items they consume. As a result, it became obligatory for multi-ingredient food products, including dietary supplements, to provide nutrition labeling that adheres to specific requirements. This Act not only established guidelines for food standards and nutrition labeling but also regulated health claims. It introduced terms such as "low-fat" and "high-fat," as well as requirements for food ingredient panels, serving sizes, daily values, and ingredient listings. This law provides information regarding various types of claims, as follows:

- **Health claims.** It shows the relationship between food substance and a reduced risk of disease or health-related function. Health claims are sub-categorized into three types:
- **SSA (significant scientific agreement) claims.** Used for conventional food and dietary supplements. SSA shows that the nutrient–disease relationship is well-established.
- **The FDA Modernization Act.** This type of claim is only applicable to conventional foods and not for dietary supplements.
- **Qualified health claims.** These claims are applicable to both regular foods and dietary supplements. If any party is interested, they can request the FDA to issue a regulation regarding a health claim (refer to 21 CFR 101.70).
- **Nutrient health claims.** This type of claim helps determine the level of nutrients present in a food product. It permits the use of words such as *free*, *high*, *low*, etc. Nutrient content claim rules usually only cover nutrients with a set daily values in Food Labeling Guide VII.
- **Structure/function claims.** When it comes to structure/function claims, they illustrate how a nutrient or dietary ingredient can influence the normal structure or function of the human body. An example of this is the claim that calcium is essential for building strong bones.

Furthermore, an essential aspect of this Act is the obligatory declaration of a wide range of elements, such as total fat, saturated fat, cholesterol, sodium, total carbohydrates, dietary fiber, protein, vitamins A and C, calcium, and iron.

1.3.6 The Dietary Supplement Health and Education Act of 1994

The DSHEA 1994 was the biggest milestone achieved in regulating nutraceutical products, including diet supplements. This Act provides the main definition of *dietary supplements* and *dietary ingredients*. Under this provision, manufacturers and distributors of dietary products are not allowed to sell products considered as misbranded or adulterated. It includes various parameters, like the Dietary Supplement Ingredient Directory, how to report problems with dietary supplements, information for consumers on using dietary supplements, and a new dietary ingredient notification process.

The DSHEA also imposes labeling requirements on dietary supplement manufacturers. These obligations necessitate the inclusion of crucial information, such as the supplements facts panel, which must encompass details regarding serving size, nutrient quantities, and other ingredients. In addition, the Act contains provisions that govern the regulation of health claims displayed on supplement labels. It is imperative that these claims are not misleading or fabricated. The DSHEA has allowed the FDA to set guidelines for Good Manufacturing Practices (GMP) for dietary supplements. These guidelines include quality control and manufacturing standards that manufacturers of dietary supplements must follow.

1.3.6.1 The Establishment of the Office of Dietary Supplements (ODS)

The DSHEA played a crucial role in the formation of the Office of Dietary Supplements (ODS), in collaboration with the National Institute of Health (NIH). The primary aim of the ODS is to actively support scientific research and to ensure the distribution of precise and trustworthy information about dietary supplements. The DSHEA established a clear distinction between the regulation of dietary supplement ingredients and food additives. Instead of treating them as food additives, the legislation introduced two distinct categories for dietary supplement ingredients: "old" or "grandfathered" ingredients and "new dietary ingredients." According to the DSHEA, any dietary supplement ingredients that were sold in the United States before October 15, 1994, are considered grandfathered.

1.3.6.2 cGMP for Dietary Supplements 2007

The nutraceutical sector relies on Good Manufacturing Practices (GMPs) to maintain high-quality standards during the manufacturing, packaging, and distribution of dietary supplements. The regulatory compliance of GMPs is ensured by the Dietary Supplement Health and Education Act (DSHEA), which also compels the Food and Drug Administration (FDA) to develop separate "current Good Manufacturing Practices" (cGMP) specifically for dietary supplements.

The main parameters covered in this Act are facility requirements, personnel training, quality control, documentation and recordkeeping, master manufacturing records, ingredient testing, labeling compliance, complaint handling, recall procedures, vendor qualification, and adverse event reporting. Under the revised cGMP regulations of 2015, dietary ingredients were incorporated into the domain of food regulations, thereby expanding the scope of oversight.

1.3.7 The Food Safety Modernization Act of 2011

This Act, enforced in 2011, effectively strengthened the food safety system. It includes seven key points regarding safe food supply. It includes thus:

- **Prevention control for human food.** To maintain the integrity of human food, it is imperative to comply with the minimum requirements for its manufacturing, processing, and packing. These requirements encompass a range of factors, including employee training, adherence to Good Manufacturing Practices (GMPs), conducting hazard analysis, and implementing risk-based preventive controls.
- **Preventive control for food for animals.** It includes the minimum requirements of manufacturing, processing, and packing of animal food.
- **Produce safety.** It provides regulations for growing, harvesting, packing, and storing dietary supplements.
- **Foreign supplier verification program.** This program is majorly focused on the import of dietary supplements that must comply with FDA regulations.
- **Third-party certification.** The establishment of external organizations to ensure the quality and standards of dietary supplements through consistent consultation and regulatory audits aids manufacturers in adequately preparing for official audits conducted by regulatory bodies to obtain approval and certifications.
- **Food defense.** According to the legislation, food establishments that are registered under the FDA are required to create a comprehensive strategy for evaluating the presence of contamination and adulteration in their products, as well as in maintaining detailed records of these assessments.

- **Sanitary transport.** This covers regulation regarding the transport of food products and dietary supplements their shipping and receiving requirements.

1.4 HISTORICAL PERSPECTIVE OF NUTRACEUTICALS IN EUROPE

Europe uses the term *food supplement* for nutraceutical products. As dietary supplements are considered as food, it is the responsibility of the manufacturer/importer/supplier to achieve the required standards for all kinds of food supplements. The 1990s witnessed an increase in cases of the bovine spongiform encephalopathy (BSE) crisis, a neurological disease caused by contaminated animal feed and the consumption of meat and bonemeal from infected animals. To address this issue, the EU Parliament introduced new regulations in 2002 aimed at enhancing food safety and traceability, known as the General Food Law Regulation (Regulation EC No. 1478/2002). This event led to the reinforcement of food safety regulations in the EU. However, it is important to highlight that non-therapeutic claims are regulated independently by each member state.

To initiate the process of harmonization in food supplement regulations and to provide a synchronized regulatory framework, a document was published in 1997 called the General Principles of Food Law in the European Union. Most EU members gave a favorable opinion toward the drafting of the basic framework directive for dietary supplements. Finally, the regulation for food supplements in EU was established for the first time in 2002, known as Directive 2002/46/EC.

1.4.1 Important Highlights of Directive 2002/46/EC

The enforcement of this directive in Europe has resulted in the creation of consistent regulations for food supplements. This has led to a unified approach to ensure that these products are safe and of high quality. This directive provides the accurate definition of *food supplement* which is applicable throughout the European Union. The effective implementation of this regulation outlines the safety and quality requirements of food supplements in EU, which ultimately shows a positive impact on consumer safety. This regulation outlines the vitamins and minerals that can be used in food supplements and sets the standards for their maximum and minimum levels. Another important parameter covered under the directive are labeling requirements for food supplements, which include necessary information like daily intake, warnings and precautions, recommended daily values, and the sentence "The product should not be used as a substitute for a varied diet" on the label of food supplements.

1.4.1.1 Regulation (EC) No. 178/2002

Through the formation of EFSA, that is, the European Food Safety Authority, this regulatory framework aims to ensure the safety of food products within the European Union. The primary aim of this regulation is to improve the quality of food supplements through rigorous scientific testing and evaluation of both food supplements and feed products. Food supplement manufacturers must provide assurance regarding quality, safety, and effectiveness to achieve the standards. As per regulation, the implementation of the Rapid Alert System for Food and Feed (RASFF) allows food safety authorities to exchange safety information between EU member states, and the European Commission to raise issues from food supplements to take further appropriate actions. This regulation also provides the provision and regulatory requirements of traceability of adulterated, unsafe food supplements and its recall, which prevent the entry of substandard and counterfeit food supplements into the market.

1.4.1.2 Regulation (EC) No. 852/2004

This regulation was drafted in 2004 and adopted in 2006, primarily concerned with the hygiene of food supplements and foodstuffs. It provides regulatory requirements of hygiene for FBOs, that is, food business operators, to ensure consumer safety. Every food business operator bears a significant responsibility to meet all the essential quality and safety standards for their products. The hygiene requirement is fulfilled by the implementation of good hygienic practices given under Annexure II and procedures based on the hazard analysis and critical control points (HACCP) principles provided under Codex Alimentarius. In case of the import and export of food products of animal origin, it must comply with Regulation (EC) 853/2004, which gives specific hygiene rules for food of animal origin, like meat and fishery products.

1.4.1.3 Regulation (EC) No. 1924/2006

This regulation outlines the necessary guidelines and standards that must be adhered to when making nutrition and health claims on food supplements. The accurate functional and health claims on labels give all the necessary information regarding the product which helps consumer select the appropriate product as per the requirement. The label must include important information, like fatty acids, saturated fatty acids, trans fatty acids, sugar, and salt. The statements used for the claims must not create any kind of confusion and should not encourage excess consumption of the product. The claims must ensure that health would not be affected by using the product.

1.4.1.4 Regulation (EC) 1925/2006

This regulation emphasizes the inclusion of essential nutrients, such as vitamins and minerals, in food supplements. It provides a list of vitamins and minerals to be used in food supplement products. Substances other than those on the list are not permissible. This regulation also mandates the inclusion of vitamins and minerals on the label, with its quantity used and its maximum limit. As per the regulation, it is not advisable to add vitamins and minerals in unprocessed foodstuffs, like fruits, meat, fish, etc. The addition of vitamins and minerals is also strictly prohibited in case of beverages containing more than 1.2% alcohol by volume. This regulation is also applicable to food for specific nutritional use, new food and food ingredients, GMO (genetically modified food), food additives and flavoring agents, etc. This regulation does not cover food supplements, which are covered under Directive 2002/46/EC.

1.4.1.5 Regulation (EC) No. 1881/2006

This regulation provides the maximum acceptable limit for contaminants present in foodstuffs, with a wide range, including fruits, vegetables, cereals, meat, and various processed food products. It covers contaminants, like mycotoxins, including aflatoxins, ochratoxin A, *Fusarium* toxins, patulin, citrinin, and ergot alkaloids; metals, like lead, cadmium, mercury, and arsenic; compounds like 3-monochloropropane-1,2-diol (3-MCPD) and its fatty acid esters and glycidyl fatty acid esters; dioxins and polychlorinated biphenyls (PCBs); and polycyclic aromatic hydrocarbons (PAHs). If all these ingredients are present in excess amount, the product is not permissible to be sold in the market. For example, in cases of oilseeds, dried nuts, dried fruits, rice, and prepacked groundnuts, which include the process of sorting and other physical treatments required to be mentioned on a line on the label: "Product shall be subjected to sorting or other physical treatment to reduce the risk of aflatoxin contamination before human consumption." Contaminants with a higher limit must not be mixed with food products that exceed the prescribed limit provided in the Annexure.

1.4.1.6 Regulation (EC) No. 1333/2008

This regulation established the regulatory framework of food additives used in food supplements and provides the list of approved food additives to be used in the formulation of food supplements. Only food additives that have been authorized by the European Union (EU) are permissible for use in food supplements. It is imperative that these additives do not comprise any detrimental components that may pose a risk to consumers. It prioritizes minimal usage of food additives in the supplements to ensure the highest therapeutic benefits. Food additives should be avoided in case of unprocessed food and supplements used for infants and young children.

In 2010, the EFSA NDA Panel (Nutrition, Novel Foods, and Food Allergens Panel) played a crucial role in formulating the scientific opinion that offers guidance on the establishment of dietary reference values (DRV) within the European Union (EU). The dietary reference values are used to guide the intake of essential nutrients taken as a part of a healthy diet. The key aspects of dietary reference values (DRV) include the recommended dietary allowances (RDA), adequate intakes (AI), and tolerable upper intake levels (UL). These DRVs serve as a valuable tool in setting dietary recommendations, creating diet charts, and shaping health policies. The guidelines provided by the Panel include methodologies and criteria that aid in determining reference values for energy.

In 2013, EMA released a guidance document called EU No. 609/2013, which came into effect in 2016, providing the regulatory requirements for various categories:

- **Infant formula and follow-on formula.** Provides the nutritional requirements for infants during the first month of life and to complement the diet for infants over 6 months old (FOF).
- **Processed cereal-based food and baby food.** Provides information to set appropriate nutrition for this particular age group, labeling, and composition requirements for infants and young children.
- **Food for special medical purpose (FSMP).** This type of product is mainly used for specific medical conditions or requirements that are not fulfilled by diet alone.
- **Total diet replacement for weight control (TDR).** This type of diet is mainly designed to replace a complete meal for the purpose of weight loss.

1.4.2 Novel Food Supplements Legislation in the EU

Novel food in the EU means food that has not been significantly consumed by humans in the EU prior to May 15, 1997. As per the principles designed by the EU for novel food, the product should comply with essential requirements, like how it must be safe for consumers. The accurate labeling of the product is of utmost importance, and it should not mislead the consumer. The consumption of novel food items does not have any negative impact on the nutritional well-being of consumers. The applicable regulations and guidelines for novel food are mentioned in the following:

- **Regulation (EC) No. 258/1997.** This regulation tried to establish harmonization in the regulation of novel food and novel food ingredients. The detailed scientific and technical evaluation of novel food became mandatory to ensure that the product does not possess any kind of risk. This regulation is not applicable in case of food additives, flavoring agents, extraction solvents, food enzymes, and GMOs (i.e., genetically modified organisms), as all these categories are regulated under Regulation (EC) 1823/2003.
- **Regulation (EC) No. 2283/2015.** Regulation (EC) No. 258/1997 was amended as Regulation (EU) No. 1169/2011 and further replaced by this regulation and laid down new regulations in marketing novel food supplements in the EU; it came into force in 2018. As per regulation, to obtain marketing approval for food supplements within the EU, a centralized procedure was implemented which is governed by the EFSA.
- **Regulation (EU) 2017/2470.** This regulation serves as an amendment to Regulation (EC) 2015/2283. It provides the Union list of authorized novel food in Article I. The list of authorized novel food is published in the *Official Journal of the European Union.*

In 2020, the most notable incident was the event known as BREXIT, which led to the UK's departure from the European Union. This separation resulted in substantial alterations to regulations and laws in all sectors, including pharmaceuticals and nutraceuticals, within the country. UK's regulatory authority, the Medicines and Healthcare Products Regulatory Agency (MHRA), got the power to build their own regulations for nutraceuticals, dietary supplements, functional foods, and beverages with claims of health benefits.

1.5 HISTORICAL PERSPECTIVE OF NUTRACEUTICALS IN CANADA

1.5.1 The Natural Health Product (NHP) Regulation of 2004

The establishment of the Office of Natural Health Products in Canada in 1999 marked a significant milestone in the country's regulatory landscape. This development paved the way for the creation of a comprehensive framework to regulate natural health products, a first of its kind in Canada. Nutraceutical products are handled as natural health products. As per the Natural Health Product (NHP) Regulation of 2004, natural products cover a wide range products, like probiotics, herbal remedies, vitamins and minerals, homeopathic medicine, traditional medicine, and essential amino acids and essential fatty acids.

Health Canada published in *Canada Gazette*, Part I, its first amendment in 2001, which proposed the necessity of building a well-defined regulatory network for natural health products in Canada. Another part of *Canada Gazette*, that is, Part II, was released in 2003. The final NHP regulations came into force in 2004, which mainly emphasized essential information, like NHP definitions, product licensing, adverse reaction reporting, GMP requirements, and labeling and packaging requirements for natural health products to be marketed in Canada.

According to the archived records found on the regulatory authority's website of Canada, the Therapeutic Products Programme and the Food Directorate played a significant role in 1996 in establishing the foundation for nutraceutical products and their regulation in Canada. The primary objective of these bodies was to develop a regulatory framework concerning the health claims attributed to nutraceuticals, functional food, and other interconnected products. The involvement of external advisory committees and working groups from both the Therapeutic Product Programme and the Food Directorate ensured a comprehensive review of the latest health claim policies for nutraceutical products. A major outcome of the review and discussion is the availability of physiologically beneficial products in Canada catering to the needs of health-conscious individuals. To guarantee the quality, safety, and effectiveness of nutraceutical products, it is essential to rely on valid and scientifically supported evidence. Information regarding products is clearly stated, without causing any confusion to or misleading consumers.

Before the NHP Act of 2004, nutraceutical products were sold either as food or drugs under the Food and Drug Act. If a nutraceutical product is categorized under the food category, the product is marketed with limited health claims and does not provide much safety information on the label. If the product is categorized under the drug category, the product must follow the drug review process and must have a drug identification number (DIN) before being placed in market.

Until 2003, only disease risk reduction–related health claims were permitted in Canada. Five such claims were approved by Health Canada, as follows.

TABLE 1.3
Claims Approved by Health Canada

Food Product Element	Reduction of Risk in Case Of
Low sodium and high potassium	High blood pressure
Adequate calcium and vitamin D	Osteoporosis
Low saturated and trans fat	Heart disease
Consumption of fruits and vegetables	Various types of cancer
Reduced dietary sugar alcohols	Dental caries (tooth carries)

The Natural Health Products Regulations (NHPR), enacted in 2004, play a pivotal role in the oversight of natural health products. These regulations establish a robust framework that ensures the quality, safety, and effectiveness of these products. Moreover, they delineate the procedures for evaluating finished products, granting licenses, and conducting post-marketing surveillance to promptly detect any adverse events associated with the use of these products.

NHPD stands for the Natural Health Product Directorate. It is a product license system that allows the applicant to use reference monographs as supporting document to demonstrate the safety and efficacy of a product. The monographs are available on the Natural Health Products Ingredients Database. This list provides the acceptable medicinal and non-medicinal ingredients, standard terminologies, and ingredient monographs.

Note: NHPD has now been replaced by the Natural and Non-Prescription Health Products Directorate.

Health Canada has released a comprehensive collection of monographs for natural health products (NHP), which serves as a reliable reference to establish the safety and efficacy of these products. To ensure the well-being of consumers, manufacturers of NHPs must obtain pre-market authorization from Health Canada and furnish all the necessary evidence and supporting documents regarding the safety and effectiveness of their products before they can be sold in the market.

The regulations governing clinical trials for natural health products were officially published in 2003, in *Canada Gazette*, Part II. The clinical trial requirements for NHPs closely mirror the requirements outlined in Division 5 of the Food and Drug Regulations for pharmaceutical drug products. However, it is essential to consider Part IV of the NHPR when strategizing for a clinical trial involving NHPs in Canada.

In 2005, Health Canada took a significant step by publishing guidelines in Part III of the Natural Health Products Regulations. These guidelines emphasized the importance of Good Manufacturing Practices (GMPs) for natural health products. The enforcement of GMP guidelines in the manufacturing, labeling, importing, and storage of these products has played a crucial role in elevating their quality, safety, and effectiveness.

To effectively regulate the labeling of natural health products eligible for sale in Canada, it is necessary to adhere to the labeling guidelines established by Health Canada in 2006. These guidelines encompass various aspects, such as artwork and label specifications, compulsory information for the product fact table (PFT), and additional requirements outlined in Part V of the regulations.

In the past, many food items with extra vitamins, minerals, amino acids, and other ingredients that did not meet FDR standards were launched as "natural health products" (NHPs) under the Natural Health Products Regulations (NHPR). Individual pre-market assessment and licensing were required for products classified as natural health products (NHPs). However, Health Canada determined that many of these products could be classified as food based on various factors, including public perception, historical use, product representation to consumers, product composition, and product format. This conclusion was reached according to the guidance document regarding the classification of products at the intersection of food and natural health products. In October 2011, the Minister of Health announced that caffeinated energy drinks, once labeled as natural health products, would now be regulated as food. Subsequently, stakeholders were informed that other NHPs meeting the definition of food would also be regulated as food products.

1.5.2 Introduction of Novel Food and Related Regulations in Canada

In 2006, the concept of novel food emerged in Canada with the introduction of the guidelines for the safety assessment of such food products. The term *novel food* was coined to describe food items that were previously not consumed or produced on a large scale. The guidelines aimed to ensure the safety of these novel food products before they could be introduced into the market. There are various factors behind the introduction of novel foods in Canada, like enhancing the influx of novel merchandise and diverse species employed as a source of sustenance.

Understanding the significance of utilizing GMOs (i.e., genetically modified organisms) and the integration of emerging food processing techniques is becoming essential. The development of such technologies has paved the way to a rise in the importation of a wide range of food products in Canada.

The regulatory authority Health Canada defines *novel food* as:

- A substance, including a microorganism (a living thing so small you need a microscope to see it), that does not yet have a history of safe use as food.
- Food that has been manufactured, prepared, preserved, or packaged by a process that
  - has not been previously used for that food and
  - causes the food to undergo a major change.
- Food that comes from a plant, animal, or microorganism that has been genetically modified so that the plant, animal, or microorganism
  - shows characteristics that it did not before,
  - does not show characteristics that it did before, or
  - has one or more characteristics that no longer fall within the expected range.

In compliance with Division 28 of the Food and Drug Regulation, manufacturers are required to notify the regulatory authority before placing a novel food product on the market. The marketing of novel food products is strictly prohibited, unless the manufacturer has obtained a NOC (notice of compliance) from the regulatory authority.

1.6 HISTORICAL PERSPECTIVE OF NUTRACEUTICALS IN JAPAN

In Japan, the term *nutraceuticals* is used to describe "health foods," which can be further classified into two categories: food products with and without functional claims. For generations, Japan has embraced the use of natural remedies to promote wellness, and the oversight of health foods is in place to safeguard the public by ensuring the safety and effectiveness of these remedies.

Japan has an extensive and rich heritage in the field of traditional medicine, commonly referred to as Kampo, which has been practiced for more than a millennium. Kampo, deeply rooted in Japanese culture, harnesses the healing properties of natural elements, like herbs, plants, and fungi, to foster well-being and address various health conditions.

The significant growth of the nutraceutical market comes from the quest for a healthy lifestyle, particularly among those who are 65 years old and above. The demand for dietary supplements that assist in supporting healthy aging, cognitive function, and joint health is extremely high in Japan.

To establish the market of healthy foods in Japan, the initiatives taken by the government of Japan are as follows.

TABLE 1.4
Initiatives Taken by the Government of Japan

Basic Law of Food Education 2005 (Shokuiku):	Matebo Law 2008:	Healthy Japan 21
The main objective behind program is to create the awareness about food, how to choose healthy food that will help to improve the health and promotion of healthy lifestyle in high schools of Japan.	The legislation includes provisions for the establishment of obesity management plans and programs, specifically designed for individuals whose examination reports indicate results surpassing the defined limits	The core elements of this program focus on promoting healthy eating habits, encouraging physical activity, aiding in smoking cessation, and curbing alcohol consumption.

1.6.1 Health Food Regulations in Japan

The functional food regulations cover three important terms, namely, "functional health claim" (FHC), "food for specified health uses" (FOSHU), and "food with nutrient function claim" (FNFC). In Japan, prior to the 1990s, the manufacturing of nutraceutical products suffered from a lack of regulations and guidance documents. To regulate health foods effectively, the MHLW introduced the term "food with nutrient function claim" (FHC) in 2001, which covers two categories, namely, FOSHU and FNFC. This initiative demonstrated the Ministry's dedication to establishing regulatory guidelines and ensuring the safety of these products. The term FOSHU stands for "food for specified health uses." Food that includes ingredients claiming to have nutritional benefits and improve human health falls under the category of FOSHU.

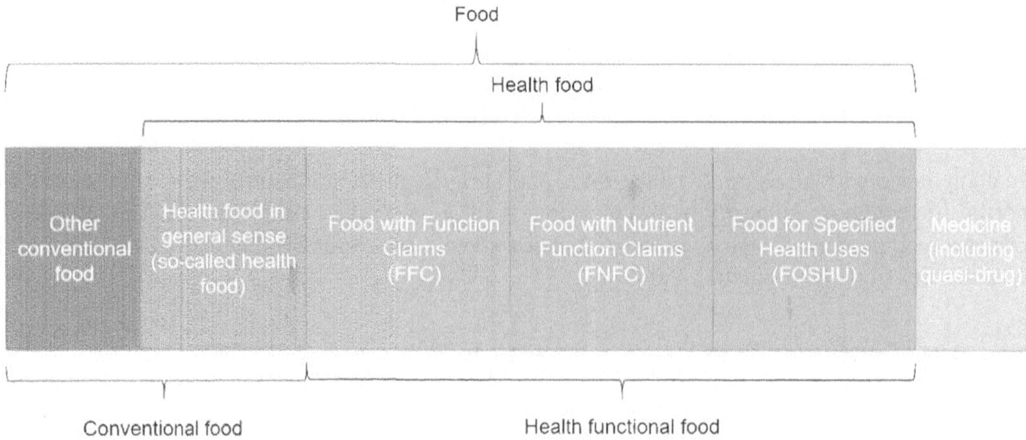

FIGURE 1.2 Classification of health functional food and conventional food in Japan.

FIGURE 1.3 FOSHU-certified logo.

TABLE 1.5
Applicable Health Food Laws in Japan

Food Sanitation Act	**Applicable to all food and drinks and food-related products.**
Pharmaceutical Affairs Law	According to regulations, health food is prohibited from making claims about its efficacy that are similar to pharmaceutical products.
Health Promotion Act	The spotlight is on false and exaggerated advertising.

The adoption of FOSHU shows a major impact on the nutraceutical market in Japan. The implementation of a safety and effectiveness assessment of ingredients used in food products that claim nutritional benefits became mandatory. FOSHU specifies the limits of ingredients used in the formulation. It also includes the provision of quality control techniques and methods of analysis for ingredients to be used in formulations. As per FOSHU, it is advisable to add a claim on the label of a product: "This product has a beneficial effect on health."

Note: The CAA (Consumers Affairs Agency) is responsible for the registration and approval of FOSHU products in Japan.

In 2005, FOSHU included four new categories, namely, "standardized FOSHU," "FOSHU with risk reduction claims," "qualified FOSHU," and "ordinary FOSHU." The benefit of qualified FOSHU is lesser documentation requirement as compared to regular FOSHU. The primary characteristic of standardized FOSHU is its ability to expedite the approval process when the utilized ingredients adhere to the prescribed standards, criteria, and specifications.

Another important category, food with nutrient function claim (FNFC), was implemented by MHLW. It permits manufacturers to put a nutritional claim on labels. As per the regulations set forth in the Food Sanitation Act, any product aiming to be eligible for FNFC must incorporate 1 of 17 crucial nutrients, consisting of 12 vitamins and 5 minerals. These nutrients must be present in quantities that align with the prescribed maximum daily intakes, thereby guaranteeing the product's safety and enhancing its nutritional value. The FNFC used to fulfill the daily requirements of vitamins and minerals in the diet.

A glimpse of applicable laws regarding health food in Japan follows.

1.7 HISTORICAL PERSPECTIVE OF NUTRACEUTICALS IN AUSTRALIA

In Australia, the term *nutraceutical* is synonymous with *complementary medicine*. The complementary medicine industry is experiencing a rapid growth, making it one of the fastest-growing sectors in the country. A significant portion, approximately 70%, of the Australian population actively incorporates complementary medicines into their daily health regimen. The regulatory body of Australia, the Therapeutic Goods Administration (TGA), with the assistance of the Health Product Regulation Group (HPRG) under the Ministry of Health, is responsible for the regulation of complementary medicines in Australia. Also, Complementary Medicines Australia (CMA) plays a crucial role in the Australian market by advocating for the interests of various stakeholders involved in the production and distribution of complementary medicines. These include manufacturers, distributors, raw material suppliers, consultants, and retailers.

As per TGA, *complementary medicine* is defined as non-prescription medicine containing complementary ingredients, such as vitamins, minerals, herbal materials, essential oils, microorganisms, and aromatherapy and homeopathic preparations. Complementary medicines are regulated as medicines under the Therapeutic Goods Acts of 1989. The federal parliament of Australia passed the TGA Act of 1989, and it has two parts: one for registered goods, and the other for listed goods. As per the TGA Act of 1989, complementary medicines are either registered or listed in the Australian Regulatory Guidelines for Complementary Medicines (ARGCM). Approved medicines by the TGA are allocated with an AUST R or AUST L number on the label.

- **AUST R.** This applies to registered medications. The TGA has reviewed the product for safety, quality, and efficacy.
- **AUST L.** Listed medicines, applicable to low-risk medicines, like vitamin products, for which only safety and quality information is required.

The TGA took a significant step in 1991 by introducing fees and charges that were to be imposed on different industries and services. These fees and charges encompassed areas such as annual licensing, GMP guidelines, and inspections that were specifically targeted toward the nutraceutical industry.

In 2003, a treaty was established between Australia and New Zealand to govern the regulations pertaining to a wide range of pharmaceutical categories, including prescription medicines, over-the-counter drugs, medical devices, and complementary medicines. This treaty led to the formation of the Australia New Zealand Therapeutic Products Authority (ANZTPA), which aimed to ensure the safety and efficacy of therapeutic products in both countries.

Between 1996 and 2006, the Australian Register of Therapeutic Goods documented the inclusion of more than 1,000 "weight loss" products. However, a significant number of these products consisted of various unevaluated ingredients, such as herbs, vitamins, and minerals, casting doubt on their effectiveness. In contrast, only 10 conventional medicines were registered during the same timeframe, each containing a single evaluated ingredient that had been proven to be effective.

Other applicable guidelines for complementary medicines in Australia are as follows:

- Food Standards Australia New Zealand Act 1991
- Customs Acts 1901 and Customs Regulations 1956
- Industrial Chemicals (Notification and Assessment) Act 1989 and the National Industrial Chemicals Notification and Assessment Scheme
- Competition and Consumer Act 2010 and the Australian Consumer Law
- Australian Dangerous Goods Code
- Agricultural and Veterinary Chemicals Code Act 1994
- Environment Protection and Biodiversity Conservation Act 1999

Recently, in 2020, the Therapeutic Goods Administration (TGA) provided clarification regarding the regulation of specific sports supplements as therapeutic goods or medicines. Furthermore, from 2023, sports supplements in the form of tablets, capsules, or pills that make claims of enhancing performance in sports or recreational activities will fall under the regulation of therapeutic goods.

1.8 HISTORICAL PERSPECTIVE OF NUTRACEUTICALS IN CHINA

The term *nutraceutical* is called *nutritional supplement* or *dietary supplement* or *health food* in China. The regulation of these product categories falls under the jurisdiction of the CFDA. Another regulatory authority is the State Administration for Market Regulation (SAMR), which is responsible for implementing health food regulation, approval, labeling, and advertisement in China. Because of the increased demand for nutritional supplements, China has emerged as the second largest nutraceutical market globally.

In China, health food is categorized into three parts:

- Dietary supplements/nutritional supplement
- Functional foods
- Traditional Chinese medicines

Nutritional supplements contain vitamins and minerals without providing energy or other active ingredients. *Functional health food* refers to food products that make specific claims about their

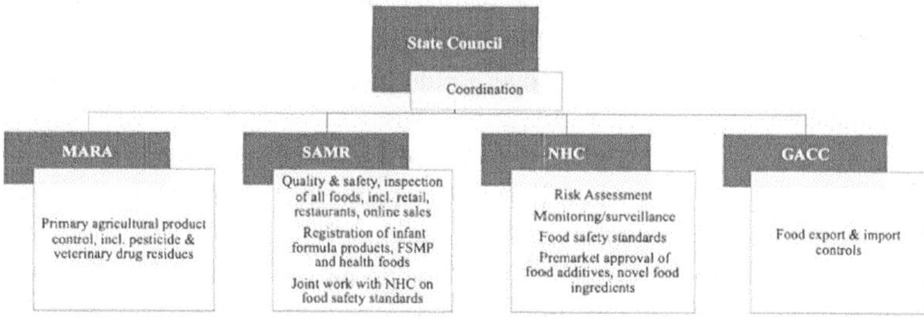

GACC, General Administration of Customs; MARA, Ministry of Agriculture and Rural Affairs; NHC, National Health Commission; SAMR, State Administration of Market Regulation.

FIGURE 1.4 Structure of the Chinese national food safety control program.

FIGURE 1.5 Blue Hat–certified logo.

impact on the body's physiological functions. As of now, 27 health claims are listed to be used for health foods in China. Deep-sea fish oil and ginseng tablets are considered prime examples of functional health foods. All health food products must be approved by SFDA, China.

In 1996, the first regulations regarding health foods in China were known as the Blue Hat, which provided regulations for the registration of health food products in China. After the successful registration of health food products, manufacturers could put the Blue Hat logo on product packaging. Under the registration procedure, vigorous scientific evaluation had been conducted to ensure the quality, safety, and efficacy of health foods.

In 2003, the Chinese government issued a public notice known as SFDA Notice No. 266 and 13, which outlined the regulatory guidelines for health food manufacturers in China. This notice established a comprehensive framework that governs the documentation and procedures to be adhered to by these manufacturers.

In 2008, an alarming scandal unfolded with the discovery of melamine being unlawfully mixed into milk and infant formula. The presence of melamine in these products proved fatal, resulting in the tragic loss of the lives of multiple infants. Such incident led to the formation of the Food Safety Law of the People's Republic of China, passed during the 7th Meeting of the Standing Committee of the 11th National People's Congress of the People's Republic of China on February 28, 2009. The latest amendment in the regulations was made in 2015.

In 2011, regulation regarding the nutritional labeling of prepackaged food, called GB 28050–2011, was implemented in China to increase the accuracy of information on the label. It mandates the inclusion of necessary information, such energy, fat, protein, carbohydrates, vitamins, and minerals on the label. GB2760–2011 includes standards for the use of food additives with its higher permissible limits in health food in China.

GB 16740–2014: General Standards for Natural and Health Foods plays a crucial role in providing direction for the production, labeling, and marketing of natural and health foods. Furthermore, these standards establish clear guidelines for the composition of ingredients in health food products.

Incorporating unauthorized and contaminated food additives in the blend can have a detrimental impact on the health of consumers. Considering this concern, the CFDA has issued standards for the control of food additives through the implementation of GB 2760–2014. This directive presents an approved catalogue of food additives that are suitable for inclusion in health food products. The additives specified include antioxidants, preservatives, coloring agents, emulsifiers, and stabilizers that are commonly employed in the formulation process.

The CFDA published regulations in 2016 for health food registration, titled Health Food Registration Application Service Guideline 2016 version. It covers dossier requirements for new products, renewals, and guidelines for domestic products containing vitamins and other nutrients.

The CFDA released the Health Food Raw Materials Directory (First Batch) in 2017. This directory lists the approved raw materials for health foods in China. Updates to the directory may include new raw materials or changes based on scientific evidence or regulations.

In China, the formula of a product is different for infants and young children. The age groups are categorized as Stage I for infants of age group 0 to 6 months; Stage II, older infants of age group 6 to 12 months; and Stage III, for young children, including the age group of 12 to 36 months. Recent changes were observed in infant formulas, published in 2021, which replaced the older version of 2010. Major changes include the separation of standards for older infants and young children, called GB10766–2021 and GB10767–2021, respectively.

1.9 HISTORICAL PERSPECTIVE OF NUTRACEUTICALS IN NEW ZEALAND

The term *nutraceutical* as used in New Zealand refers to *dietary supplement.* Dietary supplements are regulated under the Dietary Supplements Regulations 1985. The regulatory authority of New Zealand, the New Zealand Food Safety Authority (NZFSA), looks after the implementation of regulations provided under the Food Act 1981. Dietary supplements are categorized as food, but they must offer additional physiological advantages. For example, dairy products, like milk or yogurt, with the addition of folic acid, can be treated as dietary supplements in New Zealand. The ingredients used in the formulation must comply with the Animal Act 1999 or the Food Act 2014.

As per the Dietary Supplements Regulations 1985, *dietary supplements* are defined as:

- Amino acid, edible substance, herb, minerals, synthetic nutrients, or vitamins
- Sold by itself or in combination
- Sold in controlled dosage form as liquid, powder, or tablet (which might be described on the label as a cachet, capsule, lozenge, or pastille instead of as a tablet)
- Intended to be ingested orally
- Intended to supplement the amount of amino acid, edible substance, herb, mineral, synthetic nutrient, or vitamin normally derived from food

Another term used is *supplemented food,* which is food that has been enhanced with additional substances or modified to serve a physiological purpose beyond simple nutrition. But note that some categories are not categorized under supplemented food, like dietary supplements, medicines, controlled drugs or restricted substances, formulated meal replacement or formulated supplementary food, and formulated caffeinated beverages.

The regulation of additives and ingredients derived from animals plays a vital role in managing the potential risk associated with bovine spongiform encephalopathy (BSE) and/or transmissible spongiform encephalopathy (TSE), which are neurological diseases. To safeguard the well-being of

TABLE 1.6
List of Requirements and Eligibility Criteria

Purpose of Product	Composition of Product	Form of Product
Labeling of product	Effects after consumption of product	Applicable laws and regulations
Previous history of use or history of already existing product	Product usage patterns	Factors related to stakeholders of product

consumers and maintain product safety, the Animal Products Act 1999 was implemented. This Act stipulates that animal-based products must adhere to the prescribed requirements outlined in the Animal Products Act 1999.

In 2004, a joint committee, the Australia New Zealand Food Regulation Ministerial Council (ANZFRMC), was formed by the ministries of Australia and New Zealand. This committee aims to establish guidelines for fortifying food with vitamins and minerals, which were regulated under the Dietary Supplements Act 1985.

To determine the classification of dietary supplements in this joint scheme, NZFSA compiled a list of requirements and eligibility criteria. This list helps identify whether a dietary supplement is regulated as a therapeutic product (medicine) or falls under the category of fortified food-type products. This list consists of a total of nine categories.

In 2007, NZFSA proposed a recommendation which focuses on the separation of food-type and therapeutic-type dietary supplements regulations in New Zealand. The classification of supplements into food type and therapeutic type allows for a regulatory framework that is specifically designed to address the individual risks and benefits associated with each category.

Up until 2010, dietary supplements included both food and therapeutic products. With the Dietary Supplements Amendment Regulations in 2010, only therapeutic products are now classified as dietary supplements. Food products are now regulated under the New Zealand (Supplemented Food) Standard 2010. The food regulated under SFS 2010 is known as "supplemented food." The main interesting fact is that SFS is only applicable in New Zealand, not for the joint standard with Australia. Another important feature of the 2010 amendment was the authorization of a 300 mcg maximum daily dose of folic acid in dietary supplements manufactured, following the New Zealand Code of GMP for therapeutic goods. The next amendments in Supplement Food Standard were made in 2013 and 2016. In order to ensure the continued availability of dietary supplements for consumers until fit-to-purpose is completely realized, the amendment bill, which was declared in 2020, will come into effect from March 2021 until March 2026.

1.10 CONCLUSION

The significance of nutraceutical products lies in their ability to fulfill essential nutrient needs on a daily basis. These products are widely embraced across different categories on a global scale. Hippocrates clarified the connection of food with medicine: "Let food be your medicine and medicine be your food." The field of nutraceuticals fills a significant gap in providing health support to individuals. Extensive research articles have solidly established the efficiency of nutraceuticals in treating and preventing diverse disease conditions. Various countries are focusing on establishing well regulatory frameworks for nutraceutical products to maintain their safety, quality, and effectiveness. In India, nutraceutical categories are governed by the Food Safety Standards India (FSSAI). It mainly regulates the manufacturing, storage, distribution, sale, import, and export of food supplements, of nutraceutical products.

In USA, the term *nutraceutical* is used to refer to dietary supplements, regulated under the DSHEA 1994. Prior to this Act, there were no strict regulations in place for these supplements. In

Europe, nutraceutical products are referred to as food supplements. Since dietary supplements are considered food, it is the responsibility of the manufacturer/importer/supplier to meet the necessary standards for all types of food supplements. In Canada, nutraceutical products are handled as natural health products. In 1996, Canada's Therapeutic Products Programme and Food Directorate played a vital role in setting up the groundwork for nutraceutical products and their regulation.

In Japan, the term *nutraceuticals* is used for "health foods," which are categorized into two types: food products with functional claims and those without. Functional food regulations include functional health claim (FHC), food for specific health uses (FOSHU), and food with nutrient function claim (FNFC). In Australia, the term *nutraceutical* is synonymous with *complementary medicine*. Australia's Therapeutic Goods Administration (TGA) collaborates with the Health Product Regulation Group (HPRG) under the Ministry of Health to ensure the proper regulation of complementary medicines in the country. In China, they refer to nutritional supplements or dietary supplements or health food as nutraceuticals. The CFDA is responsible for regulating these product categories. In New Zealand, the 1985 Dietary Supplements Regulations oversee dietary supplements. The New Zealand Food Safety Authority (NZFSA) is responsible for enforcing the regulations outlined in the Food Act 1981.

REFERENCES

[1] https://applications.csjmu.ac.in/naacfiles/criteria3/1263.pdf (Chapter introduction)

[2] https://www.jetir.org/papers/JETIR2203472.pdf (Nutraceuticals Indian perspective)

[3] https://www.mdpi.com/2304–8158/12/2/427 (Investigating the Regulatory Process, Safety, Efficacy and Product Transparency for Nutraceuticals in the USA, Europe and Australia)

[4] https://ajcn.nutrition.org/article/S0002–9165(22)03993–4/pdf (NHP act 2004 introduction, Canada)

[5] https://epub.uni-bayreuth.de/id/eprint/5178/1/wp-3.pdf (Regulatory and Legislative Framework for Novel Foods in US, EU and other jurisdictions)

[6] https://fssai.gov.in (FSSAI official website)

[7] https://www.fssai.gov.in/upload/uploadfiles/files/Compendium_Nutra_29_09_2021.pdf (FSSAI 2016 guidance document)

[8] https://www.fssai.gov.in/upload/advisories/2022/05/627a160339d32Order_FSS_Nutra_10_05_2022.pdf (FSSAI 2022 enzymes addition document)

[9] https://www.thepharmajournal.com/archives/2018/vol7issue7/PartM/7–7–91–540.pdf (Health claims in USA)

[10] https://www.efsa.europa.eu/en/topics/topic/food-supplements (European food supplements introduction)

[11] https://food.ec.europa.eu/safety/labelling-and-nutrition/food-supplements_en (Amendment on list of vitamins and minerals list).

[12] https://eur-lex.europa.eu/legal-content/EN/ALL/?uri=celex%3A32002L0046 (Directive 2002/46/EC)

[13] https://www.eumonitor.eu/9353000/1/j4nvk6yhcbpeywk_j9vvik7m1c3gyxp/vitgbgibjuwo (Regulation (EC) No. 178/2002)

[14] https://eurlex.europa.eu/LexUriServ/LexUriServ.do?uri=CONSLEG:2004R0852:20090420:EN:PDF (EC) No. 852/2004

[15] https://eur-lex.europa.eu/legal-content/EN/ALL/?uri=celex%3A32006R1924 (Regulation (EC) No. 1924/2006)

[16] https://eur-lex.europa.eu/legal-content/EN/ALL/?uri=celex%3A32006R1925 (Regulation (EC) No. 1925/2006)

[17] https://eurlex.europa.eu/LexUriServ/LexUriServ.do?uri=OJ:L:2006:364:0005:0024:EN:PDF (Regulation (EC) No. 1881/2006)

[18] https://eur-lex.europa.eu/eli/reg/2008/1333/oj (Regulation (EC) No. 1333/2008)

[19] https://eur-lex.europa.eu/eli/reg/1997/258/oj (Regulation (EC) No. 258/1997)

[20] https://www.eumonitor.eu/9353000/1/j4nvk6yhcbpeywk_j9vvik7m1c3gyxp/vjzsag4e0num:~:text=It%20lays%20down%20rules%20for,human%20health%20and%20consumers'%20interest (Regulation (EC) No. 2283/2015)

[21] https://eur-lex.europa.eu/eli/reg_impl/2017/2470/oj (Regulation (EU) 2017/2470)

[22] https://www.canada.ca/en/health-canada/services/drugs-health-products/natural-non-prescription/legislation-guidelines/guidance-documents/labelling.html (Canada NHP labelling guidance)

[23] https://www.canada.ca/en/health-canada/services/drugs-health-products/natural-non-prescription/legislation-guidelines/guidance-documents/clinical-trials.html (NHP clinical trial guidance)

[24] https://www.canada.ca/en/health-canada/services/food-nutrition/food-labelling/health-claims/nutraceuticals-functional-foods-health-claims-foods-policy-paper.html (Health Canada policy paper 2002)
[25] https://www.canada.ca/en/health-canada/services/food-nutrition/legislation-guidelines/guidance-documents/guidelines-safety-assessment-novel-foods-derived-plants-microorganisms/guidelines-safety-assessment-novel-foods–2006.html (Novel food guidance Canada)
[26] https://www.eu-japan.eu/sites/default/files/publications/docs/2016–03-nutraceuticals-japan-min.pdf (Japan 3 programs initiative)
[27] https://treasury.gov.au/sites/default/files/202105/complementary_medicines_australia.pdf (Australia introduction)
[28] https://www.tga.gov.au/news/news/changes-regulation-sports-supplements-medicinal-dosage-forms (sports supplements in Australia)
[29] https://www.tga.gov.au/sites/default/files/australian-regulatory-guidelines complementary-medicines-argcm.pdf (Australia complementary medicine guidance documents)
[30] https://www.raps.org/News-and-Articles/News-Articles/2022/6/An-update-of-Chinas-food-safety-regulatory-framework (China chinese national food safety control system)
[31] https://www.mpi.govt.nz/food-business/labelling-composition-food-drinks/health-and-nutrition-content-claims-for-food-and-drink/supplemented-food/ (New Zealand intro)
[32] https://www.beehive.govt.nz/sites/default/files/Dietary%20Supplements%20Discussion%20Paper.pdf (ANZFRMC)
[33] https://bills.parliament.nz/v/6/a07c40a6-0f0b-4c03-a4da-8a44d165e873?Tab=history (New Zealand 2020 amendment bill)
[34] https://www.legislation.govt.nz/regulation/public/2010/0005/latest/whole.html (D.S regulation 2010)
[35] FSSAI [Internet]. https://fssai.gov.in/

2 Understanding the Different Systems of Diverse Terminologies for Dietary Supplements and Related Products

Khushi Pandya, Niyati Acharya, and Yashwant Pathak

2.1 INTRODUCTION

Although nutraceuticals have been around for about 3,000 years, it was not until 1989 that the term was coined. Stephen DeFelice, who leads the Foundation for Innovation in Medicine in Cranford, New Jersey, referred to the term as "a food (or part of a food) that provides medical or health benefits, including the prevention and/or treatment of a disease." The marketing of nutraceutical products is happening worldwide, and according to current trends, that figure will only increase in the future. Despite this, the lack of a harmonized definition for these products leads to a new set of concerns.

Nutraceuticals are classified into various categories by each nation's own set of terminologies and norms. What is a *dietary supplement* as per the USFDA (United States) becomes *food for specified health use* (FOSHU) for the PMDA (Japan). Differences in terminologies like these are often misleading and cause ambiguity, especially when it comes to marketing a product internationally. Each nation has its regulatory procedure to market products like this, but it is necessary to classify such products according to the categories dealt by that regulatory agency. Thus, it is quite essential to get a clear understanding regarding the categories provided by each regulatory agency and which type of products fall under them.

2.2 UNITED STATES (USFDA) [1–3]

Oversight for nutraceuticals in the United States is conducted by the United States Food and Drug Administration (USFDA), specifically through the Center for Food Safety and Applied Nutrition (CFSAN). Definitions established by the DSHEA categorize dietary supplements as "product[s] intended for ingestion that, among other requirements, contains a dietary ingredient intended to supplement the diet." These supplements typically comprise dietary ingredients, such as vitamins, minerals, herbs, amino acids, and enzymes, as well as other adjuncts, like fillers, binders, excipients, preservatives, sweeteners, and flavorings. The FDA maintains a periodically updated, non-comprehensive list of dietary ingredients along with their alternative names. "Notably, the DSHEA restricts the FDA from pre-approving the safety and effectiveness of dietary supplements before they enter the market." Generally, dietary supplements are considered akin to "food," unless they assert claims to mitigate, treat, prevent, cure, or alleviate symptoms of a disease, at which point they

DOI: 10.1201/9781003465041-2

are classified as drugs. In such instances, the stringent regulatory requirements applicable to drugs would govern their oversight. FDA has provided a non-comprehensive list of dietary ingredients along with their other known names, and this list is updated periodically.

Few such examples are listed in what follows. [3]

TABLE 2.1
Organizational Hierarchy of the USFDA

"Center for Biologics Evaluation and Research"
"Center for Devices and Radiological Health"
"Center for Drug Evaluation and Research"
"Center for Food Safety and Applied Nutrition"
"Center for Tobacco Products"
"Center for Veterinary Medicine"
"National Center for Toxicological Research"
"Office of Regulatory Affairs"
"Office of Operations"

TABLE 2.2
Dietary Ingredient Names Used Interchangeably

Ingredient Name	Alternatively Used Names
"Biotin"	Vitamin H, vitamin B7
"Anatabine"	Anatabloc
"Calcifediol"	"25-hydroxycholecalciferol"
	"25-hydroxy vitamin D3"
	"Calcifediol anhydrous"
	"Dedrogyl"
	"Delakmin"
	"Didrogyl"
	"Rayaldee"

2.3 EUROPEAN UNION (EFSA) [4–5]

Ideally, *food supplements* are defined as "concentrated sources of nutrients (i.e. mineral and vitamins) or other substances with a nutritional or physiological effect that are marketed in "dose" form (e.g. pills, tablets, capsules, liquids in measured doses). A wide range of nutrients and other ingredients might be present in food supplements, including, but not limited to, vitamins, minerals, amino acids, essential fatty acids, fibre and various plants and herbal extracts". They are regulated as food under the EU.

Novel food is defined as food that has not been consumed to any significant degree in the EU before 15 May 1997 (when the first novel food legislation entered into force). This can be newly developed, innovative food or food produced using new technologies and production processes, as well as food traditionally eaten outside of the EU.Novel food will only be authorised for use in the EU if it does not present a risk to public health, is not nutritionally disadvantageous when replacing

TABLE 2.3
Organizational Hierarchy of the EMA

EUROPEAN MEDICINES AGENCY SCIENCE MEDICINES HEALTH	"Committee for Medicinal Products for Human Use"
	"Pharmacovigilance Risk Assessment Committee"
	"Committee for Veterinary Medicinal Products"
	"Committee for Orphan Medicinal Products"
	"Committee on Herbal Medicinal Products"
	"Committee for Advanced Therapies"
	"Paediatric Committee"
	"Working parties and other groups"

a similar food and does not misleading to the consumer. It must undergo a scientific assessment prior to authorisation to ensure its safety. The authorisation sets out the conditions for its use, its designation as a food, specific labeling requirements (where appropriate) and post-market monitoring requirements (where appropriate). [6]

Genetically modified food means "food containing, consisting of or produced from GMOs [genetically modified organisms]." [7]

Infant formula means "food intended for use by infants during the first months of life and satisfying by itself the nutritional requirements of such infants until the introduction of appropriate complementary feeding." [8]

Follow-on formula means "food intended for use by infants when appropriate complementary feeding is introduced and which constitutes the principal liquid element in a progressively diversified diet of such infants."

Processed cereal-based food means food:

1. intended to fulfil the particular requirements of infants in good health while they are being weaned, and of young children in good health as a supplement to their diet and/or for their progressive adaptation, to ordinary food; and
2. pertaining to one of the following categories:
   a. simple cereals which are or have to be reconstituted with milk or other appropriate nutritious liquids,
   b. cereals with an added high protein food which are or have to be reconstituted with water or other protein-free liquid,
   c. pastas which are to be used after cooking in boiling water or other appropriate liquids
   d. rusks and biscuits which are to be used either directly or, after pulverisation, with the addition of water, milk or other suitable liquids. [8]

Baby food means food intended to fulfil the particular requirements of infants in good health while they are being weaned, and of young children in good health as a supplement to their diet and/or for their progressive adaptation to ordinary food, excluding:

1. processed cereal-based food; and
2. milk-based drinks and similar products intended for young children. [8]

Food for special medical purposes means food specially processed or formulated and intended for the dietary management of patients, including infants, to be used under medical

TABLE 2.4
Classification of Food as per PMDA

Food				
Health Functional Food			**Conventional Food**	
Health Foods				
"Food for specified health use"	"Food with nutrient function claims"	"Food with functional claims"	Health Food in general sense (so-called Health Food)	Other conventional food

supervision; it is intended for the exclusive or partial feeding of patients with a limited, impaired or disturbed capacity to take, digest, absorb, metabolise or excrete ordinary food or certain nutrients contained therein, or metabolites, or with other medically-determined nutrient requirements, whose dietary management cannot be achieved by modification of the normal diet alone. [8]

Total diet replacement for weight control means "food specially formulated for use in energy restricted diets for weight reduction which, when used as instructed by the food business operator, replaces the whole daily diet." [8]

2.4 JAPAN (PMDA) [9–11]

In Japan, the Ministry of Health, Labour, and Welfare regulates nutraceuticals under the name "health foods." They are classified into categories:

- Food with nutritional function claims (FNFC)
- Food with specified health uses (FOSHU)
- Food with functional claims (FFC)

Food for special health use (FOSHU) refers to "foods containing ingredients with functions for health, physiological functions, and biological activity on the body and is officially approved to claim its physiological effects on the human body."[11]

Foods for special dietary use (FOSDU) "products can use medical and nutritional expressions to indicate that they are appropriate to promote growth or maintain or recover health among infants, children, the infirm, and pregnant women."

Foods with nutrient function claims (FNFC) refers to "foods labeled with the functions of their nutritional ingredients. They are considered to be the 'foods with function claims.'"

2.5 INDIA (FSSAI) [12, 13]

In India, the Food Safety and Standards (Health Supplements, Nutraceuticals, Food for Special Dietary Use, Food for Special Medical Purpose, Functional Food, and Novel Food) Regulations of 2016 classify products into various categories based on their characteristics.

Food for special medical purpose means food intended for:

(i) particular dietary use specially processed or formulated;
(ii) the dietary management of persons and used only under medical advice;

(iii) the exclusive or partial feeding of persons with a limited, impaired or disturbed capacity to take, digest, absorb, metabolize or excrete ordinary foodstuffs or certain nutrients contained therein or metabolites; or
(iv) other medically determined nutrient requirements, whose dietary management cannot be achieved only by modification of the normal diet, by food for specific nutritional use, or a combination of them.

Food with added prebiotic ingredients means "food that contains added prebiotic ingredients which are non-viable food components that confer health benefits to the consumer by modulation of gut microbiota."

Food with added probiotic ingredients means "food with live micro-organisms beneficial to human health, which when ingested in adequate numbers as a single strain or as a combination of cultures, confer one or more specified or demonstrated health benefits in human beings."

Food for special dietary use/functional foods/nutraceuticals/health supplements mean foods which are specially processed or formulated to satisfy particular dietary requirements which exist because of a particular physical or physiological condition or specific diseases and disorders and which are presented as such, wherein the composition of these foodstuffs must differ significantly from the composition of ordinary foods of comparable nature, if such ordinary foods exist, and may contain one or more of the following ingredients, namely:-

1. plants or botanicals or their parts in the form of powder, concentrate or extract in water, ethyl alcohol or hydro alcoholic extract, single or in combination;
2. minerals or vitamins or proteins or metals or their compounds or amino acids (in amounts not exceeding the Recommended Daily Allowance for Indians) or enzymes (within permissible limits);
3. substances from animal origin;
4. a dietary substance for use by human beings to supplement the diet by increasing the total dietary intake.

Novel food means "an article of food for which standards have not been specified but is not unsafe, provided that such food does not contain any of the foods and ingredients prohibited under Food Safety and Standards Act, 2006 and regulations made there under."

Nutritional ingredients mean "the ingredients specified in Schedules other than the food additives specified in Schedule VA to Schedule VF, packed and made available in a form not for retail-consumer use, but meant for use in formulating a product falling under various categories of these regulations or other categories specified in the Food Safety and Standards (Food Products Standards and Food Additives) Regulations, 2016."

Nutrient ingredients mean "vitamins, minerals, and amino acids as specified in Schedule I and Schedule II of Food Safety and Standards Regulations 2016."

2.6 AUSTRALIA (TGA) [14–16]

In Australia, nutraceuticals are regulated as complementary medicines. The Medicines Regulation Division defines *complementary medicines* as "non-prescription medicines. They are medicines that primarily contain complementary medicine ingredients that have an established identity and tradition of use." Examples of complementary medicine ingredients are:

- Vitamins
- Minerals

- Plant material
- Nutritional substances
- Essential oils
- Microorganisms
- Animal material

Complementary medicines are divided further into three categories. [16]

Herbal medicines are "therapeutic goods that are, or contain as the major active ingredient(s), herbal substances. Herbal substances are preparations of plants, and other organisms that are treated as plants in the International Code of Botanical Nomenclature, such as fungi, algae and yeast."

Traditional medicines include "a diverse range of health practices, approaches, knowledge and beliefs incorporating medicines of plant, animal and/or mineral origin. Examples of traditional paradigms include: Traditional Chinese medicine, Ayurvedic medicine, Aboriginal and Torres Strait Islander medicine and Western herbal medicine."

Homeopathic medicines, or "homoeopathic preparations," are based upon the central tenet of homoeopathy, "let like cure like," and the principles of homoeopathic pharmacy—"potentization," being the serial dilution and succussion of a stock. Homoeopathic medicines are derived from a wide variety of natural source materials, mostly plants and minerals. Some of these source materials are poisonous, for example: *Atropa belladonna*. The highly diluted nature of homoeopathic preparations is considered to render starting materials non-toxic and therefore safe for therapeutic use.

TABLE 2.5
Organizational Hierarchy of the Therapeutic Goods Administration

	Medicines Regulation Division	• Prescription Medicines Authorisation • Complimentary and Over-the-Counter Medicines • Scientific Evaluation • Medicines Regulation Division
Australian Government Department of Health Therapeutic Goods Administration	"Medical Devices and Product Quality Division"	• "Medical Devices Branch" • "Laboratories Branch" • "Manufacturing Quality Branch"
	Regulatory Practice and Support Division	• "Regulatory Services and Drug Control Branch" • "Regulatory Education and Compliance Branch" • "Regulatory Engagement and Planning Branch"

2.7 CANADA (HEALTH CANADA) [17–19]

In Canada, various terminologies are used in place of *dietary supplements*. "The Natural and Non-Prescription Health Products Directorate is responsible for regulating the nutraceuticals in general." The definition of these terminologies are as follows.

A *nutraceutical* is a "product isolated or purified from foods that is generally sold in medicinal forms not usually associated with food. A nutraceutical is demonstrated to have a physiological benefit or provide protection against chronic disease."

Functional food is "similar in appearance to, or may be, a conventional food, is consumed as part of a usual diet, and is demonstrated to have physiological benefits and/or reduce the risk of chronic disease beyond basic nutritional functions."

According to Health Canada, "there are no regulations specifically dealing with nutraceuticals or functional foods. All foods and drugs fall under the provision of the Food and Drugs Act 1985. They may be considered drug if the claims made by the manufacturer meet the definition of a drug, in which case they will be regulated as a subset of drugs."

TABLE 2.6
Organizational Hierarchy of Health Canada

Health Canada

Office of Audit and Evaluation	• Departmental Audit Committee • Director General/Chief Audit Executive's Office • Internal Audit and Special Examinations • Program Evaluation Division • Performance Measurement Planning and Integration • Practice Management
Chief Financial Officer Branch	• Departmental Submissions and Performance Measurement Directorate • Financial Operations Directorate • Procurement and Investment Management Directorate • Resource Management and Advisory Services Directorate
Communications and Public Affairs Branch	• Ethics and Internal Ombudsman Services • Marketing and Communications Services Directorate • Planning and Operations Division • Public Affairs and Strategic Communications Directorate • Stakeholder Relations and Consultation Directorate
Controlled Substances and Cannabis Branch	• Cannabis Directorates • Controlled Substances and Overdose Crisis Response Directorates • Horizontal Policy, Planning, and Program Directorate • Tobacco Control Directorate • Branch Operational Support Services
Corporate Services Branch	
Departmental Secretariat	
Digital Transformation Branch	

TABLE 2.6 (*Continued*)
Organizational Hierarchy of Health Canada

Health Products and Food Branch	• Biologic and Radiopharmaceutical Drugs Directorate • Food Directorate • Marketed Health Products Directorate • Medical Devices Directorate • Natural and Non-Prescription Health Products Directorate • Office of Nutrition Policy and Promotion • Pharmaceutical Drugs Directorate • Policy, Planning, and International Affairs Directorate • Resource Management and Operations Directorate • Veterinary Drugs Directorate
Healthy Environments and Consumer Safety Branch	• Consumer and Hazardous Products Safety Directorate • Environmental and Radiation Health Sciences Directorate • Policy, Planning, and Integration Directorate • Safe Environments Directorate
Legal Services	
Opioid Response Team	• Controlled Substances Directorate • Opioid Response Team Directorate
Pest Management Regulatory Agency	
Regulatory Operations and Enforcement Branch	• Cannabis Directorate • Laboratories Directorate • Planning and Operations Directorate • Health Product Compliance Directorate • Policy and Regulatory Strategies Directorate • Medical Devices and Clinical Compliance Directorate • Consumer Product Safety, Tobacco and Pesticides Directorate • Controlled Substances and Environmental Health Directorate
Strategic Policy Branch	• Applied Research and Analysis Directorate (ARAD) • Canada Health Act Division (CHAD) • Federal/Provincial Relations Division (FPRD) • Official Language Health Program (OLHP) • Policy Planning and Integration Directorate (PPID) • Science Advisory Board

Natural health products are defined as a substance set out in Schedule 1 or a combination of substances in which all the medicinal ingredients are substances set out in Schedule 1, a homeopathic medicine or a traditional medicine, that is manufactured, sold or represented for use in (a) the diagnosis, treatment, mitigation or prevention of a disease, disorder or abnormal physical state or its symptoms in humans; (b) restoring or correcting organic functions in humans; or (c) modifying organic functions in humans, such as modifying those functions in a manner that maintains or promotes health.

This includes:

- Vitamins and minerals
- Herbal remedies
- Homeopathic medicines
- Traditional medicines, such as traditional Chinese medicines
- Probiotics
- Other products, like amino acids and essential fatty acids

Novel foods are described by Health Canada as:

- Foods resulting from a process not previously used for food;
- Products that do not have a history of safe use as a food (that is, products that have never been used as food); or
- Foods that have been modified by genetic manipulation, also known as genetically modified foods, GM foods, genetically engineered foods or biotechnology-derived foods

2.8 CHINA (NMPA) [20–24]

In China, nutrient supplements are regulated as one category under the umbrella of *health food* by the National Medical Products Administration (NMPA). They are defined as, A food with health function claims or for supplementing nutrients (such as vitamins/minerals), for regulating body functions, not intended to cure diseases, containing specific functional ingredients, suited for specific groups of people, and with fixed amount consumption.

Health food is divided into 2 categories, one is functional health food and the other is nutrient supplements:

- *Functional health food* refers to "health food with one or several of 27 functions."
- *Nutrient supplements* refer to "products whose purpose is to supplement nutrients such as vitamins and minerals."

2.9 ASEAN COUNTRIES [25–26]

The health supplement industry of ASEAN countries is represented by the ASEAN Alliance of Health Supplement Associations (AAHSA). According to them, *health supplement* can be defined as any product that is used to supplement a diet and to maintain, enhance and improve the healthy function of human body and contains one or more, or a combination of the following:

a. Vitamins, minerals, amino acids, fatty acids, enzymes, probiotics and other bioactive substances
b. Substances derived from natural sources, including animal, mineral and botanical materials in the forms of extracts, isolates, concentrates, metabolite
c. Synthetic sources of ingredients mentioned in (a) and (b).

2.10 RUSSIA (ROSPOTREBNADZOR) 27–28

In Russia, dietary supplements are termed *biologically active supplements* (BASs).

Biologically active supplements are [n]atural and/or identical to natural biologically active ingredients, and also probiotic microorganisms, intended for consumption with food or inclusion in products to enrich the diet with them.

TABLE 2.7
ASEAN Countries with Their Regulatory Authorities and Sub-Divisions Dealing with Nutraceuticals

Country (Authority)	Sub-Divisions
Malaysia (Ministry of Health)	• National Pharmaceutical Regulatory Agency (NPRA) • Food Safety and Quality Division
Singapore (Health Sciences Authority)	• Health Products Regulation Group • Health Products Regulation Department
Thailand (Food and Drug Administration)	• Division of Food Control • Division of Dietary Supplements Control
Indonesia (Badan Pengawas Obat dan Makanan)	• Directorate of Food Supervision and Consumption • Directorate of Traditional Medicine and Health Supplements
Philippines (Food and Drug Administration)	• Center for Food Regulation and Research (CFRR) • Center for Food and Drug Regulation and Research (CFDRR)
Vietnam (Ministry of Health)	• Food Safety Department • Department of Food Hygiene and Safety
Cambodia (Ministry of Health)	• Department of Drug, Food, and Cosmetic Quality and Safety • Department of Drug, Food, and Cosmetics Control
Laos (Ministry of Health)	• Food and Drug Department • Traditional Medicine Department
Myanmar (Food and Drug Administration)	• Food and Drug Administration • Department of Food and Drug Quality Control

According to the above definition, the following products can be identified as biologically active supplements:

- Polysaturated acids
- Mineral substances, micro and macro elements
- Some amino acids
- Some mono and disaccharides
- Food fibres
- Microorganisms present in human body
- Vitamins, whose daily dose does not exceed the one needed for the human body

However, under the Custom Union Legislation, BASs are considered as food for special dietary uses, that is, "[f]ood products with respect to which requirements have been established as to the content and/or proportion of certain or all substances and components and/or content and/or proportion of certain substances has been changed as compared to natural content of such substances, and/or into the content have been introduced substances or components (except for food additives and flavourings) not originally present therein, and/or manufacturer declares medicinal and/or preventive properties thereof and which is intended for the purposes of safely consuming this food products by certain categories of people."

Baby food is inclusive of drinking water; this is further classified into three age-wise categories: infants (3 years or less), preschool (3–4 years), and school-age (6 years and above).

Medical foods are defined as "foods with controlled nutrition and energy values and physical and organoleptic properties used in medical diets."

Dietetic foods for disease prevention are defined as "those used for correction of carbohydrate, fat and protein metabolism, with modified ratio of ingredients and/or containing foreign substances (components), or those used to reduce risk of diseases."

Mineral water is defined as "Mineral water is water that contains dissolved salts, microelements, and some biologically active components."

Sports foods are defined as "foods of a specific chemical composition, with increased nutrition value and/or tailored effect, which improves the adaptive human response to physical and emotional stress."

Food for pregnant and breastfeeding women is defined as "specialized food products in which the content and/or ratio of certain substances is modified in relation to their natural levels, and/or which contain substances or components not originally present, intended to meet the physiological needs of pregnant and breastfeeding women."

Biologically active food supplements are defined as "natural (or identical to natural) biologically active substances, including probiotic microorganisms, designed to be taken with food or made a part of food products."

2.11 GCC [29]

GCC countries have provided a harmonized definition of *nutraceuticals*, which they define as, "[f]ood supplements food/products that are used to supplement the normal diet which contain ingredients, alone or in combination, which may have a nutritional or physiological effect and often is concentrated and taken in doses. Food supplements consist of one or more of the following components: vitamins, minerals, fatty acids, amino acids, enzymes, prebiotics and probiotics, collagen, dietary fibres, melatonin, propolis, pollen, herbs or food herbal extracts, etc."

TABLE 2.8
GCC Countries along with Their Regulatory Authorities and Sub-Division Dealing with Nutraceuticals

Country (Regulatory Authority)	Sub-Division
Saudi Arabia (Saudi Food and Drug Authority)	• Food Sector • Drug Sector
United Arab Emirates (Ministry of Health and Prevention)	• Drug Control Department • Food Safety Department
Qatar (Ministry of Public Health)	• Pharmacy and Drug Control Department • Food Safety and Environmental Health Department
Bahrain (National Health Regulatory Authority)	• Drug Control Department • Food Safety Department
Oman (Ministry of Health)	• Directorate General of Pharmaceutical Affairs and Drug Control • Directorate General of Food Safety

2.12 UNITED KINGDOM (MHRA) [30–32]

In the United Kingdom, nutraceuticals are labeled as food supplements and are regulated under the authority of the Food Standards Agency, and they are defined as any food the purpose of which is to supplement the normal diet and which is a concentrated source of a vitamin or mineral or other substance with a nutritional or physiological effect, alone or in combination and is sold in dose form.

A wide range of nutrients and other ingredients might be present in food supplements. These can include thus:

- Vitamins
- Minerals
- Amino acids
- Essential fatty acids
- Fiber
- Various plants and herbal extracts

Novel foods are "any food that was not used for human consumption to a significant degree within the United Kingdom (UK) or the European Union (EU) before 15 May 1997. This means that the foods don't have a 'history of consumption.'"

2.13 THAILAND (TFDA) 33–34

The Thai Food and Drug Administration (Thai FDA) is the authority which regulates food and nutraceuticals in Thailand as per the Food Act BE 2522 (1979). Nutraceuticals are regulated under the name *food supplements* in Thailand. Its definition has been provided in the notification 293 B.E.2548 (2005) as follows.

Food supplement can be defined as products taken for consumption other than conventional foods which contain nutrients or other substances as ingredients and are in forms of tablets, capsules, powders, flakes, liquids or others, which are not conventional foods for consumers who expect for the benefit of health promotion.

Here, *other substances* mainly refer to thus:

(i) Vitamins, amino acids, fatty acids, minerals and produces obtained from plants or animals.
(ii) Concentrates, metabolites composition or extracts of substances of (i).
(iii) Artificial substances imitated like substances of (i) or (ii).
(iv) Mixtures of one or more of substances of (i) (ii) or (iii).

2.14 INDONESIA (BPOM) [35]

The nutraceutical products of Indonesia are regulated by the Indonesian National Agency for Drug and Food Control (BPOM). Under this, nutraceutical products are evaluated under the name *health supplements*.

A *health supplement* means a product designed to supplement nutrient intake, to maintain, to increase, and/or to improve health functions, to have nutritional value and/or a physiological effect, to contain one or more materials in the form of vitamins, minerals, amino acids and/or other non-plant materials combinable with plants.

Functional foods can be defined as "processed food that contains one or more functional components which, based on scientific studies, have certain physiological functions, are proven to be harmless and beneficial to health."

2.15 SRI LANKA (NMP) [36–37]

Functional food is "food with physiological benefits and/or reduce the risk of chronic diseases beyond basic nutritional functions, and may be similar in appearance to conventional food and consumed as part of a regular diet."

> Food is classified as functional when it contains biologically active substances like nutrients (such as vitamins, minerals, proteins, and omega-3 fatty acids), as well as compounds like pro-vitamin A, dietary fiber (such as inulin and fructooligosaccharides), plant sterols (for instance, phytoestrogens found in soy), organosulfur compounds, polyacetylene compounds, phenolic compounds, and prebiotics/probiotics.

2.16 MALDIVES (MFDA) [38]

Under the MFDA (Maldives Food and Drug Authority), *nutraceuticals* are defined as "[p]roducts containing vitamins, and minerals (natural& synthetic) with physiological process or specific claims to maintain or enhance a specific body function or structure."

2.17 NEW ZEALAND (MEDSAFE) [39]

Medsafe oversees the regulation of dietary supplements through the Dietary Supplements Regulations of 1985, which are governed by the Food Act 2014. Meanwhile, the Ministry for Primary Industries (MPI) is responsible for enforcing the Food Act.

Any product is considered a *dietary supplement* if the following clauses are applicable to it:

1. "It is an amino acid, edible substance, herb, mineral, synthetic nutrient, or vitamin."
2. "It is sold by itself or in a mixture."
3. "It is sold in a controlled dosage form as a liquid, powder, or tablet (which might be described on the label as a cachet, capsule, lozenge, or pastille instead of as a tablet)."
4. "It is intended to be ingested orally."
5. "It is intended to supplement the amount of the amino acid, edible substance, herb, mineral, synthetic nutrient, or vitamin normally derived from food."

2.18 MEXICO (COFEPRIS) [40]

Food supplements are regulated as food in Mexico. Article 215 of the General Health Law and Title 19 of the Regulation of Sanitary Control of Products and Services are responsible for regulating food supplements.

Food supplements can be defined as products based on herbs, plant extracts, traditional food, dehydrated or concentrated fruits, added or not, of vitamins and minerals, that can be presented in a pharmaceutical form and whose purpose of use is to increase the total dietary intake, supplement it or supply more components. These supplements can contain carbohydrates, plants, algae, dehydrated traditional food, metabolites, fatty acids, amino acids and protein.

2.19 BRAZIL (ANVISA) [41]

Nutraceuticals are considered under the name *food supplements* under ANVISA, which can be defined as "[p]roduct for oral administration, presented under pharmaceutical dosage forms

TABLE 2.9
Organizational Hierarchy of INVIMA

Planning Advisory Office
Legal Advisory Office
Internal Control Office
Information Technology Office
Citizen Service Office
International Affairs Office
Laboratory and Quality Control Office
General Secretary
Directorate of Medicines and Biological Products
Directorate of Medical Devices and Other Technologies
Food and Beverages Directorate
Directorate of Cosmetics, Toiletries, Pesticides and Domestic Hygiene Products
Directorate of Health Operations
Health Responsibility Directorate

intended to supplement the diet of healthy individuals with nutrients, bioactive substances, enzymes or probiotics, alone or in combination."

This mainly includes thus:

1. Vitamin and mineral supplements
2. Bioactive substances and probiotics
3. Novel foods
4. Foods with claims of functional properties
5. Athlete supplements
6. Supplements for pregnant women and nursing mothers
7. Specific non-prescription medicines

2.20 COLOMBIA (INVIMA) [42–44]

The regulation of dietary supplements, known as "suplementos dietarios" in Colombia, is governed by Decree (Decreto) No. 3249/2006 and Decree (Decreto) No. 3863/2008. They are categorized independently and are not classified as either food or drugs.

A *dietary supplement* is defined as a product whose purpose is to add to the normal diet and that is a concentrated source of nutrients and other substances with a physiological or nutritional effect that may contain vitamins, minerals, proteins, amino acids, other nutrients and derivatives of nutrients, plants, concentrates and plant extracts, alone or in combination.

TABLE 2.10
A Comparative View of Diversity in Terminologies Used across Various Countries

	Product							
Country	Omega-3 Fish Oil Capsules	Ginseng Extract Tablets	Probiotic Yogurt	Green Tea Extract Pills	Turmeric Curcumin Capsules	Garlic Extract Capsules	Vitamin C Tablets	Collagen Powder
United States	Classified as a dietary supplement	Classified as a dietary supplement	Classified as a functional food	Classified as a dietary supplement	Classified as a dietary supplement	Classified as a dietary supplement	Classified as a dietary supplement	Classified as a dietary supplement
European Union	Regulated as a food supplement	Regulated as a traditional herbal medicine	Regulated as a food supplement	Regulated as a food supplement	Regulated as a food supplement	Regulated as a food supplement	Regulated as a food supplement	Regulated as a food supplement
Japan	Regulated as food with nutritional function claims	Regulated as food with specified health uses	Regulated as food with functional claims	Regulated as food with nutrient function claims	Regulated as food with specified health uses	Regulated as food with nutrient function claims	Regulated as food with nutrient function claims	Regulated as food with functional claims
India	Regulated as a health supplement	Regulated as a health supplement	Regulated as a health supplement	Regulated as a health supplement	Regulated as a health supplement	Regulated as a health supplement	Regulated as a health supplement	Regulated as a health supplement
Australia	Regulated as a complementary medicine	Regulated as a complementary medicine	Regulated as a complementary medicine	Regulated as a complementary medicine	Regulated as a complementary medicine	Regulated as a complementary medicine	Regulated as a complementary medicine	Regulated as a complementary medicine
Canada	Regulated as a natural health product	Regulated as a natural health product	Regulated as a natural health product	Regulated as a natural health product	Regulated as a natural health product	Regulated as a natural health product	Regulated as a natural health product	Regulated as a natural health product
China	Regulated as a health food	Regulated as a health food	Regulated as a health food	Regulated as a health food	Regulated as a health food	Regulated as a health food	Regulated as a health food	Regulated as a health food
Russia	Regulated as a biologically active supplement	Regulated as a biologically active supplement	Regulated as a biologically active supplement	Regulated as a biologically active supplement	Regulated as a biologically active supplement	Regulated as a biologically active supplement	Regulated as a biologically active supplement	Regulated as a biologically active supplement

GCC	Regulated as a food supplement	Regulated as a food supplement	Regulated as a food supplement	Regulated as a food supplement	Regulated as a food supplement	Regulated as a food supplement	Regulated as a food supplement	Regulated as a food supplement
United Kingdom	Classified as a food supplement	Classified as a food supplement	Classified as a food supplement	Classified as a food supplement	Classified as a food supplement	Classified as a food supplement	Classified as a food supplement
Thailand	Regulated as a food supplement	Regulated as a food supplement	Regulated as a food supplement	Regulated as a food supplement	Regulated as a food supplement	Regulated as a food supplement	Regulated as a food supplement
Indonesia	Regulated as a health supplement	Regulated as a health supplement	Regulated as a health supplement	Regulated as a health supplement	Regulated as a health supplement	Regulated as a health supplement	Regulated as a health supplement	Regulated as a health supplement
Sri Lanka	Regulated as a functional food	Regulated as a functional food	Regulated as a functional food	Regulated as a functional food	Regulated as a functional food	Regulated as a functional food	Regulated as a functional food
Maldives	Regulated as products with specific claims	Regulated as products with specific claims	Regulated as products with specific claims	Regulated as products with specific claims	Regulated as products with specific claims	Regulated as products with specific claims	Regulated as products with specific claims
New Zealand	Regulated as a dietary supplement	Regulated as a dietary supplement	Regulated as a dietary supplement	Regulated as a dietary supplement	Regulated as a dietary supplement	Regulated as a dietary supplement	Regulated as a dietary supplement
Mexico	Regulated as food supplement	Regulated as food supplement	Regulated as food supplement	Regulated as food supplement	Regulated as food supplement	Regulated as food supplement	Regulated as food supplement
Brazil	Regulated as food supplement	Regulated as food supplement	Regulated as food supplement	Regulated as food supplement	Regulated as food supplement	Regulated as food supplement	Regulated as food supplement
Colombia	Regulated as dietary supplement	Regulated as dietary supplement	Regulated as dietary supplement	Regulated as dietary supplement	Regulated as dietary supplement	Regulated as dietary supplement	Regulated as dietary supplement

2.21 CONCLUSION

The landscape of nutraceutical regulation is diverse and complex, with each country employing its own terminology and classification system. While these variations may reflect cultural, historical, and regulatory differences, they also pose challenges for international trade and consumer understanding. The lack of harmonization in terminology and categorization can lead to confusion, ambiguity, and potential misinterpretation of products, especially in global marketing efforts. For instance, what may be termed a *dietary supplement* in the United States could be classified as *health food* in Japan or *food supplement* in Brazil. Such discrepancies may not only hinder market access but also impact consumer trust and safety. Efforts to standardize terminologies and harmonize regulatory frameworks across countries could mitigate these challenges. Establishing clear definitions and classification criteria for nutraceuticals would promote transparency, facilitate cross-border trade, and ensure consistent consumer protection worldwide. Furthermore, enhancing collaboration and information sharing among regulatory authorities can foster greater alignment and convergence in nutraceutical regulations. Platforms for international cooperation, such as the Codex Alimentarius Commission, provide valuable forums for harmonizing standards and guidelines, thus promoting global consistency and coherence in the regulation of nutraceuticals. In conclusion, while diverse terminologies in nutraceutical regulations may present obstacles, concerted efforts toward harmonization and standardization hold the promise of promoting global health and facilitating the sustainable growth of the nutraceutical industry.

REFERENCES

[1] Commissioner, Office of the. *FDA Organization Charts.* U.S. Food and Drug Administration (FDA). https://www.fda.gov/about-fda/fda-organization/fda-organization-charts (accessed March 12, 2024).

[2] Commissioner, Office of the. FDA 101: Dietary Supplements. U.S. Food and Drug Administration (FDA), 2022.

[3] Nutrition, Center for Food Safety and Applied Nutrition. *Dietary Supplement Ingredient Directory.* U.S. Food and Drug Administration (FDA). https://www.fda.gov/food/dietary-supplements/dietary-supplement-ingredient-directory (accessed February 12, 2024).

[4] European Food Safety Authority (EFSA). *Food Supplements.* https://www.efsa.europa.eu/en/topics/topic/food-supplements (accessed February 13, 2024).

[5] European Medicines Agency (EMA). *Committees, Working Parties, and Other Groups.* https://www.ema.europa.eu/en/committees (accessed March 11, 2024).

[6] European Commission. Q and As—Novel Food Legislation. https://food.ec.europa.eu/safety/novel-food/legislation/q-and_en (accessed February 13, 2024).

[7] Regulation (EC) No. 1829/2003 of the European Parliament and of the Council of 22 September 2003 on Genetically Modified Food and Feed (Text with EEA Relevance); 2008. http://data.europa.eu/eli/reg/2003/1829/2008-04-10/eng (accessed February 13, 2024).

[8] Regulation (EU) No. 609/2013 of the European Parliament and of the Council of 12 June 2013 on Food Intended for Infants and Young Children, Food for Special Medical Purposes, and Total Diet Replacement for Weight Control, and Repealing Certain Directives. http://data.europa.eu/eli/reg/2013/609/oj/eng (accessed February 13, 2024).

[9] Global Regulatory Partners, Inc. Japan Dietary Supplements Regulations. https://globalregulatorypartners.com/resource-center/regulatory-intelligence-platform/japan-regulatory-intelligence/japan-dietary-supplements-regulations/ (accessed February 11, 2024).

[10] khushi560. *Nutraceutical Regulation in Japan.* Artixio. https://www.artixio.com/post/nutraceutical-regulation-in-japan (accessed February 11, 2024).

[11] Expert, T. *Nutraceutical Approval Process in Japan.* Credevo Articles. https://credevo.com/articles/2020/11/15/japan-nutraceutical-approval-process/ (accessed February 14, 2024).

[12] Compendium of Nutraceutical Regulations. FSSAI. https://www.fssai.gov.in/upload/uploadfiles/files/Compendium_Nutra_29_09_2021.pdf (accessed March 12, 2024).

[13] FOOD-ACT. FSSAI. https://fssai.gov.in/upload/uploadfiles/files/FOOD-ACT.pdf (accessed February 11, 2024).

[14] Consulting, A. *Nutraceutical Regulation and Registration in Australia (TGA).* Artixio. https://www.artixio.com/post/nutraceutical-regulation-and-registration-in-australia-tga (accessed February 14, 2024).

[15] Therapeutic Goods Administration (TGA). TGA Structure. https://www.tga.gov.au/about-tga/corporate-information/tga-structure (accessed March 12, 2024).
[16] Australian Regulatory Guidelines for Complementary Medicines. 2018.
[17] Canada, Government of Canada, National Research Council (NRC). *Regulatory Information Resources for Nutraceuticals and Functional Foods: Resource Guide.* NRC Publications Archive. https://nrc-publications.canada.ca/eng/view/object/?id=a231db83-618e-4078-ab88-24582134c7ab (accessed February 14, 2024).
[18] Health Canada. *ARCHIVED—Policy Paper—Nutraceuticals/Functional Foods and Health Claims on Foods.* https://www.canada.ca/en/health-canada/services/food-nutrition/food-labelling/health-claims/nutraceuticals-functional-foods-health-claims-foods-policy-paper.html (accessed February 12, 2024).
[19] Health Canada. *Health Canada's Organizational Structure.* https://www.canada.ca/en/health-canada/corporate/organizational-structure.html (accessed March 12, 2024).
[20] Davis, M. *An Introduction to Registration of Food Supplements in China.* Global Regulatory Partners, Inc. https://globalregulatorypartners.com/registration-of-food-supplements-in-china/ (accessed February 14, 2024).
[21] National Medical Products Administration. Wikipedia. 2023.
[22] Ottaway, P. B., Ed. Copyright. In *Food Fortification and Supplementation.* Woodhead Publishing Series in Food Science, Technology and Nutrition; Woodhead Publishing, 2008; p iv. https://doi.org/10.1016/B978-1-84569-144-8.50016-X.
[23] Ottaway, P. B. 1—Principles of Food Fortification and Supplementation. In *Food Fortification and Supplementation.* Woodhead Publishing Series in Food Science, Technology and Nutrition; Woodhead Publishing, 2008; pp. 1–10. https://doi.org/10.1533/9781845694265.1.
[24] Blaze, J. A Comparison of Current Regulatory Frameworks for Nutraceuticals in Australia, Canada, Japan, and the United States. *Innovative Pharmacy*, 2021, 12 (2). https://doi.org/10.24926/iip.v12i2.3694.
[25] ASEAN Alliance of Health Supplement Associations (AAHSA). https://aahsa.org.sg/ (accessed February 17, 2024).
[26] ASEAN. *ASEAN Guidelines on Labelling Requirements for Health Supplements.* https://asean.org/wp-content/uploads/2017/09/ASEAN-Guidelines-on-Labelling-Req.-for-HS-2-Oct-2015-rev-with-discla....pdf (accessed February 19, 2024).
[27] Ottaway, P. B., Ed. Index. In *Food Fortification and Supplementation*, Woodhead Publishing Series in Food Science, Technology and Nutrition; Woodhead Publishing, 2008; pp. 271–281. https://doi.org/10.1533/9781845694265.index.
[28] Rand, G. Russia: Biologically Active Supplements. 2014.
[29] هيئة التقييس لدول مجلس التعاون لدول الخليج العربية. ة, المكملات ;.د ,مشروع
[30] Food Standards Agency. *Food Supplements.* https://www.food.gov.uk/business-guidance/food-supplements (accessed February 29, 2024).
[31] GOV.UK. Nutrition Legislation Information Sheet. https://www.gov.uk/government/publications/nutrition-legislation-information-sources/nutrition-legislation-information-sheet-2 (accessed February 29, 2024).
[32] Food Standards Agency. Novel Foods Authorisation Guidance. https://www.food.gov.uk/business-guidance/regulated-products/novel-foods-guidance (accessed February 29, 2024).
[33] khushi560. *Nutraceutical Regulation in Thailand.* Artixio. https://www.artixio.com/post/nutraceutical-regulation-in-thailand (accessed March 9, 2024).
[34] Food Compliance International. *Thailand—Food Compliance.* https://foodcomplianceinternational.com/expertise/thailand (accessed March 1, 2024).
[35] Peraturan-Bpom No. 27 Tahun 2023. https://peraturan.go.id/files2/peraturan-bpom-no-27-tahun-2023_terjemah.pdf (accessed March 10, 2024).
[36] FBDG Practitioners Handbook. English. https://nutrition.health.gov.lk/wp-content/uploads/2021/10/FBDG-Practitioners-Handbook-English.pdf (accessed March 11, 2024).
[37] National Medicines Regulatory Authority (NMRA). https://nmra.gov.lk/index.php?option=com_content&view=article&id=87&Itemid=85&lang=en (accessed March 11, 2024).
[38] Pj9req7v.Pdf. https://health.gov.mv/storage/uploads/8roO4KwP/pj9req7v.pdf (accessed March 11, 2024).
[39] New Zealand Legislation. *Dietary Supplements Regulations 1985* (SR 1985/208) (as at 01 March 2016) 2A Meaning of Dietary Supplement. https://www.legislation.govt.nz/regulation/public/1985/0208/latest/DLM2856310.html (accessed March 11, 2024).
[40] Food Compliance International. *Mexico—Food Compliance.* https://foodcomplianceinternational.com/expertise/mexico (accessed March 11, 2024).
[41] Global Regulatory Partners, Inc. *Brazil Food Supplements.* https://globalregulatorypartners.com/resource-center/regulatory-intelligence-platform/brazil-anvisa-regulatory-intelligence/brazil-food-supplements/ (accessed March 11, 2024).

[42] Food Compliance International. *Colombia—Food Compliance.* https://foodcomplianceinternational.com/expertise/colombia (accessed March 11, 2024).
[43] INVIMA. Who We Are. https://www.invima.gov.co/el-instituto/quienes-somos (accessed March 12, 2024).
[44] INVIMA. *Dietary Supplement.* https://www.invima.gov.co/atencion-al-ciudadano/suplemento-dietario (accessed March 12, 2024).

3 Regulatory Framework for Nutraceutical Registration in CARICOM

Sandeep B. Maharaj, Darren Dookeeram, and Nila-Marie Ramdeo

3.1 INTRODUCTION

Nutraceuticals are broadly defined as products derived from food sources with extra health benefits in addition to their basic nutritional value. In a direct-to-consumer form, they are a combination of "nutrition" and "pharmaceutical" and can include dietary supplements, isolated nutrients, and herbal products that may help prevent diseases, improve health conditions, or enhance health and well-being [1]. Nutraceuticals have gained importance for several reasons, such as their role in health maintenance, disease prevention, and immunity modulation, and in some instances, it can be more cost-effective compared to its pharmaceutical counterparts [2]. Nutraceuticals are strongly marketed as a bridge between food and pharmaceuticals, offering benefits derived from their nutritional as well as medicinal properties. It is suggested that they form an integral part of a comprehensive approach to health and wellness [3]. Currently, in the global market, there are excessive numbers of nutraceuticals claiming to be "natural" and offering therapeutic benefits by major online retailers.

However, their safety is at question due to the lack of regulation. The literature reflects that there is variability with the current robustness of regulation in the global landscape, and the burgeoning industries in nutraceuticals have the potential to find loopholes in bringing inadequately tested products to market [4]. This superimposes another layer of potential drug effects and interactions that may be devastating to patients [5]. The Caribbean Community (CARICOM) represents a partial governance integration of primarily Small Island Developing States (SIDS) in the Caribbean region that share similar post-colonial backgrounds and current-day challenges. While there are some health services under this umbrella, the regulation of health is largely at the individual state level and may lack uniformity [6]. It is noted that the territories of the CARICOM have great biodiversity and, resultantly, an anecdotally significant use of nutraceuticals, but with little regulatory frameworks [7]. With the knowledge that regulation of nutraceuticals inherently implies implementing stringent quality assurance measures, there is a need to strengthen this process in the Caribbean region as a benefit to health industries and as a risk mitigation strategy for end users [8].

3.2 CARICOM

CARICOM, the Caribbean Community, is a regional organization dedicated to fostering economic integration and cooperation among its member states. The Caribbean Community and Common Market was established by the Treaty of Chaguaramas, which was initially signed by Barbados, Guyana, Jamaica, and Trinidad and Tobago on July 4, 1973, and came into effect on August 1, 1973 [6]. CARICOM recognizes the critical importance of health and wellness for the development of the Caribbean region. Dr. C. James Hospedales, the executive director of the Caribbean Public Health Agency (CARPHA), emphasizes the significance of public health, stating that it is crucial for the resilience and development of the region. The region faces challenges related to chronic

DOI: 10.1201/9781003465041-3

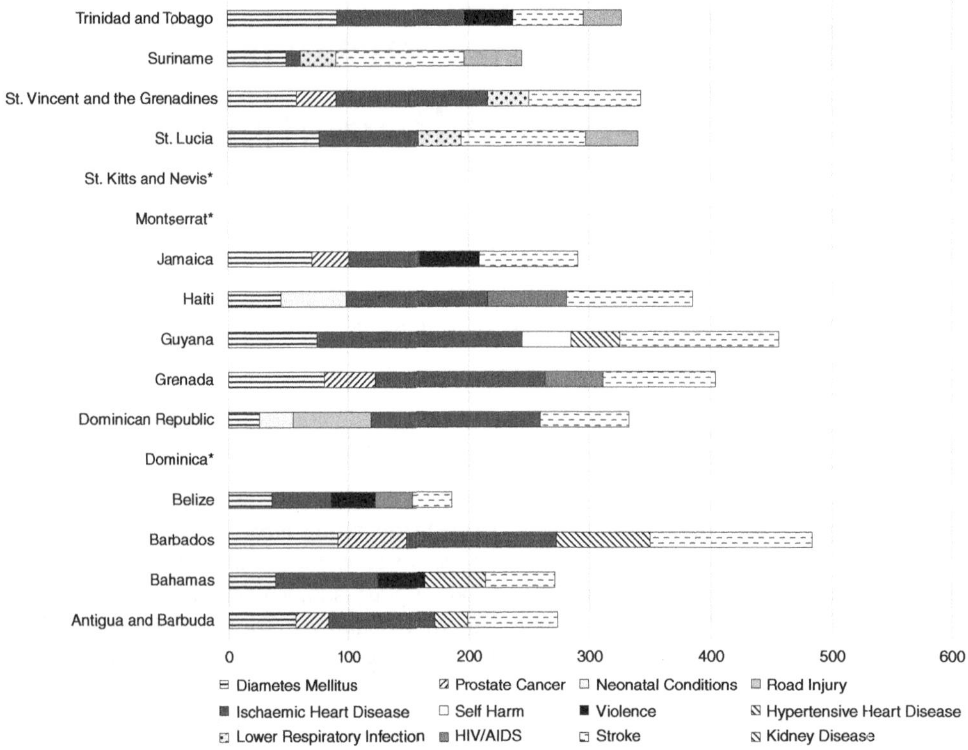

FIGURE 3.1 Top five causes of death per 100,000 population in the year 2019.

Source: [10].

non-communicable diseases (NCDs) [9], such as diabetes, heart disease, and hypertension, which have a negative impact on economic growth. Figure 3.1 identifies the top five causes of death in the region in 2019.

Nutraceuticals have a long history in CARICOM countries, playing a role in both traditional health practices and contemporary health and wellness industries. Many Caribbean countries have a rich tradition of using local herbs and botanicals for health and wellness. These traditional remedies often fall under the category of nutraceuticals and are used for their therapeutic benefits. Some CARICOM countries have local industries that produce nutraceuticals derived from indigenous plants and traditional recipes. However, the extent of local production and registration varies among the member states. While there is interest in the potential health benefits of nutraceuticals, the regulatory frameworks for these products can differ from country to country within the CARICOM area [11].

Despite the potential, there can be challenges related to quality control, standardization, and international acceptance as the region works toward a more harmonized approach to the regulation of nutraceuticals.

3.3 NUTRACEUTICALS IN THE GENERAL CONTEXT

The projected value of the global nutraceutical market is set to exceed $620 billion by 2030, with a historical market growth rate (CAGR) of 5.3% from 2018 to 2022. Currently, Europe, the USA, and Japan collectively represent more than 90% of the total global nutraceutical market share [12].

The nutraceutical industry is witnessing a significant growth on a global scale, fueled by factors like increasing demand for dietary supplements, functional foods, and beverages that promote health and well-being. Consumer awareness of the benefits of nutraceuticals in preventing and managing various health conditions is a key driver of this market expansion [12].

Nutraceuticals are increasingly viewed as a cost-effective means to support preventive health care, particularly in the context of the COVID-19 pandemic, where they are utilized as complementary to conventional medicine [13]. This trend has contributed to the industry's growth and acceptance.

In emerging economies like India, the nutraceutical market is positioned to become a global frontrunner, with a projected market value of $18 billion by 2025, growing at a remarkable CAGR of 22% [14]. This presents a significant economic opportunity for nations to harness the potential of the nutraceutical industry for both public health enhancement and economic development.

Jamaica has a significant opportunity to leverage the expanding global nutraceutical market, positioning itself to create a profitable channel for sustained foreign exchange earnings through increased production and export of nutraceutical goods [15].

The government is actively working on regulatory enhancements and the introduction of new laws to empower local products and services to gain international recognition and achieve optimal market reach [15].

Progress is evident in Jamaica's nutraceutical industry. In 2017, the Bio-Tech R&D Institute achieved a milestone by securing the inaugural Food and Drug Administration (FDA) Orphan Drug Status for a Caribbean entity or nation. This recognition was granted for their novel drug chrysoeriol, developed from cannabis (ganja), designed for leukemia treatment [15].

Through leveraging its natural resources, scientific knowledge, and regulatory advancements, Jamaica aims to unlock substantial economic and societal advantages from the global nutraceutical market, valued at over $150 billion [16]. The country's investment prospects in the nutraceutical and wellness sector rank among the most promising and active areas for commercial development worldwide.

While the global nutraceutical market offers substantial economic benefits, such as increased revenue and job creation, along with the potential to enhance public health outcomes in a cost-effective manner, it does face challenges, like stringent regulatory requirements and high costs associated with product development.

3.4 RISKS: DRUG INTERACTION, QUALITY ASSURANCE

Nutraceuticals possess the capacity to potentially offer partial prevention or complementary treatment for a range of health issues. Yet there frequently exists insufficient data to fully validate the efficacy, safety, and overall effectiveness of nutraceuticals [10]. Nutraceuticals are a broad category encompassing nutrients, herbals, and dietary supplements that offer potential health benefits beyond basic nutrition [17]. However, the majority of nutraceutical products fall under the category of "prospective" ones due to insufficient clinical evidence to support their claimed advantages [17]. This lack of substantiated information raises uncertainties surrounding their use, including challenges in categorizing them when utilized for medicinal purposes [10]. Many claims regarding their medicinal or nutritional benefits remain unverified due to a lack of data on possible drug interactions, mechanisms of action, and regulations with quality control.

3.4.1 Drug Interaction

Drug–nutrient interactions occur when the presence of a drug affects the pharmacokinetics (how the drug is absorbed, distributed, metabolized, and excreted in the body) or pharmacodynamics (the drug's effects on the body) of a dietary component, or when a drug compromises a person's nutritional status [18]. Without proper recognition and management of these interactions, severe outcomes can arise. For example, interactions between food and nutrients can diminish the absorption of certain oral antibiotics, possibly leading to insufficient levels of the drug at the infection site and increasing the risk of unsuccessful treatment [18]. Interactions involving food and drugs can have two primary clinical consequences: (1) a reduction in drug bioavailability, which may result in therapeutic failure, or (2) an increase in bioavailability, which may heighten the possibility of side effects and can lead to toxicity.

3.4.2 Quality Control Issues

Questionable quality control in manufacturing processes can result in products that have variable or unknown concentrations of active ingredients, contaminants, or even substances not declared on the label [19]. Potential contamination can occur during manufacturing, including the solvents/additives used, the purification techniques, and the storage conditions. These contaminants can pose serious health risks to consumers. The major contaminants found in nutraceuticals include heavy metals (such as mercury, arsenic, cadmium, and lead), pesticides, mycotoxins, microbial contaminants, and potentially harmful chemical substances [20]. If these are present in high amounts and one is exposed to these contaminants, serious adverse effects can occur, such as renal failure, carcinogenicity, and hepatotoxicity. Bodybuilding dietary supplements, for instance, have been found to sometimes contain anabolic steroids without this being stated on the label, posing significant health risks, especially to naïve users [21]. Defining and enacting strict, uniform manufacturing processes, along with quality assurance and control methods, are crucial.

3.5 NUTRACEUTICALS IN CARICOM

The Caribbean region is recognized as a global biodiversity hot spot, with a rich history of traditional medicinal plant use by Indigenous peoples, post-Columbian settlers, and more recent immigrant communities. This diverse ethnobotanical heritage has led to the development of a vibrant Caribbean folk medicine tradition, often based on the oral transmission of knowledge about the medicinal properties of local plant species [22].

This traditional herbal medicine system has retained its informal and less-structured nature in comparison to the ancient healing practices of China and India [22]. While there was a period when herbal medicine lost popularity as conventional pharmaceuticals gained worldwide recognition, a resurgence in the use of herbal remedies and self-medication with natural products has been observed in the Caribbean since the 1980s. The growing interest in alternative health practices in the Caribbean Community (CARICOM) region is reflected in the proliferation of herbal shops, clinics, and spas, indicating a shift toward more holistic approaches to health and wellness.

Despite this growing interest, there is a lack of scientific data on the plant species used in Caribbean folk remedies, and these plants are often not extensively documented in mainstream medical literature. Many folk remedies are created as home treatments using freshly harvested wild plants, especially in rural areas [22]. The absence of essential infrastructure, such as processing facilities and consistent herbal resources, has impeded the establishment of sustainable herbal medicine industries in the region.

In some Caribbean countries, these traditional herbal preparations are known as "bush teas," which are typically consumed as either infusions or decoctions. It is important to note that while herbal medicinal products are often perceived as inherently safe due to their natural origins, the distinction between poisonous and medicinal plants is not always clear.

During Traditional Medicines in the Islands (TRAMIL) surveys in the Caribbean, data on the use of herbal remedies was collected. TRAMIL, a research program established in 1982, focuses on popular medicine to validate and enhance primary health-care practices based on medicinal plants in the region [23]. The program aims to validate plant species, parts, and preparation methods commonly used for specific ailments (at least 20%) but not documented in scientific literature, following

TABLE 3.1
Medicinal Herbs Used in CARICOM Countries

Country	Scientific Name of Herb/ Botanical	Vernacular Names	Uses
Antigua and Barbuda	*Zingiber officinale* Roscoe	Ginger	Flatulence, cough, flu, fever, asthma, stomach pain
The Bahamas	*Lantana involucrate*	White sage	Itching from allergic reaction
Barbados	*Allium sativum* L.	Garlic	Digestive conditions, intestinal parasites, earache, skin conditions, fever, nausea, stress
Belize	*Sida rhombifolia* L.	Maluba	Conjunctivitis
Dominica	*Carica papaya* L.	Pawpaw	Urethritis, boil
Dominican Republic	*Acalypha alopecuroidea* Jacq.	Ajito con pelo	Flatulence
Grenada	*Citrus aurantium* L.	Sour orange	Cough, fever, conjunctivitis, colic, flu
Guyana	*Moringa oleifera*	Moringa	Diabetes, inflammation, asthma
Haiti	*Chamissoa altissima* (Jacq.) Kunth	Lyann panye	Asthenia, weakness
Jamaica	*Momordica charantia*	Bitter melon	Diabetes, stomach ailments, parasitic worm removal
Montserrat	*Justicia pectoralis Jacq*	Garden balsam	Colds, menopause
St. Kitts and Nevis	*Aloe vera* (L.) Burm.	Aloe vera	Diabetes, as antioxidant, constipation
St. Lucia	*Curcuma longa* L.	Indian saffron, turmeric	Jaundice, liver disfunction, abscess
St. Vincent and the Grenadines	*Capsicum annuum* L.	Pepper bush	Boil, ganglionar inflammation
Suriname	*Renealmia alpinia* (Rottb.) Maas	Masoesa	Heart problems, stomachache, swelling, sores, snakebites
Trinidad and Tobago	*Cymbopogon citratus* (DC.) Stapf	Fever grass, lemongrass	Flatulence, cough, flu, fever

Source: [25–28].

criteria of safety and effectiveness. Table 3.1 highlights some of the most commonly used medicinal herbs in CARICOM countries, along with their primary therapeutic applications.

Annona muricata, known as soursop in Trinidad and Tobago, is notable for containing flavonoids and apomorphinic alkaloids that may impact the central nervous system. Ethnopharmacological studies highlight the use of dried leaf infusions of *A. muricata* for treating anxiety, insomnia, and headaches [24].

In Souza et al.'s 2018 study, the pharmacological and toxicological effects of a hydroalcoholic extract from *A. muricata* leaves were evaluated. Animals treated with varying doses of the extract showed central nervous system depressant effects, like sedation and reduced motility. Mortality rates increased with higher doses, with an LD50 of 1091.7 mg/kg, determined using the probit method. Folk medicine often employs natural plants for psychiatric disorders under the misconception that they are safer than synthetic psychotropic drugs [24].

3.6 REGULATION IN THE CARIBBEAN FRAMEWORK

Medications are a vital component in maintaining public health, and significant financial resources are allocated by governments and individuals for their procurement. For medications to be beneficial, they must meet standards of safety, efficacy, and quality. However, neither consumers nor prescribers possess the necessary information or expertise to verify whether a specific product meets these criteria. Therefore, it is crucial for governments to regulate the medicines market to ensure

TABLE 3.2
Pharmaceutical Registration Status and Legislative Compliance across CARICOM

Country	Registration Present? Y/N	Covered in Legislation? Y/N										
		Manufacturers	Importers	Wholesaler/ Distributer	Retailers	Establishment of Quality Control Laboratory	Control of Clinical Trials	Control of Imports and Exports	Scope of Regulated Products	Power to Make Regulations	Regulations Made under the Act	Regulations of Product Promotion and Advertisement
Antigua and Barbuda	No	Yes	Yes	Yes	Yes	Yes	No	Yes	Yes*	Yes	Yes	No
The Bahamas	No	No	Yes	Yes	Yes	N/A	N/A	N/A	N/A	N/A	No	N/A
Barbados	No	Yes	Yes	Yes	Yes	Yes**	No	Yes	Yes*	Yes	Yes	No
Belize	No	Yes	Yes	Yes	Yes	Yes	No	Yes	Yes*	Yes	Yes	No
Dominica*	No											
Dominican Republic	No	Yes	Yes	Yes	Yes	Yes	No	Yes	Yes*	Yes	Yes	Yes
Grenada	No	Yes	Yes	Yes	Yes	Yes	No	Yes	Yes*	Yes	Yes	No
Guyana	Yes	Yes	Yes	Yes	Yes	Yes**	No	N/A	Yes*	Yes	Yes	Yes
Haiti	Yes	Yes	Yes	Yes	Yes	No**	No	Yes	Yes*	No	No	Yes
Jamaica	Yes	Yes	Yes	Yes	Yes	Yes	No	Yes*	Yes	Yes	Yes	Yes
Montserrat*	No											
St. Kitts and Nevis*	No											

St. Lucia	No	No	No	No	Yes	No	No	N/A	Yes	Yes	Yes	No
St. Vincent and the Grenadines	No	Yes	Yes	Yes	Yes	Yes**	No	Yes*	Yes	Yes	N/A	No
Suriname	Yes	Yes	Yes	No	Yes	No	No	N/A	Yes	N/A	No	Yes
Trinidad and Tobago	Yes	Yes	Yes	Yes	Yes	Yes**	No	Yes	Yes	Yes	Yes	Yes

Note: **Antigua and Barbuda:** *Human and veterinary drugs and poisons.
Barbados: *Human and veterinary drugs and poisons. **Not in the country's legislation, but the Caribbean Regional Drug Testing Laboratory Agreement establishes regional quality control in force, which includes Barbados.
Belize: *Food, drugs, and poisons.
Dominica: *Apart from the Dangerous Drugs Act, there are no legal provisions available that regulate the marketing, distribution, and sale of medicines or the pharmacy profession. The only pharmaceutical regulatory activity performed by the Ministry of Health–Pharmacy Division is the submission of samples of suspect quality to the Caribbean Regional Drug Testing Laboratory in Jamaica via OECS/PPS.
Dominican Republic: *Medicines regulations.
Grenada: *Human and veterinary medicines and poisons.
Guyana: *Food, drugs, and poisons; cosmetics and therapeutic devices. **Laboratory at Food and Drug Department, materials management unit minilab.
Jamaica: *imports, N/A exports
Haiti: *Human and veterinary drugs, medicated dressings, drains, and sterilized sutures. **Only required for manufacturing companies.
Montserrat: *Not aware of existing medicines legislation. Pharmacists are registered at the Magistrate Court, and their registration is published in the government gazette.
St. Kitts and Nevis: *Existing medicines legislation includes the Medical Act of 1938 and the Drug Act of 1986. Apart from these, there are no legal provisions available that regulate the marketing, distribution, and sale of medicines or the pharmacy profession. Pharmacists are, however, registered with the Medical Board under the Ministry of Health.
St. Vincent and the Grenadines: *Imports, N/A export. **Indirectly via signatory agreement in CRDTL establishment.
Trinidad and Tobago: **No explicit reference to regulatory QC laboratory.
Source: [11].

public health. International consensus emphasizes that medicines regulation involves several critical functions that should be included in national legislation, such as licensing of pharmaceutical products, premises, and distribution channels; quality control laboratory testing; adverse drug reaction monitoring; overseeing advertising and promotional activities; and regulating clinical trials, which are essential regulatory tasks. The National Regulatory Authority (NRA) is tasked with executing these functions and ensuring adherence to legal mandates [11]. Although all CARICOM members have some level of pharmaceutical legislation, none of the current laws are entirely exhaustive. Critical elements, often absent, encompass the supervision of clinical trials, monitoring of adverse drug reactions, control of product promotion and advertising, and explicit prohibition of counterfeit medications.

Regarding the medicines regulatory landscape in CARICOM countries, a distinction can be made among the member states. Only five countries have more comprehensive medicines regulatory systems, which encompass medicines registration, collectively representing about 91% of the total CARICOM population [11]. Two countries are in the process of planning the implementation of medicines registration in the near future. Eight countries, including seven from the Organization of Eastern Caribbean States (OECS), have limited regulatory systems and do not have immediate plans for medicines registration due to challenges like limited market size and constraints in human and financial resources [11].

Despite these variations, the overarching goal is to safeguard public health by ensuring that safe, effective, and high-quality medicines are accessible to patients. Therefore, it is recommended to establish a CARICOM medicines quality assurance policy and a harmonized structure to ensure timely availability of pharmaceutical products addressing prevalent health issues in all CARICOM member states, with a focus on proven safety, efficacy, and quality.

3.7 GLOBAL OVERVIEW OF NUTRACEUTICAL REGISTRATION REQUIREMENTS

The regulatory landscape for nutraceuticals varies significantly across the globe, with each country adopting its own distinct approach. Some nations, such as Australia and China, categorize nutraceuticals as a type of food [29], while others, like Colombia, Brazil, and Argentina, employ a straightforward registration-based system. However, certain countries, including Brazil, China, and Taiwan, have more stringent requirements, necessitating animal or human clinical studies before granting registration approval [29].

Japan has implemented a "foods for specified health use" (FOSHU) registration process, while China has established a "health food" (HF) registration system. These frameworks share similarities and differences in terms of the scientific evidence required, application process, evaluation criteria, legal categories, labeling specifications, and associated costs and time involved [4]. In the United States, nutraceuticals are primarily regulated as dietary supplements under the Dietary Supplement Health and Education Act (DSHEA) and the Federal Food, Drug, and Cosmetic Act. India regulates the manufacture, storage, distribution, sale, and import of nutraceuticals through the Food Safety and Standards Act (FSSA) [30]. The regulatory guidelines are less stringent than for pharmaceuticals. Canada is exploring policy options to address regulatory challenges around nutraceuticals/functional foods [31], such as defining them precisely, determining if they should be regulated as foods or drugs, and establishing standards of evidence for health claims.

3.7.1 Harmonization Challenges in Nutraceutical Regulations

There is a lack of harmonization in the registration requirements for nutraceuticals across different countries, which poses challenges for companies looking to market their products globally. The key issues hindering harmonization include varying definitions and terminologies around what constitutes a nutraceutical product; differing requirements for safety data, efficacy data, and post-market surveillance across countries; lack of consensus on technical requirements, like claims substantiation, maximum nutrient levels, and labeling. Additionally, small and medium enterprises face challenges in meeting the rigorous quality standards necessary for harmonization.

The ASEAN region has been working on developing common technical requirements and a harmonized registration process, but progress has been slowed by the COVID-19 pandemic [30]. International bodies like the ICH and APEC are also driving initiatives to standardize pharmaceutical regulations [32], which could extend to nutraceuticals.

Harmonization would provide benefits, such as faster market access for products, reduced duplication of clinical trials, and improved quality and safety standards for consumers. However, significant work remains to reconcile the diverse regulatory frameworks currently in place globally for vitamins, minerals, herbal supplements, probiotics, and other nutraceuticals.

3.7.2 Future Directions and Recommendations for Nutraceutical Regulation in CARICOM

Future directions and recommendations for nutraceutical regulation in CARICOM should focus on enhancing pharmacovigilance systems, establishing poison information centers, and harmonizing regulations to ensure consumer safety, promote public health, and support industry growth.

- **Food Security and Nutritional Needs.** CARICOM aims to achieve a 25% reduction in the regional food bill by 2025, emphasizing the importance of regulatory initiatives to support food security and nutritional needs [33]. Nutraceutical regulations should be aligned with this goal, encouraging the development and consumption of nutrient-rich foods and supplements.
- **Global Regulatory Trends and Harmonization.** Understanding global regulatory trends, such as the lack of uniform requirements for nutraceuticals in different regions, can inform CARICOM's approach to harmonizing regulations [34]. Collaborating with other regions like the ASEAN working on harmonized regulations for dietary supplements can provide beneficial insights and potential models for CARICOM to follow.
- **Pharmacovigilance and Safety.** Implementing pharmacovigilance systems is important for monitoring and addressing the adverse effects of nutraceutical products to safeguard public health. This includes establishing poison information centers to provide expert guidance and emergency assistance in cases of nutraceutical-related toxicity or adverse reactions [35].
- **Compliance and Classification.** Ensuring compliance with country-specific health authorities and ministries of health, as well as correctly classifying products, is essential for navigating the regulatory landscape and meeting the requirements in different markets [36]. Clear guidelines and streamlined processes should be developed to facilitate compliance.
- **Consumer Demand and Market Growth.** Recognizing the growing global demand for dietary supplements and the need for regulatory frameworks to support market growth can guide CARICOM in developing regulations that balance consumer needs with safety considerations [37, 38]. Regulations should promote innovation and investment while prioritizing consumer protection.
- **Preventive Health and Regulatory Restructuring.** Emphasizing the importance of nutraceuticals via product promotion and advertisement in preventive health measures can influence CARICOM's regulatory focus on promoting health and well-being through dietary supplements. This may require a restructuring of the regulatory framework, as only some CARICOM countries cover product promotion and advertisement in their legislative system [39, 40]. Doing so can assess the specific role of nutraceuticals in prevention and treatment.

- **Safety and Efficacy Assessments.** Prioritizing safety and efficacy assessments, in line with global standards and regulations, is crucial for ensuring the quality and effectiveness of nutraceutical products in CARICOM countries. Regulations should mandate clinical trials and post-market surveillance to guarantee the safety and efficacy of these products [41].

By integrating these insights and focusing on areas such as pharmacovigilance, regional collaboration, compliance, consumer demand, and preventive health, CARICOM can establish a forward-looking regulatory framework for nutraceuticals that promotes public health, supports industry growth, and ensures regulatory compliance across the region.

REFERENCES

[1] Zhou, Y.; Tian, Y.; Yang, B. Root Vegetable Side Streams as Sources of Functional Ingredients for Food, Nutraceutical, and Pharmaceutical Applications: The Current Status and Future Prospects. *Trends in Food Science & Technology*, 2023, May 11; 137:1–16.

[2] Nasri, H.; Baradaran, A.; Shirzad, H.; Rafieian-Kopaei, M. New Concepts in Nutraceuticals as Alternatives for Pharmaceuticals. *International Journal of Preventive Medicine*, 2014, Dec; 5(12):1487.

[3] Das, L.; Bhaumik, E.; Raychaudhuri, U.; Chakraborty, R. Role of Nutraceuticals in Human Health. *Journal of Food Science & Technology*, 2012; 49(2):173–83.

[4] Blaze, J. A Comparison of Current Regulatory Frameworks for Nutraceuticals in Australia, Canada, Japan, and the United States. *Innovations in Pharmacy*, 2021; 12(2).

[5] Dookeeram, D.; Bidaisee, S.; Paul, J. F.; Nunes, P.; Robertson, P.; Maharaj, V. R.; Sammy, I. Polypharmacy and Potential Drug–Drug Interactions in Emergency Department Patients in the Caribbean. *International Journal of Clinical Pharmacy*, 2017, Oct; 39:1119–27.

[6] Caribbean Community. Who We Are. Available from: https://caricom.org/our-community/who-we-are/.

[7] Vujicic, T.; Cohall, D. Knowledge, Attitudes, and Practices on the Use of Botanical Medicines in a Rural Caribbean Territory. *Frontiers in Pharmacology*, 2021, Oct 27; 12:713855.

[8] Thakur, S.; Gupta, M. M.; Sharma, D. Nutraceuticals Regulation: An Overview of the Regulatory Frameworks in the USA, EU, and Japan. In *Nutraceutical Fruits and Foods for Neurodegenerative Disorders*; 2024, Jan 1:421–40.

[9] Jones, W.; Murphy, M. M.; Henry, F.; Dunn, L.; Samuels, T. Improving Household Nutrition Security and Public Health in the CARICOM, 2018–2022. *Revista Panamericana de Salud Pública*, 2023, May 19; 46:e88.

[10] World Health Organization. Global Health Estimates: Leading Causes of Death. Available from: https://www.who.int/data/gho/data/themes/mortality-and-global-health-estimates/ghe-leading-causes-of-death

[11] Abbott, R.; Bannenberg, W. Regional Assessment of Patent and Related Issues and Access to Medicines: CARICOM Member States and the Dominican Republic (Final Report: Volume II-Country Studies). Health Research for Action, Dec 31, 2009.

[12] Puri, V. ; Nagpal, M.; Singh, I.; Singh, M.; Dhingra, G. A.; Huanbutta, K.; Dheer, D.; Sharma, A.; Sangnim, T. A Comprehensive Review on Nutraceuticals: Therapy Support and Formulation Challenges. *Nutrients*. Available from: https://www.ncbi.nlm.nih.gov/pmc/articles/PMC9654660/.

[13] Naresh, K. Impact of Nutraceutical Sector in Post COVID Era. Available from: https://fssai.gov.in/upload/media/FSSAI_News_Nutri_HealthCare_22_06_2021.pdf

[14] Naresh, K. India's Nutraceutical Market Growth. *Food Processing India Newsletter*. Available from: https://foodprocessingindia.gov.in/newsletter/emailer/two

[15] Ministry of Agriculture and Fisheries, Jamaica. Jamaica Needs to Take Full Advantage of Nutraceutical Industry. Available from: https://www.moa.gov.jm/content/jamaica-needs-take-full-advantage-nutraceutical-industry.

[16] Do Business Jamaica. Jamaica's Investment Opportunities in the Nutraceutical/Wellness Sector. Available from: https://dobusinessjamaica.com/blogs/jamaicas-investment-opportunities-in-the-nutraceuticalwellness-sector/.

[17] Chopra, A. S.; Lordan, R.; Horbańczuk, O. K.; Atanasov, A. G.; Chopra, I.; Horbańczuk, J. O.; Jóźwik, A.; Huang, L.; Pirgozliev, V.; Banach, M.; Battino, M. The Current Use and Evolving Landscape of Nutraceuticals. *Pharmacological Research*, 2022, Jan 1; 175:106001.

[18] Anadón A, Martínez-Larrañaga MR, Arés I, Martínez MA. Prebiotics and Probiotics: An Assessment of Their Safety and Health Benefits. In: Watson RR, Preedy VR, editors. *Probiotics, Prebiotics, and Synbiotics: Bioactive Foods in Health Promotion*. Cambridge, MA: Academic Press; 2016. p. 3–23.

[19] AlAli, M.; Alqubaisy, M.; Aljaafari, M. N.; AlAli, A. O.; Baqais, L.; Molouki, A.; Abushelaibi, A.; Lai, K. S.; Lim, S. E. Nutraceuticals: Transformation of Conventional Foods into Health Promoters/Disease Preventers and Safety Considerations. *Molecules*, 2021; 26(9):2540.

[20] Gil, F.; Hernández, A.; Martín-Domingo, M. *Toxic Contamination of Nutraceuticals and Food Ingredients*. Academic Press, 2016.

[21] Abbate, V.; Kicman, A. T.; Evans-Brown, M.; McVeigh, J.; Cowan, D. A.; Wilson, C.; Coles, S. J.; Walker, C. J. Anabolic Steroids Detected in Bodybuilding Dietary Supplements—A Significant Risk to Public Health. *Drug Testing and Analysis*, 2015, Jul; 7(7):609–18.

[22] Paul, J. H.; Seaforth, C. E. Harmful Plants in Caribbean Folk Medicine. *Focus on Alternative and Complementary Therapies*, 2011, Dec; 16(4):261–5.

[23] Luciano-Montalvo, C.; Boulogne, I.; Gavillán-Suárez, J. A Screening for Antimicrobial Activities of Caribbean Herbal Remedies. *BMC Complementary and Alternative Medicine*, 2013, Dec; 13:1–9.

[24] Souza, D. O.; dos Santos Sales, V.; de Souza Rodrigues, C. K.; de Oliveira, L. R.; Lemos, I. C.; de Araújo Delmondes, G.; Monteiro, Á. B.; do Nascimento, E. P.; da Costa, J. G.; da Cruz, G. M.; de Barros Viana, G. S. Phytochemical Analysis and Central Effects of *Annona muricata* Linnaeus: Possible Involvement of the Gabaergic and Monoaminergic Systems. *Iranian Journal of Pharmaceutical Research: IJPR*, 2018; 17(4):1306.

[25] Tramil. *Tramilotheque* [Internet]. Tramil.net. Available from: https://www.tramil.net/es/tramilotheque/a.

[26] Fern, K. Y. *Renealmia alpinia* [Internet]. Tropical.theferns.info, 2022. Available from: https://tropical.theferns.info/viewtropical.php?id=Renealmia+alpinia.

[27] Otero, R.; Núñez, V.; Jiménez, S. L.; Fonnegra, R.; Osorio, R. G.; García, M. E.; et al. Snakebites and Ethnobotany in the Northwest Region of Colombia. Part III: Neutralization of the Haemorrhagic Effect of *Bothrops atrox* Venom. *Journal of Ethnopharmacology*, 2000, Dec; 73(1–2):233–41.

[28] Maas, P. J. M. Alpinieae (Zingiberaceae). *Flora Neotropica*, 2004, Monograph 89: S203–S203.

[29] Mestrovic, T. Nutraceutical Regulation. *News Medical Life Sciences*, 2022. Available from: https://www.news-medical.net/health/Nutraceutical-Regulation.aspx.

[30] Patel, K.; Patel, U.; Vaghela, K.; Kanki, N.; Zaveri, M.; Movaliya, V. Comparisons of Registration Requirements of Nutraceuticals in the Philippines, Tanzania, Cambodia, and India. *International Journal of Drug Regulatory Affairs*, 2022; 10(4):18–23.

[31] Santini, A.; Cammarata, S. M.; Capone, G.; Ianaro, A.; Tenore, G. C.; Pani, L.; Novellino, E. Nutraceuticals: Opening the Debate for a Regulatory Framework. *British Journal of Clinical Pharmacology*, 2018, Apr; 84(4):659–72.

[32] Badjatya, J. K.; Chaudhari, P. Current Regulatory Requirements for Registration of Nutraceuticals in ASEAN. *International Journal of Drug Regulatory Affairs*, 2021; 9(4):37–45.

[33] CARICOM. *25% by 2025: Reduction in the Regional Food Bill* [Internet]. Georgetown, Guyana. CARICOM, 2022. Available from: https://caricom.org/25-by-2025-reduction-in-the-regional-food-bill/.

[34] Tomar, S.; Musyuni, P.; Aggarwal, G. An Overview of Regulation for Nutraceuticals and the Concept of Personalized Nutraceuticals. *Journal of Generic Medicines*, 2023, Jun; 19(2):66–74.

[35] Walker, K. E.; Bankay, R.; Jankie, S.; Dhingra, S. Pharmacovigilance in the Caribbean Countries: An Overview. *Current Pharmacology Reports*, 2023, Aug; 9(4):217–27.

[36] Athalye, M.; Vaghela, S.; Bhavsar, N. The Study of the Registration Guideline of Nutraceutical Products in ASEAN Countries. *Applied Clinical Research, Clinical Trials and Regulatory Affairs*, 2020, Aug 1; 7(2):104–16.

[37] Persistence Market Research. *Nutraceuticals Market* [Internet]. New York: Persistence Market Research, 2022. Available from: https://www.persistencemarketresearch.com/market-research/nutraceuticals-market.asp.

[38] Nutraceuticals Global Market Report. *BCC Research*, 2023. Available from: https://www.bccresearch.com/partners/tbrc-market-briefs/nutraceuticals-global-market-report.html.

[39] Gómez, C. V.; Cortéz, I. G.; Montenegro, I. M.; San Martín, O. L. The Regulatory Status of Functional Foods in the Economic Integration Organizations of Latin America and the Caribbean. *Archivos Latinoamericanos de Nutrición (ALAN)*, 2023; 73(4):297–312.

[40] Use of Nutraceuticals As A Medicine and Food. *IIMTU Blog*, 2024. Available from: https://iimtu.edu.in/blog/use-of-nutraceuticals-as-a-medicine-and-food/.

[41] Rossi, D.; Guerrini, A.; Bruni, R.; Brognara, E.; Borgatti, M.; Gambari, R.; Maietti, S.; Sacchetti, G. trans-Resveratrol in Nutraceuticals: Issues in Retail Quality and Effectiveness. *Molecules*, 2012; 17(10):12393–405.

4 Safety and Toxicity Aspects for Nutraceuticals and Related Products

Regulatory Provisions across the Globe

Jiya Patel, Aabir Pramanik, Sonali Maheshwari, Niyati Acharya, and Yashwant Pathak

4.1 INTRODUCTION

Hippocrates, the father of medicine, was a Greek physician who lived about 2,000 years ago. He is credited with coining the phrase "Let food be thy medicine and medicine be thy food." Later, it was redefined as *nutraceutical*, implying any food product or ingredient aiding in pharmaceutical benefit but not having been scientifically approved to provide clinical benefits. Such products or ingredients are also termed as dietary supplements, food additives, health foods, fortified foods, etc. across the globe, with specific regulatory bodies governing their manufacturing, sale, and distribution. Any food item or ingredient consumed as extracts or concentrates to supplement one's diet is referred to as a dietary supplement. In the United States, the Dietary Supplement Health and Education Act of 1994 included dietary supplements with vitamins, minerals, botanicals, amino acids, enzymes, metabolites, and herbs. Similarly, Japan supports the concept of functional food. As per Japan, for any food product or ingredient to be classified as functional food, it must be present in natural form, must be consumed in the diet regularly, and must regulate physiological processes and the prevention or mitigation of a chronic disease.

The nutraceutical industry is estimated to be worth USD 488.41 billion in 2024 and is expected to increase by 5.09% CAGR to USD 626.02 billion in 2029 [1]. The USA does not define the word *nutraceutical* properly. Instead, it accepts the term *dietary supplement*, which describes processed foods obtained from human food sources with the objective of enhancing health above the minimum nutritional requirements. The United States Food and Drug Administration, also known as the USFDA, has several approaches to regulate the dietary supplement market, even though regulatory standards for their safety are less stringent than those for any drug substance or drug product. Based on the historical use, dietary ingredients or dietary supplements sold in the United States prior to October 15, 1994, are exempt from being subject to a safety review [2]. Following this date in 1994, however, manufacturers are required according to the 1994 Dietary Supplement Health and Education Act, widely known as the DSHEA, to provide reasonable evidence of the safety and toxicity levels of new dietary ingredients (NDI), which need not be classified as GRAS (generally recognized as safe), approved food additives, or otherwise approved for use in conventional food types intended for human consumption. Additionally, the Act has the authority to amend the Federal Food, Drug, and Cosmetic Act in relation to dietary supplements, mandating the submission of information 75 days prior to the dietary supplement or ingredient being commercialized. Further, the FDA specifies that manufacturers should adhere to current Good Manufacturing Practices

DOI: 10.1201/9781003465041-4

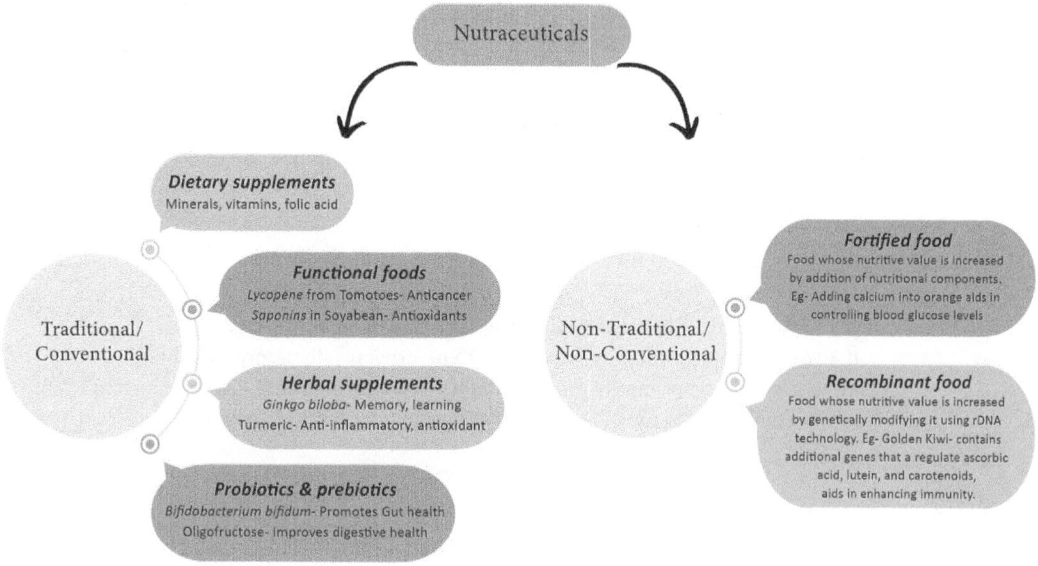

FIGURE 4.1 Comprehensive classification of nutraceuticals into two main divisions based on use and function.

(cGMP), with the aim to ensure that dietary supplements comply with the necessary standards for quality, purity, strength, composition, and identity [3].

4.2 HISTORY OF REGULATORY BODIES AND SEVERAL ACTS TO REGULATE NUTRACEUTICALS WORLDWIDE

The European Parliament and Council implemented Regulation (EC) No. 178/2002 in the year 2002, which lays down guidelines for food law, including safety, production, and distribution. This legislation also created the EU Food Safety Authority (EFSA), which aids in providing scientific advice and risk assessment. Depending on the type of nutraceutical, different processes exist to obtain marketing authorization, including EU Directive 2001/82/EC and EU Regulations 1924/2006 and 609/2013, which ensure safety and efficacy, especially for products having medical claims. Foods with medical claims have to submit biological and toxicological data, as determined by the European Food Safety Authority's (EFSA) Scientific Committees [4, 5]. Through the Food Standards Australia New Zealand (FSANZ), which is a trans-Tasman regulatory system, Australia and New Zealand collaborated to develop standards for food addressing microbiological limits, processing standards, safety, and guidelines for food fortification. This system ensures a clear division between therapeutic goods, overseen by the Therapeutic Goods Administration, and food products, regulated by FSANZ [6, 7].

However, under the Food Safety and Standards Act 2006, the Food Safety and Standards Authority of India (FSSAI) is responsible for enforcing regulations pertaining to food for special dietary use (FSDU), functional foods, food or health supplements, nutraceuticals, novel food, and food for special medical use (FSMU) in India. Likewise, countries across the globe differ in regulatory guidelines and rules on nutraceuticals, which will be explored in the later part of the chapter. The growing popularity of food supplements raises concerns about their quality, efficacy, and safety from a regulatory standpoint. Due to their concentrated nature and potential physiological effects, there is a risk of overdose and contamination with harmful substances, such as mycotoxins, heavy metals, pesticides, and environmental contaminants like polychlorinated biphenyls and dioxins.

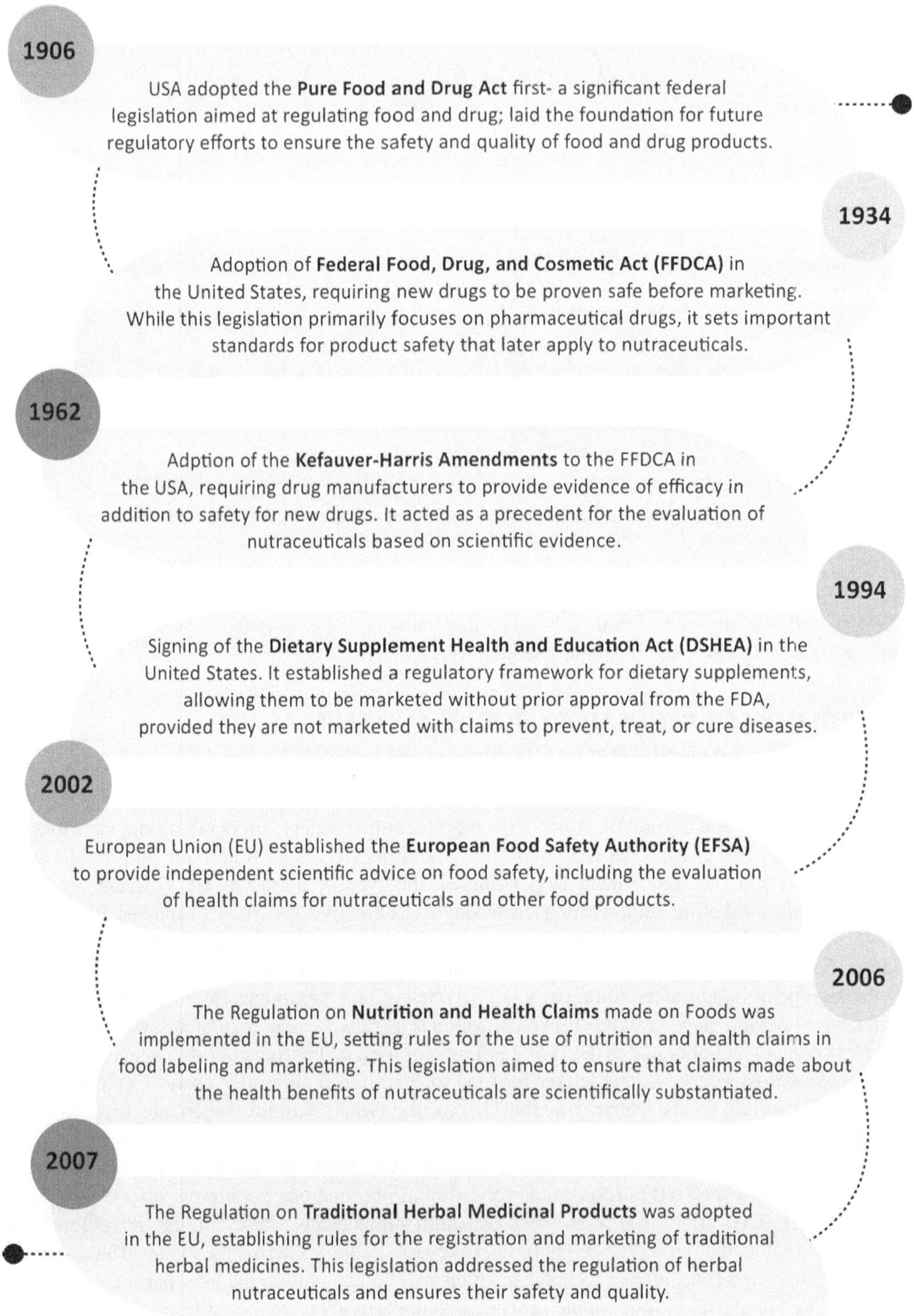

FIGURE 4.2 Historical review of the regulatory framework for safety and toxicity aspects of nutraceuticals.

TABLE 4.1
Regulatory Bodies and Acts Governing the Regulation of Nutraceuticals and Related Supplements in Various Countries across the Globe

Country	Definition	Regulatory Body	Regulation/Act
USA	*Dietary supplements* refer to products (other than tobacco) containing one or more vitamins, minerals, herbs, amino acids, dietary substances, or dietary ingredients in the form of cape, powder, soft gel, or gelcap intended to enhance the health.	Food and Drug Administration (FDA)	Dietary Supplement Health and Education Act (DSHEA) 1994 [11]
Europe	*Food supplements* refer to food products intended to enhance the beneficial effects on one or more target functions in the body, beyond adequate nutritional effects, thus aiding in improving physical conditions or reducing the risk of diseases.	European Food Safety Authority (EFSA)	Food Supplement Directive 2002/46/EC [12]
India	*Nutraceutical* refers to food or food components that aid in providing health benefits, including the treatment or prevention of any disease.	Food Safety and Standards Authority of India	Food Safety and Standards Act 2006 [10]
Canada	*Nutraceuticals* refer to food products isolated or purified from food having physiological benefits or protection against chronic diseases. *Functional food* refers to food items consumed as usual diet having physiological benefits, nutritional functions, or protection against chronic diseases.	Natural Health Products Regulations	Food and Drugs Act and Regulations [13]
Australia	*Complementary medicine* refers to food extracts containing herbs, vitamins, minerals, nutritional supplements, homeopathics, etc. with specific therapeutic potential.	Therapeutic Goods Administration (TGA)	Therapeutic Goods Act of 1989 [14]
Japan	*Health food* can be categorized as food with a functional claim that satisfies the daily requirements of consumption and food with a specific health claim about a physiological function.	Ministry of Health, Labour, and Welfare (MHLW)	Food with health claims (FHC) [15]
China		State Food and Drug Administration (SFDA)	General Administration of Quality Supervision, Inspection, and Quarantine (AQSIQ) [16]

Despite claims of antioxidant properties in supplements, these contaminants can have pro-oxidant effects, contributing to toxicity [8–10].

4.3 ADR REPORTING SYSTEMS IN VARIOUS COUNTRIES

Dietary supplements or health foods must be free from any contamination caused by contaminating agents, such as microorganisms, pesticides, residue, toxicity due to heavy metals, etc. to avoid any adverse reaction. With the increasing demand for nutraceuticals, the toxicological analysis of any functional food or dietary supplement is mandatory. The analysis and submission of records for studies indicating reproductive toxicity, teratogenicity, mutagenicity, cytotoxicity, immunotoxicity,

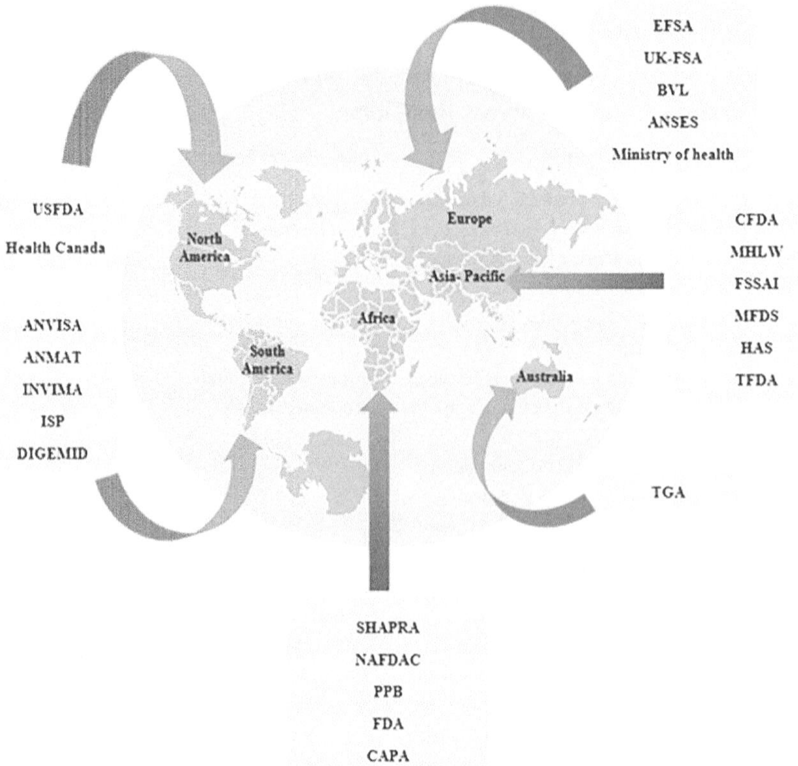

FIGURE 4.3 Various regulatory bodies that actively regulate nutraceuticals worldwide.

Note: USFDA, the United States Food and Drug Administration; ANVISA, Agência Nacional de Vigilância Sanitária; ANMAT, Administración Nacional de Medicamentos; INVIMA, Instituto Nacional de Vigilancia de Medicamentos y Alimentos; ISP, International Specialty Product; DIGEMID, La Dirección General de Medicamentos Insumos y Drogas; SHAPRA, South African Health Products Regulatory Authority; NAFDAC, National Agency for Food and Drug Administration and Control; PPB, the Pharmacy and Poisons Board; FDA, Food and Drug Administration; CAPA, Central Administration of Pharmaceutical Affairs; TGA, Therapeutic Goods Administration; CFDA, China Food and Drug Administration; MHLW, the Ministry of Health, Labour, and Welfare; FSSAI, the Food Safety and Standards Authority of India; MFDS, the Ministry of Food and Drug Safety; HSA, Health Science Authority; TFDA, Thailand Food and Drug Administration; EFSA, the European Food Safety Authority; UKFSA, United Kingdom Food Safety Authority; BVL, the Federal Office of Consumer Protection and Food Safety; ANSES, Agency for Food, Environmental, and Occupational Health and Safety.

cardiovascular toxicity, neurotoxicity, and others are now in increasing demand via regulatory bodies to audit the nutraceuticals in the market. *Nutrivigilance* is a term used to describe adverse drug reaction systems for nutraceuticals, which is comparable to *pharmacovigilance* [17]. Regulatory bodies around the world have different ways to regulate the reporting and documentation of the safety and toxicity of nutraceuticals. For instance, the CFSAN Adverse Event Reporting System is a platform that enables the reporting of adverse events and the collection of safety information for any marketed dietary supplements, prescription drugs, medical products, OTC medications, etc., which is under the direct supervision of the FDA. Another such example is the Yellow Card Scheme administered by the regulatory body of the UK (MHRA), which allows the reporting of side effects, interactions, and safety concerns in the country. Another gateway for food safety resources in the event that the food chain poses a concern to the public's health is the European Medicines Agency's Rapid Alert

System for Food and Feed, or RASFF, in collaboration with the WHO's International Food Safety Authorities Network (INFOSAN) [18, 19]. Furthermore, the pandemic is also expected to be the reason for the production and false claims of nutraceuticals with substandard quality. Therefore, countries are amending laws, acts, and regulations to protect consumers. The increasing demand for validated clinical trials of products by the Indian Council of Medical Research, commonly referred to as ICMR, and the FSSAI is one example of this. This chapter shall serve to familiarize readers with the regulatory aspects of the safety and toxicity of nutraceuticals in various countries [10].

Manufacturers of all dietary supplements in the USA are obliged by the Dietary Supplement and Non-Prescription Drug Consumer Protection Act of 2006 (which includes over-the-counter medications) to notify the FDA of any adverse reaction or adverse event occurrences within 15 days of receiving reports or complaints. Under the Natural Health Product Regulation, licensed manufacturers of health products must notify Health Canada of any major adverse reactions, similar to that in the United States. Australia gives an inclination on requirements for pharmacovigilance for analyzing and reporting adverse reactions under the Australian Regulatory Guidelines for Complementary Medicine. On the other hand, adverse events are required to be reported to the National Authority for Pharmaceuticals, Medical Devices, and Health Products in Peru. Through the National Nutrivigilance Scheme, the French Agency for Food, Environmental, and Occupational Health and Safety, better known as ANSES, established guidelines for the reporting of adverse drug reactions (ADRs) involving dietary supplements and fortified foods [20, 21].

4.4 INTERACTION OF NUTRACEUTICALS WITH VARIOUS DRUGS

4.5 Warning Letters

4.5.1 Warning Letters Issued by Various Countries for Nutraceuticals

Warning letters issued for nutraceuticals in various countries are critical regulatory actions taken by government agencies to address violations of safety, labeling, or marketing regulations. Stringent regulations are in place for nutraceuticals, a broad category of product that includes functional foods, herbal products, and dietary supplements, in order to safeguard consumers and prevent misleading claims. Here is an overview of warning letters issued in various countries, along with references, to highlight the importance of regulatory compliance in the nutraceutical industry.

4.5.1.1 United States (FDA)

Companies found to be in violation of the provisions of the Federal Food, Drug, and Cosmetic Act receive warning letters from the USFDA. Adulteration, misbranding, and making unsubstantiated health claims are a few examples of these infractions. For instance, a warning letter might be issued if a company markets a product as a cure for a disease without adequate scientific evidence. The FDA maintains a database of warning letters issued to nutraceutical companies on its website, providing transparency and serving as a deterrent against non-compliance [35].

4.5.1.2 European Union (EU)

The European Food Safety Authority's (EFSA) framework and the regulations of each member state regulate the laws governing the sale of nutraceuticals within the EU. Companies marketing nutraceuticals in the EU must comply with regulations regarding product safety, labeling, and health claims. Violations may result in warning letters from national regulatory authorities or actions taken by the European Commission to ensure compliance with EU regulations [36].

4.5.1.3 Canada

Health Canada regulates nutraceuticals under the Natural Health Products Regulations. Similar to the FDA in the United States, Health Canada issues warning letters to companies found in

TABLE 4.2
Common Drugs/Ingredients That Interact with Dietary Supplements

Name of Dietary Supplement	Name of Drug	Type of Interaction
Hypericum perforatum (Saint-John's-wort)	Antidepressants, oral contraceptives, and blood thinners	Potentially reducing their effectiveness or increasing the risk of bleeding [22]
Ginkgo biloba (maidenhair tree)	Interacts with blood thinners, such as warfarin	Increasing the risk of bleeding [23]
Vitamin K	Interacts with blood thinners	Affecting their efficacy in controlling blood clotting [24]
Coenzyme Q10 (CoQ10)	Interacts with blood pressure medications, such as beta-blockers, and certain cancer treatments, including chemotherapy drugs	Due to coenzyme Q10 reducing the efficiency of anticoagulant warfarin, could place patients at higher risk of blood clots [25]
Garlic supplements	Interact with blood thinners, such as warfarin, and HIV medications	Potentially affecting their effectiveness [26]
Fish oil supplements	Interact with blood thinners	Increasing the risk of bleeding [27]
Calcium supplements	Interfere with the absorption of certain antibiotics, thyroid medications, and iron supplements	Causing damage to kidneys and lungs, decreased blood level [28]
Vitamin E	Interacts with blood thinners, such as warfarin, and cholesterol-lowering medications	Potentially increasing the risk of bleeding [29]
Ginseng	Interacts with blood thinners and certain medications for diabetes	Affecting their efficacy [30]
Iron supplements	Interact with certain antibiotics, thyroid medications, and proton pump inhibitors	Reducing their absorption [31]
Digoxin	Hawthorn leaf and the flower	Both with positive inotropic and electrophysiological effects on the heart, thus leading to excessive effect [32]
Serenoa repens (saw palmetto)	Interacts with warfarin and other blood thinners	Prolonging bleeding time [33]
Salvia miltiorrhiza Bunge (danshen)	Interacts with blood thinners, such as warfarin	Decreasing the elimination rate of warfarin [33]
Echinacea purpura L. (purple coneflower)	Statins, fibrates, niacins, or amiodarone	Inducing hepatotoxic effects [33]
Aconitum napellus L. (aconite)		Causing bradycardia and hypotension, poisoning, paralysis [33]
Pausinystalia yohimbe (yohimbine)	Interacts with antihypertensive, cardiac, and renal medication	Increasing blood pressure, heart rate, and norepinephrine levels [33]
Glycyrrhiza glabra (licorice)	Reduces the effectiveness of antihypertensive drugs	Inducing hypokalemia, hypertension, and edema [33]
Grapefruit/grapefruit products	Interacts with antiarrhythmic-hypertensive and antiarrhythmic drugs	[34]

violation of regulations. These violations may include selling products without proper licensing, making unauthorized health claims, or failing to comply with labeling requirements. The issuance of warning letters serves to protect Canadian consumers and uphold the integrity of the regulatory system [37].

4.5.1.4 Australia

In Australia, the Therapeutic Goods Administration (TGA) oversees the regulation of nutraceuticals. Companies marketing these products are required to adhere to the Therapeutic Goods Act and its associated regulations. The TGA may issue warning letters to companies found to be non-compliant with regulatory requirements, including those related to product quality, safety, and advertising [38].

4.5.1.5 Japan

Nutraceuticals are regulated in Japan by the Ministry of Health, Labour, and Welfare (MHLW) in accordance with the Food Sanitation Act and related regulations. Warning letters or other enforcement actions by the MHLW to ascertain the safety and efficacy of nutraceuticals marketed in Japan may result from the infringement of these rules and regulations [39].

The crucial importance of complying to nutraceutical standards is made clear by these warning letters and regulatory measures in order to safeguard the public's health and maintain consumer confidence in these products. Companies operating in the nutraceutical industry must stay abreast of evolving regulations and adhere to best practices to avoid regulatory enforcement actions and maintain market integrity.

4.5.2 Various Reasons for Warning Letters Issued

Sports drinks have become increasingly popular among athletes and fitness enthusiasts for their perceived benefits in hydration and performance enhancement. However, the growing market for these beverages has also raised concerns about their safety and the accuracy of their marketing claims. Regulatory bodies such as the Food and Drug Administration (FDA) in the United States play a crucial role in ensuring that sports drinks meet safety standards and are appropriately labeled. In cases where manufacturers fail to comply with regulations, warning letters are issued to address violations and protect consumers.

The FDA regulates sports drinks under the category of dietary supplements or food products, depending on their ingredients and intended use. As such, manufacturers must adhere to labeling requirements, including the accurate representation of ingredients, nutritional content, and health claims. Failure to comply with these regulations can result in warning letters and, in severe cases, legal action [40].

Warning letters related to sports drinks include thus:

1. **Misleading health claims.** In 2019, the FDA issued warning letters to several manufacturers of sports drinks for making unsubstantiated health claims. One prominent example involved a company claiming that its sports drink could prevent or treat concussions, which prompted swift regulatory action due to the lack of scientific evidence supporting such assertions. The warning letters demanded the removal of these misleading claims from product labeling and marketing materials.
2. **Unsafe ingredients.** Another common reason for warning letters is the presence of unsafe or unapproved ingredients in sports drinks. For instance, if a manufacturer includes prohibited substances or exceeds allowable limits of certain additives, it may prompt regulatory intervention. Such violations jeopardize consumer safety and violate regulatory standards. Warning letters typically require the immediate removal of the offending ingredient and corrective action to prevent future occurrences [41].
3. **Labeling inaccuracies.** Accurate labeling is paramount to ensure consumers make informed choices about the products they consume. Warning letters may be issued if sports drinks contain inaccuracies in ingredient lists, nutritional information, or allergen warnings. These discrepancies not only undermine consumer trust but also pose risks to

individuals with allergies or dietary restrictions. Manufacturers receiving such letters must rectify labeling errors promptly and implement measures to prevent recurrence.

4.5.3 Implications and Consumer Awareness

Warning letters serve as a mechanism to hold manufacturers accountable and to safeguard consumer interest. However, their effectiveness relies on public awareness and regulatory follow-up. Consumers should remain vigilant when purchasing sports drinks, scrutinizing labels for accurate information and avoiding products with dubious claims or safety concerns. Additionally, regulatory agencies must ensure the timely enforcement of corrective actions outlined in warning letters to maintain industry compliance and uphold standards of product safety.

Warning letters play a vital role in regulating the sports drink industry, addressing violations of labeling, safety, and marketing regulations. By holding manufacturers accountable for their products' accuracy and safety, these letters contribute to consumer confidence and informed decision-making. Through continued vigilance and collaboration between regulatory bodies and manufacturers, the sports drink market can thrive while prioritizing consumer safety and regulatory compliance [42].

4.5.4 Warning Letters on Caffeine, Sports/Energy Drinks, Protein Powder

4.5.4.1 FDA Warning Letter to Eagle Energy (RE: 617712)

On December 1, 2021, the FDA issued a warning letter to Eagle Energy regarding the marketing practices of their "Caffeine Pen" products, specifically the "Lychee Ice" and "Guarana Berry" variants. The letter addresses concerns regarding the unapproved claims made about the products' efficacy, safety, and intended use. It brings attention to the fact that Eagle Energy's products are marketed as unapproved drugs, which is prohibited under the Federal FD&C Act, according to claims made on the company's website, social media profiles, and product packaging. The warning letter additionally raises safety concerns with caffeine inhalation and its possible adverse effects on respiratory health. Since the products are intended to affect body functions or prevent disease, the FDA claims that they violate the definition of *dietary supplements* according to the FD&C Act and classifies them as pharmaceuticals. Following are the strength of the case:

1. **Clarity of concerns.** The letter provides a clear outline of the specific violations and concerns identified by the FDA, including unapproved drug claims and safety issues related to the inhalation of caffeine.
2. **Detailed examples.** The FDA includes specific examples of claims made by Eagle Energy on their website, social media, and product packaging, providing concrete evidence to support their findings.
3. **Directive for compliance.** The letter outlines the necessary steps for Eagle Energy to address the violations, including ceasing the dissemination of unapproved claims, taking corrective action, and providing a written response to the FDA within a specified timeframe.

The FDA warning letter to Eagle Energy effectively communicates regulatory concerns regarding the marketing and safety of their "Caffeine Pen" products. By providing specific examples of violations and directives for compliance, the FDA aims to ensure that Eagle Energy takes appropriate corrective action to address the identified issues and protect consumer safety [41].

4.5.4.2 FDA Warning Letter to Mr. Mancuso

The FDA has issued a warning letter to Mr. Mancuso regarding significant violations of the Federal Food, Drug, and Cosmetic Act (the Act) identified on various websites and social media platforms associated with his companies. The violations primarily relate to the marketing of dietary supplements with unapproved drug claims, which classify the products as unapproved new drugs and misbranded drugs under the Act. Following are the key points that were addressed:

1. **Unapproved new drugs and misbranding.** The FDA identified multiple products marketed on Mr. Mancuso's websites and social media platforms with claims indicating their intended use as drugs for the treatment, mitigation, or prevention of diseases. Examples include CardioBurn for Her Powder (green coffee extract), Rhino Rampage Capsules™ (niacin), and Joint Revolution™ Capsules (glucosamine sulfate, chondroitin sulfate, MSM, bromelain, glucosamine with chondroitin, hyaluronic acid). Additionally, claims suggesting treatment or prevention of COVID-19 were found on the product IMMUNITY + Powder (vitamin D, zinc, elderberry, olive leaf extract, L-lysine, echinacea), raising serious concerns.
2. **Violative claims.** The FDA provided specific examples of website claims promoting the products' alleged therapeutic effects, such as reducing blood sugar levels, managing cholesterol, and alleviating joint pain. These claims indicate an intent to market the products as drugs, subjecting them to regulatory requirements for safety and efficacy.
3. **Misbranded products.** Some products, including CardioBurn for Her Powder, IMMUNITY + Powder, and HYPEX Powder (citrulline malate, Kanna, *Rauwolfia vomitoria*), were found to lack adequate directions for use, rendering them misbranded under the Act. Since these products are intended to be used in the treatment of conditions requiring a licensed practitioner's supervision, it is not possible to provide consumers sufficient instructions for safe usage.

The Food and Drug Administration's warning letter underscores the agency's commitment to enforcing regulatory standards for dietary supplements and ensuring that products marketed to consumers meet safety and efficacy requirements. The violations identified pose potential risks to public health, particularly if consumers rely on these products for the treatment of serious medical conditions.

The FDA's warning letter explains the breaches that Mr. Mancuso and his affiliated companies need to resolve immediately. This includes halting the marketing of products with unapproved drug claims, offering sufficient usage instructions where necessary, and abiding by all applicable Federal FD&C Act requirements. Failure in doing so may lead the FDA to take legal action, which could possibly involve seizure and injunction [43].

4.5.4.3 FDA Warning Letter to Mr. Harold J. Leighton

During the months of April and May of 2017, the FDA inspected Mr. Harold J. Leighton's Warren, Maine, company operations. The investigations revealed breaches of the CGMP requirements, which are set forth in 21 CFR 111 and pertain to the production, storage, labeling, and packaging operations of nutritional supplements. More specifically, significant CGMP regulation violations were documented during the manufacture of the Vanilla Recovery Whey Protein as well as Peppermint SO3 + D3 Fish Oil products. A number of misbranding violations pursuant to provisions 403 of the Act (21 USC § 343) were additionally found during the inspection. These included false labeling, failure to meet nutrient content claim requirements, improper ingredient declaration, incorrect nutrition information presentation, and non-compliance with labeling regulations across various dietary supplement products.

The claims provided on the aforementioned products on the website www.sfh.com seemed extremely concerning. Due to the products' claimed purpose of curing, mitigating, treating, or preventing illness, the FDA determined that these claims qualified the products as pharmaceuticals under section 201(g)(1)(B) of the Act. Such claims included assertions regarding heart health, inflammation reduction, and the treatment of various health conditions.

Mr. Leighton submitted letters on May 17 and September 6 in the year 2017 in response to the FDA's observations. Both of these letters addressed the violations listed in the FDA form 483,

"Inspectional Observations," which was issued during the inspection. However, upon review, the FDA found the response inadequate in addressing the identified violations.

The promotional marketing of dietary supplements for unapproved uses as pharmaceuticals, which requires prior approval from the FDA based on scientific evidence proving their efficacy and safety, was one major violation that was recently brought to attention. Furthermore, the products were misbranded subject to section 502(f)(1) of the Act, owing to the lack of adequate instructions for use.

In a nutshell, major violations of CGMP guidelines and dietary supplement labeling standards were found during the inspection, along with misleading promotion of products for use as drugs that have not been approved. The FDA has urged Mr. Leighton to take corrective actions to address these violations and ensure compliance with federal regulations governing dietary supplements and drug products. If this fails to occur, additional regulatory action—like seizure or an injunction—may be necessary.

Several CGMP violations were observed during the inspection, leading to the determination that these dietary supplements are adulterated in accordance with section 402(g)(1) of the Act [21 U.S.C. § 342(g)(1)]. Key violations include:

1. **Failure to establish specifications.** The facility neglected to establish necessary specifications for components used in the manufacturing process, such as identity specifications and component specifications, ensuring purity and strength. Moreover, product specifications for finished batches were found to be inadequate, lacking essential details to ensure product quality.
2. **Absence of finished product sampling.** Before releasing finished batches of dietary supplements, the company neglected to collect representative samples for verification with established product specifications. Despite commitments to implement corrective measures, no documented evidence of such actions has been provided to date.
3. **Inadequate handling of out-of-specification results.** The quality control personnel failed to reject the products or components when dietary supplements fell short of established standards, nor did they approve necessary changes or treatments for ensuring product quality. Additionally, the firm lacked adequate documentation of material review and disposition decisions, contravening regulatory requirements.
4. **Deficiencies in master manufacturing records.** As mandated by regulations, the facility did not prepare and follow documented key manufacturing records associated with each individual dietary supplement composition. Moreover, crucial details like the degree of strength, weight, quantity, or concentration of every single nutritional component for various batch sizes were lacking from the master manufacturing records that were already in place.

While the firm provided responses to the FDA's findings, dated May 17, 2017, indicating intentions to address the identified deficiencies, the agency deemed these responses inadequate. Despite commitments to update master manufacturing records and implement corrective actions, documented evidence of compliance has not been provided.

Overall, the inspection findings highlight serious lapses in manufacturing practices and regulatory compliance, posing potential risks to consumer safety and product quality. The FDA urges the facility to take immediate corrective actions to rectify these violations and ensure compliance with CGMP regulations for dietary supplements. Refraining from doing so results in more regulatory actions, which include enforcement measures, like product seizures and injunctions.

1. **False and misleading labeling.** The inspection revealed false statements on the website regarding antioxidant properties of certain ingredients in the Strong Strength + Muscle and Push Pre-Workout products. The claims regarding antioxidant activity attributed to

ingredients like serum protein concentrate, creatinine, and glutamine are misleading, as these ingredients do not possess antioxidant properties.

2. **Failure to meet nutrient content claim requirements.** Numerous products, such as Fuel Whey Protein and Pure Whey Protein, contain claims about their nutritional content that are not compliant with regulations. Claims like "low in cholesterol" and "high in free calcium" on Pure Whey Protein are not supported by the product's composition as per 21 CFR 101.62(d)(2) and 21 CFR 101.54(b). Similarly, the claim regarding the richness of apple pectin in Fuel Whey Protein is not in compliance with regulations.
3. **Improper declaration of ingredients.** Several products, including Peppermint SO3 + D3 Fish Oil and Coconut Fuel Whey Protein, fail to declare all ingredients by their common or usual names, as required by 21 CFR 101.4 and 21 CFR 101.36. Sub-ingredients like corn oil in vitamin D3 and rebaudioside A are not properly listed on the labels.
4. **Nutrition information presented incorrectly.** Several product labels provide nutrition information incorrectly, in contrary to FDA guidelines. Errors include incorrect order of listing for certain nutrients, incorrect percentages of daily value, incorrect rounding of values, and improper declaration of certain nutrients as "zero."
5. **Failure to list serving sizes and servings per container.** According to 21 CFR 101.36(b), several kinds of products fail to disclose serving sizes, including servings per container, on their supplement facts labeling in the correct format.
6. **Failure to list manufacturer information.** As specified by 21 CFR 101.5, certain product labels exclude the manufacturing, packers, or distributor's name and address.
7. **Failure to declare net content quantity.** A few products fail to disclose the US standard of measurement (such as pounds or ounces), as stipulated by 21 CFR 101.7(b)(1), or they indicate the net content quantity in incorrect units on the label, specifically the principal display panel.

Overall, the inspection findings indicate significant deficiencies in labeling practices, which could mislead consumers and pose risks to public health. The FDA advises the facility to promptly address these violations to ensure compliance with the Act and its implementing regulations. Failure to do so may result in enforcement action, including seizure and/or injunction [44].

4.6 RECALL PROCEDURE BY REGULATORY AGENCIES

4.6.1 Responsibilities

4.6.1.1 Quality Assurance Department

1. Initiation of product/mock recall
2. Investigation of product failure after marketed
3. Maintaining documentation as well as logs for mock recalls and product recalls

4.6.1.2 Head Quality Assurance Department

1. Investigation of post-market product failure
2. Assessment and approval for a mock or product recall
3. Intimation to regulatory authorities
4. Taking the required steps to ensure a prompt and effective mock recall and product recall

4.6.2 Process for Product Recall

4.6.2.1 Investigation Ahead of Product Recall

Any communication in writing concerning product quality, regardless of purity, efficacy, toxicity, and adverse reactions, involving physical appearance, labeling, packaging, etc., received from the complainants, which mean retailers, distributors, wholesalers, field employees, customers, and health-care professionals, shall be investigated following SOP. The head of quality assurance makes the required decisions, such as recalling the product, according to feedback provided by the market complaint investigator.

4.6.2.2 Reason for Product Recall

The product should be recalled when official notification is received from drug regulatory authorities. Apart from DRA, the head QA and company can also initiate a recall procedure if they observe any abnormality.

4.6.2.3 Classification of Defects

1. Class I: A life-threatening or related situation caused by any marketed product falls under this category. In such cases, recall of the product can extend to the consumer level. Some of the examples are:

    a. Product failure (difference between a label and the composition)
    b. Right product, incorrect label
    c. Hazardous contamination with chemicals
    d. Mixed-up product
    e. Microbial contamination in sterile products
    f. Wrong active ingredients

2. Class II: An illness or a temporary medically reversible situation caused by marketed products or mistreatment by the products falls under this category. When this occurs, a quick alert notification needs to be forwarded to all those contacts wherever the batches are placed. Some of the examples are:

    a. Mislabeling
    b. Failure to meet specifications for factors like stability, assay, dissolution, and disintegration
    c. Non-sterile products contaminated with microbes
    d. Chemical and physical contamination
    e. Leakage or sealing defects
    f. Physical/visual deterioration, like discoloration, sedimentation, crystallization, etc.

3. Class III: Defective products that do not cause any significant hazard to health but whose withdrawal should be done fall under this category. In this case, a rapid alert system is not required. Some examples are:

    a. Faulty closure and faulty packing, like missing/wrong batch number, expiry date, etc.
    b. Any physical damage, like broken or chipped dose, missing dose
    c. Contamination by microbiological spoilage, dirt, gritty/chunky particle

4.6.3 Product Recall Type

4.6.3.1 Statuary Recall of Product

The drug control authority/regulator, or any other similar entity in that nation, is responsible for this recall procedure. These authorities take action after receiving reports from government analysts where the product does not meet standard quality. They have the right to ban the product.

4.6.3.2 Voluntary Recall of Product

This recall process is initiated by the company when the company finds an abnormal observation in any marketed products. This process can be triggered by some incident, like strength, identity, safety quality, and purity of the product or batch.

4.6.3.3 Mock Recall

The purpose of this recall procedure is to verify the recall's efficacy. At least one product obtained from a distributor should be handled. Generally, it should be carried out once every two years. Mock recall is performed only up to the stock level.

The Annexure 1: Product Recall Initiation Form's section C should be followed while establishing the recall's details [45]. If, following an investigation, it becomes mandatory to recall any of the products, the QA technical personnel will do so in accordance with LOG, which is maintained in accordance with Annexure 4: Logbook for Product/Mock Recall, along with Annexure 1 [45, 46]. A recall number consisting of 12 alphanumeric characters, such as XX/AA/YY/ZZZ, will be assigned to each product recall. The letters *AA* represent product recall and mock recall (PR/MR), *YY* for current year, *ZZZ* for serial number, and *XX* for location code (manufacturing facility) [47].

4.7 USFDA RECALLS OF VARIOUS DIETARY SUPPLEMENTS: OVERVIEW AND CASE STUDIES

Any marketed products that are contaminated with substances that are hazardous to health, mislabeled products, and foods that can cause foodborne illness outbreaks are recalled from the marketplace by their manufacturers. Whenever a manufacturing company announces a recall, safety alert, or market withdrawal, this announcement is posted on the website by the USFDA for the public. A list of dietary supplements and food and beverage product recalls is available on the USFDA website [48]. This website provides recall information gathered through public notices and press releases for FDA-regulated products. In addition, the FoodSafety.gov website provides information on food recalls and food safety from the FDA and US Department of Agriculture [49]. Most meat and poultry-related products are regulated by the USDA. The FDA has a network, referred to as Coordinated Outbreak Response and Evaluation (CORE), whose purpose is to oversee surveillance and post-response procedures related to incidents that involve multiple illnesses primarily associated with FDA-regulated dietary supplements and food for humans. It also conducts investigations in case of outbreaks. The FDA search engine has filters for various FDA-regulated product recalls, which are categorized into different domains that include dietary supplements as well as food and beverages. The product recall can be filtered specifically using "terminated" and "as ongoing" or "completed." A terminated recall occurs when the FDA has determined that every necessary step has been taken to eliminate or rectify the potentially hazardous item in accordance with the recall plan, and that the disposal has been appropriately carried out in light of the degree of risk.

4.7.1 Ongoing FDA Recall Products

4.7.1.1 Recall of Magnum XXL 9800 Capsules

The Meta Herbal company voluntarily announced a recall for the product sold under the brand name Magnum XXL 9800 capsule on November 29, 2023. A pack size of 500 blisters was being recalled at the consumer level. This dietary supplement product was marketed for male sexual enhancement and was available as a blister pack of one pill. Various batches of products were distributed and sold online on Amazon. The phosphodiesterase-5 inhibitor sildenafil tainted the Magnum capsules. FDA-regulated products that are used to treat male erectile dysfunction contain sildenafil. Magnum capsules from Meta Herbal are a novel, unapproved pharmaceutical that was recalled off the market because its safety and effectiveness could not be confirmed.

A trace amount of undeclared sildenafil in the capsule results in a number of additional health problems since there is interaction between sildenafil and the nitrates found in certain prescription medicines (like nitroglycerin) that causes a sharp, potentially life-threatening decrease in blood pressure. Adult males with diabetes, high cholesterol, or high blood pressure levels, as well as heart disease, frequently consume products containing nitrate. Any report of adverse events for the

Magnum capsule has not been received to date, but the company notified consumers via Amazon regarding the recall of the product and asked consumers who had the recalled product to immediately stop using it [50].

4.7.1.2 Recall of Dr. Ergin's Advanced Glucose Support Dietary Supplement Capsules

On November 8, 2023, Advanced Glucose Support Supplement capsules of 60-, 120-, and 180-pack size bottles were announced for voluntary recall by Sugar MDs LLC. Glyburide and metformin were found in the product through FDA analysis. Since metformin and glyburide are active components in a number of FDA-regulated type 2 diabetes treatment pharmaceutical medications, FDA-unapproved products containing them are prohibited from being sold as nutritional supplements. This tainted dietary supplement product was distributed throughout the United States through online sales to maintain blood glucose levels. No adverse events have been reported for this product, although glyburide causes a significant decrease in blood sugar levels, leading to hypoglycemia and, when severe, resulting in seizure, coma, or death, which is a life-threatening risk to consumers. Along with glyburide, metformin also causes hypoglycemia and a serious rare effect called lactic acidosis, which, if untreated, leads to death [51].

4.7.1.3 Recall of Kuka Flex Forte Caplets, Reumo Flex Caplets, and Artri King Tablets

The Botanical-Be company announced the Kuka Flex Forte caplets', Reumo Flex caplets', and Artri King tablets' recall on November 7, 2023, due to the undeclared presence of the drug diclofenac, which was revealed by FDA analysis. These products were sold on online platforms, and the manufacturer notified consumers via email about the return of recalled products. These adulterated dietary supplements were prescribed to treat arthritis-related inflammation as well as pain. To date, no adverse event has been reported to the Botanical-Be company related to this recalled product. The FDA claims that the aforementioned products are unapproved for safety and include undisclosed diclofenac, a non-steroidal anti-inflammatory active ingredient that is an FDA-authorized drug and sold in a number of FDA-regulated drug products. When consumers consume these undeclared diclofenac products, there is the risk of developing serious adverse reactions, including anaphylaxis and renal, gastrointestinal, and cardiovascular problems, in patients proactively taking anticoagulants (for instance, warfarin), as well as those with hepatic or renal disorders and who have allergic reactions to diclofenac [52].

4.7.1.4 Recall of WEFUN Capsules

The WEFUN-branded dietary supplement capsules are also a male sexual enhancement product that was recalled from the market due to the presence of undeclared sildenafil in the product. This recall, dated August 25, 2023, was conducted by the manufacturing company Hua Da Trading Inc. dba Wefun Inc. Products that the FDA has granted approval for the medical treatment of erectile problems contain sildenafil. Due to the serious, life-threatening adverse events and interaction with nitrates, dietary supplement products containing undeclared sildenafil are recalled from the online market [53].

4.7.1.5 Recall of Dr. Rima's Nano Silver 10 ppm Dietary Supplement

The Natural Solutions Foundation company announced an urgent notice to customers regarding the Dr. Rima's Nano Silver 10 ppm dietary supplement product recall, dated March 7, 2023. Marketing of this product took place between January 2020 and December 2021. The dietary supplement products were used as a natural product for immune system support. The nutritional supplement's label declared that it would mitigate, alleviate, or prevent COVID-19, which the FDA believed to be an unsupported health claim, and this was the reason for the recall announcement. Considering the product's safety and efficacy were not established or approved, utilizing it exposed customers at risk to major, perhaps life-threatening, health effects. The firm had arranged the conduct of a consumer-level product recall by providing them with a notice copy, requesting consumers to participate in

TABLE 4.3
Terminated Recall Products by the FDA

FDA Publish Date	Company Name	Brand Name	Product Type	Product Description	Recall Reason
April 22, 2023	EuroPharma Inc.	Terry Naturally, EuroMedica	Dietary supplements, vitamin	Vitamin B, vitamin B complex	Undeclared milk as an allergen
May 27, 2022	Indiana Botanic Gardens Inc.	Botanic Choice	Dietary supplements	Prune and senna soft gels	Undeclared peanuts as allergen
August 2, 2021	Je Dois L'avoir Boutique	365 Skinny	Dietary supplements	High-intensity pills and emergency boutique	Contains sibutramine, which is a controlled substance by the DEA (Drug Enforcement Administration)
May 19, 2021	Miracle 8989	PremierZen Gold 7000, PremierZen Platinum 8000, and MaxxZEN Platinum 12000 capsules	Dietary supplements	Male sexual enhancement	Undeclared sildenafil and tadalafil found in FDA-approved products for the treatment of male erectile dysfunction
April 20, 2021	Church & Dwight Co. Inc.	Vitafusion	Dietary supplements, vitamin	Gummy vitamin products	Presence of metallic mesh material possibly leading to damage of the digestive tract
November 23, 2020	Fusion Health and Vitality LLC	CORE, Immune Boost	Dietary supplements, drugs	CORE, essential nutrients; Immune Boost, sublingual vitamin D3	CORE: presence of unapproved food additive hordenine HCl; Immune Boost: statements in labeling causing it to become an unapproved new drug
February 24, 2020	Med-Man Distribution	Up2, Bow & Arrow	Dietary supplements	Up2: no other all-natural libido for men and women dietary supplement; Bow & Arrow: libido enhancer for men	Tainted with sildenafil, which is FDA-approved prescription drug for erectile dysfunction
May 7, 2019	DBP Distribution	Titanium 4000	Dietary supplements, nutritional supplement	Titanium 4000 capsules	Undeclared sildenafil and tadalafil: both FDA-approved active ingredients used for the treatment of erectile dysfunction

(Continued)

TABLE 4.3 *(Continued)*
Terminated Recall Products by the FDA

FDA Publish Date	Company Name	Brand Name	Product Type	Product Description	Recall Reason
May 2, 2019	Life Rising Corporation	Life Rising, HoliCare	Dietary supplements, nutritional supplement	Dietary supplements	Contain excessive levels of lead, which causes serious and adverse health consequences
March 21, 2019	Ata Int. Inc.	BLUE FUSION	Dietary supplements, nutritional supplement	Male enhancement dietary supplements	Undeclared sildenafil, tadalafil, desmethyl carbodenafil, dithiodesmethylcarbodenafil, scutellarin, and daidzein
June 21, 2018	Gaia Ethnobotanical LLC	Gaia Ethnobotanical	Dietary supplements, nutritional supplements, drugs, generic drugs	Kratom products	Salmonella contamination

the product recall by completing the response within 22 days of the notice date, and returning any unused portion of the product to the company. Within 14 days, products were received by the firm, and reimbursements were given [54].

4.7.1.6 Recall of Various Artri Ajo King Joint Supplements

All batches of Artri Ajo King Joint dietary supplements, which were distributed by PDX Supply Warehouse LLC and Innovacion Naturals, were recalled by Walmart Inc. on May 28, 2022, from both their online and retail locations. Through laboratory investigation, the FDA found that the product's omega-3, which was a joint support supplement, includes undeclared diclofenac, which was not disclosed on the package's label. This commercialized dietary supplement was used in bone cancer, osteoporosis, muscular stiffness, and arthritis treatment. An NSAID called diclofenac elevates the risk of gastrointestinal issues and cardiovascular conditions, including stroke, as well as heart attacks. Diclofenac, an unlisted prescription ingredient in the dietary supplement, has the potential to interact with various other drugs and thus significantly raises the risk of adverse reactions, particularly when consuming numerous NSAID-containing products [55].

4.7.1.7 Recall of Certain FDA-Regulated Products in Six States, Including Dietary Supplements, Foods and Beverages, Drugs, Medical Devices, and Cosmetics

A retail-level recall for certain dietary supplements and food products by the firm Family Dollar Inc. was announced on February 18, 2022. The recall was due to the presence of rodents and rodent activity at the distribution center of the company. Because these supplements may contain salmonella, eating them causes customers to be at risk for a number of health problems. *Salmonella* is a bacterium which could infect pregnant women, newborns, the elderly, young children, and other people with compromised immune systems or pre-existing disorders, resulting in serious and often life-threatening illnesses. Infrequently, salmonella infection results in the organism spreading into the bloodstream and causes more serious complications, such as endocarditis, arthritis, and arterial infections. Salmonella also affects healthy persons, causing diarrhea, abdominal pain, fever, vomiting, and nausea [56].

4.7.2 Other Terminated FDA Recalls

The USFDA's database for information on adverse events and product complaints reported by consumers and health-care practitioners submitted to the FDA is called CAERS (CFSAN Adverse Event Reporting System). CFSAN stands for Center for Food Safety and Applied Nutrition. This database supports CFSAN's safety surveillance for dietary supplements, foods, and cosmetics [48].

4.8 TOXICOLOGICAL ASPECT OF NUTRACEUTICALS

A wide range of nutraceuticals and multi-herbal supplements is available over the counter in the market, and consumers are using it without medical advice or supervision. Many of these products have unproven safety and hidden toxicity associated with their active components. Several hospitals and medical institutes around the globe are receiving cases of patients suffering from side effects due to the toxicity and many adverse events of supplements that have been reported to date. The toxicity may be due to overconsumption of nutraceuticals, use of banned or illegal ingredients in the supplement, and lack of awareness about its safety. Therefore, these nutraceuticals should be cautiously used as per the recommended daily intake (RDA), and they require serious attention for their potential side effects.

4.8.1 Ephedra (Chinese Ma Huang)

The US Food and Drug Administration (FDA) restricted the marketing and sale of dietary supplements comprising ephedrine alkaloids, or ephedra, on February 9, 2004, because of the unjustified toxicity and significant risk involved. A naturally occurring substance, ephedra is obtained from the Chinese herb ma huang, which has three prominent species, *Ephedra sinica*, *Ephedra intermedia*, and *Ephedra equisetina*. The most prominent alkaloid in ephedra, ephedrine, contributes to 90% of the plant's total alkaloid composition. Other alkaloids that compose ephedra include pseudoephedrine, norephedrine (phenylpropanolamine), and norpseudoephedrine (cathine) [57].

According to an evidence-based review sponsored by the Office of Dietary Supplements (ODS), prior to the ban, the Chinese botanical ephedra (ma huang), alone or in combination with caffeine, was sold as a dietary supplement in the United States and showed some short-term (less than or equal to six months) improvements in energy, weight reduction, and sports performance. However, there is still a lack of scientific evidence on the long-term (greater than six months) benefits of ephedra-containing supplements on weight loss and athletic performance. Ephedrine has not been authorized for use in the United States as a medication to aid in weight loss or improve sports performance. However, ephedra alkaloids ephedrine and pseudoephedrine are available as OTC medications, as decongestants and to treat asthma [58]. The ban on ephedra use in supplements by the USFDA was due to safety-related risks, such as heart attack, stroke, seizures, psychiatric risks, autonomic risks, gastrointestinal symptoms, and sudden death reported in many individuals consuming ephedrine supplements [59]. It also leads to other disorders, like insomnia, anxiety, personality changes, irritability, dry mouth, dizziness, and other symptoms. Due to all these serious effects of ephedrine, the FDA considered it strictly unsafe for use in pregnant and breastfeeding women [60].

Despite the ephedra use ban by the FDA, it is still used in many countries for weight loss and for boosting energy. A case of acute bilateral mydriasis was reported in Korea of a 42-year-old healthy woman who experienced sudden blurred vision in both eyes and was diagnosed with dilated pupils. According to the National Research Foundation (NRF) of Korea, this drug-induced bilateral mydriasis was due to the sympathomimetic effect of ephedrine, which the woman had taken a day before the onset of her symptoms [61]. Another case, a first, of bilateral acute angle closure and myopic shift was reported when a 52-year-old woman started consuming ma huang–containing herbal supplements for weight loss. A day after consuming the supplement, she visited the emergency room complaining about acute bilateral ocular pain associated with vomiting, headache, nausea, and painful red eyes, as well as diminished vision. On evidence-based examinations, the observations were high intraocular pressure, edematous cornea, myopic shift, thickened choroid, and shallow anterior chamber in both eyes [62].

A 37-year-old African American bodybuilder had been using Hydroxycut, a MuscleTech nutritional supplement containing ephedra, every day for the previous three years in an effort to lose weight and increase energy prior to the ban. He was brought to a US hospital in May 2003 due to an erratic pulse, diminished palpitations, impaired vision, and a worsening headache. A right middle cerebral artery (also known as MCA) infarction with mild hemorrhagic transformation was discovered during an MRI scan. Following research, it was determined that the bodybuilder's usage of ephedra may have produced a dilated cardiomyopathy that was brought on by tachyarrhythmia or may have functioned as a direct toxin to the myocardium [63]. Supplements for diet that include ephedrine may be bought illegally online or via fitness facilities, since they are banned in Brazil. A 28-year-old gentleman professional marital arts fighter was noticed in Brazil consuming Therma-Pro™, a nutritional supplement intended to help with weight loss. The product included 250 mg of ma huang extract, which is comparable to 20 mg of ephedrine. He had consumed four tablets per day, which was double the amount recommended by the manufacturer, and this led him to a serious myocardial infarction associated with hemiparesis, insomnia, tremors, and chest pain [64].

The ephedrine alkaloid has α-1, β-1, and β-2 adrenergic receptor agonist properties, and it also instigates the release of endogenous catecholamine, which can result in blood vessel constriction and can possibly change the heart's electrical activity. The mechanism backing the ephedra toxicity of myocardial infarction and myocarditis is triggered by vasospasm due to α and β adrenergic stimulation as well as ephedrine's adrenergic effects, which lower the myocardial refractory time, which further contributes to arrhythmias. Being sympathomimetic, ephedrine has also been linked to temporary left ventricular dysfunction due to its stimulant effects on the heart. On the other hand, pseudoephedrine exhibits β-2 adrenergic agonist properties, triggering bronchodilation, whilst its α agonist characteristics account for relieving nasal congestion. Potential adverse effects of ephedra-containing supplements include cardiovascular complications, such as myocardial infarction, myocarditis, tachyarrhythmia, arterial hypertension, palpitations, and sudden cardiac death. Additionally, they have an impact on the central nervous system, which raises the risk of seizures, stroke, along with transient ischemic attacks [62, 64, 65].

4.8.2 *Citrus aurantium* (Bitter Orange)

After the ban on ephedrine by the FDA, multi-ingredient supplements and pre-workout supplements containing bitter orange extract gained ample popularity worldwide [66]. *Citrus aurantium* L., belonging to a genus of family Rutaceae, commonly known as bitter or sour orange, Seville orange, or bigarade, has various phytochemicals mainly comprising flavonoids and alkaloids (p-synephrine). Bitter orange has various biological properties, namely, antioxidant, antimicrobial, anticancer, antidiabetic, antiobesity, and anxiolytic effects [67]. Apart from these, bitter orange extract, specifically p-synephrine, which is the primary active constituent of bitter orange, is used in nutraceuticals along with other dietary ingredients for weight management, improving energy as well as sports performance, and in controlling appetite [68]. *Synephrine* is a sympathetic stimulator and has adrenergic characteristics similar to ephedra [69]. The principal mechanism behind the increased metabolic rate, lipolysis (breakdown of fat), and thermogenesis (production of heat) is its adrenergic activity, particularly its ability to stimulate adrenergic receptors, such as β-3 adrenergic receptors [69]. P-synephrine, the main active component [70], is a protoalkaloid phenylethylamine structurally similar to ephedrine [68].

P-synephrine, being an α and β adrenoreceptor stimulator, particularly for the stimulation of β receptors, causes cardiovascular system stimulation by increasing the heart rate and blood pressure, along with other adverse events [69]. The FDA reviewed 22 reports of adverse events (AER) related to bitter orange extract supplements between April 2004 and October 2009. Ten clinical adverse event case reports have been published in a variety of scientific medical publications over the same time period. The adverse events evaluated by the Center for Food Safety and Applied Nutrition (CFSAN) of the United States FDA include heart palpitations, tachycardia, angina, interstitial nephritis, inflammatory liver disease, chest pain, neck tightness, muscle cramps, shortness of breath and diaphoresis, acute delirium, anxiety, depression, hallucinations, suicidal thoughts, and violent behavior. The ten published clinical case reports included a variety of confounding factors linked to adverse events, such as acute lateral-wall infarction of the heart experienced by a White woman aged 55 years who took Edita's Skinny Pill, a multi-ingredient bitter orange dietary supplement for losing weight [71]; syncope induced by exercise, which was associated with QT prolongation in a 22-year-old healthy female who misused the Xenadrine EFX ephedra-free weight reduction supplement that contained bitter orange [72]; ischemic stroke reported by a 38-year-old man after consuming Stacker 2, which is an ephedra-free supplement containing caffeine and synephrine [73]; a 57-year-old man reporting variant angina developed due to bitter orange supplements [74]; ischemic colitis in a 52-year-old woman after the initiation of bitter orange supplement for weight loss [75]; a bodybuilder reporting thrombosis and coronary spasm after using synephrine and caffeine supplements [76]; another adverse event of left middle cerebral artery vasospasm and stroke for the ephedra-free product Xenadrine EFX supplement reported by a 36-year-old healthy woman [77]; the use of the Metabolift ephedra-free weight loss supplement by a 16-year-old woman leading to adverse effects of anorexia nervosa [72]; and a case report of cardiac fibrillation in a young Air Force–serving female aged 27 years linked with synephrine supplement resulting in a diagnosis of ST segment elevation myocardial infarction (STEMI) in a 24-year-old healthy Black man [63, 76]. Although all the supplements were multi-herbal products, none of the reports involved products with bitter orange extract as a single ingredient, which lack clear supportive evidence for adverse events associated with it [78].

4.8.3 Vitamin D

Many supplements and food products are fortified with vitamins to enhance the nutrition of the product beyond its basic nutritive value. Owing to the rising number of individuals who are deficient in vitamin D, a large number of unregulated nutraceuticals comprising the vitamin are sold and distributed to the general public with minimal proof of their safety. These vitamin D–containing supplements are easily available as OTCs in pharmacies and in internet sales. Vitamin D is essential in maintaining mineral metabolism, immune system, muscle function, respiratory health, and bone health. Hypervitaminosis D, or vitamin D toxicity, can result from taking excessive amounts of vitamin D supplements. An unintentional increase in vitamin D poisoning can be due to over-fortification of foods with vitamin D, like fortified milk. One instance was an elderly man, aged 72, who was regularly ingesting high doses of vitamin D through a combination calcium and ergocalciferol pill, which resulted in a disturbed mental state, tachycardia, and severe hypoxemic breathing difficulties [79].

Vitamin D, also known as calciferol, is a fat-soluble vitamin that comes in two forms: vitamin D3 from sun exposure and vitamin D2 from food. 25(OH)D, the active form of vitamin D, is produced by the enzymatic metabolization of vitamin D by cytochrome P450 mixed-function oxidases (CYP450s). Serum levels of the active form 25(OH)D are used to calculate vitamin D safe and dangerous levels; the usual range is ±30–100 ng/mL (varying for adults and kids). When blood 25(OH)D level is less than 20 ng/mL, it is considered a vitamin D deficiency; insufficiency occurs when the level is between 20 and 29 ng/mL. Vitamin D toxicity, also known as hypervitaminosis D, is typically linked to hypercalcemia and a reduction in parathyroid hormone production. It arises when blood 25(OH)D levels are beyond the normal limit or daily intake. High dosages or prolonged usage of vitamin D supplements might result in vitamin D toxicity.

Adults can become acutely toxic to vitamin D if their blood 25(OH)D level is more than 50 ng/mL (10,000 IU/day), and chronically toxic if their serum 25(OH)D level is between 50 and 150 ng/mL (4,000 IU/day for many weeks). It is advised to regularly check the nutritional vitamin D status of babies, particularly when the mother has vitamin D insufficiency. Given that children's digestive systems are still growing, it is important to take caution while administering vitamin D supplements, both in terms of dosage and time. In children, vitamin D poisoning is not frequently seen. Pediatric patients' blood 25(OH)D levels rise to more than 32 ng/mL after receiving a 700 IU/day vitamin D3 dosage for 12 weeks. This improves the patients' performance and health-related quality of life (HRQoL). When 50,000 IU of vitamin D is given to children for more than nine weeks, the blood 25(OH)D level rises to about 27 ng/mL, indicating vitamin D toxicity. For babies under 12 months of age who are breastfed or partially breastfed, a daily dose of 400 IU of vitamin is advised; in contrast, children aged 1 to 18 years should take 600 IU of vitamin D in the absence of sun exposure. Excess consumption of vitamin D causes various adverse effects, such as hypercalciuria, hypercalcemia, psychosis, kidney stones, weight loss, parathyroid hormone imbalance, and imbalance in bone metabolism. It may also include conditions such as eosinophilia (an increase in eosinophils, a type of white blood cell) and neutropenia (a decrease in neutrophils, another type of white blood cell). In order to minimize the ongoing issue about toxicity of vitamin D, upper limits and recommended dietary allowance (RDA) for vitamin D should be established for manufacturing vitamin supplements, and safety parameters of the supplement should be ensured [80].

4.8.4 Green Tea Extract

Green tea extract (GTE) is a non-fermented, non-oxidized product of the fresh leaves of *Camellia sinensis*, which belongs to the plant family Theaceae. Active constituents of green tea include polyphenols, several minerals, proteins, and pectin. The major polyphenols are catechins, namely, epigallocatechin-3-gallate (EGCG), being the highest in concentration, along with epicatechin (EC), epicatechin-3-gallate (ECG), and epigallocatechin (EGC). Green tea use has become increasingly popular worldwide either in the form of beverages or its extract in many herbal supplements for various medicinal uses, such as anticancer, antioxidant, as well as weight loss [81]. The antioxidants in green tea catechins, notably EGCG, lower oxidative stress in the body by scavenging free radicals. It is believed that this antioxidant action plays a role in the protective benefits of green tea extract against a number of ailments, including cancer and cardiovascular conditions [82].

In Germany, in February 2011, a health-care department admitted a 63-year-old woman who had breast cancer and was receiving treatment with green tea capsules recommended by a cancer support group. She had elevated transaminase levels (ALT, AST) and elevated levels of total bilirubin, along with jaundice, mild pruritus, and discoloration of urine and stool. According to the findings, the patient had drug-induced acute hepatitis due to the intake of concentrated green tea capsules [83]. A White female of age 44, in good health, was admitted to an intensive care unit in Canada in October of 2005. She had presented with sudden-onset abdominal pain, progressive malaise, and jaundice; later, she was diagnosed with acute liver failure and was under treatment. Following the institution's investigation, it was discovered that the patient's habit of taking green tea concentrate dietary supplements was primarily to blame for the liver's failure [83]. The first report of hepatotoxicity from the green tea extract product Exolise® was published in 2001, and later, it was withdrawn from the market [84]. Though the mechanism behind hepatotoxicity, liver damage, and hepatitis is unclear, it is believed to be mainly due to the EGCG presence in the supplements [85]. Numerous incidents of hepatotoxicity have been linked with high doses of green tea extract nutraceuticals, due to which many marketed products have been suspended by regulatory authorities. As a consequence, immediate discontinuation of green tea supplements must be done in cases of elevated liver enzymes [86], and these products must be used carefully in patients with liver failure or acute hepatitis [87].

4.9 TOXIC CONTAMINANTS

In the case that during production heavy metals, metalloids, mycotoxins, allergies, fertilizers, pesticides, or non product residues from hazardous plants are introduced into the product, customers may be exposed to harmful nutraceuticals [88]. The FSSAI has established the Contaminants, Toxins, and Residues Regulations 2011, which provide limits for metal contaminants, such as lead, copper, arsenic, mercury, etc., along with other crop contaminants and natural toxins, namely, aflatoxin, ochratoxin A, patulin, and other naturally occurring toxins. This regulation also specifies the maximum residue limits for insecticides and antibiotics and tolerance limits of pharmacologically active substances for fishery products, poultry, egg, meat products, and other sea foods, including a limit on antibiotics in honey products [89].

To significantly minimize or eradicate chemical hazards in food, the FDA in the United States sets guidelines, maintains track of the levels of contaminants in food, and advises food manufacturers on preventative actions. The goal of the FDA's Toxic Elements Working Group is to minimize exposure to contaminants such as arsenic, lead, cadmium, and mercury in dietary supplements, food, and cosmetics [90]. The Chemical, Metals, Natural Toxins, and Pesticides Regulations provide guidance documents for chemical contaminants, including radionuclides, melamine, acrylamide, dioxin, and PCBs; arsenic and lead as metal contaminants; pesticide residues; methyl parathion residues; and vinclozolin residues. Additional guidelines for bottled water and carbonated soft drinks are also established by the USFDA [91].

In order to protect the health of EU individuals, particularly the most vulnerable demographics, like children and pregnant women, Regulation (EU) 2023/915 on maximum levels for certain contaminants in food establishes the maximum values for certain pollutants in food. The legislation addresses additional contaminants, such as nitrates, perchlorate, melamine, and other metal and processing contaminants, mycotoxins and plant toxins, and halogenated persistent organic pollutants [92].

4.9.1 The National Toxicology Program (NTP) and the National Center for Toxicological Research (NCTR)

The National Toxicology Program (also referred to as NTP) contributes to the scientific foundation of initiatives, policies, and programs that promote well-being or prevent diseases. NTP, which was established in 1978, is vital to the generation, interpretation, and dissemination of toxicological data on compounds that may be toxic in our surroundings. As a global pioneer in toxicological research, NTP has assessed the possible health impacts of almost 3,000 environmental substances on humans [93]. NTP is an interagency initiative operated by the Department of Health and Human Services which is made up of three federal agencies, namely, FDA's National Center for Toxicological Research (NCTR), National Institutes of Health's (NIH) National Institute of Environmental Health Sciences (NIEHS), and the Centers for Disease Control and Prevention's National Institute for Occupational Safety and Health (NIOSH). The director of the NTP is concurrently the director of NIEHS. A number of external groups often consult the director and report to the assistant secretary for health [94]. The NTP's goal is to collaborate with others to develop toxicological sciences and to safeguard and enhance human health. In order to achieve the goal of innovative and reliable toxicological science that safeguards human health, federal NTP partners create and implement novel strategies, technologies, and research techniques [95].

The National Center for Toxicological Research (NCTR) of the US Food and Drug Administration (FDA) is a worldwide collaboration platform offering advice, guidance, and inventive scientific solutions to support the FDA's goal of enhancing public health. In addition to developing and promoting advanced tools and strategies that the FDA employs to safeguard and advance public and individual health, the NCTR carries out scientific research to provide evidence for FDA decision-making.

4.9.2 New Dietary Ingredient (NDI) Warning Letter

Six firms received warnings from the FDA in March 2016 due to the presence of acacia rigidly in their dietary supplements, a new dietary ingredient that lacked the necessary safety screening procedure. The FDA considers these to be adulterated supplements. The organizations faced enforcement proceedings, such as product withdrawal from the market, if they did not provide an explanation within 15 days of an issue arising [96]. While Nubreed Nutrition's product UNDISPUTED contains *Acacia rigidula*, a new dietary ingredient that has yet to obtain FDA approval, the FDA has written the company a warning letter on March 7, 2016. The FDA considers this to be adulteration of the product. The FDA is urging that Nubreed Nutrition discontinue selling the product and take corrective measures to address the breaches. They must also outline their plans for preventing future occurrences of the same offenses. If Nubreed Nutrition does not comply, the FDA has the authority to sue them in court [97].

Because Carbon Fire LLC's product 213° Metabolism Boosting Complex is an adulterated nutritional supplement, the FDA issued a warning letter to the company in March 2024. The component known as N-methyltyramine is the source of the issue. The factors that make the product deemed adulterated are that N-methyltyramine is a new dietary ingredient not accessible for sale in the United States prior to October 15, 1994, and according to FDA requirements, new dietary ingredients must be reported. Before employing such ingredients in their supplements, companies are required to submit a notification regarding new dietary ingredients. Since N-methyltyramine was not notified by Carbon Fire LLC, the FDA considered the product as adulterated, in accordance with the Federal Food, Drug, and Cosmetic Act. There is no evidence that N-methyltyramine is safe to use in dietary supplements that the FDA has been informed of. The letter makes it clear that Carbon Fire LLC is in charge of making sure its products abide by FDA guidelines. Carbon Fire LLC has been given by the FDA fifteen working days to reply with the actions that they plan to take to rectify the violations and ensure that they do not happen again. If Carbon Fire LLC rejects the FDA's evaluation, it can provide evidence to back up its position.

The FDA further cautions that if this issue is not resolved, there may be legal measures, including product seizure and injunction [98].

4.10 FUTURE ASPECTS OF NUTRACEUTICALS REVOLVING AROUND THEIR SAFETY AND TOXICITY

Since consumers inadvertently mistake nutraceuticals for organic and naturally derived products, they thereby believe they are safe for human consumption and are thus exponentially growing in popularity globally. Over-the-counter (OTC) nutritional supplements are getting more common specifically in developed countries for the beneficial reasons of managing essential body requirements and treating and preventing an ever-expanding range of medical ailments. Surprisingly, these readily accessible nutraceuticals are not thoroughly examined for their safety and efficacy characteristics due to the misconception of their being natural and safe products. Unfortunately, consumers are unaware of the toxicity linked to nutraceuticals. Marketed nutraceuticals fail to possess the safety and quality aspects of nutraceuticals established by manufacturers and regulatory bodies. To evaluate a nutraceutical for its safety and quality, nutraceutical manufacturers must closely adhere to the rules and guidelines outlined by the regulatory authorities throughout the manufacturing and marketing of their products for sale.

Dietary supplements are considered safe when used within the recommended dietary allowance (RDA) level instead of breaching the RDA limits. Overuse or misuse of nutraceuticals may end up in an array of serious adverse reactions. Numerous adverse events reported previously have been put forward on the nation's regulatory portals and published in scientific journals as well. Beyond reporting, the FDA takes up enforcement toward manufacturing companies that use prohibited and unlawful ingredients in their nutritional supplements. Also, some nutraceutical products have been recalled from sale in the marketplace. Besides this, nutraceutical products are likely to be poisoned with heavy metals, pesticide residue, mycotoxins, microbial and excipient impurities, undeclared allergens, pharmacologically active ingredients, and other adulterants, all of which contribute to its potentially toxic effects. Long-term exposure to these contaminants at higher concentrations may result in serious side effects and even to fatality. There are multiple food–drug interactions that nutraceutical users ought to refrain from causing. Consumers must additionally prevent self-medication with nutraceuticals and ingestion of miscellaneous products, and adequate health-care consultation should be achieved before the consumption of nutraceuticals. To safeguard against adverse consequences from the concurrent use of nutritional products and therapeutic pharmaceuticals, medical professionals and pharmacists must acquire the necessary training and education in the discipline of polypharmacy.

To assess a nutraceutical's safety, efficacy, and toxicity, various models can be used depending on its intended use, and biomarkers can also be used in identifying early exposure and toxicity events. On the other hand, nutraceutical products must undergo a thorough assessment for toxicity and safety before public distribution. More importantly, regulatory agencies must implement more stringent and standardized regulations and procedures to ensure the safety and quality of nutraceuticals before they become available on the shelves. To conclude, dietary supplements have the potential to foster human health and prevent disease, yet collaboration between health professionals, nutritionists, toxicologists, and regulatory authorities is essential for ensuring the most effective medical advantages and regulations.

REFERENCES

[1] Size NM. Nutraceuticals Market Size, Share, Growth, Trends, Analysis 2023–2028. Published online 2023:1–8.

[2] Brownie, S. The development of the US and Australian dietary supplement regulations: What are the implications for product quality? 2005;191–8.

[3] 103rd Congress. Dietary Supplement Health and Education Act 1994, Public Law. Published online 1994:103–417.
[4] Dietary Reference Values for Nutrients: Summary Report. *EFSA Support Publ* [Internet]. 2017 Dec 11 [cited 2024 Mar 5];14(12). Available from: https://www.efsa.europa.eu/en/topics/topic/food-supplements
[5] Regulation - 1924/2006 - EN - EUR-Lex. Accessed April 15, 2025. Available from: https://eur-lex.europa.eu/eli/reg/2006/1924/oj/eng
[6] Komala, M.G., Ong, S.G., Qadri, M.U., Elshafie, L.M., Pollock, C.A., Saad, S. Investigating the Regulatory Process, Safety, Efficacy. Published online 2023.
[7] Guston, D. Food Standards Australia New Zealand. *Encycl Nanosci Soc.* 2012(118).
[8] Santini, A., Cammarata, S.M., Capone, G., Ianaro, A., Tenore, G.C., Pani, L., et al. Nutraceuticals: Opening the Debate for a Regulatory Framework. *Br J Clin Pharmacol.* 2018;84(4):659–72.
[9] Low, T.Y., Wong, K.O., Yap, A.L.L., De Haan, L.H.J., Rietjens, I.M.C.M. The Regulatory Framework Across International Jurisdictions for Risks Associated with Consumption of Botanical Food Supplements. *Compr Rev Food Sci Food Saf.* 2017;16(5):821–34.
[10] Putta, D.S. FSSAI Guidance and Notification on Nutraceuticals—An Insight. *FnbnewsCom* [Internet]. 2020;(c):1–3. Available from: https://www.fssai.gov.in/upload/media/FSSAI_News_Guidance_FNB_09_06_2020.pdf
[11] 103rd Congress. Dietary Supplement Health and Education Act. *Public Law.* 1994;103–417.
[12] Hermans, S., Roosen, M., Vansant, G. Food Supplements. *Tijdschr Geneeskd.* 2004;60(14–15):1085–6.
[13] Programme, T.P., Branch, P. ARCHIVED—Policy Paper—Nutraceuticals/Functional Foods and Health Claims on Foods. We have archived this page and will not be updating it: 1–38.
[14] Regulatory, N. Nutraceutical Regulation and Registration in Australia (TGA). 2023;24–6.
[15] Sugiarto. 2016. Minangkabau Natural Culture Learning. 4(1):1–23.
[16] Babar, U., Xu, R. Agricultural Genetically Modified Organisms (GMOs) Regulation in China. *GMOs Polit Stance Glob GMO Regul Certif Labeling, Consum Prefer.* 2022 Jan 1;53–74.
[17] Luthra, V.R., Toklu, H.Z. Nutrivigilance: The Road Less Traveled. *Front Pharmacol.* 2023 Oct;14:1–4.
[18] EC, E.C. Rapid Alert System for Food and Feed (RASFF). Gov Webpage. 2023 Aug:1–3.
[19] Bragazzi, N.L., Martini, M., Saporita, T.C., Nucci, D., Gianfredi, V., Maddalo, F., et al. Nutraceutical and Functional Food Regulations in the European Union [Internet]. In: *Developing New Functional Food and Nutraceutical Products.* Elsevier Inc.; 2016. p. 309–322. Available from: http://dx.doi.org/10.1016/B978-0-12-802780-6/00017-1
[20] Rabbur, R.S.M., Emmerton, L. An Introduction to Adverse Drug Reaction Reporting Systems in Different Countries. *Int J Pharm Pract.* 2010;13(1):91–100.
[21] Shao, A. Global Adverse Event Reporting Regulations for Nutraceuticals, Functional Foods, and Dietary/Food/Health Supplements. *Nutraceutical Funct Food Regul United States around World.* 2019 Jan 1;619–24.
[22] Borrelli, F., Izzo, A.A. Herb-Drug Interactions with St John's Wort (Hypericum Perforatum): An Update on Clinical Observations. *AAPS J.* 2009;11(4):710–27.
[23] Ginkgo Biloba and Warfarin Interactions—Drugs.com [Internet]. [cited 2024 Mar 5]. Available from: https://www.drugs.com/drug-interactions/ginkgo-biloba-with-warfarin-1174-10621-2311-0.html
[24] Drug Interaction Report: Vitamin K, Warfarin [Internet]. [cited 2024 Mar 5]. Available from: https://www.drugs.com/interactions-check.php?drug_list=1868–16217,2311–0
[25] Coenzyme Q10—Special Subjects—MSD Manual Consumer Version [Internet]. [cited 2024 Mar 5]. Available from: https://www.msdmanuals.com/en-in/home/special-subjects/dietary-supplements-and-vitamins/coenzyme-q10
[26] Interaction Between Garlic and Anticoagulant Warfarin | Vinmec [Internet]. [cited 2024 Mar 5]. Available from: https://www.vinmec.com/en/pharmaceutical-information/use-medicines-safely/interaction-between-garlic-and-anticoagulant-warfarin/
[27] Buckley, M.S., Goff, A.D., Knapp, W.E. Fish Oil Interaction with Warfarin. *Ann Pharmacother.* 2004;38(1):50–3.
[28] Name, O., Vitamin, S.B.C., B.C.D.V., More, F.S. Riboflavin—Uses, Side Effects, and More Side Effects:22–24.
[29] Keltner, D., Sauter, D., Tracy, J., Cowen, A. 乳鼠心肌提取 HHS Public Access. *J Nonverbal Behav.* 2019;43(2):133–60.
[30] Anonim. Drug Interactions Between Amoxicillin and Warfarin. 2021;3–5. Available from: https://www.drugs.com/drug-interactions/amoxicillin-with-warfarin-187-0-2311-0.html
[31] IRON: Overview, Uses, Side Effects, Precautions, Interactions, Dosing and Reviews. Accessed April 18, 2025. Available from: https://www.webmd.com/vitamins/ai/ingredientmono-912/iron

[32] Isbn, C., Pdf, T., Press, N.A., Academy, N., Panel, S., Interactions, E., et al. Engineering in Society Copyright © National Academy of Sciences. All rights reserved. Unless otherwise indicated, all materials in this PDF File are copyrighted by the National Academy of Sciences. Distribution, posting, or copying is strictly prohibited. *Society.* 1985;II:790 p.
[33] Anadón, A., Martínez-Larrañaga, M.R., Ares, I., Martínez, M.A. Interactions Between Nutraceuticals/Nutrients and Therapeutic Drugs [Internet]. *Nutraceuticals: Efficacy, Safety and Toxicity.* Elsevier Inc.; 2016. p. 855–874. Available from: http://dx.doi.org/10.1016/B978-0-12-802147-7.00060-7
[34] What Medication Can Grapefruit Interact With?—Ben's Natural Health [Internet]. [cited 2024 Mar 6]. Available from: https://www.bensnaturalhealth.com/blog/general-health/grapefruit-medications-interactions-cautious-drug-list/
[35] Lane, F. Warning Letters:3–5.
[36] Safety—European Commission.
[37] Facts, Q., Products, N.H. Natural Health Products:1–4.
[38] Therapeutic Goods Administration (TGA) | Australian Government Department of Health [Internet]. [cited 2024 Mar 24]. Available from: https://www.tga.gov.au/
[39] Food Safety Ministry of Health, Labour and Welfare [Internet]. [cited 2024 Mar 24]. Available from: https://www.mhlw.go.jp/stf/seisakunitsuite/bunya/kenkou_iryou/shokuhin/index_00006.html
[40] Warning Letters | FDA [Internet]. [cited 2024 Mar 25]. Available from: https://www.fda.gov/inspections-compliance-enforcement-and-criminal-investigations/compliance-actions-and-activities/warning-letters
[41] Eagle Energy USA, Inc.—617712–12/01/2021 | FDA [Internet]. [cited 2024 Mar 25]. Available from: https://www.fda.gov/inspections-compliance-enforcement-and-criminal-investigations/warning-letters/eagle-energy-usa-inc-617712–12012021
[42] Dietary Supplements | FDA [Internet]. [cited 2024 Mar 25]. Available from: https://www.fda.gov/food/dietary-supplements
[43] Muscle Sports Products, LLC—625731–09/23/2022 | FDA [Internet]. [cited 2024 Mar 25]. Available from: https://www.fda.gov/inspections-compliance-enforcement-and-criminal-investigations/warning-letters/muscle-sports-products-llc-625731–09232022
[44] Maine Natural Health, Inc.—525870–12/19/2017 | FDA [Internet]. [cited 2024 Mar 25]. Available from: https://www.fda.gov/inspections-compliance-enforcement-and-criminal-investigations/warning-letters/maine-natural-health-inc-525870–12192017
[45] mceclip0.
[46] Logbook for Product/Mock Recall:*1.*
[47] SOP for Drug Product Recall & Mock Recall—Pharma Beginners [Internet]. [cited 2024 Mar 25]. Available from: https://pharmabeginners.com/sop-for-product-recall/
[48] Recalls, Market Withdrawals, & Safety Alerts | FDA [Internet]. [cited 2024 Mar 6]. Available from: https://www.fda.gov/safety/recalls-market-withdrawals-safety-alerts
[49] Sanchez, E., Simpson, R.B., Zhang, Y., Sallade, L.E., Naumova, E.N. Exploring Risk Factors of Recall-Associated Foodborne Disease Outbreaks in the United States, 2009–2019. *Int J Environ Res Public Health.* 2022;19(9):4947. doi:10.3390/IJERPH19094947/S1
[50] Meta Herbal Issues Voluntary Nationwide Recall of Magnum XXL 9800 Capsules Due to Presence of Undeclared Sildenafil | FDA. Accessed April 18, 2025. Available from: https://www.fda.gov/safety/recalls-market-withdrawals-safety-alerts/meta-herbal-issues-voluntary-nationwide-recall-magnum-xxl-9800-capsules-due-presence-undeclared
[51] SugarMDs LLC Issues Voluntary Nationwide Recall of Advanced Glucose Support Supplements Capsules Due to Presence of Undeclared Glyburide and Metformin | FDA. Accessed April 18, 2025. Available from: https://www.fda.gov/safety/recalls-market-withdrawals-safety-alerts/sugarmds-llc-issues-voluntary-nationwide-recall-advanced-glucose-support-supplements-capsules-due
[52] Botanical-Be Issues Voluntary Nationwide Recall of Kuka Flex Forte, Reumo Flex (Caplets) and Artri King (Tablets). Due to Presence of Undeclared Diclofenac | FDA. Accessed April 18, 2025. Available from: https://www.fda.gov/safety/recalls-market-withdrawals-safety-alerts/botanical-be-issues-voluntary-nationwide-recall-kuka-flex-forte-reumo-flex-caplets-and-artri-king
[53] Announcement C. Hua Da Trading Inc dba Wefun Inc Issues Voluntary Nationwide Recall of WEFUN Capsules Due to Presence of Undeclared Sildenafil:1–5.
[54] Announcement C. Urgent Notice of Product Recall:24–26.
[55] Announcement C. Walmart Inc. Issues Voluntary Nationwide Recall of Various Artri Ajo King Joint Supplements Due to Potential Hidden Drug Ingredients:1–8.

[56] Announcement C. Family Dollar Stores Issues Voluntary Recall of Certain FDA-Regulated Products in Six States Including Drugs, Devices, Cosmetics, Foods:2–4.
[57] Sheet F, This HP, Center-rand SCEP, Services H, Supplements D, Medicine A, et al. Ephedra and Ephedrine Alkaloids for Weight Loss and Athletic Performance. *2004*;2004.
[58] National Institutes of Health. FDA Prohibits Sales of Dietary Supplements Containing Ephedra. 2004;36915. Available from: https://ods.od.nih.gov/HealthInformation/Ephedra.aspx
[59] Shekelle, P.G., Hardy, M.L., Morton, S.C., Maglione, M., Mojica, W.A., Suttorp, M.J., et al. Efficacy and Safety of Ephedra and Ephedrine for Weight Loss and Athletic Performance: A Meta-analysis. *JAMA* [Internet]. 2003 Mar 26 [cited 2024 Mar 7];289(12):1537–45. Available from: https://jamanetwork.com/journals/jama/fullarticle/196247
[60] Ephedra | NCCIH [Internet]. [cited 2024 Mar 7]. Available from: https://www.nccih.nih.gov/health/ephedra
[61] Kang, Y.E., Chae, J., Nguyen, T.T., Kang, J.J., Oh, S.Y. A Case Report of Acute Bilateral Mydriasis Induced by Herbal Medication. *Neuro-Ophthalmology* [Internet]. 2024 Jan 22 [cited 2024 Mar 7]; Available from: https://www.tandfonline.com/doi/abs/10.1080/01658107.2023.2301363
[62] Ryu, S.J., Shin, Y.U., Kang, M.H., Cho, H.Y., Seong, M. Bilateral Acute Myopia and Angle Closure Glaucoma Induced by Ma-huang (Ephedra): A Case Report. *Med* (United States). 2017 Dec 1;96(50).
[63] Libman, R.B., Menna, B.L., Gulati, S. Case Report: Consequences of Ephedra Use in an Athlete. *Lancet*. 2005;366(Suppl. 1).
[64] Forte, R.Y., Precoma-Neto, D., Chiminacio Neto, N., Maia, F., Faria-Neto, J.R. Myocardial Infarction Associated with the Use of a Dietary Supplement Rich in Ephedrine in a Young Athlete. *Arq Bras Cardiol*. 2006;87(5):2005–2007.
[65] Libman, R.B., Menna, B.L., Gulati, S. Case Report: Consequences of Ephedra Use in an Athlete. *Lancet* [Internet]. 2005 Dec 1 [cited 2024 Mar 7];366(SUPPL. 1):S22. Available from: http://www.thelancet.com/article/S0140673605678324/fulltext
[66] Stohs, S.J., Ray, S.D. Review of Published Bitter Orange Extract and p-Synephrine Adverse Event Clinical Study Case Reports. *J Diet Suppl* [Internet]. 2020 May 3 [cited 2024 Mar 7];17(3):355–63. Available from: https://www.tandfonline.com/doi/abs/10.1080/19390211.2019.1577936
[67] Maksoud, S., Abdel-Massih, R.M., Rajha, H.N., Louka, N., Chemat, F., Barba, F.J., et al. Citrus aurantium L. Active Constituents, Biological Effects and Extraction Methods. *An Updated Review*. [Internet]. 2021 Sep 26 [cited 2024 Mar 7];26(19):5832. Available from: https://www.mdpi.com/1420–3049/26/19/5832/htm
[68] Stohs, S.J., Ray, S.D. Review of Published Bitter Orange Extract and p-Synephrine Adverse Event Clinical Study Case Reports. *J Diet Suppl* [Internet]. 2020 May 3 [cited 2024 Mar 8];17(3):355–63. Available from: https://www.tandfonline.com/doi/abs/10.1080/19390211.2019.1577936
[69] de Jonge, M.L.L., Kieviet, L.C., Sierts, M., Egberink, L.B., van der Heyden, M.A.G. Review of Case Reports on Adverse Events Related to Pre-workout Supplements Containing Synephrine. *Cardiovasc Toxicol* [Internet]. 2023 Jan 1 [cited 2024 Mar 8];23(1):1–9. Available from: https://link.springer.com/article/10.1007/s12012-022-09777-z
[70] Deshmukh, N.S., Stohs, S.J., Magar, C.C., Kadam, S.B. Citrus aurantium (Bitter Orange) Extract: Safety Assessment by Acute and 14-day Oral Toxicity Studies in Rats and the Ames Test for Mutagenicity. *Regul Toxicol Pharmacol* [Internet]. 2017;90:318–27. Available from: https://doi.org/10.1016/j.yrtph.2017.09.027
[71] Nykamp, D.L., Fackih, M.N., Compton, A.L. Possible Association of Acute Lateral-Wall Myocardial Infarction and Bitter Orange Supplement. 2004 May 1 [cited 2024 Mar 8];38(5):812–6. http://dx.doi.org/101345/aph1D473 [Internet]. Available from: https://journals.sagepub.com/doi/10.1345/aph.1D473
[72] Stephensen, T.A., Sarlay, R. Ventricular Fibrillation Associated with Use of Synephrine Containing Dietary Supplement. *Mil Med*. 2009;174(12):1313–9.
[73] Bouchard, N.C., Howland, M.A., Greller, H.A., Hoffman, R.S., Nelson, L.S. Ischemic Stroke Associated with Use of an Ephedra-Free Dietary Supplement Containing Synephrine. *Mayo Clin Proc*. 2005 Apr 1;80(4):541–5.
[74] Gange, C.A., Madias, C., Felix-Getzik, E.M., Weintraub, A.R., Estes, N.A.M. Variant Angina Associated with Bitter Orange in a Dietary Supplement. *Mayo Clin Proc*. 2006;81(4):545–8.
[75] Sultan, S., Spector, J., Mitchell, R.M. Ischemic Colitis Associated with Use of a Bitter Orange–Containing Dietary Weight-Loss Supplement. *Mayo Clin Proc*. 2006;81(12):1630–1.
[76] Kieviet, M.L.L.D.J.L.C., Egberink, M.S.L.B. Review of Case Reports on Adverse Events Related to Pre-workout Supplements Containing Synephrine. *Cardiovasc Toxicol* [Internet]. 2023;23(1):1–9. Available from: https://doi.org/10.1007/s12012-022-09777-z

[77] Holmes, R.O., Tavee, J. Vasospasm and Stroke Attributable to Ephedra-Free Xenadrine: Case Report. *Mil Med.* 2008;173(7):708–10.
[78] Stohs, S.J. Assessment of the Adverse Event Reports Associated with *Citrus aurantium* (Bitter Orange) from April 2004 to October 2009. *J Funct Foods* [Internet]. 2010;2(4):235–8. Available from: http://dx.doi.org/10.1016/j.jff.2010.10.003
[79] Tan, X., Wang, Y., Wu, Z., Zhou, Q., Tang, Y., Liu, Z., Yuan, G., Luo, S., Zou, Y., Guo, S., Han, N., Yao, K. The role of Her-2 in penile squamous cell carcinoma progression and cisplatin chemoresistance and potential for antibody-drug conjugate-based therapy. *Eur J Cancer.* 2023 Nov:194:113360. Available from: https://doi.org/10.1016/j.ejca.2023.113360
[80] Levita, J., Wilar, G., Wahyuni, I., Bawono, L.C., Ramadaini, T., Rohani, R., et al. Clinical Toxicology of Vitamin D in Pediatrics: A Review and Case Reports. *Toxics.* 2023;11(7):1–15.
[81] Cavalieri, L., Arcucci, S. Hepatitis Grave Producida por Intoxicación con Té Verde en un Niño. Presentación de un Caso. *Arch Argent Pediatr.* 2019;117(6):655–8.
[82] NTP. NTP Technical Report on the Toxicology and Carcinogenesis Studies of Perfluorooctanoic Acid. 2020;(May).
[83] Pillukat, M.H., Bester, C., Hensel, A., Lechtenberg, M., Petereit, F., Beckebaum, S., et al. Concentrated Green Tea Extract Induces Severe Acute Hepatitis in a 63-Year-Old Woman—A Case Report with Pharmaceutical Analysis. *J Ethnopharmacol.* 2014;155(1):165–70.
[84] Jaundice as a Misadventure of a Green Tea (Camellia Sinensis) Lover: A Case Report. *PubMed* [Internet]. [cited 2024 Mar 8]. Available from: https://pubmed.ncbi.nlm.nih.gov/19317284/
[85] Jimenez-Saenz, M., Martinez-Sanchez, M.D.C. Acute Hepatitis Associated with the Use of Green Tea Infusions. *J Hepatol.* 2006;44(3):616–7.
[86] Pillukat, M.H., Bester, C., Hensel, A., Lechtenberg, M., Petereit, F., Beckebaum, S., et al. Concentrated Green Tea Extract Induces Severe Acute Hepatitis in a 63-Year-Old Woman—A Case Report with Pharmaceutical Analysis. *J Ethnopharmacol* [Internet]. 2014;155(1):165–70. Available from: http://dx.doi.org/10.1016/j.jep.2014.05.015
[87] Perkins, J.D. Techniques to Ensure Adequate Portal Flow in the Presence of Splenorenal Shunts. *Liver Transplant.* 2007;13(5):767–8.
[88] Chopra, A.S., Lordan, R., Horbańczuk, O.K., Atanasov, A.G., Chopra, I., Horbańczuk, J.O., et al. The Current Use and Evolving Landscape of Nutraceuticals. *Pharmacol Res.* 2022;175(October 2021).
[89] MoH, I. Food Safety and Standard (Contaminants, Toxins and Residues) Regulation. *Minist Heal Fam Welfare Food Saf Stand Auth India.* 2011;0(3):1–19.
[90] Testing, C.M. Environmental Contaminants in Food: FDA's Toxic Elements Working Group. *FDA*:27–9.
[91] U.S. Food and Drug Administration. Chemical Contaminants, Metals, Natural Toxins & Pesticides Guidance Documents & Regulations. *Guid Doc Contain Nonbinding Recommendations* [Internet]. 2015(January 2017):28–9. Available from: http://www.fda.gov/Food/GuidanceRegulation/GuidanceDocumentsRegulatoryInformation/ChemicalContaminantsMetalsNaturalToxinsPesticides/default.htm
[92] Commission Regulation. Maximum Levels for Certain Contaminants in Foodstuffs. *Comm Regul No 1881/2006* [Internet]. 2006;1–33. Available from: https://eur-lex.europa.eu/legal-content/EN/ALL/?uri=CELEX:02006R1881-20210831
[93] Who We Are [Internet]. [cited 2024 May 29]. Available from: https://ntp.niehs.nih.gov/whoweare
[94] Organization [Internet]. [cited 2024 May 29]. Available from: https://ntp.niehs.nih.gov/whoweare/organization
[95] Mission & Goals [Internet]. [cited 2024 May 29]. Available from: https://ntp.niehs.nih.gov/whoweare/about
[96] Recent FDA Action on Products Marketed as Dietary Supplements Labeled as Containing *Acacia rigidula.* 2016;46.
[97] Service PH. Nubreed Nutrition, IncInc 3/7/16. 2016;201(1):5–7.
[98] Woods, D. Carbon Fire, LLC. 2024;413(d):3–5.

5 Regulatory Requirements and Challenges for Nutraceuticals in the United States

Anita Jacob and Yashwant Pathak

5.1 INTRODUCTION

Nutraceuticals, more commonly known as dietary supplements in the United States, became a topic of discussion after the US Food and Drug Administration sent warning letters to 14 US-based companies that were illegally sending more than 65 products that fraudulently claimed to prevent, diagnose, treat, or cure cancer in April of 2017. These products included pills, ointments, oils, and teas seemingly harmless to consumers. Such sales and commercialization are considered a violation of the Federal Food, Drug, and Cosmetic Act, especially since the manufacturers did not demonstrate to the FDA that they were safe and effective for labeled uses [1]. In the United States, various nutraceuticals, or dietary supplements, such as vitamins and botanicals, are available as tablets, soft gels, capsules, powders, and liquids. Many consume them without a physician's recommendation in hopes that they will treat, cure, or prevent diseases.

5.2 DIETARY SUPPLEMENT USE AMONG ADULTS: UNITED STATES, 2017–2018

The Centers for Disease Control and Prevention conducted a National Health and Nutrition Examination Survey including questions on nutraceuticals, commonly called dietary supplements. The definition of *dietary supplement* that was used was

> a product (other than tobacco) that is intended to supplement the diet; contains one or more dietary ingredients (including vitamins, minerals, herbs or other botanicals, amino acids, and other substances) or their constituents; is intended to be taken by mouth as a pill, capsule, tablet or liquid, and is labeled on the front panel as being a dietary supplement [2].

It was found that from 2007–2008 through 2017–2018, the prevalence of dietary supplement use increased in all age groups among US adults. It was also found that 57.6% of US adults aged 20 and over who took the survey used a dietary supplement in the past 30 days [2]. Research has been published on how diets with varied foods are enough to provide adequate nutrients for the body to develop and to reduce the risk of chronic disease, yet the demand for dietary supplements is still great [3].

5.3 COMMON NUTRACEUTICALS IN THE UNITED STATES

5.3.1 American Ginseng

Panax quinquefolius, commonly known as American ginseng, is an herb that grows in North America and is endangered in some states due to the high demand. This adaptogen is believed to

DOI: 10.1201/9781003465041-5

affect insulin levels and lower blood sugar. The recommended dose for adults is 200–400 mg by mouth twice daily for three to six months [4].

Despite the supposed benefits, there are also patient risks associated with ginseng abuse and misuse. Some long-term symptoms found in a study of 133 patients ingesting high doses of ginseng for two years include morning diarrhea, skin eruption, sleeplessness, and blood pressure fluctuation [5]. It is the display of such symptoms that pushes for medical tracking of dietary supplements that are consumed by patients, especially if the patients need medications for maintenance.

5.3.2 Green Tea

Green tea is another dietary supplement that has increased in popularity in the United States, specifically for weight loss. However, it has been found that despite labeling on fine green tea supplements, it is still difficult for the consumer to determine the true quality of the products used. There have also been reports of increased liver trouble with the consumption of green tea extracts [6].

5.3.3 Folic Acid

Folic acid is a supplement taken by women of reproductive age who would like to have children, in order to prevent birth defects such as anencephaly, where parts of the child's brain and skull may not form, or spina bifida, where there is no development of the backbone to protect the spinal cord [7]. Despite the immense benefits the supplement, with proper dosing, has for pregnant women and their babies, high doses may increase the risk of colorectal cancer [8].

5.3.4 Echinacea

Echinacea has been used for common colds and infections based on the belief that the supplement stimulates the immune system to fight off infection more effectively [9]. In 1998, it was reported that echinacea represented 10% of the total herbal medicine market in the United States [10]. Research under the National Center for Complementary and Integrative Health suggests that this dietary supplement may help with infections because of the types and amounts of bacteria in the plants themselves when looking at just isolated cells. Again, though there are believed benefits, side effects may include digestive tract issues, including nausea and stomach pain [9].

5.4 TRENDS IN NUTRACEUTICAL USE IN THE UNITED STATES

In 1944, courts dismissed FDA reports that some vitamin B capsules were misbranded as foods and as drugs and later declared that a USP (the official *Homeopathic Pharmacopoeia of the United States*) listing was not enough to deny the drug status of a product. To move forward with controlling dietary supplement production and sale, the FDA then wished to apply the Food Additives Amendment of 1958 to botanical products under the definition of a food additive. However, the amendment allowed for the proof of safety of new substances added to food to be provided by the manufacturers themselves rather than the FDA having to search for proof. This process was changed, so the FDA is responsible for determining whether a new dietary ingredient has any possible risk of harm prior to being marketed as a dietary supplement [11].

5.5 THE DIETARY SUPPLEMENT HEALTH AND EDUCATION ACT OF 1994 (DSHEA)

The Act focused on the benefits of dietary supplements, including the prevention of chronic diseases, including cancer and heart disease. Despite the many benefits that these nutraceuticals or dietary supplements may have, it was proved necessary that regulations be placed on products being sold as dietary supplements, as well as on the information that linked nutrition and long-term good health [12]. Dietary supplements have been promoted to the point that nearly 50% of 260,000,000 surveyed Americans reported consuming dietary supplements, such as vitamins, herbs, or minerals, to improve their health [12].

The DSHEA 1994 specifies that a *dietary supplement* is defined as a product that supplements the diet and contains one or more dietary ingredients. These ingredients can include a vitamin; a mineral; an herb or another botanical; an amino acid; a dietary substance for use by man to supplement the diet; a concentrate, metabolite constituent, or extract; or a combination of the previously mentioned. The DSHEA definition also included ingredients such as botanicals and hormones, to be labeled as supplement ingredients rather than drug ingredients. These dietary supplements may be consumed as a powder, soft gel, gelcap, or capsule [13].

The Act also addresses the possible misinformation or mislabeling that could occur through publications, articles, chapters in a book, or peer-reviewed scientific publications. The Act specifies that statements can be made on the benefit of a dietary supplement and how it affects the structure or function of humans, but manufacturers must explicitly write, "This statement has not been evaluated by the Food and Drug Administration. This product is not intended to diagnose, treat, cure, or prevent any disease" [13].

To prevent misbranded supplements or falsely labeled products, the supplement manufacturer must be able to list the name and quantity of each of its ingredients. If the supplement manufacturer decides to include an expiration date, the FDA requires current Good Manufacturing Practices (cGMPs) to provide stability testing data. The DSHEA did not put a mandate on supplements displaying expiration dates, but some manufacturers do include a "use by" date. The FDA themselves must also determine whether the dietary supplement poses a "significant or unreasonable risk of illness or injury," rather than just the manufacturer providing data to prove its safety. Such procedures prevent manufacturers from falsifying data and possibly harming individuals who may choose to consume those dietary supplements. Despite having the ability to protect the public from harm, the FDA does not have the authority to approve dietary supplements for safety and effectiveness before they are marketed [14].

5.6 DIETARY SUPPLEMENT LABELING

On the containers and packages of dietary supplements, there are requirements for labels. These required statements include thus:

- Statement of identity/name of the dietary supplement
- The net quantity of contents of the amount of the dietary supplement within the product
- Nutrition labeling
- Ingredient list
- Name and place of business of the manufacturer, packer, or distributor [15]

These must all be on the front label panel, which is what is most likely to be seen by the consumer, or on the information panel. If there are alternate principal display panels, the same information must be placed on each different principal display panel. If everything does not fit on the principal display panel, there may be a "supplement facts" panel. If there are other ingredients, such as flavors, preservatives, starches, gelatin, or stabilizers, they should be listed in an "other ingredients" list [16]. The expiration date of the dietary supplement is not required according to the FDA. Unfortunately, this may cause consumers to continue to ingest supplements that have gone bad.

In September of 2018, the US FDA released a draft guidance for those who manufacture, market, or distribute dietary supplements with live microbial ingredients. Such action would enforce firms to declare in the supplement facts label quantitative data on live microbials in colony forming units and by weight required by regulation. The International Probiotics Association did attempt to change 21 CFR 101.36 to allow probiotic ingredients to be labeled by CFUs instead of weight, but it was denied in hopes of keeping consumers and health-care professionals updated [17].

5.7 SAFETY, RISK, AND QUALITY REGULATIONS

It is common knowledge that too much of any substance may be harmful to one's body. Similarly, there are side effects for dietary supplements when taken at high doses or taken with certain prescribed medicines. For example, taking vitamin K, which is available over the counter, may interact with warfarin, a blood thinner, and prevent it from working properly. As the popularity of nutraceuticals/dietary supplements grew, the FDA worked with the Institute of Medicine and the National Research Council of the National Academies to establish a framework for categorizing dietary supplement ingredients and how to determine the safety and efficacy of such nutraceuticals [18].

The safety of nutraceuticals was particularly important, as multiple readily available dietary supplements contain active ingredients that can have strong adverse effects on the body. Intake greater than the recommended dosage can increase the severity of the side effects of prescribed maintenance medications as well.

To make sure that such products were made properly and safely, Good Manufacturing Practices (GMPs) were adopted by the World Health Assembly in 1969. This ensured quality standards for medicinal products so that they were safe for use. The policies under the GMP ensure that the production and testing processes are clearly "defined, validated, reviewed and documented" [19].

In September 2015, the FDA established the "Current Good Manufacturing Practice, Hazard Analysis, and Risk-Based Preventive Controls for Human Food" rule in the Code of Federal Regulations Part 117. Part 117 was specifically developed in order to establish requirements for cGMPs and for hazard analysis and risk-based preventive controls for human food. Part 111 was specifically developed to outline manufacturing, packaging, labeling, and holding operations for dietary supplements that are manufactured, packaged, or labeled by a company when they are manufactured by oneself and then packaged/labeled by another person or if the dietary supplement is imported in any state or territory of the USA, District of Columbia, or the Commonwealth of Puerto Rico. The definitions and interpretations of the terms included in the Act were provided originally from Section 201. In September 2015, the FDA established the "Current Good Manufacturing Practice, Hazard Analysis, and Risk-Based Preventive Controls for Human Food" rule in the Code of Federal Regulations of the Federal Food, Drug, and Cosmetic Act [19].

Subpart B of the CGMP addresses requirements applied for preventing microbial contamination from sick or infected personnel and hygienic practices that should be used. Measures for preventing microbial contamination include excluding any person confirmed by medical examination that has or appears to have an "illness, infection, open lesion or any other abnormal source of microbial contamination" that could contaminate components of dietary supplements or contact surfaces. Employees should also be notified if they are believed to have a health condition that may lead to microbial contamination [19].

Hygienic practices that should be followed to prevent microbial contamination of dietary supplements are stated in Title 21, Chapter I, Subchapter B, Part 111, as follows.

This section also addresses the personnel qualification requirements applied to those who work in the dietary supplement industry, including having qualified employees with the proper education, training, and/or experience and having people designated to quality control operations. The manufacturing company/corporation must also keep a record of the requirements mentioned previously and documentation of training [19].

TABLE 5.1
Hygienic Practices to Prevent Microbial Contamination of Dietary Supplements

Wearing outer garments in a manner that protects against the contamination of components, dietary supplements, or any contact surface.

Maintaining adequate personal cleanliness.

Washing hands thoroughly (and sanitizing, if necessary, to protect against contamination with microorganisms) in an adequate handwashing facility:

1. Before starting work
2. At any time when the hands may have become soiled or contaminated

Removing all unsecured jewelry and other objects that might fall into components, dietary supplements, equipment, or packaging, and removing hand jewelry that cannot be adequately sanitized during periods in which components or dietary supplements are manipulated by hand. If hand jewelry cannot be removed, it must be covered by material that is maintained in an intact, clean, and sanitary condition and that effectively protects against contamination of components, dietary supplements, or contact surfaces.

Maintaining gloves used in handling components or dietary supplements in an intact, clean, and sanitary condition. The gloves must be of an impermeable material.

Wearing, where appropriate, in an effective manner, hairnets, caps, beard covers, or other effective hair restraints.

Not storing clothing or other personal belongings in areas where components, dietary supplements, or any contact surfaces are exposed or where contact surfaces are washed.

Not eating food, chewing gum, drinking beverages, or using tobacco products in areas where components, dietary supplements, or any contact surfaces are exposed, or where contact surfaces are washed.

Taking any other precautions necessary to protect against the contamination of components, dietary supplements, or contact surfaces with microorganisms, filth, or any other extraneous materials, including perspiration, hair, cosmetics, tobacco, chemicals, and medicines applied to the skin.

5.8 REGULATIONS FOR LOCAL MANUFACTURING [20]

Regulations for physical plants and grounds that apply to those that manufacture dietary supplements are addressed in Subpart C of the "Current Good Manufacturing Practice, Hazard Analysis, and Risk-Based Preventive Controls for Human Food" Title 21, Chapter I.

The grounds of a physical plant must be organized in a manner that prevents attracting pests or allowing them to breed. The grounds must also have proper inspection and extermination plans (with no insecticides, fumigants, fungicides, or rodenticides) in place to avoid any extraneous materials that may lead to contamination. The water supply used during the manufacturing process must be safe and sanitary and be distributed through a plumbing system with proper drainage [20].

In order to reduce the possibility of mix-ups or contamination of components, dietary supplements, or contact surfaces with "microorganisms, chemicals, filth or other extraneous material," the plant must be organized with the dietary supplements, packaging, or labels in separate areas of the plant. The section also outlines what aspects of the architecture are required for the plant to function properly.

Subpart D is focused on the usage of equipment and utensils within manufacturing plants. Procedures are included for calibrating instruments and controls for automated, mechanical, and electronic equipment that are used in the manufacturing or testing of a component or dietary supplement. It is specified that the equipment and utensils must be made in a way that minimizes the accumulation of dirt or any other extraneous materials.

Subpart E specifies what the requirements are for establishing a production and process control system. Specifications that must be established include component specifications to ensure the "purity, strength, and composition of dietary supplements." There must also be established limits

TABLE 5.2
Production Processes

Subpart H	Requirements for the master manufacturing record
Subpart I	Requirements for the batch production record
Subpart J	Requirements for laboratory operations
Subpart K	Requirements for manufacturing operations
Subpart L	Requirements for packaging and labeling operations

on the types of contamination that may lead to the adulteration of a finished batch of dietary supplements. Tests or examinations must be conducted in order to make sure that the set specifications are met, along with collecting representative samples of unique lot components, packaging, labels, in-process materials, unique shipments, and lots of packaged and labeled dietary supplements. Based on the samples, quality control personnel conduct a material review and make decisions that may cause in-process adjustments to make the product suitable for use. This subsection also specifies what documentation is required for each part of the process in order to ensure the product is manufactured safely.

Subpart F specifies the requirements for the quality control operations in the production and process control system, while Subpart G focuses on the requirements for components, packaging, and labels of the dietary supplement being manufactured. Subpart H through Subpart L focus on the requirements for the following production processes.

Subpart M directs attention to the holding and distributing requirements for dietary supplements, which include requirements for holding components, dietary supplements, packaging, and labels. Conditions such as temperature, humidity, and light are essential to the purity and composition of the dietary supplements and their components. Holding process conditions should protect against possible mix-ups, contamination, or deterioration. Similar requirements and conditions apply to holding reserve samples as well.

Requirements and processes for returned dietary supplements are outlined in Subpart N. Before salvaging a returned dietary supplement, the quality control personnel must conduct a material review and make sure that the reprocessed product meets all product specifications. Processes in response to product complaints are specified in Subpart O. Any complaints are to be reviewed by the qualified personnel. After complaints are reviewed, the quality control personnel will then decide on whether the product complaint needs to be investigated and reviewed. This section is then followed by Subpart P, with information on the requirements for records and recordkeeping, with all records available for inspection or copying by the FDA when requested [20].

5.8.1 FDA Facility Registration [21]

Any foreign or domestic food facility that wishes to market and sell their products in the United States must register with the Food and Drug Administration. This can be done via the completion of the FDA Food Facility Registration Form. The sections required by this form include:

1. Type of registration
2. Facility name and address information
3. Parent company name and address information
4. Trade names
5. US agent if it is the registration for foreign food facilities
6. Seasonal facility dates of operation (optional)
7. General product categories

   a. Regarding whether the food/product is for human consumption or for animal consumption
   b. Selecting what type of food product for human consumption the product would be classified as

8. The dietary supplement categories:

   a. Proteins, amino acids, fats, and lipid substances
   b. Vitamins and minerals
   c. Animal by-products and extracts
   d. Herbals and botanicals

9. Owner, operator, or agent-in-charge information
10. Inspection statement
11. Certification statement [21]

5.9 IMPORT REGULATIONS

5.9.1 Prior Notice

A prior notice must be filed with the FDA before the product arrives at its port of entry, adhering to the following guidelines.

Prior notice must be submitted:

- Road, two hours before the shipment's arrival
- Rail, four hours before the shipment's arrival
- Air, four hours before the shipment's arrival
- Sea, eight hours before the shipment's arrival

Prior notice must include thus:

- The origin of the shipment
- Expected Customs and Border Protection (CBP) port of entry
- CBP Automated Commercial System (ACS) used to track date and type
- The manufacturer's name and address
- Any associated email address
- Working telephone and fax numbers
- Standard Carrier code [22]

5.9.2 Foreign Supplier Verification Programs for Importers of Food for Humans and Animals (FSVP Regulation)

The Foreign Supplier Verification Program for Importers of Food for Humans and Animals is essential to ensuring that both food and beverage products meet the standards for public health protection [23]. The program is mandatory for importers of food to the United States and ensures that companies/producers follow safety regulations and labeling rules. Guidelines provided by the FSVP follow.

Food and beverages that are exempt from the FSVP guidelines are as follows.

5.10 THE PUBLIC HEALTH SECURITY AND BIOTERRORISM PREPAREDNESS AND RESPONSE ACT OF 2002

The Public Health Security and Bioterrorism Preparedness and Response Act of 2002 applies to certain dietary supplement manufacturers that need to register with the FDA. The Act "requires the biennial review and republication of the HHS list of select agents and toxins." This was done

TABLE 5.3
FSVP Guidelines

Use a qualified individual to develop an FSVP and to perform FSVP activities.

Perform a hazard analysis that includes identifying known or reasonably foreseeable hazards associated for each type of food and determining whether they require a control.

Evaluate risks posed by the food and the performance of the foreign supplier, considering thus:

- The hazard analysis for the food
- The entity that will be applying hazard controls, such as the foreign supplier or the foreign supplier's ingredient supplier
- The foreign supplier's food safety practices and procedures
- Applicable US food safety regulations and information regarding the foreign supplier's compliance with those regulations, including whether the foreign supplier is the subject of an FDA warning letter or import alert
- The foreign supplier's food safety performance history, including results from testing, audit results, and the supplier's record of correcting problems

Conduct appropriate supplier verification activities to ensure that the hazards requiring control in the food you import have been significantly minimized or prevented. These activities may include:

- Annual on-site audits (must be performed by a qualified auditor)
- Sampling and testing of food
- A review of the supplier's relevant food safety records and/or other appropriate activities

Take corrective actions (if necessary), and investigate the adequacy of the FSVP (when appropriate).

Re-evaluate the food and foreign supplier every three years or sooner if the FSVP importer becomes aware of new information about the hazards in the food or the foreign supplier's performance.

Identify the FSVP importer when filing for entry with US Customs and Border Protection using the FSVP importer's name, electronic mailing address, and unique facility identifier (UFI) recognized as acceptable to the FDA.

TABLE 5.4
Food and Beverages Exempt from FSVP Guidelines

Juice and seafood from foreign suppliers that comply with the respective HACCP regulations (21 CFR Part 120 or 123) and certain ingredients that are used by the importer in the manufacturing or processing of juice and seafood products in accordance with the respective HACCP regulations.

Small quantities of food are imported for research and evaluation purposes that are not intended for retail sale and are not sold or distributed to the public.

Small quantities of food imported for personal consumption that are not intended for retail sale and are not sold or distributed to the public.

Food produced in compliance with the FDA's low-acid canned food requirements in 21 CFR Part 113 (except with respect to microbiological hazards controlled by 21 CFR Part 113 only), as well as certain ingredients for use in LACF products, but only with respect to microbiological hazards).

Certain alcoholic beverages.

Food that is transshipped through the United States or that is imported for future export and not sold or distributed in the United States.

Food that is manufactured/processed, raised, or grown in the United States, exported, and returned to the United States without further manufacturing/processing.

Certain meat, poultry, and egg products.

Source: [24, 25].

so that the FDA could easily trace the manufacturer of a recalled product. This became especially important when there were dietary supplements that had prescription drugs or other potentially dangerous substances [26].

TABLE 5.5
Steps in Monitoring the Food System from Harvesting to Consumption

1. **Conduct a hazardous analysis.** Create a list of hazards that could possibly cause injury or illness and determine whether there are factors that may influence the occurrence and severity of the hazard being controlled. Measures to control the hazard must also be included.
2. **Determine critical control points (CCPs).** A *critical point* is any step in which hazards can be prevented, eliminated, or reduced to acceptable levels. Ignorance of critical points, such as cooking, cooling, re-heating, or holding, typically leads to foodborne illness outbreaks or other hazards.
3. **Establish critical limits.** Critical limits ensure that CCPs have control over possible biological, chemical, or physical hazards. These limits would be monitored by measurement or observation, such as temperature or water activity.
4. **Establish monitoring procedures.** These procedures would outline how to measure or observe the "flow of food" through an establishment. Monitoring procedures would also outline which individuals would be authorized to monitor actions and what training must be conducted.
5. **Establish corrective actions.** Corrective actions should be in place in case a critical control point is not met. The plan should also include what the corrective action would be and who is responsible for carrying out the corrective action.
6. **Establish verification procedures.** Such procedures make sure that the operations of the plan are scientifically and technically sound. Through expert advice, scientific studies, observations of the flow of food, measurements, evaluations, and on-site review of established critical limits, verification can be carried out. Verification then allows an establishment to modify plans if necessary.
7. **Establish recordkeeping and documentation procedures.** Recordkeeping and documentation procedures would be essential to ensuring that all the established standards are met. Such procedures would also include information on training employees as well.

Source: [27].

5.10.1 Possible Use of Hazard Analysis and Critical Control Guidance

The Hazard Analysis and Critical Control Points system guides the monitoring of the food system from harvesting to consumption. Such a system reduces the risk of foodborne illnesses. Seven formal steps can be applied to both commercial and noncommercial settings.

5.11 RECENT UPDATES ON NUTRACEUTICAL GUIDELINES IN THE UNITED STATES

5.11.1 Updates from the FDA Regarding Dietary Supplements from 2023 to May 2024

5.11.1.1 March 2023: FDA Launches the New Directory of Ingredients Used in Products Marketed as Dietary Supplements

The Dietary Supplement Ingredient Directory is a webpage where the public can search more on ingredients in dietary supplements. Such a resource allows them to learn more about the FDA's stance on the ingredient and whether any other agencies have taken any action regarding the ingredient. This directory is in place of the FDA Dietary Supplement Ingredient Advisory List, which quickly alerted the public when the agency identified ingredients [28].

5.11.1.2 February 2024: Update on the FDA's Launch of the New Directory of Ingredients Used in Products Marketed as Dietary Supplements

The Dietary Supplement Ingredient Directory has now been renamed Information on Select Dietary Supplement Ingredients and Other Substances. Updates also include new categories to classify the type of FDA action or communication for ingredients and when the listed ingredient was added to the directory [28].

5.11.1.3 March 2024: The FDA Issues Final Guidance on New Dietary Ingredient Notification Procedures and Timeframes

On March 6, 2024, the FDA finalized the complicated and controversial pre-market safety notification process titled "NDI Notification Procedures and Timeframes." Such guidance is important for manufacturers and distributors of new dietary ingredients (NDIs) and dietary supplements to prepare and submit notifications to the FDA [28].

The topics included in the section are as follows.

TABLE 5.6
Topics Included in the Section

Who needs to submit an NDIN and when?

Under Section 413(a)(2) of the FD&C Act, "the manufacturer or distributor of an NDI, or of the dietary supplement that contains the NDI, must submit a premarket notification to FDA at least 75 days before introducing the product into interstate commerce or delivering it for introduction into interstate commerce, unless the NDI and any other dietary ingredients in the dietary supplement 'have been present in the food supply as an article used for food in a form in which the food has not been chemically altered' (21 U.S.C. 350b(a)(1))."

According to Section 201 of the FD&C Act, *dietary ingredients* include thus:

- Vitamins
- Minerals
- Herbs or other botanicals
- Amino acids
- Dietary substances for use by man to supplement the diet by increasing the total dietary intake
- A concentrate, metabolite, constituent, extract, or combination of any dietary ingredient from the preceding categories

What to include in an NDIN?

- Citation in published articles with reasonable expectation of safety
- Name/complete address of the manufacturer or distributor submitting the notification
- Name of the NDI (if a botanical, the Latin binomial name must be included)
- The level of the NDI in the dietary supplement and its recommended use included in the labeling of the supplement
- History of use or another evidence that suggests that proper usage, as specified by the label, is safe
- Signature of the authorized individual from the manufacturer or distributor to sign the notification.

How to file an NDIN?

The FDA recommends that the NDIN be filed through the CFSAN Online Submission Module to allow for data validation. It also helps prevent incomplete notifications.

How should the information be organized and presented?

The CFSAN Online Submission Module offers a template with recommendations on NDIN organization.

What happens after an NDIN is submitted?

The FDA will send an acknowledgment to the notifier with the date of the receipt and NDIN number.

FDA response types commonly sent by the FDA:

- Letter of acknowledgement without objection
- Letter listing deficiencies that make the NDIN incomplete under 21 CFR 190.6
- Objection letter raising concerns with the adequacy of the identity information or safety information (e.g., identifying gaps in the history of use or that the safety information on the NDI does not support the conditions of use)
- Letter raising other regulatory issues with the ingredient or product (e.g., the ingredient is not a dietary ingredient under 21 USC 321(ff)(1), or the product does not satisfy all parts of the dietary supplement definition under 21 USC 321(ff)).

Source: [29].

5.11.1.4 April 2024: FDA Issues Draft Guidance on New Dietary Notification Master Files for Dietary Supplements

A draft guidance was issued to provide specific guidance on recommendations about master files for new dietary ingredient notifications. A master file usually includes the identity, manufacturer, and safety information of the new dietary ingredient. This would be helpful for industrial compliance with the NDIN requirements, as recommendations on content and submission would be included. Comments made on the draft guidance are to be submitted 60 days after it is published in the Federal Register, to ensure that the FDA considers their comments on the draft guidance before working on the final draft [28].

5.12 REGULATIONS OF DIETARY SUPPLEMENT ADVERTISEMENTS

In 1998, the Federal Trade Commission (FTC) staff had guidance on confirming or dismissing claims about the effectiveness and safety of health-related products through the *Dietary Supplements: An Advertising Guide for Industry* [30].

The document clarifies the difference between the responsibilities of the FDA and those of the FTC, stating that the FTC's responsibility is to regulate advertising, "including print and broadcast ads, infomercials, catalogs, and similar direct marketing materials." The FTC aims to make sure that advertising is truthful, not misleading, and substantiated. They also will take actions against

TABLE 5.7
How the FTC Law Applies to Dietary Supplement Advertising

Dietary supplement advertisements are analyzed by:

- *Identifying claims and interpreting ad meaning.*
- *Substantiating claims.* Substantiating a claim can be done by determining thus:
- The type of product
- The type of claim
- The benefits of a truthful claim, and the cost/feasibility of developing the substantiation for the claim
- The consequences of a false claim
- The amount of substantiation that experts in the field believe is reasonable

Other issues relating to dietary supplement advertising:

- *Claims based on consumer testimonials or expert endorsements.* The FTC law requires for there to also be appropriate scientific evidence behind the claim.
- *Claims based on traditional use.* Products that are advertised with claims referring to traditional use will portray themselves as already being backed by scientific evidence. If such claims are found to be false and present a notable risk of harm to the consumer's health and safety, they are subject to a higher level of scientific proof.

Use of a DSHEA disclaimer in advertising:

- The DSHEA disclaimer typically includes that the statement has not been evaluated by the FDA and that the product is not intended to "diagnose, treat, cure, or prevent any disease." Such a disclaimer may lead consumers to think that there is no other sort of deception in the advertisement, which may be false.
- *Using third party literature.* The FTC law does prohibit the deceptive use of newspaper articles, abstracts, or scientific studies to promote brands or products. This can be applied to dietary supplement manufacturers that may use such literature to promote products without actually having scientific proof to back up their claims.

Source: [31].

supplement manufacturers, distributors, retailers, catalog companies, infomercial producers, or others involved in deceptive promotions. The FTC makes it clear that anyone who is involved in this process is obligated to present the truth and to check the adequacy of the support behind those claims. They want to ensure a clear language, by avoiding small types, placing any qualifying information close to the claim being qualified, and avoiding making inconsistent statements or distracting elements [31].

Overall, to ensure that the FTC laws are followed, it is recommended that supplement advertisers follow two steps:

> (1) careful drafting of advertising claims with particular attention to how claims are qualified and what express and implied messages are conveyed to consumers; and (2) careful review of the support for a claim to make sure it is scientifically sound, adequate in the context of the surrounding body of evidence, and relevant to the specific product and claim advertised. [31]

5.13 FUTURE TRENDS FOR NUTRACEUTICALS IN THE UNITED STATES

In 2021, the Council for Responsible Nutrition found that the most consumed supplements were vitamins and minerals, specialty supplements (like omega-3 fatty acids or probiotics), botanicals and herbs, sports supplements, and weight management supplements. The most cited reasons for taking nutraceuticals were to improve immune health or maintain general well-being. Especially after the COVID-19 pandemic, there was increased public awareness of the importance of supplements in preventing and reducing symptoms. Vitamins C, D, and B, zinc, or probiotics were the supplements taken in hopes of boosting immunity [32].

There has also been an increased interest in dietary supplements due to the rise of social media personalities promoting products such as Bloom, a digestive supplement, or berberine, a compound derived from various plants which has been promoted as "nature's Ozempic." A Pew Research Center report stated that "80% of internet users have looked for health information online" [33].

A study analyzed various dermatologic supplements promoted on Instagram through posts from March 2021. These posts were evaluated for account verification status, endorsements, ingredients, and health claims. Out of the 100 posts evaluated, only "4% of posts were made by verified accounts, and 1% of posts contained a visible Supplement Facts label." Advertising through social media platforms has become more of an issue for dermatology supplements, especially since the FDA does not really evaluate the safety or effectiveness of the dietary supplements that are available on the market. The research found that 19% of posts had no information on what ingredients were in a supplement, and almost 31% mentioned vitamins and minerals without providing specifics. These statistics are concerning because there are issues with adverse reactions and the possibility of overdoses of said vitamins and minerals [34].

Such a spike in marketing would cause an increase in sales but also bring the rapid spread of misinformation, causing concern for the FDA and the FTC. If such trends continue, there may be an increase in the rules and regulations published by the FDA and the FTC regarding nutraceuticals within the next decade due to false information and increased production of supposed dietary supplements [35]. There is also the possibility of public health officials promoting initiatives to educate consumers on dietary supplements but also to recommend consulting a physician before purchasing and consuming them.

5.14 CONCLUSION

The world of nutraceuticals is expanding in the United States with the influence of regulating factors and frameworks, consumer trends, and medical professional opinion. What initially started as the consumption of supplements like ginseng, green tea, folic acid, and echinacea has become the consumption of supposed supplements like Bloom without any scientific evidence. Since the enactment of the Dietary Supplement Health and Education Act of 1994, various regulations have been published to ensure the quality of dietary supplements and safety for consumers. There are constant efforts to regulate the labeling, safety and risk, manufacturing, importing, quality, and advertising of dietary supplements. As diet culture heightens in popularity with influencers promoting products with little evidence, the FDA and the FTC will likely release regulatory frameworks to prevent the public from consuming harmful or ineffective products, by looking at the manufacturing, importing, and sale of nutraceuticals.

REFERENCES

[1] U.S. Food and Drug Administration (FDA). *FDA Takes Action Against 14 Companies for Selling Illegal Cancer Treatments* [Internet]. 2020. Available from: https://www.fda.gov/news-events/press-announcements/fda-takes-action-against-14-companies-selling-illegal-cancer-treatments

[2] Centers for Disease Control and Prevention (CDC). *Products—Data Briefs—Number 399—February 2021* [Internet]. 2021. Available from: https://www.cdc.gov/nchs/products/databriefs/db399.htm

[3] Hunt, J., Dwyer, J. Position of the American Dietetic Association: Food Fortification and Dietary Supplements. *Am Diet Assoc.* 2001 Jan 1;101(115–132).

[4] MedlinePlus. *American Ginseng: MedlinePlus Supplements* [Internet]. Available from: https://medlineplus.gov/druginfo/natural/967.html#:~:text=People%20take%20American%20ginseng%20for

[5] Paik, D.J., Lee, C.H. Review of Cases of Patient Risk Associated with Ginseng Abuse and Misuse. *J Ginseng Res.* 2015 Apr;39(2):89–93.

[6] National Center for Complementary and Integrative Health. *Green Tea* [Internet]. 2020. Available from: https://www.nccih.nih.gov/health/green-tea

[7] Centers for Disease Control and Prevention (CDC). *Folic Acid* [Internet]. 2022. Available from: https://www.cdc.gov/ncbddd/folicacid/about.html

[8] National Institutes of Health. *Office of Dietary Supplements—Folate* [Internet]. 2022. Available from: https://ods.od.nih.gov/factsheets/Folate-Consumer/

[9] National Center for Complementary and Integrative Health. *Echinacea* [Internet]. 2020. Available from: https://www.nccih.nih.gov/health/echinacea

[10] Brevoort, P. "Booming U.S. Botanical Market: A New Overview." *HerbalGram.* 1998.

[11] Committee on the Framework for Evaluating the Safety of Dietary Supplements. *Dietary Supplements: A Framework for Evaluating Safety.* Washington, DC: National Academies Press; 2005.

[12] National Institutes of Health. *Dietary Supplement Health and Education Act of 1994 Public Law 103–417, 103rd Congress* [Internet]. 2017. Available from: https://ods.od.nih.gov/About/DSHEA_Wording.aspx

[13] Institute of Medicine (US) and National Research Council (US). *Committee on the Framework for Evaluating the Safety of Dietary Supplements* [Internet]. National Academies Press (US); 2013. Available from: https://www.ncbi.nlm.nih.gov/books/NBK216048/

[14] U.S. Food and Drug Administration (FDA). *Current Good Manufacturing Practice (CGMP) Regulations* [Internet]. 2019. Available from: https://www.fda.gov/drugs/pharmaceutical-quality-resources/current-good-manufacturing-practice-cgmp-regulations

[15] Center for Food Safety and Applied Nutrition. *Dietary Supplement Labeling Guide: Chapter I* [Internet]. 2018. Available from: https://www.fda.gov/food/dietary-supplements-guidance-documents-regulatory-information/dietary-supplement-labeling-guide-chapter-i-general-dietary-supplement-labeling

[16. Center for Food Safety and Applied Nutrition. *Questions and Answers on Dietary Supplements* [Internet]. U.S. Food and Drug Administration; 2022 May 6. Available from: https://www.fda.gov/food/information-consumers-using-dietary-supplements/questions-and-answers-dietary-supplements#:~:text=FDA%20generally%20does%20not%20approve

[17] Center for Food Safety and Applied Nutrition. *FDA Issues Draft Guidance on the Labeling of Dietary Supplements Containing Live Microbials* [Internet]. 2020 Jul 7. Available from: https://www.fda.gov/food/cfsan-constituent-updates/fda-issues-draft-guidance-labeling-dietary-supplements-containing-live-microbials

[18] Institute of Medicine (US) and National Research Council (US). *Dietary Supplements: A Framework for Evaluating Safety* [Internet]. National Academies Press (US); 2013. Available from: https://www.ncbi.nlm.nih.gov/books/NBK216048/

[19] World Health Organization (WHO). *Good Manufacturing Practices* [Internet]. 2024. Available from: https://www.who.int/teams/health-product-policy-and-standards/standards-and-specifications/gmp

[20] Department of Health and Human Services, Food and Drug Administration. *PART 111—Current Good Manufacturing Practice in Manufacturing, Packaging, Labeling, or Holding Operations For Dietary*

Supplements [Internet]. National Archives; 2007. Available from: https://www.ecfr.gov/current/title-21/chapter-I/subchapter-B/part-111

[21] Liberty Management Group. *FDA Food Facility Registration Form* [Internet]. Available from: https://www.fdahelp.us/food-form.html [cited 2024 May 6].

[22] U.S. Customs Clearance. *The Essential Guide to Importing Dietary Supplements* [Internet]. USA Customs Clearance; 2022. Available from: https://usacustomsclearance.com/process/importing-dietary-supplements/#:~:text=Health%20Inspection%20Service [cited 2024 May 6].

[23] Det Norske Veritas. *FSVP—Foreign Supplier Verification Program* [Internet]. Available from: https://www.dnv.us/services/fsvp-foreign-supplier-verification-program-190893/#:~:text=The%20Foreign%20Supplier%20Verification%20Program%20is%20an%20FDA%20program%20aimed [cited 2024 May 6].

[24] U.S. Food and Drug Administration (FDA). *Final Rule on Foreign Supplier Verification Programs* [Internet]. 2023 Jun. Available from: https://www.fda.gov/media/97893/download.

[25] U.S. Department of Health and Human Services, Food and Drug Administration, Center for Food Safety and Applied Nutrition. *Foreign Supplier Verification Programs for Importers of Food for Humans and Animals: What You Need to Know About the FDA Regulation: Guidance for Industry Small Entity Compliance Guide* [Internet]. 2024 May. Available from: https://www.fda.gov/media/110327/download [cited 2024 May 6].

[26] U.S. Department of Health and Human Services. *Public Health Security and Bioterrorism Preparedness and Response Act of 2002; Biennial Review and Republication of the Select Agent and Toxin List* [Internet]. Federal Register. 2010. Available from: https://www.federalregister.gov/documents/2010/07/21/2010–17728/public-health-security-and-bioterrorism-preparedness-and-response-act-of-2002-biennial-review-and.

[27] Cooperative Extension Food Safety Education. *Hazard Analysis of Critical Control Points Principles* [Internet]. University of Rhode Island. Available from: https://web.uri.edu/foodsafety/hazard-analysis-of-critical-control-points-principles/#:~:text=Hazard%20Analysis%20Critical%20Control%20Points.

[28] Center for Food Safety and Applied Nutrition. *Guidance for Industry: New Dietary Ingredient Notification Procedures and Timeframes—Dietary Supplements* [Internet]. U.S. Food and Drug Administration; 2024. Available from: https://www.fda.gov/regulatory-information/search-fda-guidance-documents/guidance-industry-new-dietary-ingredient-notification-procedures-and-timeframes-dietary-supplements.

[29] Center for Food Safety and Applied Nutrition. *What's New in Dietary Supplements* [Internet]. U.S. Food and Drug Administration; 2024. Available from: https://www.fda.gov/food/dietary-supplements/whats-new-dietary-supplements.

[30] Federal Trade Commission. *Health Products Compliance Guidance* [Internet]. Federal Trade Commission; 2022. Available from: https://www.ftc.gov/business-guidance/resources/health-products-compliance-guidance.

[31] Federal Trade Commission Bureau of Consumer Protection. *Dietary Supplements: An Advertising Guide for Industry.* DIANE Publishing; 2001.

[32] Djaoudene, O., Romano, A., Bradai, Y.D., Zebiri, F., Ouchene, A., Yousfi, Y., et al. A Global Overview of Dietary Supplements: Regulation, Market Trends, Usage during the COVID-19 Pandemic, and Health Effects. *Nutrients* [Internet]. 2023 Jul 26;15(15):3320. Available from: https://www.ncbi.nlm.nih.gov/pmc/articles/PMC10421343/.

[33] Pew Research Center. *Health Information Online* [Internet]. Pew Research Center; 2005. Available from: http://www.pewinternet.org/2005/05/17/health-information-online/ [cited 2024 May 6].

[34] Zamil, D.H., Ameri, M., Fu, S., Abughosh, F.M., Katta, R. Skin, Hair, and Nail Supplements Advertised on Instagram. *Baylor University Medical Center Proceedings*. 2022 Sep 23;36(1):38–40.

[35] Klein, J., Schweikart, S. Does Regulating Dietary Supplements as Food in a World of Social Media Influencers Promote Public Safety? *AMA Journal of Ethics* [Internet]. 2022 May;24(5):E396–401. Available from: https://journalofethics.ama-assn.org/sites/joedb/files/2022–04/hlaw1-peer-2205_0.pdf.

6 Nutraceuticals and Regulatory Compliance in Latin America

Juan Montilla, Niyati Acharya, and Yashwant Pathak

6.1 INTRODUCTION

The term *nutraceutical* is an amalgamation of the words "nutrition" and "pharmaceuticals." They are food products that provide health benefits in addition to essential nutrition. These products come from natural sources and are enriched with bioactive components, vitamins, minerals, herbs, or other physiologically beneficial substances. The main objective of nutraceuticals is to improve health, prevent diseases, support biological processes, and contribute to general well-being. The different categories of nutraceuticals include medicinal products, nutritional supplements, and fortified beverages. They are made to help avoid or minimize specific symptoms and offer additional health benefits over traditional diets. Nutraceuticals, such as vitamins, minerals, probiotics, and omega-3 fatty acids, are very popular in a specific population segment because they complement the usual diet by providing a series of benefits that contribute to health and well-being.

Nutrient supplementation offers a practical and effective means to compensate for dietary deficiencies or insufficient intake of certain nutrients. Specific nutraceuticals, such as functional foods or herbal supplements, are designed to support specific health problems, improve immunity, promote better digestion, or help with cognitive function. As part of a comprehensive approach to health, regular consumption of nutraceuticals can help increase vitality, improve quality of life, and achieve higher energy levels.

The growing awareness of Latin American consumers about health, well-being, and quality of life has turned the region into a potential market for nutraceutical products. The market for natural and functional products has increased significantly in the region, which is indicative of a movement toward preventive health-care practices. The different nutraceutical products found in the Latin American market are mainly composed of botanical supplements, functional foods and beverages, fortified products, and dietary supplements.

As consumers become more aware of the connection between nutrition, food, and overall health, there is a greater need for nutraceuticals. Due to dietary imbalances caused by different factors, such as poverty, the difficulty in certain population sectors of accessing the health system, or an unbalanced diet and lifestyle changes, a growing market for products that can compensate for the dietary deficiencies has been generated. Another thing to keep in mind is that as the population ages, products that help improve joint health, vitality, healthy aging, and cognitive health become increasingly desirable. The accessibility of nutraceutical products has facilitated this market's expansion through various retail channels, such as pharmacies, health food stores, supermarkets, and internet platforms. Multinational corporations and regional producers are prominent participants in the Latin American nutraceutical sector, offering a variety of products that may satisfy a wide range of customer tastes.

Government regulations based on objective scientific, commercial, and public health criteria fundamentally guarantee the safety, efficacy, and quality of nutraceutical products in Latin America. These regulations are designed to safeguard consumer health and maintain product standards. The approval process typically involves strict evaluations and compliance with established guidelines before products can be marketed. Regulatory approval procedures require safety assessments to determine any potential hazards related to nutraceutical components. This process includes

 DOI: 10.1201/9781003465041-6

evaluating toxicity, allergenicity, and possible interactions with other chemicals. Nutraceutical manufacturers must support their health claims and the effectiveness of their products with scientific data. Regulatory agencies evaluate these statements to ensure truthfulness and avoid providing false information. To ensure the consistent quality, potency, and purity of nutraceutical products, approval processes emphasize compliance with Good Manufacturing Practices. Nutraceutical product labels must be legible and accurate to provide consumers with the necessary information about ingredients, dosage, usage guidelines, and allergens.

Throughout this research, the regulatory framework in Latin America, country-specific regulations, labeling requirements, classification of nutraceuticals, and challenges, including future perspectives, will be discussed. *Nutraceuticals* are an essential product category that promises to improve overall health and well-being. The relevance of regulatory frameworks that guarantee the safety, efficacy, and quality of these products may also guarantee the expansion of the nutraceutical market in Latin America, which would eventually benefit customers and the industry.

6.2 GENERAL REGULATORY FRAMEWORK FOR NUTRACEUTICAL APPROVAL IN LATIN AMERICA

When it comes to the regulatory framework for nutraceutical approval in Latin America, there is a widespread situation where many countries, individuals, and even within their own regions have a different structure and balance to handling these nutraceuticals. There are approximately more than 20 nations in Latin America. Due to the lack of standardized work in the more than 20 nations in Latin America, there can be significant regional variations in the regulatory framework associated with food supplements. Even though these nations might not have the same framework, several characteristics within the region as a whole share these values of regulating nutraceutical approvals.

6.3 ROLE OF HEALTH AUTHORITIES

Some of the highlighted roles of health authorities or regulatory agencies in overseeing the approval process can be counted as an additional aspect that is different within each individual nation. Supervising the approval process for various health-related products and interventions is a critical function of regulatory agencies and health authorities. The approval process is an organized and demanding examination meant to guarantee the quality, safety, and effectiveness of these items before their release into the market or their application in medical environments. This oversight is crucial to protect the public's health and foster confidence in the health-care system. In this session, one will go over the primary components of regulatory bodies and health authorities' roles in the approval process.

The standardization and regulation of supplements and nutraceuticals are essential to the regulatory procedure. These items have to meet specific standards specified by health authorities in order to be approved for ingestion. These requirements frequently cover labeling, production procedures, ingredient quality, and health claims. These standards are created and updated by regulatory bodies, like the National Institute for Food and Drug Surveillance (INVIMA) in Colombia, the National Health Surveillance Agency (ANVISA) in Brazil, and others in the region, to reflect the most recent advancements in science and global best practices.

The manufacturer or distributor of the supplement or nutraceutical usually submits an application or dossier to start the approval process. This dossier includes extensive details about the product, such as composition, manufacturing procedures, safety research, and proof of any health claims. Health officials carefully examine this data to ensure that the product complies with requirements and does not put customers at unnecessary risk.

Essential elements of the evaluation procedure are clinical trials and safety evaluations. In order to determine the safety and effectiveness of the supplement or nutraceutical, health authorities review the scientific data submitted by the applicants. Examining information on the product's impact on human health, possible interactions with other drugs, and any documented adverse reactions may be part of this process. The objective is to ascertain whether the product fulfills its promises of health advantages without jeopardizing its users. Their goal is to ensure that the public eye is in favor of these procedures and that individuals who are consuming these pharmaceuticals are in good health.

National regulatory bodies play a crucial role in health systems by supervising medical items' efficacy, safety, and quality. Their primary roles are stated as follows on page 61 of the document "Plan for self-sufficiency in health matters in Latin America and the Caribbean: lines of action and proposals": The process of approval includes the "[r]egistration (licensing) of products, inspection and licensing of manufacturers, inspection and licensing of distributors, post-marketing surveillance, regulation of any claims made to promote the products commercially, authorization of clinical trials." These results and actions have a direct impact on the nation's economy's growth as well as the expansion and assistance of the health industry.

Facilities are inspected to ensure that good manufacturing procedures (GMP) are followed. These checks guarantee the consistency of the final product's quality and safety, as well as the dependability of the manufacturing processes. During these inspections, health authorities may evaluate elements like the source of raw materials, production techniques, quality control, and adherence to hygienic requirements.

6.4 IMPORTANCE OF SCIENTIFIC EVIDENCE

A point that remains to be mentioned that is extremely important for the safety of the nutraceutical consumer is scientific evidence. Scientific evidence plays a very important role in supporting the safety and effectiveness of products. That is why all kinds of experts are needed to verify, test, and approve the product so that it goes on the market.

> The first step to assess therapeutic efficacy should be based on positive evidence from clinical data. This step should include the contribution of different professionals and different expertise, ranging from food chemistry and food safety to structure–activity studies, with the aim of assessing the mechanism of action, nutritional aspects, pharmacology, pharmacokinetics and pharmacodynamics of each nutraceutical [1].

Determining how nutraceuticals function and their potential advantages can be helped by evaluating their pharmacokinetics, nutritional value, and mechanisms of action. This guarantees a complete grasp of the efficacy of nutraceuticals by putting a priority on clinical data and involving a variety of expertise, assisting in the making of well-informed judgments about their usage for health-related goals. These professionals make sure everything is safe for the consumer. As stated in the National Library of Medicine:

> Ultimately, the proof of efficacy, as well as of safety, should come from well-designed and well-executed human studies. Although the basic methodology of designing and conducting human studies on the health benefits of nutraceuticals is similar to that of pharmaceutical trials, there are aspects that need special consideration when investigating nutraceuticals [2–5].

These types of tests that help determine whether a product is safe or unsafe for the consumer are called clinical trials. This consists of several series of research that test and evaluate the effects of, in this case, supplements on the human body. Why are all these trials or tests important to ensure that a product is safe? Most nutraceutical products state some health benefits on their label. Scientists have to make sure that the product meets what it says on the label, and if this is met, it will help consumers easily decide which product to buy. Another example is, if a patient is undergoing treatment and is taking certain medications and starts taking nutraceuticals, it can cause serious health problems, since some nutraceuticals contain certain compounds or very concentrated forms of nutrients that can alter and interact with the medication. These are just a few examples of why it is important to keep these regulations, and of the importance of scientific evidence when it comes to supporting the safety and efficacy of nutraceutical products.

6.5 LABELING REQUIREMENTS

The Latin American nutraceutical company attaches considerable importance on labeling. When things are properly labeled, consumers may make informed decisions about what they buy, since transparency is ensured. Additionally, it assists distributors and producers in meeting the diverse regulatory standards that exist in the several Latin American nations, which can differ greatly.

To gain the trust of consumers, nutraceuticals—which include functional foods, dietary supplements, and other items that offer health advantages beyond basic nutrition—need to be labeled with clarity and accuracy. Latin American consumers are becoming more aware of the contents and claims made on nutraceutical products, and they depend on information on the label to help them make decisions.

Labeling, which includes the whole ingredient list, nutritional information, and any therapeutic or health claims, can help set premium nutraceuticals apart from inadequate or deceptive goods on the market. In an era where customers are being more discriminative and the nutraceutical sector is changing quickly, this transparency is essential. Adherence to appropriate labeling protocols not only safeguards consumer trust but also strengthens the sustained expansion and prestige of the nutraceutical industry in Latin America.

Being a very varied continent, Latin America has certain requirements for the labeling of supplements, one different from the other, depending on the country. These regulations cover a wide range of requirements meant to guarantee that customers receive accurate and honest information from product labels. Detailed component lists, nutritional data, allergen declarations, dose guidelines, expiration dates, and manufacturer information are among the common labeling requirements in the area. Furthermore, it is crucial to follow the laws controlling health claims, as they frequently impose strict requirements to verify the veracity and correctness of any claims stated on product labels. Although these standards are fundamental to all Latin American countries, differences may arise from regulations unique to each nation. This clearly emphasizes how crucial it is for producers of nutraceuticals to manage the region's variety of regulatory environment in order to guarantee consumer confidence and compliance.

6.6 DIFFERENT TYPES OF NUTRACEUTICALS

Nutraceuticals are a broad category of goods that are consumed throughout Latin America for reasons other than simple nutrition. They are divided into groups according to their components, purposes, and legal and regulatory environments. The primary categories of nutraceuticals found in Latin America are as follows:

- Dietary supplements, including those containing proteins, vitamins, minerals, and herbs
- Functional foods, including dairy products, cereals, baked goods, and other functional food categories
- Functional, sports, and energy drinks

In Latin America, these nutraceuticals are commonly used to address a wide range of health issues, from weight management and athletic performance to nutrient deficiencies and problems with digestion. The protection, labeling, and distribution of these items are regulated by regional regulatory bodies, such as COFEPRIS in Mexico and ANVISA in Brazil, to guarantee consumer protection and adherence to national regulations.

6.7 REGULATION OF APPROVAL OF NUTRACEUTICALS IN SPECIFIC COUNTRIES

Latin America is a large continent with a variety of geological features, including the wild Andes Mountains, the dense Amazon rainforest, and the vast Pampas Plain. This wide geographic diversity has a major role in the distinctive political, linguistic, and cultural environments that each nation in the region has to offer. Because of this, the legal systems that oversee the authorization of nutraceuticals differ greatly between countries.

The various traditions, customs, and food habits that are observed in the various countries of Latin America are evidence of the region's vast cultural diversity. These cultural differences affect not only the kinds of nutraceutical goods that are consumed but also the regulation strategies that the different countries have chosen. The differences in regulatory criteria and procedures for the approval of nutraceuticals are further influenced by the diversity of governance structures, which span from democratic republics to socialist states and totalitarian regimes.

Language also has a big influence on regulatory systems. In Latin America, Spanish is the official language in most countries, but Portuguese is used in Brazil, with the largest population of the region, and in many of those countries, there are a lot of autochthonous languages still being used. The regulatory environment is made more complex by this linguistic diversity, since documents and regulations need to be modified to meet the unique language requirements within each region.

Comprehension of the varied regulatory frameworks set by each country is imperative when navigating the regulatory approval procedures for nutraceuticals in Latin America. These frameworks include a wide range of topics, including safety evaluations, ingredient guidelines, product classification, and labeling regulations. The complexity of the clearance process is increased by additional elements that further influence regulatory dynamics, such as international trade agreements, political stability, and economic conditions.

Being that Latin America is such a diverse region, businesses seeking to market nutraceutical products there must carefully review each nation's regulatory environment. To ensure adherence to relevant laws and ease market access, this calls for careful study, calculated planning, and coordination with regional regulatory bodies. Through comprehension and adjustment to the nuances of regulatory mandates throughout Latin America, businesses can proficiently present and distribute nutraceuticals in this constantly evolving and quickly expanding marketplace. In what follows, processes from the most populated Latin American countries—Mexico, Brazil, Colombia, and Argentina—will be discussed.

6.7.1 Mexico

The nutraceutical market in Mexico has grown exponentially due to the public health deficiencies that exist throughout the country. The body in charge of regulating all supplements is COFEPRIS. The Federal Commission for the Protection against Sanitary Risk, as it is in Spanish, "La Comisión Federal para la Protección contra Riesgos Sanitarios (COFEPRIS)," is the top regulatory authority. This company is in charge of controlling lots of things, including food, food supplements, and herbal products. COFEPRIS also tests the importance of consumer health, which is why they regulate the health safety measures that must be carried out so that the product can be for sale. Before the regulatory processes are mentioned and processed, it is important to clarify that *nutraceuticals* are defined differently according to their regulations. Products that can be offered in a pharmaceutical form and that are intended to improve, complete, or supply specific components of a diet, such as those based on herbs, plant extracts, traditional foods, or dehydrated or concentrated fruits, are considered as nutraceuticals.

There are multiple steps in regulations to register a nutraceutical on the market. These include:

- *Examination of the product's ingredient.* This means that the manufacturer must submit a list of all chemical and natural components used in the product. Then COFEPRIS will check them one by one to make sure they are healthy.

- *Categorization of the product.* Nutraceuticals are classified as foods in Mexico. For this reason, a product made by means of a manufacturing process that utilizes substances with pharmacological effects will be categorized as a medication rather than a nutraceutical.
- *Delivery of prior consultation to COFEPRIS.* COFREPRIS will reach back to the company with a response, approval, or denial.
- *Submission of documents.* Such as product details, safety information, and labeling data.
- *Approval or rejection from COFEPRIS.* If the product is accepted, it can be distributed in the whole country.

In order to accelerate the approval of their nutraceutical products, companies must rigorously follow COFEPRIS standards and requirements during the approval process. Furthermore, it is essential that manufactures are informed about any modifications or revisions to rules to guarantee compliance with existing requirements, as they may alter over time.

The regulation of nutraceuticals in Mexico is characterized by a number of distinct difficulties and facets that are indicative of the intricacy of the public health environment and the market. The market's diversity, which covers a broad spectrum of goods, from contemporary dietary supplements to conventional herbal medicines, is one of the main challenges. Because of this variability, it is challenging to establish and implement consistent regulatory standards, particularly in light of the sizeable segment of the market that works informally and might not follow rules. The need for nutraceuticals designed to help people lose weight and control their blood sugar is fueled by public health issues, including the high prevalence of obesity and diabetes. But these circumstances also call for strict regulation to guard against misleading health claims and to guarantee the effectiveness and safety of products.

As with nutraceutical approval requirements, the Federal Commission for the Protection against Sanitary Risk (COFEPRIS) also plays a very important role in supplement labeling. They are the ones in charge of ensuring that all nutritional product companies in Mexico comply with the labeling standards set by COFEPRIS. Thanks to the requirements that will be mentioned later, COFEPRIS may guarantee consumer safety and present clear and accurate product information.

Respecting these guidelines, it is essential for the lawful advertising and distribution of nutraceuticals in the Mexican market to give customers the guarantee that the goods are fit for human use and are safe.

> In March 2020, an amendment to NOM-051-SCFI/SSA1–2010 included Front-of-Pack warning labels for foods and beverages high in energy, sugar, saturated fat, trans fat and sodium was approved. This comes into force from 1st October 2020 and applies to all prepacked foods (Food Compliance International PTE.) [6–9]

This new amendment changed the labeling requirements in Mexico. Now, food supplements need to abide by the updated labeling regulations in order to give consumers clear, accurate, and thorough information. The most important change included front-of-pack labels. A food supplement must have the relevant front-of-pack warning labels on it if it added saturated fat, trans fat, sugar, sodium, or calorie content exceeding the set criteria. Customers can rapidly detect products that might contain excessive concentrations of these components thanks to these labeling. In addition to this recent amendment, there is a strict set of standards that manufacturers must follow for their labeling to be appropriate.

Firstly, product labeling must include a clear and appropriate name along with a description of the characteristics of the product. This enables consumers to easily identify the product they are seeking. Secondly, manufacturers must provide a comprehensive list of product ingredients, arranged in descending order by weight. Ensuring completeness and accuracy in this list is vital for consumers with specific dietary requirements or allergies. Nutrition facts, including a detailed table outlining the levels of important nutrients per serving, servings per container, and serving size, must be prominently displayed on the label. This information empowers consumers to make informed dietary decisions by understanding the nutritional content of the product.

Instructions for use should be provided, offering clear guidance on dosage, administration methods, and any necessary safety precautions. This ensures that consumers can use the product effectively and safely. Contact information for the manufacturer, including address, phone number, importer, and company name, must be readily available on the label. This allows consumers to reach out for further information or assistance if needed.

To safeguard consumer health, manufacturers are required to include any contraindications or warnings on the label, ensuring that consumers are aware of potential risks associated with product use. Finally, expiration dates and lot numbers must be provided to guarantee product freshness and traceability, further contributing to consumer safety and well-being.

The Official Mexican Standard permits the incorporation of additional languages in dietary supplements but mandates that all information be displayed in Spanish. If other languages are employed, the information must, nevertheless, also be equally prominently displayed in Spanish.

It is prohibited to utilize terms or expressions that suggest the product can satisfy a person's nutritional needs on its own or serve as a meal replacement. It is also prohibited to give unclear information about the ingredients, source, effects, and other characteristics of the products, or to suggest that they are intended for therapeutic, rehabilitative, or preventative purposes.

6.7.2 Brazil

The largest country in Latin America, and one of the largest in the world, has an extensive list of nutraceutical regulations at the national level. First, it is important to mention that nutraceuticals in Brazil are not considered as drugs, because they are viewed as an element of food. "[P]roducts for oral ingestion, presented in pharmaceutical forms, intended to supplement the diet of healthy individuals with nutrients, bioactive substances, enzymes or probiotics, isolated or combined (RDC No. 243/2018)" [10–15]. This classification suggests that claims regarding therapeutic or medical effects of dietary supplements should be avoided. Rather, they ought to be promoted and supported only as goods that enhance overall health and well-being. The main company responsible for the registration of nutraceuticals in Brazil is the National Health Surveillance Agency, or the Agência Nacional de Vigilância Sanitária (ANVISA), and its job is to regulate, market, and approve supplements. Companies have to inform the state's health authority when they begin manufacturing products. To register a nutraceutical on the market, there are several steps to follow as part of the requirements of approval in Brazil. These consist of:

- The ingredients of the product must be examined.
- Claims and labels have to be reviewed.
- Papers and documents should be submitted, including:
    - Information about the company that is going to manufacture the product
    - Characteristics of the manufacturing facility
    - Evidence of fees paid
    - The type and title of the product
    - Kind of packing
- Wait for ANVISA's response.

This means manufacturers and distributors must gather extensive documentation that satisfies ANVISA's requirements in order to be approved. This paperwork addresses a number of subjects, including the labeling, manufacturing procedures, proof of payments of fees, safety, effectiveness, supporting evidence, and content of the product. After this is prepared, ANVISA carefully reviews the material to make sure it complies with regulatory requirements. This evaluation looks at the safety of the ingredients, the effectiveness of the product, its intended use, and the correctness of the labeling. Furthermore, ANVISA has the authority to inspect manufacturing facilities in order to confirm that Good Manufacturing Practices (GMP) and other quality requirements are being followed. ANVISA approves the product after a satisfactory evaluation, enabling it to be sold in Brazil. On the other hand, rejected products are given thorough feedback and given the chance to fix every mistake and submit their application again. The form can be found online, and all papers should be printed and physically delivered to the National Agency of Sanitary Surveillance.

With the purpose of protecting the public's health, ANVISA continues to monitor nutraceutical goods through post-market surveillance after approval. The result of the inspection depends on several things: most recent inspection date, company history, and the guarantee that the product is 100% safe to consume. If any of these requirements do not meet the standards set by ANVISA, it is considered a failed inspection; therefore, production will be suspended, and the product removed from the market.

ANVISA in Brazil regulates the labeling standards for nutraceuticals to make sure that customers obtain complete details about the products people acquire. These policies aim at encouraging consumer decision-making that is well-informed, transparent, and safe. For items to be legally commercialized in Brazil, including food with functional properties and dietary supplements, they must comply with strict labeling requirements. These products are known as nutraceuticals. This contains comprehensive component lists, usage guidelines, cautions, and nutritional data—all essential for preserving consumer confidence and preserving public health. Manufacturers and distributors must ensure that their products comply with the regulatory and legal standards established by Brazilian authorities by adhering to ANVISA's labeling regulations. The labeling of nutraceuticals in Brazil must contain the following steps:

1. The label should prominently feature the term "Suplemento Alimentar" (food supplement), along with the presentation format (e.g., tablets, capsules) and the primary nutrient contained within.
2. All text, including letters and numbers, must be legible to ensure clarity and ease of understanding for consumers. Product indications based on age demographics should be provided, along with clear guidance on recommended consumption amounts and frequency for each demographic group. A prominent warning boldly stating "This product is not a medicine" serves to clarify the intended use of the supplement and emphasize that it does not treat or cure any diseases. Recommendations to avoid exceeding the recommended daily consumption amount contribute to consumer safety and prevent potential adverse effects.
3. A clear warning to keep the product out of reach of children helps mitigate the risk of accidental ingestion. Instructions on proper storage after opening the package are essential for maintaining product integrity and efficacy over time.
4. Including the daily value percentage for key nutrients provides consumers with valuable information regarding the nutritional content of the supplement relative to their daily dietary needs.

In conclusion, customer health and well-informed decision-making are prioritized by Brazil's strict labeling regulations for nutraceuticals. These rules cover specific product indications, readability requirements, conspicuous warnings, instructions for correct storage, and labels that are both descriptive and easy to read. Brazil hopes to empower customers with thorough information, encourage responsible usage, and eventually improve general health and well-being by following these regulations.

6.7.3 Colombia

The National Institute for Food and Drug Surveillance, or Instituto Nacional de Vigilancia de Medicamentos y Alimentos, or INVIMA, is principally in charge of enforcing Colombia's regulations pertaining to nutraceuticals.

Like other foods and dietary supplements, nutraceuticals usually need to be registered with INVIMA to be sold and marketed in Colombia. In order to complete the registration procedure, a product's formulation, labeling, safety information, and proof that it complies with Colombian laws must all be submitted.

Colombia has become increasingly appealing to over-the-counter and prescription enterprises due to the authority's recognition globally for its stringent regulations and expedited licensing processes in recent times.

For new prescription, consumer health, and phytomedicines to be released into the market, they must comply with INVIMA regulatory standards. It makes it possible for the medications to be marketed and subsequently commercialized. Manufacturers of nutraceuticals are required to provide authoritative evidence of the product's efficacy and safety. To support the claims made regarding the health advantages of the product, this may entail carrying out clinical trials or supplying information from previously completed research.

This documentation could take the form of research findings, clinical studies, or related scientific publications that support the product's stated advantages. INVIMA assesses the scientific evidence submitted as part of the registration procedure to make sure the statements are factual and backed by trustworthy statistics.

Clinical trials include any research conducted on humans that aims to identify any adverse reaction to the investigational product(s) and to find or verify the pharmacological, clinical, or other pharmacodynamic effects of the product(s) and to investigate the experimental product's absorption, distribution, metabolism, and excretion to confirm its efficacy and/or safety.

INVIMA must register and approve any human subject research intended to determine or confirm the clinical effects of the product(s) under examination, as well as to confirm their safety and effectiveness. When evaluating the research protocols, the development of the study in Colombia is conceptualized by considering both the clinical and quality components of the research product. As per the standards outlined in this section, INVIMA keeps an eye on adherence to the conditions permitted for the study throughout its duration to safeguard the participants.

In order to guarantee the quality and safety of their products, manufacturers of nutraceuticals are required to follow Good Manufacturing Practices. This entails upholding sanitary and hygienic manufacturing facilities, putting quality control procedures into place, and meticulously documenting all production operations.

Prior to the process intended to confirm compliance with the established technical-legal requirements, the health registry is the publicly available document released by INVIMA or the delegated authority. It permits a natural or legal person to produce, market, import, export, package, process, or dispense medications and pharmaceutical preparations.

Following the completion of the safety and efficacy evaluation (pharmacological evaluation), the user is required to provide evidence of the item's quality. The group of chemically synthesized medication health records, which is responsible for issuing chemically synthesized medicine health records and related procedures, also evaluates this information.

Following is the process of legal documentation conducted by INVIMA:

- Properly signed legal representative's full application name and company details
- Brand name and name of the manufacturer
- Evidence of the petitioner organization's legal counsel and constitution
- Certificate of legal representation from the manufacturer

- Certificate of Good Manufacturing Practices
- If necessary, a power of attorney for process management
- Receipt for payment issued by the sanitary registry
- Industrial records

Nutraceuticals and other food products in Colombia are subject to strict labeling regulations that have been set forth by multiple Ministry of Health and Social Protection directives. Resolution 810 of 2021 [16] and its revisions, especially Resolution 2492 of 2022 [17–22], contain the main rules. These labeling requirements are used to provide the people information with each packet and to safely transfer and attain solid information about the product. These strategic steps and requirements are necessary for supplying nutraceuticals to be as efficient and as safe.

Products containing high levels of sugars, sodium, or saturated fats must feature front-of-pack labels. These labels are designed to provide clear nutritional information and include warning symbols to alert consumers to the presence of excessive nutrients that could pose health risks. The objective is to make it easier for consumers to make informed dietary choices quickly and easily.

A comprehensive breakdown of the macro- and micronutrient levels. This includes the serving size, calorie count, and amounts of proteins, carbohydrates, fats, vitamins, and minerals. By providing this detailed information, the regulations help consumers understand the nutritional value of the product and how it fits into their overall diet.

Certain products are exempt from some of these labeling requirements. These exemptions include children's formula, food for special medical purposes, natural foods, and bulk food products. This ensures that products intended for specific uses or populations are not subjected to unnecessary labeling requirements that might not be relevant or useful.

Product attributes and any health advantages must be listed on labels. In order to prevent deceiving customers, there are additional rules about how these claims are to be presented.

Companies have 18 months from the date of issuance to adjust to these new labeling regulations. Products that fall under this category, such as returnable beverage bottles, have five years to comply. According to the regulations, labels must be made so that the information they convey is accurate and clear and they cannot lead to customer error. To get their label designs approved, businesses must submit them to the National Institute for the Surveillance of Medications and Food Products (INVIMA).

6.7.4 Argentina

Nutraceuticals are usually categorized as food items in Argentina as opposed to medications. Products made from food sources called nutraceuticals are those that offer health advantages over and above basic nourishment. They are regulated by food authorities rather than pharmaceutical regulatory agencies because they are taken orally and frequently have forms and functions like regular food items.

Therefore, the National Administration of Medicines, Food, and Medical Technology (ANMAT), which oversees guaranteeing the security, caliber, and effectiveness of food items, medications, and medical equipment in the nation, is principally in charge of regulating and supervising nutraceuticals in Argentina. To guarantee that nutraceuticals fulfill specified safety and quality standards, ANMAT creates standards and guidelines for their manufacture, labeling, and marketing. These requirements are:

- Authorizations for raw components and packaging materials that come into contact with food
- Information about ingredients and nutritional content
- Product details
- Storage parameters

- Production procedure
- Information supporting "free-from" or nutrition claims
- If voluntary designations like "organic" or "kosher" made, certificates included
- Payment of the fees

Nutraceuticals and other food products sold in Argentina are subject to strengthened labeling laws designed to protect consumers and promote openness. These rules are part of larger initiatives to encourage wholesome eating practices and shield customers from information that can be deceptive. One of the strictest and most comprehensive food policy rules in the world was passed into law by the president of Argentina. It mandates that ultra-processed items with excessive amounts of sodium, sugar, fats, and calories have black octagonal warnings on the front of the packaging. Thanks to new food marketing rules that require front-of-package warning labels—the first phase of which began recently—consumers in Argentina can now identify dangerous food. Also, information about voltage, fire safety, child safety, and safety information must all be labeled. For nutritional reasons, black octagon labels must be present on product packaging for processed foods and non-alcoholic beverages as of March 2022. Dates of manufacture and expiration, as well as importer and manufacturer identification, are also required. Pharmaceuticals are required to list the ingredients by weight/volume and the name of the pharmaceutical laboratory's responsible technical director.

The steps of labeling in Argentina are as follows.

When considering the packaging and labeling of a dietary supplement, numerous details must be meticulously included to meet regulatory standards and provide consumers with essential information. First and foremost, the product's brand and trade name must be prominently displayed. This name, often a key element of the product's identity, is crucial for brand recognition and consumer trust. Additionally, the product must be clearly identified as a "Suplemento Dietético," which serves as its legal designation, ensuring that consumers understand its intended use as a dietary supplement.

The gross weight of the product is another critical piece of information that must be included. This measurement, usually listed in grams or ounces, informs consumers of the total quantity of the product they are purchasing. It is often a determinant in the product's pricing and can influence purchasing decisions, especially for those comparing similar products.

A comprehensive list of ingredients is also essential. This list must detail every component included in the product, from the primary active ingredients to any additives or preservatives. The ingredient list provides transparency, allowing consumers to make informed choices based on their dietary needs and preferences. It also helps individuals with dietary restrictions or allergies avoid harmful substances.

Speaking of allergens, it is imperative to highlight any allergenic substances present in the product. Common allergens, such as nuts, dairy, soy, and gluten, must be clearly identified to prevent adverse reactions in sensitive individuals. This information is crucial for consumer safety and is often mandated by food and drug regulatory bodies.

The manufacturer's or importer's name and address should also be listed. This information not only provides a point of contact for consumers who may have questions or concerns but also adds a layer of accountability for the product's quality and safety. Alongside this, the maker's or importer's National Registry of Establishment (RNE) number must be included. This number is a unique identifier assigned to registered establishments, ensuring that the product comes from a verified source that complies with regulatory standards.

The nation of origin must be declared as well. This information can influence consumer perceptions and preferences, as some may prefer products from specific countries known for their quality standards. Additionally, it can affect the product's import and export regulations.

Assembly and usage instructions are another critical component of the label. Clear, concise directions on how to assemble (if applicable) and use the product ensure that consumers can safely and effectively utilize the supplement. This section typically includes recommended dosages, the best time to take the supplement, and any preparation steps required before consumption.

The daily quantity or recommended daily intake should be specified. This information helps consumers understand the appropriate amount to take each day to achieve the desired benefits without overconsumption, which could potentially lead to adverse effects.

A best-before date or use-by date is also essential for ensuring the product's quality and safety. This indicates the timeframe the product is expected to retain its best quality and potency. Consuming the product beyond this date could reduce its effectiveness or potentially lead to health risks.

The lot identity and batch number are crucial for traceability. In the event of a recall or quality issue, these identifiers allow the manufacturer to track and isolate affected products efficiently. This information is vital for maintaining product safety and consumer trust.

Finally, warning remarks are required on the label. These warnings might include potential side effects, contraindications with other medications, or specific groups of people (such as pregnant women or individuals with certain medical conditions) who should avoid the product. These warnings are critical for consumer safety, ensuring that the product is used correctly and safely.

The following requirements are part of the specified labeling requirements that take more action. Apart from the standard labeling guidelines, nutritional supplements are subject to specific regulations as per Article 1381 of CAA: Herbal ingredients must be listed on the label together with the source plant part and their common and Latin names; each serving's nutritional information and the proportion of the RDI must be given as well. The label states that "free sale" is forbidden.

6.8 CONCLUSION

To come to an end, in several countries in Latin America, the term *nutraceutical* refers to a broad category of goods that are intended for uses other than basic nourishment and are grouped according to their components, applications, and regulatory settings. Dietary supplements (proteins, vitamins, minerals, and herbs), functional foods (dairy, cereals, baked goods), and functional drinks (energy drinks and sports drinks) are the main categories. These products target a range of health conditions, including digestive problems, weight management, sports performance, and nutrient deficiencies.

The protection, labeling, and distribution of nutraceuticals are supervised by regulatory organizations, such as ANVISA in Brazil and COFEPRIS in Mexico, to guarantee consumer safety and adherence to national laws. Because each Latin American nation has a different political, linguistic, and cultural environment, so do they all have different regulatory frameworks. Because of these variations, specific methods for product approval and market entry are required, taking into account things like safety assessments, criteria for ingredients, product classification, and labeling specifications. Businesses looking to successfully sell nutraceutical products throughout this broad and continuously evolving region must comprehend and adjust to these various regulatory mandates.

Latin American nutraceutical companies prioritize accurate labeling to guarantee openness and well-informed consumer decisions. Precise labeling is essential to complying with the many and sometimes disparate regulatory requirements in that region. With detailed ingredient lists, accurate nutritional information, and validated health claims, labels that are clear and exact aid customers in choosing and putting their trust in nutraceuticals, including functional foods and dietary supplements. In a quickly changing market where customers are becoming more perceptive, this openness is crucial. Respecting local labeling laws benefits the Latin American nutraceutical industry's growth and reputation, in addition to safeguarding customers. The distinct legal requirements of every nation emphasize how crucial it is to manage these variations to preserve industry compliance and consumer trust.

REFERENCES

[1] Nutraceuticals: opening the debate for a regulatory framework. https://bpspubs.onlinelibrary.wiley.com/doi/10.1111/bcp.13496

[2] LatinAmericamarketopportunitiesforfoodsupplements:Navigatingregulatorycomplexities.https://www.raps.org/news-and-articles/news-articles/2020/6/latin-america-market-opportunities-for-food-supple
[3] Nutraceuticals: opening the debate for a regulatory framework. https://www.ncbi.nlm.nih.gov/pmc/articles/PMC5867125/
[4] Nutraceuticals: is there good science behind the hype? https://www.ncbi.nlm.nih.gov/pmc/articles/PMC3575926/
[5] Tamaño del mercado de nutracéuticos y análisis de participación tendencias de crecimiento y pronósticos (2024–2029). https://www.mordorintelligence.com/es/industry-reports/global-nutraceuticals-market-industry
[6] Nutraceuticals & Dietary supplements regulations in Mexico. https://www.artixio.com/post/nutraceutical-dietary-supplements-regulations-in-mexico
[7] Registration of food supplements in Mexico—General description. https://foodsupplements.freyrsolutions.com/food-supplement-registration-in-mexico
[8] Mexico COFEPRIS labeling requirements for food supplements under NOM-051-SCFI/SSA1–2010. https://www.artixio.com/post/mexico-cofepris-labeling-requirements-for-food-supplements-under-nom-051-scfi-ssa1–2010#:~:text=Product%20Name%20and%20Brand%20Identity,supplement%20and%20its%20intended%20use.
[9] Mexico—Food compliance international. https://foodcomplianceinternational.com/expertise/mexico
[10] Brazil—Food compliance international. https://foodcomplianceinternational.com/expertise/brazil
[11] Food Supplements Registration in Brazil—Overview. https://foodsupplements.freyrsolutions.com/food-supplements-registration-in-brazil#:~:text=The%20Brazilian%20Regulatory%20agency%20responsible,and%20Novel%20Ingredients%20(CTCAF).
[12] ANVISA nutraceuticals and dietary supplements regulations in Brazil. https://www.artixio.com/post/nutraceutical-and-dietary-supplements-regulations-in-brazil
[13] Regulatory framework for dietary supplements and the public health challenge. https://rsp.fsp.usp.br/wp-content/uploads/articles_xml/1518–8787-rsp-53–90/1518–8787-rsp-53–90Regulatory framework for dietary supplements and the public health challenge https://rsp.fsp.usp.br/wp-content/uploads/articles_xml.x34413.pdf
[14] Food and dietary supplements launch in brazil—an overview. https://www.freyrsolutions.com/blog/food-and-dietary-supplements-launch-in-brazil-an-overview
[15] Understanding the OTC and pharma regulatory process and requirements in Colombia. https://www.chameleon-pharma.com/understanding-the-otc-and-pharma-regulatory-process-and-requirements-in-colombia/
[16] Resolución 810 de 2021. https://scj.gov.co/sites/default/files/marco-legal/resolucion-810-de–2021.pdf
[17] Resolución 2492 de 2022. https://www.minsalud.gov.co/Normatividad_Nuevo/Resolución%20No.%202492de%202022.pdf
[18] Nutraceutical Regulations and Registration in Colombia (INVIMA). https://www.artixio.com/post/nutraceutical-regulations-and-registration-in-colombia-invima#:~:text=The%20primary%20regulatory%20authority%20responsible,Drug%20Surveillance%20Institute%20(INVIMA).
[19] .Medicamentos homeopáticos, fitoterapéuticos y suplementos dietarios. https://www.invima.gov.co/productos-vigilados/medicamentos-y-productos-biologicos/medicamentos-homeopaticos-fitoterapeuticos-y-suplementos-dietarios
[20] Colombia—Food compliance international. https://foodcomplianceinternational.com/expertise/colombia
[21] Argentina now requires big food to add warning labels to ultra-processed foods. https://healthpolicy-watch.news/argentina-warning-labels-processed-foods/
[22] Argentina—Food compliance international. https://foodcomplianceinternational.com/expertise/argentina

7 The UK and the Medicines and Healthcare Products Regulatory Agency (MHRA)

Regulatory Requirements and Challenges for Nutraceuticals

Ramesh Pennamareddy, Shivkumar Somani, and Kandi Purushothama Reddy

7.1 INTRODUCTION

Nutraceuticals are substances derived from food sources that offer health benefits, including disease prevention and management. These products are used to support bodily functions, enhance overall wellness, and address various health conditions.

Often, nutraceuticals are utilized to promote health, slow down aging, prevent chronic illnesses, and maintain the body's structure and function. The regulation of nutraceuticals varies by country, and they are not always subject to the strict testing standards applied to pharmaceutical drugs.

7.1.1 Categories of Nutraceuticals

Nutraceuticals are generally divided into main categories. It includes a wide range of products, such as:

- Dietary supplements
- Functional foods
- Medicinal foods
- Probiotics
- Prebiotics

7.1.1.1 Dietary Supplements

Dietary supplements are products designed to be ingested and that contain dietary ingredients that help supplement the diet.

Many individuals use dietary supplements to boost their overall health, address specific nutritional deficiencies, enhance physical performance, or manage certain health issues. Although supplements can offer benefits, they should be taken responsibly and alongside a balanced diet. It is crucial to understand that dietary supplements are not meant to substitute whole foods or serve as a primary source of nutrition.

These supplements exist in many forms, including solid forms, gummies, powdered substances, liquids, and energy bars.

DOI: 10.1201/9781003465041-7

Frequent nutritional supplements include vitamins (e.g., calciferol and cyanocobalamin), minerals (e.g., calcium and iron), herbs (e.g., aloe vera and ginger), along with additional products, like glucosamine, probiotics, and fish oils.

Consulting a health-care professional before beginning any supplement regimen is recommended to ensure it is safe and suitable for your individual health needs.

Examples of dietary supplements include:

- Vitamins (vitamin D, vitamin C, vitamin B12, vitamin B6, vitamin A, vitamin K)
- Probiotics
- Calcium, zinc, magnesium, selenium
- Amino acids
- Omega-3 fatty acids
- Garlic
- Multivitamins
- Fish oil
- Biotin

7.1.1.2 Functional Foods

Functional foods include natural or processed foods that offer health benefits beyond basic nutrition when consumed regularly as part of a balanced diet.

These foods are defined as those that provide beneficial effects on one or more bodily functions, beyond basic nutrition, contributing to better health and a reduced risk of disease.

The primary roles of functional food include nutrient absorption, meeting psychological needs, and physical adjustment. Functional foods are often integrated into a balanced diet to help manage health and improve quality of life.

Functional foods can be categorized into two main types: conventional and modified.

Examples of functional foods follow:

- **Fortified foods.** Products that have added nutrients, like fortified cereals or milk with added vitamins and minerals.
- **Probiotic foods.** Foods containing live beneficial bacteria, such as yogurt or kefir, which support digestive health.
- **Prebiotic foods.** Foods rich in fiber or other substances that promote the growth of beneficial gut bacteria, such as bananas or garlic.
- **Enriched foods.** Foods enhanced with additional bioactive compounds, such as omega-3 fatty acids added to eggs or bread.
- **Soy protein, plant sterols, and stanols.**

7.1.1.3 Medicinal Foods

Medicinal foods are specially designed and formulated to address specific dietary needs for the management of certain diseases or medical conditions that cannot be fulfilled by a regular diet.

These foods are intended for the partial or complete feeding of patients, administered orally or via enteral feeding tubes.

Special Formulations: They are often tailored to meet the dietary needs of individuals with specific health conditions, such as diabetes, heart disease, or gastrointestinal disorders.

Clinical Use: They are typically used in clinical settings or under medical supervision to support disease management or recovery.

Nutrient Composition: They may contain a unique combination of nutrients, bioactive compounds, or other ingredients aimed at addressing specific health concerns.

Examples of medicinal foods include specialized infant formulas, products for metabolic disorders, oral rehydration solutions, and specialized nutrition formulas for diabetic patients.

Additional examples of medicinal foods include:

- Papaya—often used to aid digestion
- Turmeric—may help reduce inflammation
- Moringa leaves—consumed for their high vitamin C content

7.1.1.4 Probiotics

Probiotics are live microorganisms, often referred to as "beneficial" or "good" bacteria, which can offer health benefits when consumed in suitable quantities. Found in various foods and supplements, they support or restore a healthy balance of gut microbiota, which is essential for digestive health and overall wellness.

Here are some key points about probiotics:

- **Sources:** They are commonly found in fermented foods, like yogurt, kefir, sauerkraut, kimchi, and miso, and are also available in supplements in forms such as capsules, tablets, or powders.
- **Health Benefits:** Probiotics can support digestion, enhance immune function, alleviate symptoms of irritable bowel syndrome (IBS), and assist in preventing or treating diarrhea related to antibiotics or infections.
- **Strain-Specific Effects:** Different probiotic strains provide various benefits, so choosing the right strain for specific health issues is important.
- **Safety:** While probiotics are generally safe for most people, individuals with weakened immune systems, serious health conditions, or underlying health problems should consult with a health-care professional before starting probiotic use.

Incorporating probiotics into your diet can help maintain a healthy gut microbiome and support overall health.

7.1.1.5 Prebiotics

Prebiotics are food components that are not digested but promote the development and functioning of beneficial microbes in the gut.

Unlike probiotics, which are live bacteria, prebiotics act as nourishment for these beneficial bacteria, supporting their growth and helping maintain a balanced digestive system.

Here are some important aspects of prebiotics:

- **Sources:** Prebiotics are typically found in fiber-rich foods, such as bananas, onions, garlic, leeks, asparagus, chicory root, and whole grains. They can also be included in some dietary supplements.
- **Health Benefits:** By promoting the growth of beneficial gut bacteria, prebiotics can improve digestive health, enhance nutrient absorption, support immune function, and aid in regulating bowel movements [1].
- **Types:** Various types of prebiotics, such as inulin, fructooligosaccharides (FOS), and galactooligosaccharides (GOS), offer different benefits for gut health.
- **Safety:** Generally, prebiotics are safe for most individuals, though excessive intake may sometimes cause digestive issues, like bloating or gas.

Including prebiotic-rich foods in your diet helps in maintaining a healthy gut microbiome and supports overall well-being.

7.2 REGISTERING AND LICENSING NUTRACEUTICALS IN THE UK

In the UK, nutraceuticals are governed as food supplements under the Food Supplements Regulations 2003, which has been amended over time. Following are some key points regarding nutraceutical regulations in the UK:

- **No registration requirement.** Food supplements do not need to be registered in the UK. They can be sold as long as they adhere to the specific laws for food supplements and all other relevant food regulations. It is the responsibility of manufacturers, importers, or retailers to ensure compliance with these laws.
- **Good Manufacturing Practices.** Most supplement producers in the UK adhere to the " 'Guidelines for Good Manufacturing Practice for Manufacturers of Food Supplements," which outline the standards for the production process.
- **Regulatory oversight.** The Medicines and Healthcare Products Regulatory Agency (MHRA) is the main authority in the UK responsible for regulating nutraceuticals and ensuring their safety, quality, and efficacy.
- **Legal representation.** Manufacturers and importers of nutraceutical products must appoint a legal representative based in the UK. This representative serves as the point of contact for regulatory authorities and ensures that the products comply with relevant regulations.
- **Regulations overview.** The regulations outline specific requirements for the composition and labeling of food supplements, including the permitted vitamin and mineral substances. The document also provides detailed information on the laws applicable to the production, import, and sale of food supplements in the UK, along with links to relevant legislation and additional guidance.
- **EU directive and national legislation.** The regulations outline the allowable vitamin and mineral ingredients in food supplements, specifying the measurement units, labeling requirements, presentation, and advertising standards. *Food supplements* are defined as concentrated sources of vitamins, minerals, or other substances that have nutritional or physiological effects, provided in a dosage form.
- **Permitted substances.** Various lists of permitted vitamins, minerals, and sources have been tracked for use in dietary supplements. These listings reveal the medications that have been approved for use.

7.2.1 Water-Soluble Vitamins

1. **Biotin**
   - Average intake from food: 0.033 mg/day
   - Supplement intake: up to 2 mg/day
   - Estimated maximum intake: 2.07 mg/day
2. **Folic Acid**
   - Average intake from food: 0.26 mg/day
   - Supplement intake: males, up to 0.50 mg/day; females, up to 0.80 mg/day
   - Estimated maximum intake: males, 0.99 mg/day; females, 1.29 mg/day
3. **Niacin (Vitamin B3)**
   - Average intake from food: 34 mg/day
   - Supplement intake: up to 250 mg/day
   - Estimated maximum intake: 307 mg/day

4. **Pantothenic Acid (Vitamin B5)**
    - Average intake from food: 5.4 mg/day
    - Supplement intake: up to 550 mg/day
    - Estimated maximum intake: 560 mg/day
5. **Riboflavin (Vitamin B2)**
    - Average intake from food: 1.8 mg/day
    - Supplement intake: up to 100 mg/day
    - Estimated maximum intake: 103 mg/day
6. **Thiamin (Vitamin B1)**
    - Average intake from food: 1.50 mg/day
    - Supplement intake: up to 300 mg/day
    - Estimated maximum intake: 303 mg/day
7. **Pyridoxine (Vitamin B6)**
    - Average intake from food: 2.0 mg/day
    - Supplement intake: up to 100 mg/day
    - Estimated maximum intake: 104 mg/day
8. **Vitamin B12**
    - Average intake from food: 0.0062 mg/day
    - Supplement intake: up to 3.0 mg/day
    - Estimated maximum intake: 3.02 mg/day
9. **Vitamin C**
    - Average intake from food: 64 mg/day
    - Supplement intake: up to 3,000 mg/day
    - Estimated maximum intake: 3,160 mg/day

7.2.2 Fat-Soluble Vitamins

1. **Vitamin A (Retinol)**
    - From food: median intake is 520 μg RE/day, 97.5th percentile at 6,050 μg RE/day
    - From supplements: up to 2,400 μg RE/day
    - Estimated maximum intake: 6,050 μg RE/day (food) + 2,400 μg RE/day (supplements) = 8,450 μg RE/day
2. **ß-Carotene**
    - From food: average intake 2.3 mg/day, 97.5th percentile, at 7.0 mg/day
    - From supplements: up to 20 mg/day
    - Estimated maximum intake: 7.0 mg/day (food) + 20 mg/day (supplements) = 27 mg/day
3. **Vitamin D**
    - From food: average intake 0.003 mg/day, 97.5th percentile, at 0.009 mg/day
    - From supplements: up to 0.0125 mg/day.
    - Estimated maximum intake: 0.009 mg/day (food) + 0.0125 mg/day (supplements) = 0.022 mg/day

4. **Vitamin E**

- From food: Average intake 8.5 mg/day, 97.5th percentile, at 18 mg/day
- From supplements: up to 670 mg/day
- Estimated maximum intake: 18 mg/day (food) + 670 mg/day (supplements) = 690 mg/day

5. **Vitamin K**

- From food: 0.068 mg/day
- From supplements: up to 0.20 mg/day
- Estimated maximum intake: 0.068 mg/day (food) + 0.20 mg/day (supplements) = 0.27 mg/day

7.2.3 Trace Elements

1. **Boron**

- From food: 1.5 mg/day, 97.5th percentile, at 2.6 mg/day
- Intake from water: 0.2–0.6 mg/day
- From supplements: up to 10 mg/day
- From cosmetics and consumer products: up to 0.47 mg/day
- Estimated maximum intake: 2.6 mg/day (food) + 0.6 mg/day (water) + 10 mg/day (supplements) + 0.47 mg/day (cosmetics) = 14 mg/day

2. **Chromium**

- From food: mean intake 0.10 mg/day, 97.5th percentile, at 0.17 mg/day
- Intake from water: up to 0.002 mg/day
- From supplements: up to 0.6 mg/day
- Estimated maximum intake: 0.17 mg/day (food) + 0.002 mg/day (water) + 0.6 mg/day (supplements) = 0.77 mg/day

3. **Cobalt**

- From food: mean intake 0.012 mg/day, 97.5th percentile, at 0.019 mg/day
- Intake from water: approximately 0.02 mg/day
- From supplements: typically present as part of vitamin B12, so generally not counted separately
- Estimated maximum intake: 0.019 mg/day (food) + 0.02 mg/day (water) = 0.039 mg/day

4. **Copper**

- From food: mean intake 1.4 mg/day, 97.5th percentile, at 3.0 mg/day
- From supplements: up to 2 mg/day
- Intake from water: up to 6 mg/day
- Estimated maximum intake: 3.0 mg/day (food) + 2 mg/day (supplements) + 6 mg/day (water) = 11 mg/day

5. **Germanium**

- From food: 0.004 mg/day, 97.5th percentile, at 0.007 mg/day
- Estimated maximum intake: 0.007 mg/day

6. **Iodine**

- From food: mean intake 0.22 mg/day, 97.5th percentile, at 0.43 mg/day
- Intake from water: <0.03 mg/day

- From supplements: up to 0.49 mg/day
- Estimated maximum intake: 0.43 mg/day (food) + 0.03 mg/day (water) + 0.49 mg/day (supplements) = 0.95 mg/day

7. **Manganese**

- From food: mean intake 4.9 mg/day, 97.5th percentile, at 8.2 mg/day
- From supplements: up to 10 mg/day
- Intake from water: 0.1 mg/day
- Estimated maximum intake: 8.2 mg/day (food) + 10 mg/day (supplements) + 0.1 mg/day (water) = 18 mg/day

8. **Molybdenum**

- From food: mean intake 0.11 mg/day, 97.5th percentile, at 0.21 mg/day
- Intake from water: 0.02 mg/day
- From supplements: up to 0.33 mg/day
- Estimated maximum intake: 0.21 mg/day (food) + 0.02 mg/day (water) + 0.33 mg/day (supplements) = 0.56 mg/day

9. **Nickel**

- From food: mean intake 0.13 mg/day, 97.5th percentile, at 0.21 mg/day
- Intake from water: up to 0.04 mg/day
- From supplements: up to 0.005 mg/day
- Estimated maximum intake: 0.21 mg/day (food) + 0.04 mg/day (water) + 0.005 mg/day (supplements) = 0.26 mg/day

10. **Selenium**

- From food: mean intake 0.039 mg/day, 97.5th percentile, at 0.1 mg/day
- From supplements: up to 0.3 mg/day
- Estimated maximum intake: 0.1 mg/day (food) + 0.3 mg/day (supplements) = 0.4 mg/day

11. **Tin**

- From food: mean intake 1.8 mg/day, 97.5th percentile, at 6.3 mg/day
- From supplements: up to 0.01 mg/day
- Estimated maximum intake: 6.3 mg/day (food) + 0.01 mg/day (supplements) = 6.31 mg/day

12. **Vanadium**

- From food: mean intake 0.013 mg/day
- From supplements: up to 0.025 mg/day
- Intake from water: 0.01 mg/day
- Estimated maximum intake: 0.013 mg/day (food) + 0.025 mg/day (supplements) + 0.01 mg/day (water) = 0.05 mg/day

13. **Zinc**

- From food: mean intake 9.8 mg/day, 97.5th percentile, at 17 mg/day
- Intake from water: up to 10 mg/day
- From supplements: up to 50 mg/day
- Estimated maximum intake: 17 mg/day (food) + 10 mg/day (water) + 50 mg/day (supplements) = 77 mg/day

7.3 ADDING NEW SOURCES OF VITAMINS AND MINERALS TO THE APPROVED LIST

The European Food Safety Authority (EFSA) welcomes the submission of dossiers with safety information on vitamins, minerals, and their sources that are not currently listed in the Annexes of Directive 2002/46/EC.

Submissions can be made at any time, and guidance on this process is available at EFSA's website. If EFSA issues a favorable opinion on the dossier, the Standing Committee on the Food Chain and Animal Health will review the findings before any new legislation is enacted and the source is added to the Annex [2].

The legislation currently focuses on vitamins and minerals, with other nutrients, such as essential fatty acids, amino acids, and botanicals, still available for purchase despite not being listed. These substances, however, must comply with the general provisions of the law regarding safety, labeling, and packaging [3].

Directive 2002/46/EC also establishes minimum and maximum limits for vitamins and minerals in supplements, based on scientific and safety considerations.

7.3.1 Other Relevant Legislation

7.3.1.1 Kava-Kava

- **Definition:** *Kava-kava* refers to the entire *Piper methysticum* plant or its extracts.
- **Purpose:** The regulation is aimed at safeguarding public health due to concerns about the plant's potential liver toxicity and other health risks [4].
- **Geographic Scope:** This legislation is enforced across the UK, including Scotland, Wales, and Northern Ireland.
- **Regulation Scope:** The sale of any food containing kava-kava is prohibited in the UK under the Kava-Kava in Food (England) Regulations 2002, which include supplements containing the plant or its extracts [5].

7.3.1.2 Tryptophan

- **Supplement Use:** Tryptophan is commonly used to support sleep and mood but requires careful management due to potential drug interactions.
- **Permitted Form:** Only levorotatory tryptophan (L-tryptophan) is permitted in food supplements [6].
- **Scope:** Regulations govern the inclusion of tryptophan in foods and the sale of products containing it.
- **Medical Certification:** An appropriate medical certificate from a registered practitioner is required for products containing tryptophan to treat diagnosed conditions. These regulations came into effect on November 11, 2005.

7.3.2 Medical Claims

The Medicines and Healthcare Products Regulatory Agency (MHRA), under the Department of Health, ensures that medicines and medical devices are safe and effective. While most supplements are regulated under food laws, if a product claims to prevent, treat, or cure diseases, it must be licensed under medicines legislation [7].

- **Role of MHRA:** Responsible for ensuring safety and effectiveness of medicinal products, including supplements making medicinal claims.

- **Regulatory Framework:** Products with medicinal claims fall under stricter standards and must be licensed, whereas most supplements only need to comply with general food safety and labeling laws [8].

7.3.3 Regulations of Nutraceuticals in Europe

According to Functional Food Science in Europe (FUFOSE), functional foods must demonstrate benefits beyond basic nutrition. In the EU, nutraceuticals are classified based on their physiological effects, with products falling under either food or medicinal substance categories depending on their impact on health [9].

Regulatory Authority: The European Medicines Agency (EMA) oversees the evaluation of nutraceuticals, while the EFSA handles risk assessments and safety guidelines for foods. The General Food Law Regulation (EC No. 178/2002) outlines the principles for food safety, with specific laws ensuring that nutraceuticals meet stringent health and safety requirements.

This law serves as the cornerstone for establishing broad guidelines, standards, prerequisites, and protocols for making decisions about food safety, food production, and food distribution. Additionally, it created the independent EU Food Safety Authority (EFSA), which oversees risk assessment and scientific advice. In order to provide a high degree of health protection, it also creates emergency management protocols [10, 11].

Nutraceuticals in the EU comprise a broad range of products that can be used for feed materials, feed additives, or medicinal purposes. The EU legislation for the different types of nutraceuticals is very complex.

Different regulatory procedures apply to each of these categories for obtaining marketing authorization. For example, food products designed for special medical purposes or nutraceuticals that make medical claims or exhibit pharmacological effects must comply with Directive 2001/82/EC and are governed by Regulations (EU) No. 1924/2006 and No. 609/2013 to ensure their efficacy and safety. Foods with medical claims must have biological and toxicology data, according to the EFSA's Scientific Committees and Panels. Because nutraceuticals are subject to complicated regulations, marketing authorization is granted with a high safety margin and the precautions required for human health.

7.4 LABELING AND PACKAGING OF FOOD SUPPLEMENTS

The Food Labelling Regulations 1996 (as amended) and Food Supplements Legislation outline specific labeling requirements. These include details like the product name, ingredients list, recommended daily dose, and warnings against exceeding the dose or using the product as a substitute for a varied diet [12].

Manufacturers must also ensure compliance with health claim regulations and, when using organic raw materials, provide certification to verify this claim.

7.4.1 General Labeling Requirements

The general labeling requirements in the UK, as governed by the Food Labelling Regulations 1996 (as amended), mandate that most foods carry the following information [13]:

- **A name.** The product must have a clear and accurate name.
- **List of ingredients.** All ingredients used in the product must be listed.
- **Amount of certain ingredients.** The quantity of specific ingredients must be stated if they are emphasized on the label or are essential to characterize the product.
- **Date mark.** The product must have an appropriate date mark (e.g., "use by" or "best before" dates).

- **Special storage conditions or conditions of use.** Any particular storage instructions or usage conditions must be provided.
- **Manufacturer, packer, or EC seller details.** The name and address of the responsible party, such as the manufacturer, packer, or European Community seller, must be included.
- **Instructions for use.** If necessary, the product must include clear instructions for safe and proper use.
- **Place of origin.** The place of origin must be provided if omitting it might mislead consumers.

Food products that are prepared for delivery to a catering establishment or the "ultimate consumer" are subject to these regulations. Someone who purchases goods other than (1) for resale, (2) for a catering establishment, or (3) for a manufacturing business is considered the "ultimate consumer." You are advised that, even in cases where there are no legal requirements, it is best practice for a wholesale importer to provide as much information as possible (as long as the information is accurate), as this will assist the company that manufactures the final product in meeting its labeling requirements.

7.5 HEALTH BENEFITS OF NUTRACEUTICALS

Nutraceuticals offer a range of health benefits, often bridging the gap between food and pharmaceuticals. Here are some key health benefits [14–16]:

1. **Enhanced nutritional value.** Nutraceuticals can improve the overall nutritional profile of a diet, providing essential vitamins, minerals, and other bioactive compounds that may be lacking.
2. **Disease prevention.** Many nutraceuticals contain antioxidants, anti-inflammatory agents, and other bioactive substances that may help prevent chronic diseases, such as heart disease, diabetes, and cancer.
3. **Support for longevity.** Regular consumption of certain nutraceuticals has been linked to a longer lifespan by supporting immune function, reducing inflammation, and protecting against age-related conditions [15].
4. **Management of specific health conditions.** Nutraceuticals can be targeted to manage or alleviate symptoms of specific conditions, such as joint pain, digestive issues, or cognitive decline.
5. **Specialized nutrition.** Nutraceuticals can be tailored to meet the nutritional needs of specific populations, such as athletes, the elderly, or those with specific dietary restrictions or health conditions [16, 17].

7.6 CONCLUSION

The regulatory landscape for nutraceuticals in the UK, governed by the Medicines and Healthcare Products Regulatory Agency (MHRA), presents both opportunities and challenges. While the MHRA offers a framework aimed at ensuring safety and efficacy, the classification of nutraceuticals often leads to ambiguity in regulatory requirements. This complexity can hinder market entry and innovation, particularly for small and medium-sized enterprises. Key challenges include navigating the distinction between food supplements and medicinal products, compliance with labeling and health claims, and the necessity for robust scientific evidence. Furthermore, the dynamic nature of consumer preferences and emerging research can complicate adherence to existing regulations. Despite these obstacles, the increasing interest in health and wellness presents a significant opportunity for nutraceuticals. Stakeholders must engage with regulatory bodies early in product development to foster compliance and streamline processes. Continuous dialogue with the MHRA, along with investment in research and development, will be essential for addressing regulatory challenges and capitalizing on market potential.

In summary, while the regulatory environment for nutraceuticals in the UK is complex, proactive engagement and adherence to MHRA guidelines can pave the way for successful product development and consumer trust.

REFERENCES

[1] Thakur, Shyamli, Madan Mohan Gupta, and Deepak Sharma. "Nutraceuticals regulation: an overview of the regulatory frameworks in USA, EU, and Japan." *Nutraceutical Fruits and Foods for Neurodegenerative Disorders* (2024): 421–440.

[2] Bansal, Rohit, and Anju Dhiman. "Nutraceuticals: A comparative analysis of regulatory framework in different countries of the world." *Endocrine, Metabolic & Immune Disorders-Drug Targets (Formerly Current Drug Targets-Immune, Endocrine & Metabolic Disorders)* 20.10 (2020): 1654–1663.

[3] Komala, Muralikrishna Gangadharan, et al. "Investigating the regulatory process, safety, efficacy and product transparency for nutraceuticals in the USA, Europe and Australia." *Foods* 12.2 (2023): 427.

[4] Gupta, Ramesh C., Ajay Srivastava, and Rajiv Lall. "Toxicity potential of nutraceuticals." *Computational Toxicology: Methods and Protocols* (2018): 367–394.

[5] Schmidt, Mathias, et al. "Kava: A risk-benefit assessment." *The Essential Guide to Herbal Safety* (2005): 155–221.

[6] Sarris, Jerome, et al. "Adjunctive nutraceuticals for depression: a systematic review and meta-analyses." *American Journal of Psychiatry* 173.6 (2016): 575–587.

[7] Keservani, Raj K., Anil K. Sharma, and Rajesh K. Kesharwani, eds. *Nutraceuticals and Dietary Supplements: Applications in Health Improvement and Disease Management*. CRC Press, 2020.

[8] Santini, Antonello, et al. "Nutraceuticals: Opening the debate for a regulatory framework." *British Journal of Clinical Pharmacology* 84.4 (2018): 659–672.

[9] Gulati, Om P., and Peter Berry Ottaway. "Legislation relating to nutraceuticals in the European Union with a particular focus on botanical-sourced products." *Toxicology* 221.1 (2006): 75–87.

[10] Cheftel, J. Claude. "Food and nutrition labelling in the European Union." *Food Chemistry* 93.3 (2005): 531–550.

[11] Ruckman, Stephen A. "Regulations for nutraceuticals and functional foods in Europe and the United Kingdom." In *Nutraceutical and Functional Food Regulations in the United States and Around the World*. Academic Press, 2008: 221–238.

[12] Kaur, Sandeep, et al. "Potential health benefits of nutraceuticals for human health." *Environmental Contaminants and Natural Products; Bentham Science Publishers: Sharjah, United Arab Emirates* (2019): 193–208.

[13] Coppens, Patrick, Miguel Fernandes Da Silva, and Simon Pettman. "European regulations on nutraceuticals, dietary supplements and functional foods: a framework based on safety." *Toxicology* 221.1 (2006): 59–74.

[14] Chopra, Avijeet S., et al. "The current use and evolving landscape of nutraceuticals." *Pharmacological Research* 175 (2022): 106001.

[15] Singh, MEERA CHANDRADATT, and KISHORE N. Gujar. "Nutraceuticals: Uses, risks and regulatory scenario." *International Journal of Pharmacy and Pharmaceutical Sciences* 5.Suppl 3 (2013): 23–26.

[16] Coppens, Patrick, Miguel Fernandes Da Silva, and Simon Pettman. "European regulations on nutraceuticals, dietary supplements and functional foods: A framework based on safety." *Toxicology* 221.1 (2006): 59–74.

[17] Tedesco, Doriana Eurosia Angela, and Petra Cagnardi. "Regulatory guidelines for nutraceuticals in the European union." *Nutraceuticals in Veterinary Medicine* (2019): 793–805.

8 Regulatory Frameworks and Guidance for Food Products Registration, Safety, Labeling, and Consumer Protection Guidelines

A Comprehensive Review on German Standards

Nikunjana Patel, Nishit Parekh, and Sanjeev Acharya

8.1 INTRODUCTION

Nutraceuticals are products that offer physiological benefits, protect against chronic diseases, improve overall well-being, slow down aging, and extend life expectancy. These products are gaining popularity in Europe, primarily due to the growing prevalence of lifestyle-related illnesses and the increasing emphasis on preventive health care. As life expectancy rises and lifestyle-related diseases become more common, the significance of nutraceuticals cannot be overlooked by consumers in developed markets like Europe.

Within the market, functional foods hold the largest share, followed by functional beverages and dietary supplements. Functional beverages, in particular, are experiencing rapid growth and show potential for further expansion in developing markets. In Europe, the elderly population (aged 65 and above) is substantial, leading to an increase in health conditions such as obesity, heart disease, and arthritis. As of 2021, approximately 21.13% of Europe's population falls into this age group, driving higher demand for nutraceuticals and functional foods.

Moreover, probiotic supplements are becoming increasingly popular, targeting digestive issues, obesity, and other health concerns. In October 2022, OptiBiotix Health Plc, a global leader in microbiome modulation solutions, launched "OptiXOS," a new xylooligosaccharide prebiotic. This product claims to improve gut health by promoting the growth of beneficial microbes at low active doses.[1] This is indicative that the nutraceutical and functional food segment is emerging in Europe and developed countries.[2]

8.1.1 Market Trends of Nutraceutical in Europe with Respect to Germany

The growing number of health-conscious consumers in Europe, particularly in Germany, who are seeking products with health-promoting ingredients, has driven market growth. Nutraceutical ingredients are increasingly being incorporated into functional end products, such as dairy goods, baked items, snacks, cereals, fats and oils, and meat products. These fortified products often contain essential nutrients, like vitamins, minerals, fiber, and protein. Additionally, there is a rising

DOI: 10.1201/9781003465041-8

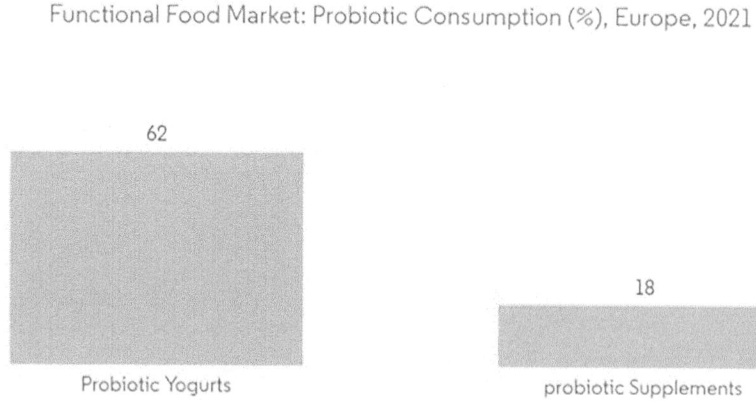

FIGURE 8.1 Functional food market consumption (%), Europe 2021.

demand for probiotic foods, such as yogurt and supplements, due to their numerous benefits, including improved gut health and immune support.(3)

In response to these trends, market players in Germany are introducing innovative probiotic products catering to various age groups. For example, in March 2023, Müller introduced a new probiotic yogurt range in Germany called "Müller Vitality," designed to support digestive health and strengthen the immune system. Similarly, in June 2023, the German company Kneipp launched "Kneipp ProBalance," a line of probiotic supplements aimed at enhancing gut health and overall well-being. Figure 8.1 indicates the functional food market consumption (%) in 2021.

To meet consumer demands for convenience and portability, many fortified products are now offered in single-serving sizes, driving demand for functional snack bars. An example is Kellogg's recent launch in Germany of two new gluten-free Nut Butter bar flavors under its Crunchy Nut range: Cocoa Hazelnut and Almond. These bars are free from palm oil, artificial colors, and preservatives, making them a healthier snack option.(4)

8.1.2 Future Prospects of the Nutraceutical Market in Germany

The nutraceutical market in Germany is poised for strong growth, driven by several key factors. One of the main drivers is the increasing consumer awareness of health and wellness, especially in the wake of the COVID-19 pandemic, which highlighted the importance of preventive health care. This trend is leading to a rising demand for dietary supplements, functional foods, and beverages that boost immunity and overall well-being.(5)

Germany, with its aging population and high consumer demand for organic and health-focused products, is one of the leading nutraceutical markets in Europe. This is due to the growing prevalence of lifestyle-related diseases, such as obesity, heart conditions, and arthritis. Functional beverages and foods, including probiotics and prebiotics, are particularly popular. Additionally, the German market benefits from a stable economy and increased interest in personalized nutrition.

However, the industry faces challenges, such as stringent regulations and high R&D costs, which can slow down product development. Despite these hurdles, innovations in product offerings and an increasing focus on sustainability are expected to shape the future of the nutraceutical market in Germany. Overall, the market shows promising growth potential as consumer demand for health-enhancing products continues to rise.(6) Figure 8.2 shows the projected European nutraceutical market.

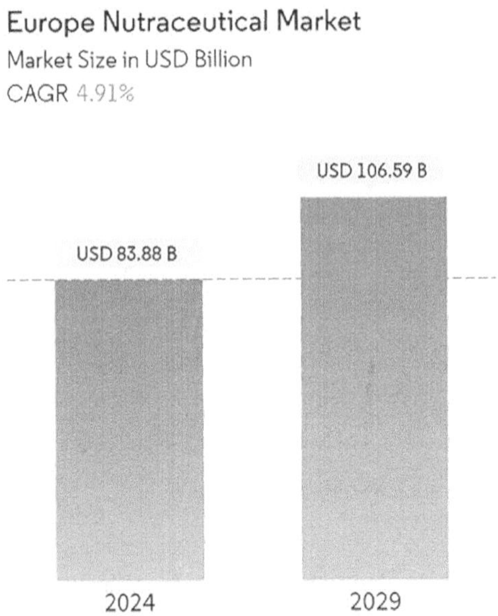

FIGURE 8.2 Europe nutraceutical market (market size in USD billion).

Study Period	2018 - 2029
Base Year For Estimation	2023
Market Size (2024)	USD 83.88 Billion
Market Size (2029)	USD 106.59 Billion
CAGR (2024 - 2029)	4.91 %
Market Concentration	Low

FIGURE 8.3 Study period and projected market size.

8.1.2.1 Germany Dominates the Market

Germany, the leading market in Europe, is closely followed by the United Kingdom and France in the nutraceutical sector. One of the key trends driving this market is the increasing popularity of high-protein fortification in breakfast cereals. With a population of around 108 million, Germany boasts the largest food and beverage retail market in Europe. Although German consumers tend to purchase less frequently, they prioritize quality and have a strong preference for organic and health-conscious products. This focus on quality has led to a notable increase in demand for functional beverages and foods over the past few years.

Responding to this growing demand, many companies are introducing new products specifically tailored for the German market. For example, in April 2021, PepsiCo launched "Rockstar Energy + Hemp," a unique beverage combining hemp seed extract with ginseng, guarana, B vitamins, caffeine, sugar, and taurine. The drink is offered in three variations: Original, Tropical Burst, and Prickly Cactus, catering to the German market's interest in energy-boosting and health-oriented drinks.[7]

Additionally, the popularity of probiotic products has been on the rise in Germany, particularly in the form of yogurts and sour milk items. Sour milk products, in fact, account for 74% of European consumption, according to the Europe International Probiotic Association's 2021 statistics. Companies like Müller have capitalized on this trend by launching the "Müller Vitality" range in Germany, designed to enhance digestive health and strengthen the immune system. Another example is Danone's introduction in Germany in late 2022 of "Activia Plus," a probiotic yogurt fortified with added vitamins and minerals to support overall health.

Germany's thriving nutraceutical market is further bolstered by the country's strong economy and low unemployment rate, which contribute to increased consumer spending on premium health products. The market's growth is also evident in the rising demand for functional snack bars. Kellogg's, for instance, expanded its Crunchy Nut range in Germany with two new gluten-free Nut Butter bar flavors: Cocoa Hazelnut and Almond. These bars, which are free from palm oil, artificial colors, and preservatives, have been well-received by health-conscious German consumers.

Moreover, the trend toward plant-based and sustainable products has seen brands like Alpro introduce a range of plant-based drinks and yogurts that appeal to the German market's eco-conscious consumers. Alpro's "Not Milk," launched in 2023, is a plant-based milk alternative that mimics the taste and texture of dairy milk, catering to the increasing number of consumers in Germany who are adopting plant-based diets.[8]

Germany's position as a leader in the European nutraceutical market is driven by a combination of high consumer demand for quality, health-focused products, and the innovative responses of companies that are continuously launching new offerings tailored to the specific needs and preferences of the German market.

8.1.2.2 Germany Nutraceutical Market

The consumer health market in Germany is flourishing, reflecting a growing emphasis on preventive care and self-treatment. Food supplements have experienced a significant expansion as Germans prioritize their well-being. In 2019, the total revenue from food supplements in Germany reached EUR 2.6 billion, representing a 5% increase from the previous year. This upward trend underscores the rising interest of consumers in health products.[9]

Pharmacies continue to be the leading distribution channel for these supplements, comprising more than 84% of total revenues, totaling EUR 2.2 billion. This can be attributed to their reputation as a reliable source for health products and individualized recommendations. This preference has remained strong even during the COVID-19 pandemic, which further increased the demand for immune-supporting supplements. In the initial quarter of 2020, sales of pure vitamin C nearly doubled, and growth rates between 40 and 60% were observed for other supplements such as zinc preparations, vitamin C combinations, and immune stimulants compared to the previous year.

The German consumer health market has also seen notable developments in several areas.

It is segmented into various categories, including vitamins, minerals, herbs, and specialty supplements, in the German food supplements market. There is a strong demand for vitamins and minerals, particularly vitamin D and B-complex vitamins. Vitamin D, for example, has been shown to support bone health and immunity, which has driven growth in these categories.

Companies are constantly introducing innovative products to meet the changing needs of health-conscious consumers. In early 2021, Biolectra, a German company, launched its "Biolectra Immun" line, which combines vitamins, minerals, and natural extracts to support immune health. In addition, Kneipp's "Wellkorn" line of functional foods includes fortified cereals and snacks designed to boost daily nutrient intake.

E-Commerce Growth: The rise of online shopping has also influenced the consumer health market. E-commerce platforms have become increasingly popular for purchasing food supplements, offering convenience and a broad range of products. As health-conscious consumers seek easy access to high-quality supplements and personalized health solutions, online retail is on the rise.

Food supplements are regulated in Germany with a robust regulatory framework, ensuring product safety and effectiveness. The regulation of these products is overseen by the German Federal Institute for Risk Assessment (BfR) and the Federal Office for Consumer Protection and Food Safety (BVL). It is through regulatory oversight that consumer trust and high standards are maintained.

German consumer health is expected to continue growing as health awareness increases, the country's population ages, and preventive health measures become increasingly popular. The market is expected to grow further with innovations in product formulations, an emphasis on natural and organic ingredients, and continued consumer demand for functional foods and supplements.

Germany's consumer health market is driven by a strong focus on preventive care, a preference for high-quality products, and a dynamic regulatory landscape. As German consumers' attitudes toward health and wellness evolve, this market continues to grow.[10]

8.1.2.3 Market Diversity Creating New Products

Germany's food supplement market experienced significant expansion with the introduction of nearly 8,000 new products, as reported by the German Federal Office of Consumer Protection and Food Safety (BVL). This influx of new products underscores the dynamic nature of the market and the growing consumer interest in health and wellness solutions.

Key market trends and data follow:

1. **Dominance of vitamins and minerals.** Vitamins and minerals have emerged as the most prominent categories in the supplement market. In 2019, these products generated substantial revenue, accounting for EUR 1.11 billion, which represents approximately 50% of the total supplement revenues from pharmacy sales. This dominance highlights the continued consumer reliance on these essential nutrients for overall health and well-being.
2. **Growth in specific supplement categories.** The market has seen notable increases in the sales of supplements targeting various health issues. Products aimed at cardiovascular health, muscle and joint pain relief, bladder and reproductive health, and digestive tract probiotics have all contributed to the market's growth. For example:
   a. **Cardiovascular health.** Supplements designed to support heart health have gained popularity as awareness of cardiovascular diseases increases.
   b. **Muscle and joint pain relief.** There is a growing demand for supplements that address muscle and joint discomfort, particularly among the aging population and active individuals.
   c. **Bladder and reproductive health.** Products targeting bladder function and reproductive health reflect a holistic approach to personal wellness.
   d. **Digestive health.** Probiotics and other digestive health supplements continue to see strong demand, driven by increasing awareness of gut health.
3. **Sales growth rates.** The growth rates for these supplement categories have ranged between 7 and 8%. This consistent growth indicates a stable and expanding market for these products, as consumers increasingly seek targeted solutions for specific health concerns.
4. **Market dynamics.** The German food supplement market is characterized by a high level of innovation and product diversification. Companies are continuously developing new formulations and addressing emerging health trends to meet consumer demands. This is reflected in the diverse range of products now available, from general wellness supplements to specialized formulations targeting specific health conditions.
5. **Consumer preferences.** The strong market performance of vitamins and minerals, along with other health-focused supplements, underscores a broader trend toward preventive care. German consumers are increasingly proactive about their health, seeking supplements that support various aspects of their well-being.

The German food supplement market is really buzzing right now, thanks to a flood of new products and a growing focus on health and wellness. Vitamins and minerals are still leading the charge, bringing in impressive revenue, but there is also a lot of excitement around other types of supplements. This all points to a market that is not only vibrant but also increasingly tuned in to what consumers want when it comes to staying healthy. It is clear that people are more interested than ever in finding effective ways to support their well-being, and the market is responding with a range of options to meet those needs.[11]

8.2 REGULATORY FRAMEWORK FOR FOOD SUPPLEMENTS AND NUTRACEUTICALS IN GERMANY

In Germany, food supplements are treated as part of the broader food category, which means they have to meet some pretty strict rules to ensure they are safe and correctly labeled. Here iss a breakdown of how things work.

8.2.1 Federal Registration

The Federal Office of Consumer Protection and Food Safety (BVL)—think of the BVL as the watchdog for food safety in Germany. Before a food supplement can be sold, it needs to be registered with the BVL. This step ensures that the product is safe and its labeling is accurate.

Website: BVL (https://www.bvl.bund.de/).

What You Need to Know: Under the German Food Supplements Regulation (NemV), manufacturers must notify the BVL of their product. This helps make sure that everything, from ingredient safety to labeling, complies with the rules.

8.2.1.1 Overview of the German Federal States

In Germany, the oversight of official food control is a responsibility entrusted to the federal states. This important task is carried out by the state ministry or the senate administration department in city-states. Local food control and veterinary offices in districts and towns with independent administration diligently conduct inspections and take samples to ensure compliance with regulations at all levels of food production, from producers to processing facilities, storage, transport, and sales. These inspections may occur periodically without prior notice or in response to complaints from consumers or third parties. Establishments with a history of non-compliance are subject to more frequent inspections in the interest of upholding public health and safety. Additionally, the authorities monitor and ensure compliance with laws related to the composition, safety, and labeling of food products, including assessing hygiene standards and overseeing the implementation of internal control systems within businesses. Samples are meticulously tested for harmful microorganisms, plant protection residue, heavy metals, and other undesirable substances, while also ensuring the accuracy of food labeling. In the event of regulatory violations, appropriate measures are taken, such as filing an official complaint and swiftly removing potentially hazardous products from the market.

In the event that consumers have concerns about a specific product's safety or conformity with regulations, they are encouraged to contact their local food control and veterinary offices, which will attentively address and resolve these issues.

8.2.1.2 Federal Ministry of Food and Agriculture

The Federal Ministry of Food and Agriculture (BMEL) in Germany plays a critical role in shaping the country's food, agricultural, and consumer policies. The Ministry's three main objectives are:

1. **Precautionary health protection for consumers.** Ensuring that the food consumed by the public is safe and that health risks are minimized.

2. **Quality assurance.** Guaranteeing that food production meets high-quality standards and that products are reliable and trustworthy.
3. **Sustainable food production.** Promoting agricultural practices that respect environmental sustainability and consider the welfare of animals.

Several key agencies and institutions operate under the BMEL's jurisdiction, each contributing to these goals:

1. **Federal Institute for Risk Assessment (BfR).** Focuses on evaluating risks to human health from food, chemicals, and other products.
2. **Federal Office of Consumer Protection and Food Safety (BVL).** Responsible for ensuring the safety of food and feed, as well as coordinating monitoring programs and acting as a point of contact between the EU and German states.
3. **Federal Office of Plant Varieties (Bundessortenamt).** Manages the registration of plant varieties and maintains the integrity of agricultural biodiversity.
4. **Federal Research Institutes.** Four research institutes under the Ministry conduct scientific studies to inform policy decisions and advance knowledge in agriculture, nutrition, and food safety.

The Ministry also oversees five legally independent institutions under public law. These include entities like the marketing promotion fund for German wines, agricultural and food products, and the Federal Office for Agriculture and Food. The Federal Institute for Risk Assessment also falls under this category, further reinforcing its role in consumer health protection.

8.2.1.3 The Federal Institute for Risk Assessment (BfR)

The Federal Institute for Risk Assessment (BfR) is a crucial entity in Germany dedicated to ensuring food safety and consumer health protection. Operating independently, the BfR provides scientific evaluations and recommendations based on rigorous risk analysis and internationally recognized criteria.

8.2.1.3.1 Key Functions and Responsibilities

Scientific independence. The BfR is renowned for its scientific independence, allowing it to conduct unbiased evaluations on a wide range of issues affecting food safety and consumer health. This independence ensures that the BfR's assessments are objective and not influenced by external pressures.

Risk analysis and proposals. The Institute conducts comprehensive risk analyses to identify and evaluate potential hazards related to food products, chemicals, and other consumer goods. Based on these analyses, the BfR makes well-informed proposals aimed at mitigating risks and improving public health.

Transparency and communication. The BfR prioritizes transparency in its work. It ensures that its evaluation results are accessible and comprehensible to the public, experts, and other stakeholders. This openness helps build trust and allows interested parties to stay informed about the safety and quality of consumer products.

Confidentiality. While the BfR strives to be transparent, it also upholds the confidentiality of sensitive data. Protected information that cannot be disclosed for legal or proprietary reasons is kept secure, ensuring that privacy and intellectual property are maintained.

Public and expert engagement. The BfR engages with both the public and experts by making its evaluations available for review. This includes publishing reports and opinions on various topics related to food safety, health risks, and regulatory measures.

By adhering to these principles, the BfR plays a vital role in protecting consumer health in Germany. Its work not only informs regulatory decisions and public policies but also fosters a safer

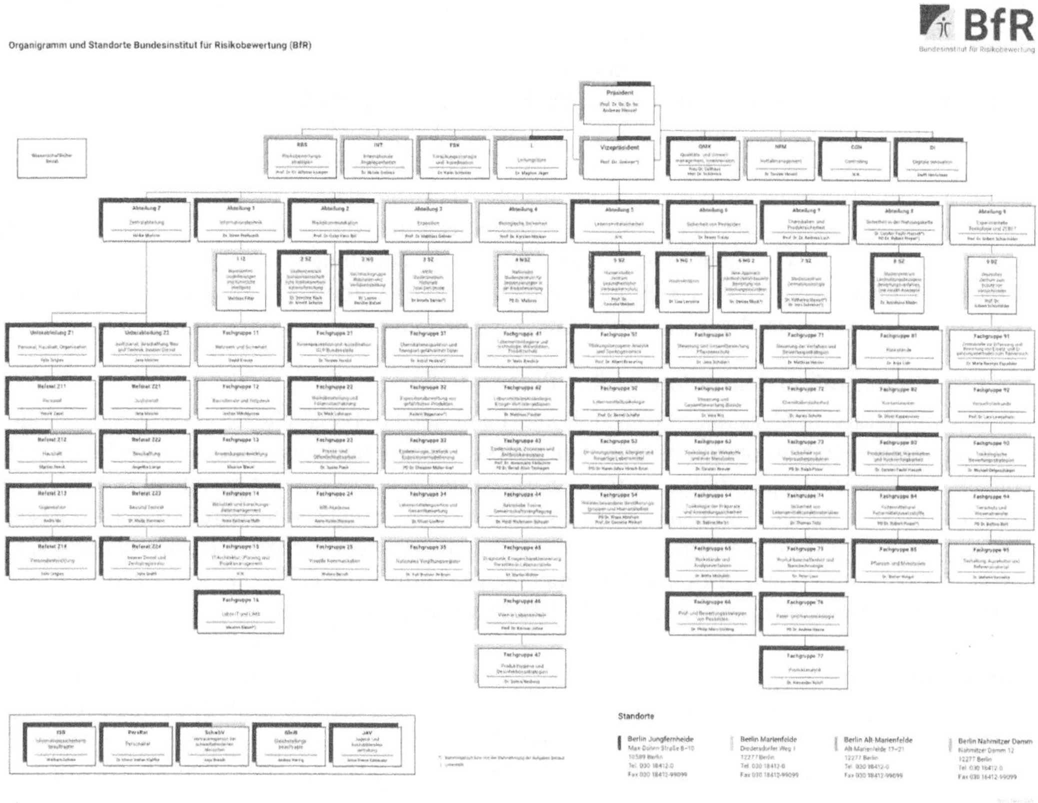

FIGURE 8.4 Organizational chart of the German Federal Institute for Risk Assessment.

environment for consumers by addressing potential risks with scientific precision and transparency. Figure 8.4 indicates the organizational chart of the German Federal Institute for Risk Assessment.

8.2.1.4 Federal Office of Consumer Protection and Food Safety (BVL)

The Federal Office of Consumer Protection and Food Safety (BVL) plays a vital role in ensuring food safety and consumer protection in Germany. Its responsibilities include developing general administrative provisions, coordinating control and monitoring programs, and supporting the federal states by acting as a central coordinating office.

The BVL serves as the administrative hub for various expert working groups. These groups specialize in different areas, including:

- Genetically modified foods
- Wine and spirits
- Diet foods
- Cosmetic products
- Food hygiene and animal-origin products (subgroups include hygiene and microbiology, immunology, serology, molecular biology, meat and meat products, fish and fishery products, and milk and dairy products)

As Germany's national contact point for the European Rapid Alert System for Food and Feed (RASFF), the BVL is responsible for managing notifications about potentially hazardous products. When the federal states identify such products, the BVL reviews and, if necessary, supplements the information before forwarding it to the European Commission. The Commission then shares this information with all EU member states. Conversely, the BVL ensures that rapid alert notifications from other EU member states are communicated to the relevant authorities in Germany.

To prevent food crises and manage risks, the BVL proactively develops measures aimed at minimizing consumer risks and limiting economic damage. They monitor potential problem areas and devise procedures to identify and mitigate risks. An internet-based technical information system has been established to facilitate swift communication between state and federal authorities.

In recent years, significant food crises have often been linked to contaminated feed. The BVL takes on numerous tasks related to feed safety, with the overarching goal of enhancing the safety of the entire food supply chain. Through these efforts, the BVL contributes to protecting consumer health and maintaining public trust in food safety regulations.

8.2.1.5 Organizational Structure of Regulatory Bodies

The Federal Office of Consumer Protection and Food Safety (BVL) plays a critical role in safeguarding food safety and consumer protection in Germany. Operating as an independent higher federal authority under the jurisdiction of the Federal Ministry of Food and Agriculture (BMEL), the BVL has offices in Braunschweig and Berlin, with its management headquartered in Braunschweig. The agency is led by Friedel Cramer, who became the president on December 1, 2019, supported by the vice president, Dr. Gerd Fricke, and a comprehensive organizational structure that includes five departments, a head office, and specialized administrative units.

The BVL's organizational structure is designed to meet a wide range of requirements while ensuring high service standards across its operations. Central to this is the Quality Management System (QMS), based on the internationally recognized DIN EN ISO 9001:2015 standard. This system ensures the legality and quality of the BVL's services, supporting continuous improvement and adherence to strict regulatory requirements. Figure 8.5 indicates the organizational structure of regulatory bodies.

8.2.1.5.1 Administrative Departments

1. **Quality management.** The BVL has established a dedicated staff office to oversee the implementation and maintenance of the QMS. The quality manager, who reports directly to the president, coordinates these efforts, ensuring that the principles of quality management are communicated and upheld throughout the organization.
2. **Internal revision.** The Internal Audit Department conducts independent, risk-based audits to support the BVL's governance and control functions. This department ensures that all operations are legally compliant, economically efficient, and safe. It provides recommendations and advises employees on improving administrative procedures based on its audit findings.
3. **International affairs.** This department manages the BVL's international projects and collaborations with sister authorities worldwide. It serves as the national contact point for several EU initiatives, including the "Better Training for Safer Food (BTSF)" program, and develops international training courses like the BVL's "International Training for Safer Food (ITS Food)" for non-EU countries.

8.2.1.5.1.1 Department Z: Central Services Department Z handles essential administrative functions, including human resources, financial management, legal affairs, real estate, procurement, and IT. This department ensures that the BVL's operations are well-supported and that the necessary infrastructure is in place to carry out its wide-ranging responsibilities effectively.

8.2.1.5.2 Specialized Departments

1. **Department 1: Food, Feed, and Commodities.** This department is instrumental in ensuring food safety across Germany. It supports the BMEL in governmental tasks and coordinates

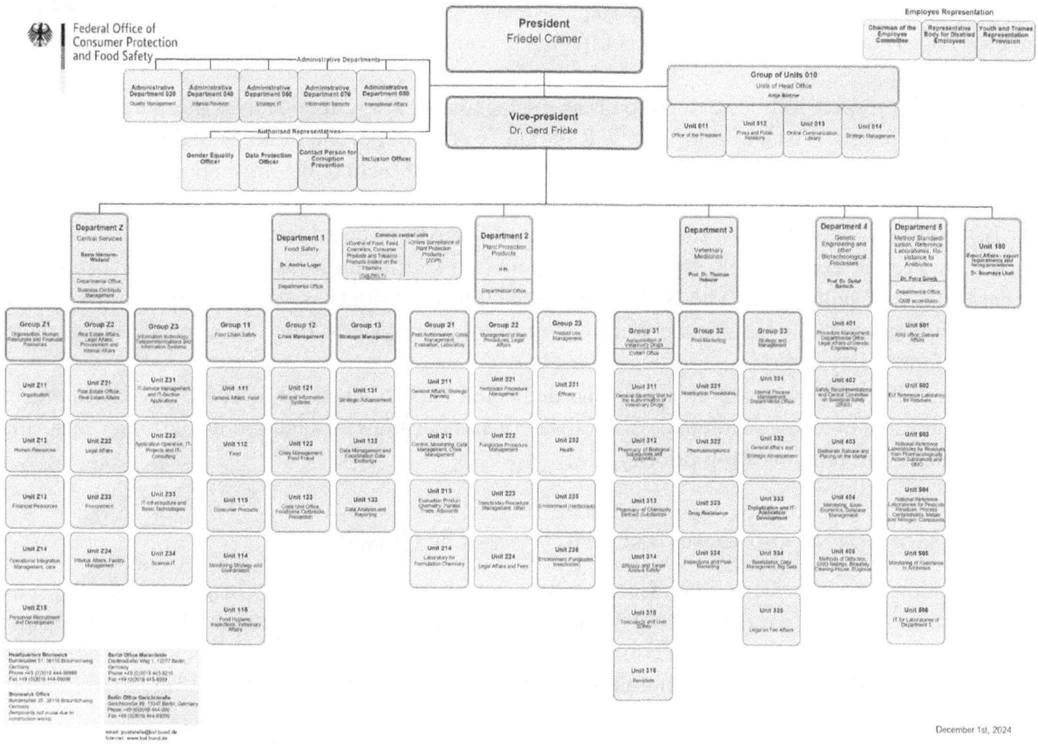

FIGURE 8.5 Organizational chart of the federal office.

the official control of food traded on the internet, ensuring consistent safety standards. It also plays a crucial role in crisis prevention and management, particularly in the food and feed sectors, and acts as the national contact point for the European Rapid Alert System.

2. **Department 2: Plant Protection Products.** This department is responsible for authorizing plant protection products in Germany and serves as the national coordination point for evaluating active substances within the EU. The department's risk management activities are designed to prevent negative impacts on human health, animals, and the environment.
3. **Department 3: Veterinary Drugs.** It handles the authorization of veterinary drugs in Germany and participates in related EU procedures. It also monitors the safety of these drugs, requiring veterinarians and pharmaceutical companies to report any adverse effects.
4. **Department 4: Genetic Engineering.** It focuses on the regulation of genetically modified organisms (GMOs) in Germany. It reviews applications for scientific trials involving GMOs and provides national opinions in the EU's approval process for commercial GMO cultivation.
5. **Department 5: Method Standardization, Reference Laboratories, Resistance to Antibiotics.** It encompasses the BVL's laboratories, including eight national and one European reference laboratory. These labs support official food monitoring efforts by developing and validating sampling and analysis methods. The department also plays a key role in monitoring antibiotic resistance in bacteria that affect animals, contributing to public health by developing measures to counteract resistance.

8.2.1.5.3 The BVL's Role in Food Safety and Consumer Protection

The BVL is deeply involved in the development of general administrative provisions, control, and monitoring programs, and it assists the federal states as a coordinating office. It also serves as the national contact point for the European Rapid Alert System for Food and Feed, ensuring that information about potential health hazards is effectively communicated between the EU and Germany's federal states.

The BVL's responsibilities extend to crisis management, where it conceives measures to prevent crises and minimize risks to consumers. It monitors potential problem areas and develops procedures to identify and address risks, using an internet-based technical information system for efficient communication with other authorities.

In the event of a food crisis, such as those caused by contaminated feed in recent years, the BVL is responsible for coordinating the response, including setting up a task force that involves both federal and state authorities. The BVL's commitment to food safety is further supported by its role as the national contact point for the worldwide information system, InfoSan, contributing to international food safety efforts.

Through its comprehensive structure and dedicated departments, the BVL effectively manages the complex and evolving challenges of food safety and consumer protection in Germany, ensuring that public health remains a top priority. [12]

8.2.2 Monitoring and Compliance Mechanisms in Germany

Food monitoring authorities—each German state has its own team of food inspectors who keep an eye on the supplements available in the market. They ensure that products meet safety standards and that any advertising or claims are truthful. These authorities perform regular checks, sampling, and inspections to ensure compliance. If a product is not up to snuff, they have the power to take action.

8.2.3 General Guidance Documents for Regulatory Compliance for Germany

The Federal Office of Consumer Protection and Food Safety (BVL) provides a range of guidance documents to assist businesses in understanding and complying with regulations. These documents are essential for ensuring that products meet all necessary legal requirements. BVL's general guidance documents offer detailed instructions on various regulatory aspects, such as application procedures, data requirements, and technical specifications. They help businesses navigate the complexities of food and product safety regulations.

For instance, the "Guidance for Applicants" document outlines how to submit dossiers for plant protection products. It provides step-by-step instructions on what information and data are needed to ensure a smooth application process.

8.2.4 Specific Procedural Guidelines

For more targeted needs, the BVL also publishes procedural guidance documents tailored to specific tasks or situations. These documents are designed to provide clear instructions and best practices for particular regulatory processes:

1. **Adverse event reporting.** Guidelines on how to report adverse events related to veterinary medicines, ensuring that any safety issues are communicated effectively.
2. **Risk management.** Documents explaining communication protocols for managing risks associated with food and veterinary products.

BVL guidelines are designed to help businesses understand and comply with German regulations for food supplements, veterinary medicines, and plant protection products. They provide comprehensive instructions and explanations, ensuring that companies can navigate the regulatory landscape effectively.

8.2.5 Federal Ministry of Food and Agriculture (BMEL)

The BMEL, or Bundesministerium für Ernährung und Landwirtschaft, is a key cabinet-level ministry in Germany responsible for overseeing all matters related to food and agriculture. The BMEL's role includes [1] *regulation*, setting policies and regulations for food safety, agriculture, and nutritional standards; [2] *support*, providing resources and support to businesses and consumers in these sectors; and [3] *oversight*, ensuring that food and agricultural products meet high standards of quality and safety.

8.2.5.1 Many Paths, One Goal: Safe Food

The Federal Office of Consumer Protection and Food Safety (BVL) plays a crucial role in ensuring food safety in Germany through various measures:

1. **Authorization and monitoring.** The BVL grants authorizations for products and coordinates monitoring programs in collaboration with the federal states. This ensures that food products meet stringent safety standards before reaching consumers.
2. **Rapid Alert System.** As part of the European Rapid Alert System, the BVL ensures efficient communication between the EU and the German federal states. This system helps in quickly addressing and managing food safety issues across Europe.
3. **European inspections.** The BVL serves as the contact point in Germany for inspections conducted by the Food and Veterinary Office of the European Commission, ensuring that the country adheres to European food safety standards.
4. **Crisis management.** In times of food safety crises, the BVL supports the Federal Ministry of Food, Agriculture, and Consumer Protection (BMEL) by providing expertise and coordinating efforts to protect consumers.

Through these activities, the BVL helps maintain high standards of food safety and consumer protection in Germany, ensuring that any potential risks are swiftly identified and managed.[13]

8.3 Regulatory Guideline for Genetically Modified Food or Feed in Germany

The general requirements include:

1. **Safety first.** Manufacturers and distributors are responsible for making sure their products are safe for consumption.
2. **Clear labels.** Labels must be clear and not misleading. Claims about curing or preventing diseases are not allowed unless officially approved.
3. **Health claims.** If a product claims to reduce the risk of a disease, it needs to be backed by scientific evidence and approved by the relevant authorities, like the European Food Safety Authority (EFSA).

Specific regulations, meanwhile, include thus:

1. **German Food Supplements Regulation (NemV).** This regulation outlines what is required for registering a supplement, including details about labeling and health claims.
2. **EU Regulations.** Germany also follows European rules, such as Regulation (EC) No. 1924/2006, which sets out conditions for making health claims and ensures consistency across EU countries.

8.3.1 Scope of the Regulation in Germany

The Regulation Implementing Union Law on the Provision of Food Information to Consumers (Food Information Implementation Regulation—LMIDV) [14] supplements the provisions of Regulation (EU) No. 1169/2011 of the European Parliament and of the Council of October 25, 2011, on the provision of food information to consumers, amending Regulations (EC) No. 1924/2006 and (EC) No. 1925/2006 of the European Parliament and of the Council, and repealing Commission Directive 87/250/EEC, Council Directive 90/496/EEC, Commission Directive 1999/10/EC, Directive 2000/13/EC of the European Parliament and of the Council, Commission Directives 2002/67/EC and 2008/5/EC, and Commission Regulation (EC) No. 608/2004 (OJ L 304, 22.11.2011, p. 18; L 331, 18.11.2014, p. 41; L 50 of 21.2.2015, p. 48; L 266 of 30.9.2016, p. 7), and the provisions of the European Union legal acts based on them on the following:

1. The labeling of foodstuffs intended for sale to thus:
   a. End consumer within the meaning of Article 3(18) of Regulation (EC) No. 178/2002 of the European Parliament and of the Council of January 28, 2002, laying down the general principles and requirements of food law, establishing the European Food Safety Authority, and laying down procedures in matters of food safety (OJ L 31, 1.2.2002, p. 1), as amended.
   b. Providers of mass catering services within the meaning of Article 2(2)(d) of Regulation (EU) No. 1169/2011.
2. The passing on of information to other food business operators when supplying food which is not intended for sale to final consumers or mass caterers.

This regulation shall not apply to the extent that special legal provisions lay down labeling requirements within the meaning of paragraph 1.

8.3.2 Packaging Requirements of Genetically Modified Food and Feed in the European Union and in Germany

8.3.2.1 Authorization of Genetically Modified Food and Feed

The approval procedure as laid down in Regulation (EC) No. 1829/2003 on genetically modified food and feed requires a safety assessment by the European Food Safety Agency (EFSA).

An authorization can only be granted for genetically modified food and feed which do not have adverse effects on human health, animal health, or the environment, and if consumers or users are not misled. Moreover, the genetically modified food and feed must not differ from the products which it is intended to replace to such an extent that its normal consumption would be nutritionally disadvantageous for consumers or animals. Genetically modified feed must not harm or mislead the consumer by impairing the distinctive features of the animal products.

The safety assessment of food and feed derived from genetically modified organisms (GMOs) is based on the comparison of the GMO in question with its conventional comparator. This comprises the molecular characterization of the genetic modification and the resulting new protein(s), as well as a comparative analysis of the nutrients, and anti-nutrients, as well as endogenous toxins and allergens that are of relevance for the specific organism. This permits a conclusion of whether the genetic modification may have caused any unintended changes. If changes in the content of compounds are found, a decision is to be made depending on the type and scope of differences, which necessitates further testing in order to provide evidence that the GMO-derived products are as safe as their conventional counterparts.

8.3.2.2 Authorization Procedure for Genetically Modified Food and Feed

Authorizations for the placing on the market of genetically modified food and feed are granted by the European Commission. It is the task of EFSA to examine, on the basis of the documents enclosed with the application, whether the product corresponds to the requirements in terms of placing on the market and labeling.

Taking into account the opinion of EFSA, the European Commission prepares a draft decision; for it to be approved, it requires a qualified majority of the member states.

The competent authorities of the EU member states are provided with the applications and can provide EFSA with their opinion on the safety assessment. In accordance with article 1, para. 14, no. 4 of the German law implementing the EC Regulations on Genetic Engineering, the German competent authority for providing EFSA with national comments is the Federal Office of Consumer Protection and Food Safety (BVL). The comments are elaborated in consultation with the Federal Office for Nature Conservation (BfN) and the Robert Koch Institute (RKI). In addition, the BVL requests statements from the German Federal Institute for Risk Assessment (BfR) and the Julius Kühn Institute (JKI).

8.3.3 Labeling Requirements for Genetically Modified Food and Feed

Regulation (EC) No. 1829/2003 requires specific labeling in order to inform consumers about any characteristic or property which renders a genetically modified food or feed different from its conventional counterpart with respect to the following:

- Composition, nutritional value or properties and nutritional effects, intended use
- Implications for the health of certain sections of the population groups or certain species or categories of animals
- Ethical or religious concerns

Moreover, it must also be evident from the label that the food or feed is genetically modified.

Irrespective of the detectability of the genetic modification, Regulation (EC) No. 1829/2003 requires the labeling of all foods and feeds produced from GMO, including those not containing any GMO-derived material, such as refined oils produced from GM oilseeds that are not distinguishable from conventional oil. The words "genetically modified" or "produced from genetically modified [e.g., corn]" shall appear on the label. Labeling is not required to foods and feeds containing GMO-derived material in a proportion no higher than 0.9%, provided that there is evidence that the presence of this material is adventitious or technically unavoidable due to traces of GMOs that might accidentally have entered the supply chain during cultivation, transport, or processing.

8.3.4 Traceability

Regulation (EC) No. 1830/2003 provides a framework for the traceability of GMO-derived food and feed with the objective of facilitating accurate labeling; monitoring the effects on the environment and, where appropriate, on health; and if necessary, withdrawing products. To this end, the Commission has adopted Regulation (EC) No. 65/2004, which lays down a system for the development and assignment of unique identifiers to all GMOs. The parties involved in placing such products on the market must establish systems and standardized procedures to ensure that the data on GMOs can be stored for a period of five years.

In addition to a labeling proposal, applications for authorization must be accompanied by methods for the identification of the GMOs as well as reference material necessary for monitoring compliance with the labeling requirements and traceability provisions, which in Germany is performed by the competent authorities of the federal states.

The testing and evaluation of the detection methods is within the remit of the Joint Research Centre of the EU Commission (JRC), which established the European Network of GMO Laboratories (ENGL) for this purpose back in December 2002. The German Federal Institute for Risk Assessment (BfR), whose staff began with the development of detection methods for genetically modified animals even earlier in 1995, is one of the founding members of this network.

The international standardization of methods for the detection of genetically modified foods and for the quantification of the GMO-derived material is performed by working groups of the German and European standardization institutes DIN and CEN under the lead and with the participation of BfR scientists.

8.3.5 Safety Assessment of Genetically Modified Food and Feed

Up-to-date guidelines for the safety assessment of genetically modified plants and derived food were published by the GMO Panel of the European Food Safety Authority (EFSA) (cf. Guidance Document for the Risk Assessment of Genetically Modified Plants and Derived Food and Feed).

The safety assessment is based on the principle of substantial equivalence, which was described by an OECD working group in 1993 and further developed by the Codex Alimentarius Commission (CAC), the international food standards organization FAO, and the WHO (cf. Codex Principles and Guidelines on Foods Derived from Biotechnology). It is based on a comparison of the genetically modified organism with the non-modified starting organism (more information can be found in the document "The Concept of Substantial Equivalence").

Genetically modified food and feed may only receive marketing authorization if they have demonstrated in comprehensive studies to be just as safe as comparable conventional products. Furthermore, in the case of GMO-containing food and feed, it must also be shown that the GMOs do not constitute a threat to the environment.

The studies required for the testing of food safety encompass the characterization of genetic modification and the resulting new proteins. Furthermore, comparative analyses are to be undertaken of the nutritive, anti-nutritive, toxic, and allergenic ingredients of relevance for the respective organism. This is in order to be able to determine any unintended changes triggered by the genetic modification. If changes are observed, then—depending on the type and scale—a decision must be taken as to which more extensive studies are needed in order to prove the safety of the product.

If new findings are available indicating that a genetically modified food or feed which has been legally placed on the market probably constitutes a serious threat to the health of humans, animals, or the environment, trade with the product may be suspended or subjected to specific requirements.

8.3.6 Labeling of Food and Food Ingredients Produced "without Genetic Engineering"

The Novel Foods Regulation permits member states to issue statutory provisions to inform consumers through labeling that the product concerned is not novel within the intendment of the Regulation or that the novel methods mentioned in the Regulation were not used to produce the food.

Based on this option, national labeling provisions for foods produced without genetic engineering were laid down in the reused national Novel Foods Regulation (NLV, new version of February 14, 2002, BgBl.I—Federal Gazette, p. 123) that entered into force in October 1998.

The application of the label "without genetic engineering" is restricted to foods and food ingredients which neither contain nor were produced from genetically modified organisms (GMOs).

Furthermore, the application of GMO-derived processing aids, such as extraction solvents or enzymes, does not comply with the labeling requirements. This exception also applies to foods from animals fed with GMO-derived feed or feed additives or from animals that have been treated with GMO-derived medicinal products. The latter are only permitted if no comparable conventionally

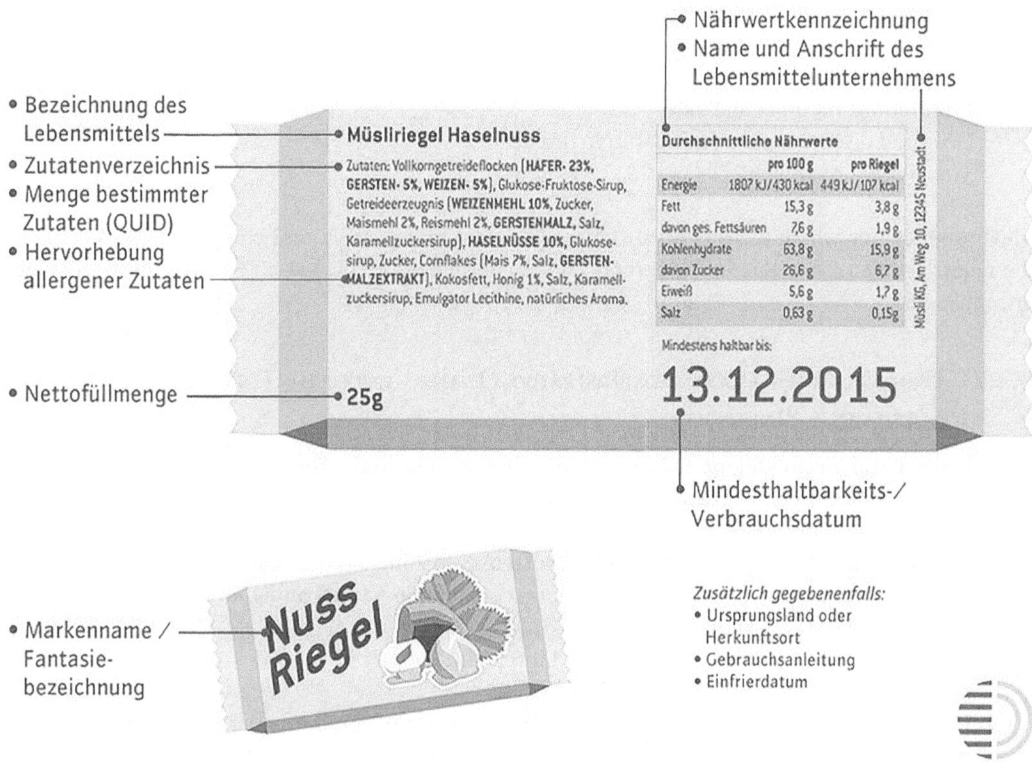

FIGURE 8.6 Overview of food labeling.

produced medicinal products, in terms of therapeutic efficacy, are available. Figure 8.6 depicts an overview of labeling requirements for Germany.

The labeling of food is regulated uniformly throughout Europe in the Food Information Regulation (LMIV)—Regulation (EU) 1169/2011.[15]

8.3.6.1 General Requirements for the Labeling of Foodstuffs When Placed on the Market

1. Foodstuffs shall be labeled in German when placed on the market if labeling is compulsory under the following:
   a. This Regulation
   b. Regulation (EU) No. 1169/2011
   c. The legal acts of the European Union based on Regulation (EU) No. 1169/2011
2. By way of derogation from paragraph 1, foods placed on the market by air may be labeled in another easily understandable language, provided that the information on ingredients and processing aids pursuant to Article 9(1)(c) of Regulation (EU) No. 1169/2011 must always also be provided in German.
3. Special requirements for the labeling of certain prepackaged foods when placed on the market. By way of derogation from Article 16(4) of Regulation (EU) No. 1169/2011, beer supplied as prepackaged food shall, when placed on the market, be labeled with a list of ingredients in accordance with Article 9(1)(b) of Regulation (EU) No. 1169/2011.
4. Special provisions for the labeling of non-prepacked foodstuffs when placing them on the market or supplying them:

a. Extended nutritional labeling
b. Origin labeling for non-prepackaged fresh, chilled, and frozen pork, sheep, goat, and poultry meat

5. Prohibitions on trade and sale.
6. Criminal offences and administrative offences.
7. Transitional provision.

Meat products for which there is no sufficient information available on February 1, 2024, to meet the requirements of Section 4b Paragraph 2 may continue to be placed on the market until stocks are exhausted.

8.3.7 Criminal Offenses and Administrative Offenses under the Food Supplements Regulation

1. **Intentional or negligent misuse of nutrients: criminal offense.** Anyone who intentionally or negligently uses a nutrient contrary to the regulations faces penalties under Section 58, paragraph 1, no. 18, paragraphs 4–6 of the Food and Feed Code.
2. **Improper market placement: criminal offense.** Placing a food supplement on the market contrary to the regulations, including incorrect labeling or packaging, is punishable under Section 59(1), no. 21(a) of the Food and Feed Code.
3. **Administrative offense.** Negligent violations are subject to penalties under Section 60, paragraph 1, no. 2 of the Food and Feed Code.
4. **Failure to report breaches: criminal offense.** Failing to report breaches of the law, or incorrect, incomplete, or late reporting, is considered an administrative offense under Section 60(2), no. 26(a) of the Food and Feed Code.
5. **Unlawful market placement: administrative offense.** Placing a food supplement on the market in violation of specific sections of the Regulation is punishable under Section 60(2), no. 26(b) of the Food and Feed Code. [16].

8.4 VARIOUS FOOD CATEGORIES/PRODUCT REGISTRATIONS AND HEALTH CLAIMS IN GERMANY

1. **BVL Guidelines:** The BVL provides guidelines to help businesses navigate the registration, labeling, and advertising of food supplements. These guidelines are essential for making sure everything is done right.
2. **EU Guidelines:** Germany aligns with broader EU guidelines to maintain consistency. This includes guidelines on health claims and food safety.
3. **Federal Institute for Risk Assessment (BfR):** Offers scientific advice on food safety issues.
4. **European Food Safety Authority (EFSA):** Provides guidance on health claims and conducts safety evaluations across Europe.
5. **European Food and Veterinary Office (EFVO):** Ensures that EU food safety laws are followed.

A NOTE ON BVL GUIDELINES

The term “BVL guideline” can refer to various specific documents issued by the Federal Office of Consumer Protection and Food Safety. The exact meaning depends on the context, but generally, it covers everything from product registration to labeling and health claims.

8.4.1 Nutritional Supplements

The regulation of dietary supplements in the EU is governed by Directive 2002/46/EC, which outlines the approved minerals and vitamins that can be included. *Dietary supplements* are defined as "[a] food product designed to complement the regular diet, providing nutrients or other substances that have a nutritional or physiological effect, either individually or in combination."

These supplements are available in measured doses, such as capsules, tablets, lozenges, powders, and liquid forms. Since these products are considered food under German law, they are subject to the Food and Feed Code (LFGB). The allowable vitamins and minerals are listed in Annex 1 of the 2004 Nutrition Regulation (NemV), and all other ingredients must comply with LFGB and the Novel Food Regulation.

Regulation (EC) No. 1924/2006 governs health claims associated with supplements, ensuring that any promotional statements are approved by the EU. Claims related to disease prevention or treatment are prohibited, just as with other food products. [17]

8.4.2 Supplementary Balanced Diets

These foods are designed for specific medical purposes and are intended for patients who cannot meet their nutritional needs through regular food intake. Regulation (EU) No. 609/2013, 2(2)(g) applies to patients with:

1. Difficulty ingesting, digesting, absorbing, or metabolizing standard foods or certain nutrients
2. Medically necessary nutritional requirements

Such medical foods must be used under medical supervision and be labeled accordingly. They are intended to meet specific nutritional needs, not for the treatment of diseases.

Under DiätV sections 4a and 14b, efficacy must be proven through generally accepted scientific evidence. The EFSA provides guidance for evaluating these foods under Regulation (EU) No. 609/2013, requiring manufacturers to submit relevant data when requested.

Regulation (EU) 2016/128, effective from February 22, 2019, supplements the earlier Regulation 609/2013 regarding composition and labeling requirements, with specific rules for foods aimed at infants taking effect from February 2020. [18, 19]

8.4.3 Supplementary Feed

Supplementary feed is intended to provide additional nutrients for animals and help achieve balanced nutrition. According to the Feed Regulation, these are compound feeds with higher levels of specific substances, such as minerals, than typical complete feeds for the respective animal category. The purpose of these feeds is to complement other animal feed products and support the overall nutritional needs of livestock. [20]

8.4.4 Functional Food

Functional foods, also referred to as nutraceuticals, are ordinary foods enhanced with additional ingredients to promote health benefits. These foods are not concentrated nutrients like dietary supplements but are consumed in typical food forms. They may also be referred to as designer foods or nutraceuticals.

The Bavarian Ministry of Food, Agriculture, and Forestry has outlined several requirements for functional foods:

1. They should be regular foods, not supplements.
2. They should differ from conventional foods, and this modification should be evident in the final product.
3. The modification should offer a clear health benefit beyond basic nutrition.

Although functional foods are not specifically defined by law, they must comply with general food regulations. Products classified as novel foods under the Novel Foods Regulation must undergo an EU approval process. Health claims related to functional foods are regulated under the European labeling directive and must avoid misleading or health-related statements unless they comply with the relevant legislation. [21]

8.4.5 Special Food

Special foods are designed to meet the specific dietary needs of individuals who require a modified diet for health reasons or to achieve particular nutritional or physiological outcomes. These include:

1. Lactose-free products
2. Salt substitutes
3. Low-protein foods
4. Gluten-free products [16]

8.4.5.1 Guidelines for the Registration of Foods Registered as Dietetic Foods

Dietetic foods are developed for particular nutritional purposes, targeted at specific population groups. They must be distinct from regular food products in composition or manufacturing process (§1 paras. 1 and 2 of the Regulation on Dietetic Foods).

The DiätVO provides additional regulations for specific categories of dietetic foods, such as those designed for infants and young children. For products not listed in Annex 8, the notification procedure outlined in §4a DiätVO must be followed, with the Federal Institute for Protection of Consumers and Food Safety (BVL) responsible for registration. The BfR advises the German government on infant nutrition and statutory regulations concerning dietetic foods. [22]

8.4.5.2 Guidelines for the Registration of Foods Registered as Infant Nutrition

Infant nutrition typically progresses through two phases:

1. A diet consisting solely of breast milk or breast milk substitutes during the first six months.
2. A diet where breast milk or substitutes are supplemented with complementary foods, introduced no earlier than 5 months and no later than 7 months.

Since 1999, the Federal Institute for Risk Assessment (BfR) has served as the headquarters for the National Committee for the Promotion of Breastfeeding, which was established in 1994. The committee provides advice to the federal government on breastfeeding promotion and overcoming challenges. In addition, the Research Institute for Child Nutrition (FKE) in Dortmund offers recommendations for the nutrition of infants and young children.

For mothers who are unable to or choose not to breastfeed, breast milk substitutes are available for healthy, full-term infants. According to practical advice from the FKE, infant formula can fully replace breast milk from birth and can be used as the only source of nutrition for the first four to six months. After this period, infant formula may be combined with weaning foods during the first year of life. In Germany and the European Union, there are four types of infant formula:

1. Formula based on cow's milk protein with lactose as the only carbohydrate
2. Formula based on cow's milk protein with additional carbohydrates besides lactose
3. Formula made from soybean protein isolate (Note: In Germany, this formula is lactose-free)
4. Formula with partially hydrolyzed protein (labeled as HA), designed for infants from families with a history of allergies

Follow-on formula, intended for use after the fifth month, can replace breast milk but is not suitable for newborns, as it does not meet the nutritional needs of very young infants. However, there is no strict requirement to switch from infant formula to follow-on formula.

8.4.5.2.1 Complementary Foods

Complementary food refers to any food given to infants in the second half of their first year, alongside breast milk or infant formula/follow-on formula. These foods can either be homemade from basic ingredients or purchased as ready-to-use products. In Germany, commercially available complementary foods are classified as dietetic foods and must comply with the "Dietetic Foods Ordinance" (Diet Regulation). When introducing complementary foods, it is recommended to follow the guidelines provided by the Research Institute for Child Nutrition in Dortmund.

Both infant formulas and complementary foods are subject to stringent regulations regarding their composition, use of additives (with bans on colorants, flavorings, and preservatives), as well as microbiological standards and limits for residues and contaminants. These regulations are consistent across the European Union.

8.4.5.3 Guidelines for the Registration of Foods Registered as Food Supplements

Food supplements are classified as foods that:

- Aim to complement a regular diet
- Contain concentrated sources of nutrients or substances with nutritional or physiological effects (such as vitamins, minerals, trace elements, amino acids, fibers, or plant/herbal extracts), either individually or in combination
- Are available in dose forms, like capsules, powder sachets, and other non-traditional food forms, consumed in small, controlled amounts

To safeguard public health, food supplements must provide clear information on the recommended daily intake. Unlike medicinal products, food supplements are required to be safe, without causing side effects, and are regulated under general food safety standards. However, they are not required to be registered with the Federal Office of Consumer Protection and Food Safety (BVL). Instead, monitoring of food supplements sold in stores and their manufacturing facilities falls under the responsibility of the federal states (Länder).

The regulatory framework for food supplements includes the provisions of the Food and Feed Code (LFGB) and the Food Supplements Ordinance. Currently, no binding maximum levels for the nutrient content of supplements (including vitamins and minerals) have been set at the national or European level, although efforts are underway to establish an EU-wide regulation on maximum permissible levels.

8.4.6 Novel Food

Novel foods are regulated under Regulation (EU) 2015/2283, which replaced earlier regulations (EC 258/97 and EC 1852/2001). According to this regulation, a *novel food* refers to any food that was not consumed to a significant degree within the European Union before May 15, 1997, and falls under one or more of the following categories:

1. Foods with a new or intentionally modified molecular structure not previously used as food in the EU before May 15, 1997
2. Foods made from microorganisms, fungi, or algae
3. Foods derived from mineral sources
4. Foods made from plants or their parts, except when there is a history of safe use in the EU or the plant was produced using traditional breeding methods
5. Foods derived from animals or their parts, except for animals obtained by traditional breeding practices that have a history of safe use in the EU
6. Foods produced from cell or tissue cultures derived from animals, plants, microorganisms, fungi, or algae
7. Foods made through production processes not previously used in the EU, which significantly alter the food's composition, structure, nutritional value, metabolism, or levels of undesirable substances
8. Foods consisting of engineered nanomaterials
9. Vitamins, minerals, and other substances regulated under Directive 2002/46/EC, Regulation (EC) No. 1925/2006, or Regulation (EU) No. 609/2013, where:
   - A new production process not used before May 15, 1997, has been applied
   - The vitamins or minerals contain or consist of engineered nanomaterials
10. Foods used exclusively in food supplements in the EU before May 15, 1997, that are now intended to be used in foods other than supplements

Traditional foods from non-EU countries that were not widely consumed in the EU before 1997, such as baobab fruit pulp, also fall under the novel food regulation.

8.4.6.1 Guidelines for the Registration of Novel Foods

Novel foods authorized and included in the European Union (EU) list can only be placed on the market or used in food products if they meet specific conditions and labeling requirements set forth in the list. This "Union list" of authorized novel foods is continuously updated by the European Commission. A critical requirement for the authorization of novel food is that it must not pose a safety risk to human health, as established by available scientific data.

8.4.6.2 Authorization Process

Two primary processes are intended for placing novel foods on the market:

1. **Authorization process (Article 10 of Novel Food Regulation EU 2015/2283).** This process involves the submission of an application for authorization to the European Commission. Applications must include scientific data to confirm the novel food's safety for human consumption. The process ensures that the novel food is included in the Union list. This list is updated either at the request of the applicant or by the Commission itself.
2. **Notification process for traditional foods from non-EU countries (Articles 14 and 16 of the Novel Food Regulation EU 2015/2283).** Applicants marketing traditional foods from third countries can opt for a notification process instead of the full authorization process. In this process, applicants must demonstrate the food's safety and provide data on its composition. Additionally, they must prove that the food has been part of a conventional diet for at least 25 years, consumed by a significant number of people. Applications are submitted electronically to the EU Commission.

8.4.6.2.1 Involvement of the Federal Institute for Risk Assessment (BfR) and the Federal Office of Consumer Protection and Food Safety (BVL)

The European Food Safety Authority (EFSA) is responsible for risk assessment in accordance with the novel food regulation. In Germany, this responsibility lies with the Federal Office of Consumer Protection and Food Safety (BVL).

Upon request by the BVL, the German Federal Institute for Risk Assessment (BfR) may conduct risk assessments to determine if any objections exist regarding the food's safety. The BfR's assessment is forwarded to the BVL during the notification process.

8.4.6.2.2 Conditions for Authorization and Union List

To include a novel food in the Union list, the following conditions must be met:

- The food must not pose any safety risk to human health based on available scientific evidence.
- Consumers must not be misled by the food's intended use, especially if it is meant to replace another food item, and there must not be significant changes in nutritional value.
- If the novel food replaces another, it must not disadvantage the consumer in terms of nutrition.

The Union list includes all novel foods authorized for use in the EU, specifying their conditions of use, labeling requirements, and other specifications. Once a food is authorized and added to the list, the authorization applies across the entire European Union. The list is continuously updated by the European Commission.

8.4.7 Functional Food

8.4.7.1 Guidelines for the Registration of Functional Foods

Functional foods are increasingly popular due to their potential health benefits, beyond basic nutritional value. These foods aim to influence consumers' health-related physiological parameters in a long-term, targeted manner. Functional foods are typically marketed in conventional food forms, unlike food supplements, and can include products like probiotic yogurt or margarine enriched with plant sterols. The terms *designer foods* and *nutraceuticals* are often used interchangeably with *functional foods*.

8.4.7.2 Legal Status of Functional Foods

There is no formal legal definition for *functional foods* in German or European Union law. These foods can be sold as:

- General consumption foods, such as probiotic yogurt
- Foods for special dietary uses, such as margarine enriched with plant sterols

However, if functional foods fall under the category of novel foods, in line with the novel foods regulation, or if they contain novel ingredients, they must undergo a European approval process before entering the market.

8.4.7.3 Advertising and Labeling

The claims made about functional foods are regulated in several ways:

- For foods for general consumption, the relevant regulations include the German Food and Feed Code (LMBG) and laws on food or nutrient labeling.
- For special dietary uses, the provisions of the diet regulation apply.

Functional food claims may present legal challenges due to bans on:

- Misleading advertising (§17 LMBG)
- Health-related advertising (§18 LMBG)

There are currently no plans to establish a statutory definition for functional foods as a separate group within the European Union. As a result, advertising statements relating to their function must comply with the European labeling directive.

The Code of Conduct for Compliance with Article 140 of Regulation (EU) 2017/625 is a document that offers practical guidance on managing reports of violations under Article 140 of Regulation (EU) 2017/625 (EU Control Regulation, OCR). It details procedures for handling reports and the treatment of whistleblowers. [23]

8.5 A USER-FRIENDLY APP FOR MOBILE DEVICES LIKE TABLETS AND SMARTPHONES

After its comprehensive relaunch, the joint federal and state consumer protection portal lebensmittelwarnung.de is launching with a new design and new functions. Also included is a user-friendly app for mobile devices, such as tablets and smartphones. [25]

The Federal Office of Consumer Protection and Food Safety (BVL), together with the Federal Ministry of Food and Agriculture (Federal Ministry of Food, Agriculture, and Consumer Protection), presented the consumer-friendly revised version of the portal on June 18, 2024.

"In the new app, consumers can be informed about new public announcements anytime and anywhere automatically via push function," says Friedel Cramer, president of the BVL. If you want to share a message with friends, you can use the social networks Facebook and X, email, or messenger services via the app. "The relaunch of the website and the launch of the app are a significant contribution to strengthening food safety in Germany and thus to preventive health protection for the population."

The app of the portal lebensmittelwarnung.de can be found in Google Play and the Apple App Store.

8.6 THE CODEX ALIMENTARIUS

The Codex Alimentarius is a collection of international food standards. It is based on the assumptions and decisions of expert bodies (committees) of the Codex Alimentarius Commission of the Food and Agriculture Organization (FAO) and the World Health Organization (WHO) of the United Nations.

The standards of the Codex Alimentarius ensure consumer health protection and fair trade practices for food. They serve the international harmonization of food law in the member states and are an important reference in trade disputes of the World Trade Organization (WTO). The Codex Alimentarius Commission currently has 188 members, as well as the European Union. Germany is also a member and is actively involved in standard-setting work in several committees on various topics in the Codex Alimentarius. [26]

8.6.1 The Codex Committee on Nutrition and Foods for Special Dietary Uses (CCNFSDU)

The Codex Committee on Nutrition and Foods for Special Dietary Uses (CCNFSDU), founded in 1960, is responsible for nutrition issues and develops standards in the field of nutrition and foods for special nutritional uses. These include, for example, dietetic foods, such as food for infants and young children, but also fortified foods and food supplements.

TABLE 8.1
Nutrition Claims and Conditions in Germany for Functional Food

Sr. No.	Claim on Label	Limit	Remark
1	Low energy	≤40 kcal (170 kJ)/100 g (solids) or ≤20 kcal (80 kJ)/100 mL (liquids)	Tabletop sweeteners: ≤4 kcal (17 kJ)/portion
2	Energy reduced	≥30% energy reduction	Must indicate the characteristic(s) responsible for reduction
3	Energy-free	≤4 kcal (17 kJ)/100 mL	Tabletop sweeteners: ≤0.4 kcal (1.7 kJ)/portion
4	Low-fat	≤3 g fat/100 g (solids) or ≤1.5 g fat/100 mL (liquids)	Semi-skimmed milk: ≤1.8 g fat/100 mL
5	Fat-free	≤0.5 g fat/100 g or 100 mL	"X% fat-free" claims prohibited
6	Low saturated fat	≤1.5 g saturated + trans fats/100 g (solids) or ≤0.75 g/100 mL (liquids)	Sum of saturated + trans fats should provide ≤10% of energy
7	Saturated fat–free	≤0.1 g saturated + trans fats/100 g or 100 mL	
8	Low-sugar	≤5 g sugar/100 g (solids) or ≤2.5 g/100 mL (liquids)	
9	Sugar-free	≤0.5 g sugar/100 g or 100 mL	
10	No added sugar	No added mono- or disaccharides or sweetening substances	Label must state: "CONTAINS NATURALLY OCCURRING SUGARS" if naturally present sugars exist
11	Low-sodium/salt	≤0.12 g sodium (or equivalent salt)/100 g or 100 mL	Waters: ≤2 mg sodium/100 mL
12	Very low sodium/salt	≤0.04 g sodium (or equivalent salt)/100 g or 100 mL	Not applicable to natural mineral waters
13	Sodium-free or salt-free	≤0.005 g sodium (or equivalent salt)/100 g or 100 mL	Nil
14	Source of fiber	≥3 g fiber/100 g or ≥1.5 g/100 kcal	Nil
15	High-fiber	≥6 g fiber/100 g or ≥3 g/100 kcal	Nil
16	Source of protein	>12% of energy value from protein	Nil
17	High-protein	≥20% of energy value from protein	Nil
18	Source of vitamins and minerals	Meets "significant amount" criteria of Directive 90/496/EEC or equivalent	Nil
19	High in vitamins and minerals	Contains ≥ twice the value of "source of" vitamin/mineral	Nil
20	Contains [nutrient]	Must comply with applicable provisions and meet "source of" conditions for vitamins/minerals	Nil
21	Increased [nutrient]	≥30% increase in content compared to a similar product	Applies to nutrients other than vitamins and minerals
22	Reduced [nutrient]	≥30% reduction in content compared to a similar product	Sodium/salt reduction ≥25%; micronutrients ≥10%
23	Light/lite	Must meet "reduced" conditions	Must indicate the characteristics that make the food "light" or "lite"
24	Naturally/natural	Food naturally meets conditions for a nutritional claim	Can be used as a prefix to the claim

Source: [24].

8.6.2 Other Committees of the Codex Alimentarius

BVL experts participate as members of the German delegation and in working groups in the work of many of the 14 other technical committees of the Codex Alimentarius. These include thus:

- Codex Committee on Contaminants in Food (CCCF)
- Codex Committee on Food Hygiene (CCFH)
- Codex Committee on Food Additives (CCFA)
- Codex Committee on Methods of Analysis and Sampling (CCMAS)
- Codex Committee on Pesticide Residues (CCPR)
- Codex Committee on Veterinary Medicinal Products Residues (CCRVDF)
- Codex Committee on Food Import and Export Control and Certification Systems (CCFICS)

8.7 FUTURE OPPORTUNITIES IN THE GERMAN NUTRACEUTICAL MARKET

With a compound annual growth rate (CAGR) of 6.55% from 2023 to 2033, the size of the Germany nutritional supplements market is expected to surpass USD 9.66 billion. A growing dietary supplement research and development is transforming the industry. The growing demand for natural and organic goods in Germany is expected to propel the growth of the nutritional supplement market in that nation. Because customers are more aware of the ingredients and manufacturing procedures of the products they use, they choose natural and organic alternatives. Given the increasing need for nutraceutical components, it is expected that the market for nutritional supplements would develop. Germany has been a major contributor to the expansion of the nutritional supplement sector in Europe. The region's rising level of disposable income and the increased level of public awareness are two factors that should help the market thrive. Furthermore, market development is likely to be fueled by government initiatives and regulations in Germany that encourage youth participation in sports and healthy lives. Healthy dietary programs for school-age children are part of the German federal government's National Action Plan "IN FORM," which aims to improve children's and young people's eating habits and nutrition [27].

The growth of the vitamins and minerals market has been significantly influenced by Germany's robust pharmaceutical sector. The nation's well-established regulatory system guarantees the quality and safety of its products. Customers' faith and confidence in the market have increased as a result of this. Moreover, vitamins and minerals are easily accessible in Germany because of the country's extensive network of pharmacies and health food stores. These retail outlets frequently provide expert counsel and direction, which help sustain the market's expansion. [28]

8.8 CONCLUSION

The German nutraceutical market has evolved into one of the most significant in Europe, driven by increasing consumer awareness about health and wellness, an aging population, and advancements in biotechnology and nutrition science. This chapter explored the key trends shaping the market, future opportunities for growth, and the intricate regulatory landscape that businesses must navigate.

Germany has witnessed a surge in demand for nutraceuticals, particularly in functional foods, dietary supplements, and fortified beverages. Health-conscious consumers are prioritizing products that offer preventive health benefits, especially those that support immune function, cognitive health, and overall well-being. Plant-based nutraceuticals, probiotics, omega-3 fatty acids, and vitamins have been standout categories, reflecting global health trends. Additionally, personalized nutrition, fueled by advancements in genetic testing and wearable health tech, is rapidly gaining ground as consumers seek tailored solutions to meet their unique health needs.

The future of the nutraceutical market in Germany is promising, with several growth opportunities on the horizon. The rising popularity of personalized nutrition opens doors for companies to innovate in product development and distribution. Furthermore, the increasing demand for sustainable and organic products will likely encourage more research and innovation in eco-friendly nutraceuticals. The aging population offers a substantial market for products addressing age-related health concerns, such as joint health, cardiovascular support, and cognitive enhancement. Digital health solutions that combine nutraceuticals with data-driven insights also present a potential growth area, as consumers embrace technology in their wellness journeys.

Navigating Germany's regulatory landscape is crucial for businesses aiming to succeed in this market. The European Food Safety Authority (EFSA) plays a pivotal role in overseeing the approval of health claims on nutraceutical products. Stringent regulations ensure that only scientifically backed claims are permitted, thus protecting consumers from misleading information. Additionally, companies must adhere to European Union (EU) regulations concerning food safety, labeling, and product registration. Although these regulatory hurdles can be challenging, they also serve to maintain high product standards and foster consumer trust in nutraceuticals.

In conclusion, while the German nutraceutical market presents a robust growth trajectory with numerous opportunities for innovation and expansion, companies must remain vigilant in addressing the evolving regulatory requirements. Success in this dynamic industry will depend on a company's ability to adapt to consumer demands for personalized, sustainable, and scientifically supported products while ensuring compliance with Germany's and the EU's stringent regulatory standards.

REFERENCES

[1] Admin. Optibiotix. 2022. OptiBiotix launches highly effective low dose prebiotic to stimulate targeted gut microbial growth. [cited 2024 Sep 7]. Available from: https://optibiotix.com/optibiotix-launches-highly-effective-low-dose-prebiotic-to-stimulate-targeted-gut-microbial-growth/

[2] ltd R and M. Europe Nutraceutical—Market Share Analysis, Industry Trends & Statistics, Growth Forecasts 2018–2029 [Internet]. [cited 2024 Sep 6]. Available from: https://www.researchandmarkets.com/reports/4775076/europe-nutraceutical-market-share-analysis

[3] Croci F. Vitamins & Dietary Supplements Market Trends Overview.

[4] Nutraceutical in Europe Market—Share & Revenue [Internet]. [cited 2024 Sep 6]. Available from: https://www.mordorintelligence.com/industry-reports/europe-nutraceutical-market

[5] Nutraceuticals Market Size, Industry Share & Forecast—2033 [Internet]. [cited 2024 Sep 7]. Available from: https://www.futuremarketinsights.com/reports/nutraceuticals-market

[6] Nutraceuticals World. Considerations for the Nutraceuticals Industry in 2024 and beyond. [Internet] [cited 2024 Sep 7]. Available from: https://www.nutraceuticalsworld.com/issues/2023-12-01/view_editorials/considerations-for-the-nutraceuticals-industry-in-2024-and-beyond/

[7] Lucas A. CNBC. 2022. PepsiCo to launch hemp seed-infused drink under Rockstar Energy [cited 2024 Sep 7]. Available from: https://www.cnbc.com/2022/02/01/pepsico-to-launch-hemp-seed-infused-drink-under-rockstar-energy.html

[8] Alpro This is not M*lk Whole | Alpro [Internet]. [cited 2024 Sep 7]. Available from: https://www.alpro.com/uk/products/not-mlk/not-mlk/not-mlk-whole

[9] Spherical Insights. Germany Nutritional Supplements Market Forecasts to 2033 [Internet]. [cited 2024 Sep 7]. Available from: https://www.sphericalinsights.com/reports/germany-nutritional-supplements-market

[10] The Food Supplements Market in Germany.

[11] Bundesportal | Notification of food supplements for sale in Germany [Internet]. [cited 2024 Sep 6]. Available from: https://verwaltung.bund.de/leistungsverzeichnis/EN/leistung/99118031261000

[12] BVL—Structure [Internet]. [cited 2024 Sep 6]. Available from: https://www.bvl.bund.de/EN/FederalOffice/01_Structure/Structure_node.html

[13] BVL—Food [Internet]. [cited 2024 Sep 6]. Available from: https://www.bvl.bund.de/EN/Tasks/01_Food/food_node.html;jsessionid=88A0291C833DE9C1EF944097E85579D9.internet991

[14] LMIDV—nichtamtliches Inhaltsverzeichnis [Internet]. [cited 2024 Sep 6]. Available from: https://www.gesetze-im-internet.de/lmidv/

[15] BVL—Kennzeichnung von Lebensmitteln [Internet]. [cited 2024 Sep 6]. Available from: https://www.bvl.bund.de/DE/Arbeitsbereiche/01_Lebensmittel/03_Verbraucher/02_KennzeichnungLM/lm_kennzeichnung_lebensmittel_node.html
[16] Laws / Norms / Guidelines Fakten NEM—NEM e. V.—Verband—Nahrungsergänzungsmittel [Internet]. [cited 2024 Sep 6]. Available from: https://nem-ev.de/laws-norms-guidelines-fakten-nem/
[17] BVL—Startseite [Internet]. [cited 2024 Sep 6]. Available from: https://www.bvl.bund.de/DE/Home/home_node.html
[18] BfArM—Home [Internet]. [cited 2024 Sep 6]. Available from: https://www.bfarm.de/DE/Home/_node.html
[19] BVMed. Bundesverband Medizintechnologie. [Internet] [cited 2024 Sep 6]. Available from: https://www.bvmed.de/startseite
[20] BVL—Official Collection of Methods of Analysis. [Internet] [cited 2024 Sep 6]. Available from: https://www.bvl.bund.de/EN/Tasks/09_Laboratories/01_Tasks/04_official_collection_methods_analysis/official_collection_methods_analysis_node.html
[21] Regulation (EC) No 258/97 of the European Parliament and of the Council of 27 January 1997 concerning novel foods and novel food ingredients. OJ L Jan 27, 1997. [Internet] Available from: http://data.europa.eu/eli/reg/1997/258/oj/eng
[22] Health assessment of dietetic foods—BfR [Internet]. [cited 2024 Sep 6]. Available from: https://www.bfr.bund.de/en/health_assessment_of_dietetic_foods–731.html
[23] BVL—HoA—Heads of Food Safety Agencies—Code of Conduct for compliance with Article 140 of Regulation (EU) 2017/625 (available in English only) [Internet]. [cited 2024 Sep 6]. Available from: https://www.bvl.bund.de/SharedDocs/Downloads/01_Lebensmittel/Internationales/HoA_WG_FF_Code_of_practice_Art_140-de.html?nn=19720086
[24] Regulation—1924/2006—EN—EUR-Lex [Internet]. [cited 2024 Sep 6]. Available from: https://eur-lex.europa.eu/eli/reg/2006/1924/oj
[25] BVL—Press releases—Food warnings via push message on your smartphone [Internet]. [cited 2024 Sep 6]. Available from: https://www.bvl.bund.de/SharedDocs/Pressemitteilungen/01_lebensmittel/2024/07_LMWarnung-Relaunch.html
[26] BVL—CCNFSDU—Codex Alimentarius-Komitee für Ernährung und Lebensmittel für besondere Ernährungszwecke [Internet]. [cited 2024 Sep 6]. Available from: https://www.bvl.bund.de/DE/Aufgaben/08_Internationales/02_Gremien/CCNFSDU/CCNFSDU_node.html
[27] Nutraceuticals | Free Full-Text | Design and Development of Novel Nutraceuticals: Current Trends and Methodologies [Internet]. [cited 2024 Sep 7]. Available from: https://www.mdpi.com/1661–3821/2/2/6
[28. Augusti, P.R.; Conterato, G.M.M.; Denardin, C.C.; Prazeres, I.D.; Serra, A.T.; Bronze, M.R.; Emanuelli, T. Bioactivity, Bioavailability, and Gut Microbiota Transformations of Dietary Phenolic Compounds: Implications for COVID–19. *J. Nutr. Biochem.* 2021, 97, 108787.

9 Regulatory Landscape of Nutraceuticals in the European Union

Compliance and Standards

Arindam Chatterjee, Mayank Bansal, Rakesh Kumar Gupta, Devid Chutia, Deepak Sharma, and Madan Mohan Gupta

9.1 IMPORTANCE OF REGULATION IN THE NUTRACEUTICAL INDUSTRY

Nutraceuticals consist of a range of substances, such as nutrients, bioactive compounds, extracts, herbs, etc. The quantities added vary from manufacturer to manufacturer. Therefore, standards should exist to maintain quality and legitimacy to the claims made. [1] As per the European Food Safety Authority, it is a basic right of the consumer to know from what and how his/her food is sourced, processed, and packaged. The European Union's food policy or regulation aims for the promotion and protection of the customer. [2]

9.1.1 Consumer Safety

Over the years, across the globe, people have started recognizing nutraceuticals as an alternative for the management and prevention of chronic diseases. The lack of rules and regulation generates confusion in consumer leading to misuse of the nutraceuticals. Well-defined rules and regulations protect consumers from false and misleading claims. [3] The EU has adopted Regulation (EU) No. 1047/2012 on the nutritional and health claims made on foods. A *nutritional claim* is defined as "any claim which states, suggests or implies that a food has particular beneficial nutritional properties due to the presence, absence, increased or reduced levels of a particular nutrient or other substance," while a *health claim* is "any claim that states, suggests or implies that a relationship exists between a food category, a food or one of its constituents and health." [4] Different countries have different regulations for nutraceuticals, but the ultimate goal is of customer safety.

9.1.2 Industry Integrity

As per the studies, consumers prefer nutraceuticals that have scientific evidence for the effectiveness and safety.[5] Nutraceuticals have displayed potential benefits for anticancer, cardiovascular diseases, diabetes, pathological disorders, etc. For maintaining industrial integrity, regulations can ensure adherence to Good Manufacturing Practices by industry members, monitoring by national authorities, protocol development, and harmonization of international regulatory regimes. [6]

9.1.3 Market Access and Global Trade Background

The global nutraceutical market was valued at over $423 billion in 2022 and is expected to reach $693.06 billion by 2028. In 2024, the European nutraceutical market size is estimated at $83.88

DOI: 10.1201/9781003465041-9

billion and will reach $106.59 billion by 2029. [7] Regulatory approval demonstrates that a product meets the necessary safety, quality, and labeling standards, allowing it to be sold legally in different jurisdictions. Non-compliance can result in market barriers, trade restrictions, and legal penalties. However, products may be placed in different categories in different countries, which results in variations in the regulatory requirements and safety assessments. So the manufacturer has to follow and stick to the guidelines of each regions where they intend to market their product. Recently, efforts have been taken by different regulatory bodies for the harmonization of regulations, thereby promoting trade. [8]

9.1.4 Objectives of the Study

Regulation in the nutraceutical industry is decisive for ensuring consumer safety, product quality, and efficacy. It establishes standards for manufacturing practices, ingredient quality, and labeling accuracy. Regulations prevent fraud and misrepresentation, promote market access and trade, and encourage evidence-based practice. They also support public health initiatives by monitoring products and contributing to disease prevention strategies. Overall, regulatory frameworks play a vital role in safeguarding public health, maintaining consumer trust, and promoting responsible industry practices.

9.2 REGULATORY FRAMEWORK IN THE EUROPEAN UNION

The utmost priority in the European Union (EU) is to maintain food safety, which is accomplished by means of a thorough regulatory structure. This structure has been meticulously crafted to ensure that food items distributed within the EU strictly comply with rigorous safety criteria, thereby safeguarding the health and welfare of consumers throughout member nations.

At the core of this system of regulations lies the European Food Safety Authority (EFSA), a separate organization with the responsibility of offering scientific counsel on the risks associated with food safety. Through conducting risk assessments, analyzing emerging dangers, and providing guidance to policymakers, the EFSA guarantees that decisions concerning food safety are grounded in the most up-to-date scientific information. [9]

Regulation (EC) No. 178/2002, commonly referred to as the General Food Law, serves as the foundation of food safety regulation within the European Union at the legislative level. This pivotal regulation outlines the fundamental principles and criteria that govern food law, while also establishing a framework for conducting risk assessment, management, and communication. Additionally, it clearly defines the roles and obligations of various stakeholders, including food business operators, competent authorities, and the European Food Safety Authority (EFSA), in upholding and safeguarding food safety across the entire supply chain. [10]

Various aspects of food production, processing, and distribution are subject to specific regulations. One such regulation is Regulation (EC) No. 852/2004, which establishes hygiene guidelines for foodstuffs. It requires food establishments to implement measures that prevent contamination and ensure cleanliness. Additionally, Regulation (EC) No. 183/2005 governs the labeling and traceability of genetically modified organisms (GMOs) in food and feed. This regulation aims to promote transparency and allow consumers to make informed choices.

It is crucial to note that the European Union has implemented a strong framework of official controls to validate adherence to food safety standards. Member states are tasked with carrying out inspections, collecting samples, and conducting laboratory tests in order to ensure that food establishments comply with regulatory obligations. Operators who fail to meet these requirements are subject to various penalties, such as fines, closure of their premises, or potential criminal charges. These consequences serve as a powerful incentive for businesses to prioritize and uphold food safety regulations. [11]

The EU also promotes global collaboration in the realm of food safety by forming alliances with various countries and organizations. Through the exchange of knowledge, implementation of effective strategies, and utilization of expertise, the EU actively contributes to worldwide initiatives aimed at protecting both food security and public health.

9.2.1 European Food Safety Authority

The European Food Safety Authority (EFSA) was established under Joint Regulation No. 178/2002 EC of the European Parliament to provide scientific advice and risk assessments related to foods and food supplements. EFSA's main tasks include the assessment of risks associated with the food chain, including food additives, contaminants, pesticides, genetically modified organisms (GMOs), and novel foods. Through rigorous scientific assessments, EFSA identifies potential hazards and assesses their effects on human health, animal health, and the environment.

In addition, EFSA plays a vital role in risk communication by informing policymakers, stakeholders, and the public on emerging food safety issues, scientific evidence, and recommendations. By promoting transparency and evidence-based decision-making, EFSA helps strengthen consumer confidence in the safety and quality of the EU's food supply. [12–14]

9.2.2 New Food Regulations

In the ever-evolving field of culinary products, the regulation of novel foods is a core part of ensuring food safety and innovation. The concept of novel foods includes products that were not traditionally consumed in large quantities in the European Union (EU) before May 15, 1997. The purpose of this regulation is to protect public health while promoting innovation and facilitating market access for new food products.

The European Commission regulates novel foods under Regulation (EU) 2015/2283, which came into force in January 2018. The regulation simplifies the approval process for novel foods and ensures a harmonized approach across all EU member states. New foods undergo rigorous scientific assessment by the European Food Safety Authority (EFSA) before being placed on the market to ensure they are safe and meet regulatory standards. [15]

The regulation describes different categories of novel foods, including newly developed foods, foods derived from new production processes, and foods traditionally consumed outside the EU. It provides a transparent framework for submitting novel food authorization applications and includes detailed information on product ingredients, manufacturing processes, intended use, and safety data.

Additionally, the regulation emphasizes the importance of appropriate labeling of novel foods to inform consumers of their novelty and any specific instructions for use or precautions. This ensures consumer awareness and promotes responsible consumption while maintaining transparency and trust in the food market. [16]

By facilitating the entry of innovative foods into the market, novel food regulation promotes research and development in the food industry, promotes economic growth, and expands consumer choice. However, it also highlights the critical importance of safety and risk assessment in protecting public health and welfare. [17–19]

9.2.3 Regulation of Health Claims

The regulation of health claims plays a vital role in preventing misleading information in the food industry and ensuring consumer protection. A *health claim* is any claim that shows a link between a food or its ingredients and a health benefit. These claims influence consumer perceptions and purchasing decisions and therefore require effective regulation.

In the EU, food health claims are regulated by Regulation (EC) No. 1924/2006, which sets out harmonized rules for the use of nutrition and health claims. The purpose of this regulation is to maintain a high level of consumer protection while promoting innovation and fair competition in the food industry. [20]

Under the regulation, only authorized health claims listed in the EU Register of Nutrition and Health Claims can be used on foods. These statements undergo rigorous scientific evaluation by the European Food Safety Authority (EFSA) to ensure that they are supported by scientific evidence and provide consumers with accurate information. [21]

Health claims are divided into general health claims and specific health claims, and each health claim has different approval processes and standards. *General health claims* refer to statements about the role of a nutrient or substance in maintaining health, while *specific health claims* refer to the relationship between a food and a specific health role.

In addition, the regulation regulates the presentation of health claims and sets strict standards to prevent ambiguity or exaggeration. The information must be clear, accurate, and based on the latest scientific evidence so consumers can make informed decisions about their diets.

Effective enforcement mechanisms are critical to ensuring compliance with health claims rules, thereby safeguarding consumer trust and the integrity of the food market. Regulators monitor the use of health claims and take appropriate steps to address any infringement or misleading behavior by food manufacturers. [22, 23]

In essence, the regulation of health claims aims to strike a balance between promoting innovation and protecting public health by regulating the use of food claims. By providing consumers with accurate and reliable information, they can make informed decisions that contribute to their overall health and well-being.

9.2.4 Labeling Requirements

Labeling requirements are a cornerstone of food regulation and serve a variety of purposes, from ensuring consumer safety and information transparency to facilitating market access and trade. Proper labeling enables consumers to make informed decisions about the products they purchase and consume, while also assisting regulatory agencies in enforcing compliance with relevant standards and regulations.

In the European Union, food labeling is regulated by Regulation (EU) No. 1169/2011 on the provision of information about food to consumers. This comprehensive legislation harmonizes labeling requirements across EU member states and promotes consistency and clarity in the presentation of information on food packaging. [24]

The regulation requires basic information to be included on food labels, including food name, ingredient list, allergen information, nutritional statement, net content and expiry date. This information enables consumers to identify and evaluate the characteristics of a food, including its ingredients, nutritional content, and potential allergens or hazards.

Allergen labeling is especially important because it can help people with food allergies or intolerances avoid substances that may cause adverse reactions. The regulations establish a list of allergenic ingredients that must be prominently displayed on food labels to ensure the safety of vulnerable consumer groups. [25]

Additionally, the regulation requires clear and legible labels as well as mandatory font size requirements to improve consumer readability and accessibility. In addition, misleading or deceptive practices, such as false claims or misleading images, are prohibited to prevent consumer confusion or misunderstanding.

Country of origin labeling is another important aspect of food labeling requirements and provides consumers with information about the geographical origin of food. This transparency enhances consumer trust and supports informed purchasing decisions, especially in the context of global food supply chains.

Additionally, labeling requirements extend to non-prepackaged foods sold in bulk or in bulk to ensure that basic information is provided to consumers at the point of sale. Retailers have a responsibility to provide this information clearly and prominently so that consumers can make informed decisions when purchasing unpackaged food.

Overall, labeling requirements promote consumer trust, promote fair competition, and ensure food safety and market information transparency.

Overall, labeling requirements play a vital role in promoting consumer trust, facilitating fair competition, and ensuring food safety and information transparency in the marketplace. By providing clear and accurate information, food labels empower consumers to make choices that align with their dietary preferences, health needs, and ethical considerations. [25, 26]

9.2.5 Maximum Levels of Contaminants

Contaminants in food pose a significant risk to public health and require strict regulation to ensure that food meets established safety standards. Maximum contaminant levels (MLCs) are designed to protect consumers from potential health hazards associated with harmful substances in food, including heavy metals, mycotoxins, pesticides, and environmental contaminants.

In the EU, maximum levels of contaminants are set and regulated by Regulation (EC) No. 1881/2006, which sets limits for various contaminants in food. The limits are based on a scientific risk assessment by the European Food Safety Authority (EFSA), which takes into account factors such as toxicological data, dietary exposure, and possible health effects. [27]

The regulation sets maximum levels of certain contaminants in various food categories, ensuring that products intended for human consumption meet safety standards. For example, maximum levels could be set for contaminants such as aflatoxins in nuts and dried fruits, lead in some fish, or cadmium in cocoa and chocolate products.

Implement a regular monitoring and surveillance program to assess compliance with maximum levels of contaminants and detect any non-compliance or contamination. Regulators sample and test food to verify its safety and compliance with established standards and take appropriate enforcement action in the event of non-compliance. [28]

In addition, the regulation provides for risk management and mitigation measures, including the implementation of Good Agricultural Practices (GAP) and Good Manufacturing Practices (GMP), to prevent or reduce contamination at different stages of the food production chain. These include practices such as soil and water management, pest control, and quality assurance to minimize the presence of contaminants in food.

Effective communication and transparency are critical to ensuring consumer confidence in food safety. The regulation requires that information about the presence of contaminants and the highest level of compliance be communicated to consumers through labeling or other means so that they can make informed decisions about the foods they consume.

To sum up, the EU contaminant limit standards play a vital role in protecting public health and ensuring food safety. Establishing strict standards and monitoring mechanisms will reduce the risks associated with food contamination and increase consumer confidence in the safety and quality of the food supply. [29]

9.3 SPECIFIED NUTRACEUTICAL INGREDIENTS: REGULATIONS AND RESTRICTIONS

Nutraceutical ingredients, including vitamins, minerals, herbal extracts, and other functional ingredients, are strictly regulated by the European Union (EU) to ensure consumer safety, product effectiveness, and transparency. The regulation of these ingredients includes a comprehensive framework, standards, and restrictions to protect the integrity of the nutraceutical industry and public health.

9.3.1 Vitamins and Minerals

Vitamins and minerals are micronutrients essential for various physiological functions of the human body. In the EU, the regulation of vitamins and minerals in food supplements is governed by the Food Supplements Directive 2002/46/EC. The purpose of this directive is to harmonize the rules for the composition, labeling, and marketing of food supplements containing vitamins and minerals within the EU member states.

Directive 2002/46/EC requires vitamins and minerals used in food supplements to be safe for human consumption and specifies maximum levels to prevent excessive intake and ensure consumer safety. [30]

The regulatory decision-making process is bolstered by the European Food Safety Authority (EFSA), which evaluates the safety and bioavailability of nutrient sources and forms utilized in food supplements. Scientific opinions are provided by the EFSA to support this process. Additionally, Directive 2002/46/EC mandates that vitamins and minerals be clearly labeled on food supplement packaging, including details such as the specific nutrient name, quantity per recommended daily dose, and percentage of the nutrient reference values (NRVs). Provided this information, consumers are empowered to make well-informed choices regarding their dietary intake while also preventing any deceptive claims or advertising practices. [31]

In accordance with Directive 2002/46/EC, Annex II outlines specific conditions and restrictions for the inclusion of certain vitamins and minerals in food supplements. To address safety concerns related to high doses, there are limitations on the use of specific forms of vitamin B6 and vitamin D, and maximum levels have been set to minimize the risks associated with excessive intake.

The primary objective of vitamin and mineral regulation in the European Union is to guarantee the safety, quality, and proper utilization of these vital nutrients in food supplements. Simultaneously, it strives to offer consumers transparent and precise information that aids in making healthy dietary decisions. Specifically, when it comes to herbal extracts, the regulation remains consistent. [32]

9.3.2 Herbal Extracts

For centuries, medicinal uses of herbal extracts, which are derived from plants and botanicals, have been well-established. These extracts are frequently included in food supplements due to their perceived health advantages. Nevertheless, regulating herbal extracts in the EU necessitates careful attention to their composition, as well as potential variations and safety issues.

Regulations within the European Union govern the use of herbal extracts in food supplements, as outlined in Directive 2002/46/EC on food supplements, and Directive 2004/24/EC on traditional herbal medicinal products. The primary objective of these regulations is to uphold the safety, quality, and effectiveness of herbal extracts, while also safeguarding the traditional wisdom and cultural significance tied to herbal remedies. [33]

Compliance with safety and labeling regulations, as outlined in Directive 2002/46/EC, is mandatory for herbal extracts used in food supplements. This includes the accurate identification of the plant species, the specific part of the plant utilized, and the extraction method employed. Furthermore, adherence to Good Manufacturing Practices (GMPs) is essential in the production of herbal extracts to guarantee uniformity, purity, and safety. [34]

The regulatory framework outlined in Directive 2004/24/EC provides guidelines for the registration and promotion of traditional herbal medicinal products that have a long-standing history of use within the European Union. Compliance with this directive necessitates a thorough evaluation of the safety, quality, and effectiveness of herbal extracts utilized in these products, utilizing both scientific research and traditional wisdom. [35]

The sale of herbal extracts as food supplements in the EU is subject to the Novel Food Regulation (Regulation (EU) 2015/2283). This regulation specifically applies to herbal extracts that were not traditionally consumed in the EU prior to May 15, 1997. In order to ensure the protection of consumer health and safety, these novel herbal extracts are required to undergo comprehensive safety assessments before they can be authorized for sale in the market.

In the European Union, the regulation of herbal extracts has a dual objective: to safeguard traditional knowledge and cultural heritage while also prioritizing consumer safety, product quality, and adherence to regulations within the nutraceutical industry. Additionally, there are other functional ingredients that fall under this regulatory framework. [36]

9.3.3 Other Functional Ingredients

Apart from vitamins, minerals, and herbal extracts, food supplements may contain various other functional ingredients that possess specific health benefits or physiological effects. These ingredients consist of a wide array of substances, such as amino acids, fatty acids, probiotics, enzymes, and plant-derived compounds that possess antioxidant properties or the ability to modulate the immune system.

Directive 2002/46/EC on food supplements is the governing regulation in the European Union for other functional ingredients. Its purpose is to establish guidelines and criteria for the composition, labeling, and promotion of food supplements that contain functional ingredients. The primary objective of this directive is to guarantee the safety, quality, and proper utilization of functional ingredients, while also ensuring that consumers are provided with precise and transparent information. [37]

In accordance with Directive 2002/46/EC, food supplements are required to utilize functional ingredients that are deemed safe for human consumption. To ensure that consumers are protected from excessive intake, maximum permitted levels have been established. Furthermore, food supplement packaging must clearly display the functional ingredients, providing the specific substance name, quantity per recommended daily dose, as well as any necessary warnings or precautions.

In addition, the European Food Safety Authority (EFSA) conducts evaluations on the safety and effectiveness of functional ingredients utilized in food supplements, offering scientific judgments to aid in regulatory decision-making. Through this scientific evaluation, functional ingredients are guaranteed to meet established safety criteria and are deemed appropriate for their designated purpose in food supplements. [38]

9.4 PRE-MARKET APPROVAL PROCESS IN THE EUROPEAN UNION

The primary objective of regulating other functional ingredients in the European Union is to guarantee the safety of consumers, maintain product quality, and ensure compliance with regulations within the nutraceutical industry. This regulatory framework also strives to offer consumers access to cutting-edge and efficient dietary supplements that promote their overall health and well-being. In the European Union, the pre-market approval process plays a crucial role in achieving these objectives. [39]

Within the European Union (EU), there exists a meticulous regulatory framework known as the pre-market approval process. This framework serves the purpose of guaranteeing the safety, effectiveness, and quality of products prior to their introduction into the market. The pre-market approval process is applicable to a wide range of products, such as pharmaceuticals, medical devices, cosmetics, and food supplements. In the context of this discussion, we will specifically explore the pre-market approval process for food supplements, with a particular emphasis on the application procedure, the assessment of safety and efficacy, and the necessary data requirements. [40]

9.4.1 Application Procedure

The process of applying for pre-market approval of food supplements in the EU requires the submission of a thorough dossier to the appropriate regulatory body, which is usually the European Food Safety Authority (EFSA) or national competent authorities. This dossier must include extensive details regarding the composition of the product, its manufacturing process, intended purpose, and safety data. These pieces of information are crucial in order to obtain authorization for the sale of the product. [41]

To initiate the application process, applicants are required to submit a comprehensive application dossier comprising crucial elements, as outlined in what follows.

In the "Product Information" section, you will find comprehensive details about the food supplement, such as its name, brand, packaging, dosage form, and the specific population it is intended for. The "Composition" section provides an overview of the ingredients utilized in the food supplement, encompassing vitamins, minerals, herbal extracts, and other functional components. Every ingredient is clearly specified, with information regarding its origin, form, and concentration. [42]

The manufacturing process is described in detail, highlighting the adherence to Good Manufacturing Practices (GMP), the implementation of quality control measures, and the conduct of stability testing. These measures are put in place to guarantee the consistency and quality of the product. Safety data, such as toxicological studies, human clinical trials, and reports on adverse events, are indispensable in demonstrating the safety of the food supplement for its intended purpose. This data plays a critical role in assessing potential risks and establishing appropriate dosage levels to ensure safety.

To meet regulatory standards, it is essential to provide comprehensive information regarding labeling and packaging. This includes details about product labeling, packaging materials, and any claims or statements featured on the packaging. Once the application dossier is submitted, regulatory authorities meticulously review the provided information to evaluate the product's safety, effectiveness, and adherence to regulatory requirements. This evaluation process may entail seeking input from scientific experts, conducting peer reviews, and engaging in public consultations to ensure a thorough assessment. [43]

9.4.2 Safety and Efficacy Assessment

The evaluation of safety and efficacy is an essential step in the EU's pre-market approval process for food supplements. Regulatory bodies carefully examine the scientific evidence provided in the application dossier to ascertain the product's suitability for human consumption and its effectiveness in fulfilling its intended purpose.

The process of safety assessment entails the examination of toxicological information, which encompasses research on immediate toxicity, genotoxicity, and prolonged exposure, with the objective of pinpointing potential risks and determining appropriate dosage thresholds. In addition, human clinical trials may be carried out to evaluate the safety and effectiveness of a particular product or treatment within specific subsets of the population, such as pregnant women, children, or the elderly. [44]

The evaluation of efficacy centers on determining whether a product can effectively provide the promised health benefits or physiological effects as stated by the manufacturer. To establish the validity of these claims, scientific evidence is thoroughly examined, including clinical studies, data on bioavailability, and mechanistic studies, all of which contribute to substantiating and confirming the product's assertions with reliable evidence.

In addition, when evaluating the effectiveness of a product and guaranteeing its desired outcomes for consumers, regulatory bodies take into account various elements, including the relationship between dosage and response, the extent to which the product is absorbed by the body, and the mechanism by which it operates. These factors are carefully considered by regulatory authorities in their assessment process. [45]

9.4.3 Data Requirements

In order to guarantee the safety and quality of food supplements, the EU has established a rigorous and thorough set of data requirements for pre-market approval. These requirements encompass various types of data that must be included in the application dossier.

The safety of the product's components, including acute toxicity, subchronic toxicity, and genotoxicity, is evaluated in toxicological studies to determine potential risks and establish appropriate dosage levels. Human clinical trials are then conducted to assess the supplement's safety, effectiveness, and tolerability in real-life scenarios, generating important information on its impact on human health and guiding regulatory decisions. [46]

The bioavailability of a product's active ingredients and their pharmacokinetic properties are evaluated through bioavailability studies. These studies examine how the ingredients are absorbed, distributed, metabolized, and excreted in order to gain insight into the product's behavior within the body. This information is crucial for optimizing dosage regimens and understanding the product's effects.

The evaluation of a product's stability over time is known as stability testing. This process involves subjecting the product to different storage conditions in order to verify that it maintains its quality, potency, and efficacy throughout its designated shelf life. The information obtained from stability testing is essential for determining the appropriate shelf-life duration and storage guidelines for the product. [47]

To evaluate the safety profile of a product and identify possible risks to consumers, it is essential to provide information on adverse events, including reported side effects or adverse reactions. The extensive data requirements for the pre-market approval of food supplements in the EU encompass various scientific studies and evidence that substantiate the product's safety, efficacy, and quality. By meeting these requirements, manufacturers can guarantee compliance with regulatory standards and streamline the authorization process for selling their products in the EU market.

To summarize, the European Union's system for approving food supplements before they enter the market is a comprehensive and meticulous regulatory structure aimed at safeguarding consumer well-being, guaranteeing product effectiveness, and promoting transparency. By following meticulous application protocols, conducting thorough safety and effectiveness evaluations, and fulfilling data prerequisites, manufacturers can successfully navigate the regulatory environment and introduce secure, top-notch products to meet consumer demands. [48]

9.5 HEALTH CLAIMS AND SCIENTIFIC SUBSTANTIATION

In the EU, scientifically verifying health claims is intricate and entails assessing the data about the connection between food and human health. An essential player in this process is the European Food Safety Authority (EFSA), which evaluates the scientific data based on standards [49]. Nevertheless, the requirement to prove cause-and-effect links in claim substantiation is controversial. Some contend for a strict evidentiary standard that prioritizes preventing misleading claims by requiring causal facts for proof. However, many who disagree with this strategy support laxer regulations, which would sell more goods with health claims. The evaluation of health claims by the EFSA has faced criticism due to its emphasis on randomized controlled trials (RCTs) and its scant attention to alternative techniques in nutrition research [50]. In general, the process of scientifically verifying health claims is a complicated and continuous discussion that requires striking a balance between the advancement of innovation and public health and the requirement to protect consumers.

9.5.1 Criteria for Health Claims

Food claims imply a connection between the consumer's health and a food product. The use of health claims is controlled in many nations, and to safeguard consumers, the scientific substantiation procedure needs to be founded on solid facts and tenable mechanisms of action. Nevertheless, when it comes to food claims, there needs to be more observance of European Union (EU) regulations, especially when disclosing the quantity and mode of consumption necessary to achieve the purportedly advantageous outcome [51]. Because individuals evaluate health claims individually and focused marketing is necessary to reach consumers for whom the stated benefit is relevant, the impact of health claims on public health, innovation, and food product sales has been limited. Plain language is crucial for health claims, since consumers may be unable to distinguish between nutrition and health claims. To make health claims about bioactive chemicals in food, one must provide evidence from rigorous clinical trials, laboratory results, and assurances of efficacy, safety, and bioavailability.

9.5.1.1 Role of EFSA on the Approval of Nutraceuticals

The European Food Safety Authority (EFSA) is essential in nutraceuticals' approval process. Legislation about animal nutrition is adopted by the European Commission using the conclusions that EFSA produces. EFSA assesses health claims by the EU's nutrition and health claims regulation. EFSA is involved in opinion formulation and decision-making processes, and its guidelines are followed during the evaluation process. Thoroughly controlled clinical trials, including European participants, are the gold standard when evaluating health claims, although EFSA takes into account all available evidence. Protecting consumers from deceptive product advertising and ensuring that rigorous scientific standards support the effects of nutraceuticals are the goals of these rules and assessments [52].

9.5.2 Scientific Substantiation Process

Examining the data to back up the claim is part of the scientific substantiation process for health claims. The European Food Safety Authority (EFSA) is a vital entity in the European Union to evaluate health claims [53]. The EFSA employs a three-step process of defining the food or element, evaluating the data, and establishing a cause-and-effect relationship to examine scientific substantiation. Nonetheless, there is disagreement about whether EFSA's evidence hierarchy is the most appropriate method for determining causal linkages in nutrition and accomplishing societal goals. The Nutri-Score system has been assessed concerning particular health claims, and although the data indicates that it is advantageous to human health, a cause-and-effect relationship could not be proved. Based on well-conducted randomized controlled trials, systematic literature reviews are valuable for verifying health claims and establishing the causal links between food consumption and observed health impacts [54].

9.5.3 Examples of Approved Health Claims

The European Union has various instances of authorized health claims. One such product is the tomato water-soluble extract known as Fruitflow, which was approved as a health claim due to its ability to decrease blood pressure and inhibit platelets [19]. Alternatively, various modified, non-alcoholic fruit juice drinks under the "Toothkind" brand have been approved as a health claim for their ability to help keep teeth healthy instead of fruit juices and beverages with added sugar [26]. Furthermore, the EU has over 300 approved claims about nutrition and health that have been governed since 2006. These claims, which include a wide range of substances and their health benefits, are intended to safeguard consumers from deceptive and fraudulent claims while encouraging innovation in the food sector.

9.6 LABELING AND PACKAGING REQUIREMENTS

Different nations have different regulations regarding the labeling and packaging of *nutraceuticals*, defined as "foods for special dietary uses or functional foods or health supplements" by India's Food

Safety and Standards Authority. Nutraceuticals are referred to as dietary supplements and health goods in Sudan [15], while they are referred to as health supplement items in Cambodia. The labels of nutritional products, which contain details on active substances, precautions, indications, storage requirements, instructions, and purposes, should be clear and easy to read. Nutraceuticals must undergo quality control procedures and standardization to guarantee uniformity, effectiveness, and safety [51–52]. In order to keep shelves stable and avoid cross-contamination, packaging is essential. Segments or trays may be used in the packaging to safeguard the food product and nutraceuticals. In general, different nations have different regulations and norms. Thus, manufacturers must adhere to each market's particular labeling and packaging specifications. [55]

9.6.1 Nutrient Content and Functionality Claims

Food and beverage products' perceived usefulness and nutritional value are major consumer perception and purchase behavior determinants. Research indicates that making claims about the presence or lack of specific nutrients or chemicals in a product can lead to a 75% rise in customer purchases. The Food and Drug Administration (FDA), the European Authority of Food Safety (EFSA), and MHLW (Japan) are among the regulatory bodies that have developed protocols for claiming health advantages from functional food items [56]. These statements are meant to enlighten customers on the possible impacts of food ingredients on the composition or operation of the human body. On the other hand, familiarity, variety, price, flavor, and packaging are some of the variables that can affect how consumers perceive content claims. Sports celebrities' endorsements and nutrient content claims can also affect children's preferences for energy-dense and nutrient-poor (EDNP) food products [57]. Generally, to prevent deception and encourage the consumption of healthier foods, the use of functionality and nutrient content claims in food marketing should be controlled.

9.6.2 Allergen Information

Nutraceuticals are foods or meals with therapeutic qualities that can be utilized to treat or avoid disease. They originate from a number of things, such as plants, dietary fiber, and enzymes. Fatty acids, proteins, alkaloids, phenolic compounds, and other diverse chemical components are all included in nutraceuticals [58]. By boosting nutrient absorption and supporting the gastrointestinal system, they are considered safe for humans and can enhance health. Conversely, the allergy data about nutraceuticals lacks clarity. More investigation and clinical testing are required to prove the effectiveness and safety of dietary supplements and nutraceuticals and any potential allergic effects. The most current research has not addressed in detail allergic reactions from nutraceuticals in clinical trials. The assessment of allergen immunotherapy (AIT) outcome measures in allergic asthma, the effect of neuro-nutraceuticals on brain health and cognitive function, the use of natural resources' mast cell stabilizers for treating allergies, suggestions for enhancing allergen immunotherapy clinical trials, and the influence of smoking on allergen-induced airway responses in asthmatics receiving inhaled corticosteroids are just a few of the topics covered [59].

9.6.3 Language and Presentation

Nutraceuticals offer medicinal and health advantages, such as illness prevention and therapy. The body can employ them to sustain functions, fend off chronic illnesses, delay ageing, and enhance health. Three general categories can be used to group nutraceuticals: nutrients, botanical products or herbs, and reagents obtained from other sources. Nutrients include minerals, vitamins, fatty acids, and amino acid. A thorough investigation discovered that herbal nutraceuticals could help treat a wide range of oxidative stress–related diseases, including allergies, Alzheimer's disease, cardiovascular diseases, cancer, diabetes, eye problems, immunological disorders, inflammation, and obesity [60]. Examples of the kind of clear, essential information that has to be included in the terminology and presentation of nutraceuticals are a succinct summary of the critical nutrients and a graphic depiction of the dietary groups and their nutritional profiles. This can assist customers in making health-related decisions with knowledge.

9.7 COMPLIANCE CHALLENGES AND INDUSTRY BEST PRACTICES

Numerous studies have been conducted on compliance difficulties across various industries, particularly software engineering, organizational management, and occupational safety and health. Despite the wealth of accessible theoretical knowledge, empirical evidence regarding the difficulties faced by industrial practitioners in maintaining regulatory compliance is still sparse. Research has been done to shed light on typical procedures and difficulties in examining and assessing regulatory compliance, especially in software engineering. Implementing compliance risk in industrial companies can guarantee stable and efficient operation, as it is strongly linked to organizations' management and control systems. The risks associated with compliance include financial loss, loss of market share, legal and regulatory penalties, and reputational harm from breaking laws, rules, and standards. It has been discovered that good compliance methods lower workplace accidents, underscoring the significance of comprehension, dedication, practice, and behavior in reaching regulatory compliance [47].

9.7.1 Common Compliance Issues

Inadequate regulation and a lack of international agreement on regulatory procedures are common problems with nutraceuticals' compliance. Concerns regarding product authenticity and the existence of tampered or counterfeit goods have resulted in problems with quality and safety [48]. Problems include poor formulation, insufficient levels of valuable components, and contamination from heavy metals, herbicides, pesticides, and adulterants. To guarantee the identification, potency, purity, and composition of supplement ingredients, adherence to current reasonable manufacturing procedures (cGMPs) is crucial. For dietary supplements, regulatory compliance is crucial for claims, labeling, advertising, and packaging [61]. Marketed nutraceutical formulations' accuracy and regulatory compliance have been found to differ, with specific formulations failing to reach suggested values and content as indicated on the label.

9.7.2 GMP

The manufacture of nutraceuticals requires adherence to Good Manufacturing Practices, or GMPs. GMPs guarantee that production procedures adhere to fundamental guidelines and can monitor and stop product contamination [62]. The food and pharmaceutical industries must adhere to the same GMP regulations; however, although independent control is required in the pharmaceutical business, it is advised in the food industry. One nutraceutical that has garnered interest is glycomacropeptide (GMP), which possesses biological and functional properties. Applications in the food and nutrition sectors can benefit from the specific amino acids found in GMP. Nutraceuticals can help prevent and treat disease in addition to providing basic nourishment. They include herbal products, functional foods, and dietary supplements. GMP offers various advantageous characteristics that can be employed in manufacturing functional foods, such as impeding bacterial adhesion, diminishing gastric secretions, and enhancing mineral absorption [63].

9.7.3 Strategies for Ensuring Compliance

Quantifying the active components and heavy metal levels in marketed formulations using sophisticated analytical techniques is one strategy for guaranteeing compliance in the nutraceutical industry. To guarantee adherence to current protocols and supply of nutraceutical goods that are both safe and effective, regulatory bodies must put strict monitoring procedures into place. Establishing worldwide risk assessment guidelines and botanical compendia and fostering greater collaboration among consumers, health-care providers, and government agencies depend on creating an efficient and ethical system. Evidence-based methods and adherence to current Good Manufacturing Practices (cGMPs) can guarantee the manufacturing of safe and high-quality nutraceuticals and give consumers and healthcare professionals patient-specific evidence [64]. Pharmaceutical firms must adhere to regulatory regulations to produce safe, high-quality products that improve health outcomes and contribute to good social change.

9.8 FUTURE TRENDS AND DEVELOPMENTS

9.8.1 Emerging Nutraceuticals

Nutraceuticals in infancy comprise substances originating from microorganisms, plants, and animals, like aloe vera, amla, ginger, turmeric, and garlic. These natural nutraceuticals provide numerous medicinal advantages, such as cardiovascular, antidiabetic, antibacterial, antifungal, and allergy-alleviation effects. The nutraceutical market is expected to grow worldwide, with a predicted value of $285 billion by 2021 [65]. The purpose of developing nano-nutraceuticals is to enhance the delivery of dietary bioactives, and this is one area of increasing nutraceutical development. With their enhanced biodistribution, regulated and targeted release, and increased tissue absorption, solid lipid nanoparticles (SLNs) are being investigated as a possible nutraceutical delivery medium. However, the production of SLNs continues to present difficulties, and regulatory agencies such as the USFDA emphasize using quality-by-design (QbD) methods [36]. The market for foods with therapeutic and nutritional benefits is known as emerging nutraceuticals, and it is expanding quickly. These products, derived from organic resources, such as microbes and plants, are acknowledged for their potential to prevent cancer, ageing, and diabetes. Because they alter the microbiome's composition, nutraceuticals have been proposed as a potential chronic illness treatment and prevention strategy. Nutraceuticals, or functional foods, have gained popularity because they can assist in managing many conditions and offer extra health benefits. Nutraceuticals comprise vitamins, minerals, and herbal remedies that help prevent disease, maintain and improve health, and enhance overall well being [37]. Proper regulation is necessary to guarantee safety and effectiveness, as there is a dearth of clear information and evidence supporting the purported health advantages of nutraceuticals.

9.8.2 Potential Regulatory Changes

The regulatory environment about nutraceuticals is intricate and differs among nations. There is not yet a consensus on the regulatory procedures for nutraceuticals around the globe. The Food Safety Standards (FSS) Act of 2006 has given laws about nutraceuticals and functional foods in India more significance [38]. The nutraceutical industry confronts several challenges, including high prices, inadequate distribution and marketing, a lack of standards, and inadequate understanding. Nutraceuticals are governed by regulations in many countries, including the United States, EU member states, China, Japan, and India. There are still not enough comprehensive regulatory frameworks in place worldwide. It might be necessary to modify nutraceutical regulations to uphold industry norms of efficacy, safety, and consistency.

REFERENCES

[1] V. Puri et al., "A comprehensive review on nutraceuticals: Therapy support and formulation challenges," *Nutrients*, vol. 14, no. 21. MDPI AG, p. 4637, Nov. 03, 2022. https://doi.org/10.3390/nu14214637.

[2] Food Supplements, "European food safety authority," https://www.efsa.europa.eu/en/topics/topic/food-supplements (accessed Apr. 26, 2024).

[3] S. Thakur, M. M. Gupta, and D. Sharma, "Nutraceuticals regulation: an overview of the regulatory frameworks in USA, EU, and Japan," in *Nutraceutical Fruits and Foods for Neurodegenerative Disorders*. Elsevier, pp. 421–440, 2024. https://doi.org/10.1016/b978-0-443-18951-7.00023-2.

[4] N. L. Bragazzi et al., "Nutraceutical and functional food regulations in the European Union," in *Developing New Functional Food and Nutraceutical Products*. Elsevier, pp. 309–322, 2017. https://doi.org/10.1016/b978-0-12-802780-6.00017-1.

[5] S. L. Teoh, S. Ngorsuraches, N. M. Lai, and N. Chaiyakunapruk, "Consumer preferences and willingness to pay for nutraceuticals: A discrete choice experiment," *Value in Health Regional Issues*, vol. 24, pp. 167–172, May 2021. https://doi.org/10.1016/j.vhri.2020.09.003

[6] M.G. Komala, S.G. Ong, M.U. Qadri, L.M. Elshafie, C.A. Pollock, S. Saad, "Investigating the regulatory process, safety, efficacy and product transparency for nutraceuticals in the USA," *Europe and Australia, Foods*, vol. 12, p. 427, 2023. https://doi.org/10.3390/foods12020427.

[7] "Nutraceutical in europe market—share & revenue," Nutraceutical in Europe Market—Share & Revenue. https://www.mordorintelligence.com/industry-reports/europe-nutraceutical-market (accessed Apr. 26, 2024).

[8] S. Thakkar et al., "Regulatory landscape of dietary supplements and herbal medicines from a global perspective," *Regulatory Toxicology and Pharmacology*, vol. 114. Elsevier BV, p. 104647, Jul. 2020. https://doi.org/10.1016/j.yrtph.2020.104647.

[9] A. Santini et al., "Nutraceuticals: Opening the debate for a regulatory framework," *British Journal of Clinical Pharmacology*, vol. 84, no. 4. Wiley, pp. 659–672, Feb. 14, 2018. https://doi.org/10.1111/bcp.13496.

[10] P. Coppens, M. F. da Silva, and S. Pettman, "European regulations on nutraceuticals, dietary supplements and functional foods: A framework based on safety," *Toxicology*, vol. 221, no. 1. Elsevier BV, pp. 59–74, Apr. 2006. https://doi.org/10.1016/j.tox.2005.12.022.

[11] O. P. Gulati and P. Berry Ottaway, "Legislation relating to nutraceuticals in the European Union with a particular focus on botanical-sourced products," *Toxicology*, vol. 221, no. 1. Elsevier BV, pp. 75–87, Apr. 2006. https://doi.org/10.1016/j.tox.2006.01.014.

[12] Md. F. Ahmad, S. A. Ashraf, F. A. Ahmad, J. A. Ansari, and Md. R. A. Siddiquee, "Nutraceutical Market and its Regulation," *American Journal of Food Technology*, vol. 6, no. 5. Science Alert, pp. 342–347, Apr. 15, 2011. https://doi.org/10.3923/ajft.2011.342.347.

[13] M. Nicoletti, "Nutraceuticals and botanicals: overview and perspectives," *International Journal of Food Sciences and Nutrition*, vol. 63, no. supl. Informa UK Limited, pp. 2–6, Feb. 24, 2012. https://doi.org/10.3109/09637486.2011.628012.

[14] A. S. Chopra et al., "The current use and evolving landscape of nutraceuticals," *Pharmacological Research*, vol. 175. Elsevier BV, p. 106001, Jan. 2022. https://doi.org/10.1016/j.phrs.2021.106001.

[15] R. Bansal and A. Dhiman, "Nutraceuticals: A comparative analysis of regulatory framework in different countries of the world," *Endocrine, Metabolic & Immune Disorders—Drug Targets*, vol. 20, no. 10. Bentham Science Publishers Ltd., pp. 1654–1663, Dec. 02, 2020. https://doi.org/10.2174/1871530320666200519084415.

[16] R. Giunta, G. Basile, and A. Tibuzzi, "Legislation on Nutraceuticals and Food Supplements: A Comparison between Regulations in USA and EU," *Advances in Experimental Medicine and Biology. Springer US*, pp. 322–328, 2010. https://doi.org/10.1007/978-1-4419-7347-4_24.

[17] D. Tedesco, P. Cagnardi, "Regulatory guidelines for nutraceuticals in the European Union," 2019, pp. 793–805. https://doi.org/10.1007/978-3-030-04624-8_58.

[18] P. B. Ottaway, "The regulation of functional foods and nutraceuticals in the European Union," in *Regulation of Functional Foods and Nutraceuticals*. Wiley, pp. 227–245, Mar. 24, 2005. https://doi.org/10.1002/9780470277676.ch14.

[19] I. Tańska, "Salient features of European Union regulations on nutraceuticals and functional foods," in *Nutraceutical and Functional Food Regulations in the United States and around the World*. Elsevier, pp. 323–336, 2019. https://doi.org/10.1016/b978-0-12-816467-9.00021-6.

[20] J.R. Lupton, "Scientific substantiation of claims in the USA: focus on functional foods," *European Journal of Nutrition*, vol. 48, pp. 27–31, 2009. https://doi.org/10.1007/S00394-009-0073-3.

[21] N.-G. Asp, "Health claims within the Swedish code," *Scandinavian Journal Nutrition*, 2016. https://doi.org/10.1080/11026480260363260.

[22] S. H. Zeisel, "Regulation of 'nutraceuticals,'" *Science*, vol. 285, no. 5435. American Association for the Advancement of Science (AAAS), pp. 1853–1855, Sep. 17, 1999. https://doi.org/10.1126/science.285.5435.1853.

[23] O. K. Nwosu and K. I. Ubaoji, "Nutraceuticals: History, classification and market demand," in *Functional Foods and Nutraceuticals*. Springer International Publishing, pp. 13–22, 2020. https://doi.org/10.1007/978-3-030-42319-3_2.

[24] L. Domínguez Díaz, V. Fernández-Ruiz, and M. Cámara, "The frontier between nutrition and pharma: The international regulatory framework of functional foods, food supplements and nutraceuticals," *Critical Reviews in Food Science and Nutrition*, vol. 60, no. 10. Informa UK Limited, pp. 1738–1746, Mar. 29, 2019. https://doi.org/10.1080/10408398.2019.1592107.

[25] S. A. Ruckman, "Regulations for nutraceuticals and functional foods in Europe and the United Kingdom," in *Nutraceutical and Functional Food Regulations in the United States and Around the World*. Elsevier, pp. 221–238, 2008. https://doi.org/10.1016/b978-012373901-8.00015-9.

[26] A. de Boer, "Fifteen years of regulating nutrition and health claims in Europe: The past, the present and the future," *Nutrients*, vol. 13, 2021, p. 1725. https://doi.org/10.3390/NU13051725.

[27] A.K. Duttaroy, "Regulation of functional foods in European Union: assessment of health," *European Food Safety Authority*, 2019, pp. 267–276. https://doi.org/10.1016/B978-0-12-816467-9.00019-8.

[28] M.J. Sadler, Authorised EU health claim for reformulated, non-alcoholic, acidic drinks, (2018) 103–111. https://doi.org/10.1016/B978-0-08-100922-2.00007-3.

[29] A. Koch, S. Brandenburger, S. Türpe, and M. Birringer, "The need for a legal distinction of nutraceuticals," *Food and Nutrition Sciences*, vol. 05, no. 10. Scientific Research Publishing, Inc., pp. 905–913, 2014. https://doi.org/10.4236/fns.2014.510100.

[30] A. Shao, "Global market entry regulations for nutraceuticals, functional foods, dietary/food/health supplements," in *Developing New Functional Food and Nutraceutical Products*. Elsevier, pp. 279–290, 2017. https://doi.org/10.1016/b978-0-12-802780-6.00015-8.

[31] O. P. Gulati, P. Berry Ottaway, and P. Coppens, "Botanical nutraceuticals, (food supplements, fortified and functional foods) in the European union with main focus on nutrition and health claims regulation," in in *Nutraceutical and Functional Food Regulations in the United States and Around the World*. Elsevier, pp. 221–256, 2014. https://doi.org/10.1016/b978-0-12-405870-5.00014-1.

[32] D. Bagchi, *Nutraceutical and Functional Food Regulations in the United States and around the World*, 2nd ed. San Diego: Academic Press, 2014.

[33] V. D. Sattigere, P. Ramesh Kumar, and V. Prakash, "Science-based regulatory approach for safe nutraceuticals," *Journal of the Science of Food and Agriculture*, vol. 100, no. 14. Wiley, pp. 5079–5082, Nov. 16, 2018. https://doi.org/10.1002/jsfa.9381.

[34] S. Girdhar, D. Pandita, A. Girdhar, and V. Lather, "Safety, quality and regulatory aspects of nutraceuticals," in *Applied Clinical Research, Clinical Trials and Regulatory Affairs*, vol. 4, no. 1. Bentham Science Publishers Ltd., pp. 36–42, Feb. 27, 2017. https://doi.org/10.2174/2213476x03666160801150607.

[35] S. R. Garud, A. G. Lamdande, and S. R. Gholap, "Regulations on functional foods and nutraceuticals," in *Industrial Application of Functional Foods, Ingredients and Nutraceuticals*. Elsevier, pp. 785–823, 2023. https://doi.org/10.1016/b978-0-12-824312-1.00022-4.

[36] J. Smith and E. Charter, Eds., *Functional Food Product Development*. Wiley, Apr. 26, 2010. https://doi.org/10.1002/9781444323351.

[37] P. Daliu, A. Santini, and E. Novellino, "From pharmaceuticals to nutraceuticals: bridging disease prevention and management," *Expert Review of Clinical Pharmacology*, vol. 12, no. 1. Informa UK Limited, pp. 1–7, Nov. 30, 2018. https://doi.org/10.1080/17512433.2019.1552135.

[38] M. T. Yeung, J. E. Hobbs, and W. A. Kerr, "Accessing US and EU markets for nutraceuticals and functional foods," *Journal of International Food & Agribusiness Marketing*, vol. 19, no. 1. Informa UK Limited, pp. 53–79, Dec. 07, 2006. https://doi.org/10.1300/j047v19n01_04.

[39] New Product Development and Regulatory Challenges in Nutraceutical Industry (2022) 1012–1035. https://doi.org/10.4018/978-1-6684-3546-5.ch049.

[40] S.R. Peters, H. Verhagen, "An evaluation of the nutri-score system along the reasoning for scientific substantiation of health claims in the EU—A narrative review," *Foods*, vol. 11, p. 2426, 2022. https://doi.org/10.3390/foods11162426.

[41] K.M.P. Patel, U. Patel, K. Vaghela, N. Kanki, M. Zaveri, V. Movaliya, "Comparisons of registration requirements of Nutraceuticals in Philippines, Tanzania, Cambodia and India," *International Journal of Drug Regulatory Affairs*, vol. 10, pp. 18–23, 2022. https://doi.org/10.22270/ijdra.v10i4.546.

[42] B. Bradford, Benefit Claims of Functional Food, 2015, pp. 1–6. https://doi.org/10.1007/978-3-642-27728-3_122-1.

[43] H. Dixon, M. Scully, P. Niven, B. Kelly, K. Chapman, R.J. Donovan, J. Martin, L.A. Baur, D. Crawford, M. Wakefield, Effects of nutrient content claims, sports celebrity endorsements and premium offers on pre-adolescent children's food preferences: experimental research, Pediatric Obesity, vol. 9, pp. 47–57, 2014. https://doi.org/10.1111/J.2047–6310.2013.00169.X.

[44] J.H. Ryu, D. Kang, "Physicochemical properties, biological activity, health benefits, and general limitations of aged black garlic: A review," *Molecules*, vol. 22, 2017 https://doi.org/10.3390/molecules22060919.

[45] I.S. Iyen, Ogofotha, G. Oghenekeno, J.O. Oko, *Advances in Nutraceuticals*, vol. 5, pp. 8–22, 2019. https://www.arcjournals.org/pdfs/ijmpnp/v5-i2/2.pdf.

[46] S. Fiala and H.B. Fleit, "Clinical and experimental treatment of allergic asthma with an emphasis on allergen immunotherapy and its mechanisms," *Clinical and Experimental Immunology*, vol. 212, no. 1. Oxford University Press, pp. 14–28, Mar. 05, 2023. https://doi.org/10.1093/cei/uxad031.

[47] D. Johnson, S. Johnson-Verma, "The use of nutraceuticals in sports and exercise medicine," *British Journal of Sports Medicine*, vol. 47 (2013). https://doi.org/10.1136/BJSPORTS-2013–092558.27.

[48] J. Dwyer, P. Coates, and M. Smith, "Dietary supplements: regulatory challenges and research resources," *Nutrients*, vol. 10, no. 1. MDPI AG, p. 41, Jan. 04, 2018. https://doi.org/10.3390/nu10010041.

[49] C.L. Cheng, Y.H.K. Yang, S.J. Lin, C.-H. Lin, Y.J. Lin, Compliance with dosing recommendations from common references in prescribing antibiotics for preterm neonates, *Pharmacoepidemiology and Drug Safety*, vol. 19, pp. 51–58, 2010. https://doi.org/10.1002/PDS.1858.

[50] J.L. Turner, S. Pharm, The implementation of European community directives concerning GMP, *Drug Information Journal*, vol. 25, pp. 537–544, 1991. https://doi.org/10.1177/009286159102500410.

[51] I. G. Télessy, "Nutraceuticals," in *The Role of Functional Food Security in Global Health*. Elsevier, pp. 409–421, 2019. https://doi.org/10.1016/b978-0-12-813148-0.00024-4.

[52] A. M. Thayer, "Europe Adopts GMP Regulations," *Chemical & Engineering News Archive*, vol. 84, no. 7. American Chemical Society (ACS), pp. 29–33, Feb. 13, 2006. https://doi.org/10.1021/cen-v084n007.p029.

[53] K. Varma A.C., S. Kuttappan, and A. Amalraj, "Regulatory aspects of nutraceuticals," in *Herbs, Spices and Their Roles in Nutraceuticals and Functional Foods*. Elsevier, pp. 337–346, 2023. https://doi.org/10.1016/b978-0-323-90794-1.00002-8.

[54] P. Dudeja and R. K. Gupta, "Nutraceuticals," in *Food Safety in the 21st Century*. Elsevier, pp. 491–496, 2017. https://doi.org/10.1016/b978-0-12-801773-9.00040-6.

[55] D. D. O'Dwyer and S. Vegiraju, "Navigating the maze of dietary supplements," *Topics in Clinical Nutrition*, vol. 35, no. 3. Ovid Technologies (Wolters Kluwer Health), pp. 248–263, Jul. 2020. https://doi.org/10.1097/tin.0000000000000207.

[56] U. Golla, "Emergence of nutraceuticals as the alternative medications for pharmaceuticals," *International Journal of Complementary & Alternative Medicine*, vol. 11, no. 3. MedCrave Group, LLC, 2018. https://doi.org/10.15406/ijcam.2018.11.00388.

[57] H. Sachdeva, S. Khaturia, and M. Chahar, "Nutraceuticals as therapeutic agents for prevention and treatment of diseases," in *Prebiotics, Probiotics and Nutraceuticals*. Springer Nature, pp. 45–68, 2022. https://doi.org/10.1007/978-981-16-8990-1_4.

[58] P. Sahu, G. Sahu, H. Sharma, and G. K. Sahu, "Preparation and characterization of nutraceutical drink," *International Journal for Multidisciplinary Research*, vol. 5, no. 3, May 12, 2023. https://doi.org/10.36948/ijfmr.2023.v05i03.2983.

[59] B. Rasuli, "Regulatory aspects of nutraceuticals and functional foods in India," 2022, pp. 155–165. https://doi.org/10.1007/978-981-19-2366-1_10.

[60] L. Inbathamizh, D. Prabavathy, and S. Sudha, "Quality assurance of nutraceuticals and their approval, registration, marketing," in *Handbook of Nutraceuticals and Natural Products*. Wiley, pp. 337–360, May 13, 2022. https://doi.org/10.1002/9781119746843.ch27.

[61] E. B.-M. Daliri and B. H. Lee, "Current trends and future perspectives on functional foods and nutraceuticals," in *Beneficial Microorganisms in Food and Nutraceuticals*. Springer International Publishing, pp. 221–244, 2015. https://doi.org/10.1007/978-3-319-23177-8_10.

[62] K. Trifković and M. Benković, "Introduction to nutraceuticals and pharmaceuticals," in *Nutraceuticals and Natural Product Pharmaceuticals*. Elsevier, pp. 1–31, 2019. https://doi.org/10.1016/b978-0-12-816450-1.00001-5.

[63] A. Anadón, I. Ares, M.-R. Martínez-Larrañaga, and M.-A. Martínez, "Evaluation and regulation of food supplements: European perspective," in *Nutraceuticals*. Elsevier, pp. 1241–1271, 2021. https://doi.org/10.1016/b978-0-12-821038-3.00073-2.

[64] K. Bhupathiraju, A. V. Krishnaraju, K. Sengupta, T. Golakoti, S. K. Akolkar, and P. Datla, "Regulations on nutraceuticals, functional foods, and dietary supplements in India," in *Nutraceutical and Functional Food Regulations in the United States and Around the World*. Elsevier, pp. 445–464, 2019. https://doi.org/10.1016/b978-0-12-816467-9.00029-0.

[65] J. P. da Costa, "A current look at nutraceuticals—Key concepts and future prospects," *Trends in Food Science & Technology*, vol. 62. Elsevier BV, pp. 68–78, Apr. 2017. https://doi.org/10.1016/j.tifs.2017.02.010.

10 Regulatory Requirements for South Africa for Marketing and Importing Nutraceuticals, Herbal Products, and Functional Foods

Pravin Semnarayan, Deon du Rand, and Megan Saffer

10.1 INTRODUCTION

There is a proliferation of nutraceuticals and herbal products in the South African market, with quantitative data on complementary medicines not captured and available for the South African market and industry, according to the Health Products Association of South Africa. This chapter discusses the regulatory framework that is still under development to regulate these products. The intention of the authorities is to ensure consistent safety, quality, and efficacy of medicines to protect the South African public (Medicines Act). However, industry participants strongly oppose regulating these products, questioning both the capacity and the authority of regulatory authorities to regulate complementary medicines in South Africa (SCA Appeal Judgement Case 256/2021; 2022).

10.2 DEFINITION OF COMPLEMENTARY MEDICINES

This section discusses how the products are defined by the authorities.

Medicine's definition is described in the Medicines Act (May 26, 2017), in Section 1, as:

> "[M]edicine"—
>
> (a) means any substance or mixture of substances used or purporting to be suitable for use or manufactured or sold for use in—
>   (1) the diagnosis, treatment, mitigation, modification or prevention of disease, abnormal physical or mental state or the symptoms thereof in humans; or
>   (2) restoring, correcting or modifying any somatic or psychic or organic function in humans; and
>
> (b) includes any veterinary medicine.

Medicines can be classified into categories A, B, C, and D, as in the Medicines Act General Regulations (August 25, 2017), in Section 9.1:

> Medicines shall be classified into categories as follows:
>
> (a) Category A = Medicines which are intended for use in humans and which are, without manipulation, ready for administration, including packaged preparations where only a vehicle is added to the effective medicine;

DOI: 10.1201/9781003465041-10

(b) Category B = Medicines intended for use in humans and animals which cannot normally be administered without further manipulation;
(c) Category C = Medicines intended for veterinary use which are, without further manipulation, ready for administration, including packaged preparations where only a vehicle is added to the effective medicine; and
(d) Category D = Complementary medicines intended for use in humans and animals which are, without further manipulation, ready for administration, including packaged preparations where only a vehicle is added to the effective medicine.

Complementary medicine (category D) is further classified in the Medicines Act General Regulations (August 26, 2017), in Section 9.2, as:

> Medicines in category D shall be classified into the following sub-categories:
>
> (a) discipline-specific medicines with such disciplines as determined by the Authority; and
> (b) health supplements.

Complementary medicine's definition is described in the Medicines Act General Regulations, Regulation 1 (August 26, 2017) as:

> "[C]omplementary medicine" means any substance or mixture of substances that—
>
> (a) originates from plants, fungi, algae, seaweeds, lichens, minerals, animals or other substance as determined by the Authority;
> (b) is used or purporting to be suitable for use or manufactured or sold for use—
>   (i) in maintaining, complementing or assisting the physical or mental state; or
>   (ii) to diagnose, treat, mitigate, modify, alleviate or prevent disease or illness or the symptoms or signs thereof or abnormal physical or mental state of a human being or animal; and
> (c) is used—
>   (i) as a health supplement; or
>   (ii) in accordance with those disciplines as determined by the Authority.

The capacity and authority of the Minister of Health and SAHPRA Africa to regulate complementary products was contended by the Alliance of Natural Health Products of South Africa in the Supreme Court of Appeal of South Africa Judgement (Case 256/2021; 2022). The Alliance argued that the Minister of Health was only empowered to regulate medicines and scheduled substances within the meaning of *medicine* as described by the Medicines Act, and that the Medicines Act General Regulations purported to regulate medicines that were not within the description of *medicine* in the Medicines Act. The dispute was whether *complementary medicine* as defined in the Medicines Act General Regulations was aligned with the definition of *medicine* in the Act and limited to therapeutic or medicinal purposes—"drinking water is clearly not a medicine under the Medicines Act, but could be construed as complementary medicine in terms of Medicines Act General Regulations." The Appeals Court affirmed the decision of the Supreme Court and granted 12 months to the Minister of Health to remedy the issue. According to SCA 256/2021:

> The definition of "medicine" in section 1 of the Medicines and Related Substances Act No.101 of 1965 is declared to apply only to substances that are used or purport to be suitable for use or are manufactured or sold for use in the diagnosis, treatment, mitigation, modification or prevention of maladies, in order to achieve a medicinal or therapeutic purpose, in human beings and animals.
>
> The General Regulations promulgated on 25 August 2017 under General Notice 859 in Government 41064 are declared unlawful to the extent that they apply to "complementary medicines" and "health

supplements" that are not "medicines" or "Scheduled substances" as defined in section 1 of the Medicines and Related Substances Act No. 101 of 1965.

The General Regulations to the Medicines Act have since been amended in April 2023 to provide a new definition for *complementary medicine*, in the Medicines Act Regulations Amendment of 2023. *Complementary medicine* is now defined as a category D medicine. Category D medicines are outlined in the Medicines Act Regulations Amendment sub-regulation (2):

> "**[C]omplementary medicine**" means a category D medicine.
>
> (Regulation 1 of the Medicines Act General Regulations Amendment, April 2023)

This definition of *complementary medicine* is highly specific to discipline-specific and health supplements as outlined in Annexure 1 for humans and Annexure 2 for animals in the General Regulations of the Medicines Act. Category D medicines are regulated according to the Medicines Act General Regulations.

Discipline-specific traditional claims are listed in Section 33 of Annexure 1 of the Medicines Act General Regulations (August 26, 2017) as "aromatherapy, homeopathy, phytotherapy, traditional Chinese medicine, Unani medicine, Western herbal medicine, combination product, and other herbal."

Health supplements are listed in Section 34 of Annexure 1 of the Medicines Act General Regulations (August 26, 2017) as "amino acids; monosaccharides; animal extracts, products, and derivatives; carotenoids; enzymes; fats, oils, and fatty acids; minerals; polyphenols (including bio-flavonoids); probiotics; saccharides (including prebiotics); vitamins; multiple substance formulation; and other."

10.3 REGULATORY AUTHORITIES

This structure of regulation in South Africa, the integration and key interactions between legislations and the authorities, and the requirements for the importer/exporter of complementary medicines, is shown in Figure 10.1. The Minister of Health and SAHPRA, through the Medicines Act, the Medicines Act General Regulations, the Medicines Act General Regulations Amendment, and the SAHPRA Guidelines, regulate medicines and related substances. The Minister of Health and SAPC, meanwhile, regulate pharmacies and pharmacists through the Pharmacy Act, the Pharmacy Act Regulations, and the Pharmacy Act Rules.

The importer of complementary medicine is required to have thus:

- A National Department of Health license for the pharmacy to import and/or export medicines
- A SAPC certificate of recording of the pharmacy
- A SAPC certificate of recording of the pharmacy owner
- A SAPC registration of the responsible pharmacist
- A SAPC pharmacy grading certificate
- A SAHPRA license to import medicines with the medicines listed
- The medicines registered (or listed in the case of complementary) with SAHPRA

Examples of the wording of these certificates are in the sections and figures that follow.

10.3.1 National Department of Health

Figure 10.2 shows an example of the wording on the license issued by the National Department of Health.

10.3.2 South African Health Products Regulatory Authority

The objectives and functions of SAHPRA are listed in Sections 2A and 2B of the Medicines Act.

Figure 10.3 shows an example of the wording on the license issued by SAHPRA.

10.3.3 South African Pharmacy Council

The objectives of the SAPC are listed in Section 3 of the Pharmacy Act and focus on the promotion of the health care of the population and the registration and conduct of pharmacists.

Figure 10.4 shows an example of the wording on a certificate of recording of a pharmacy owner issued by SAPC.

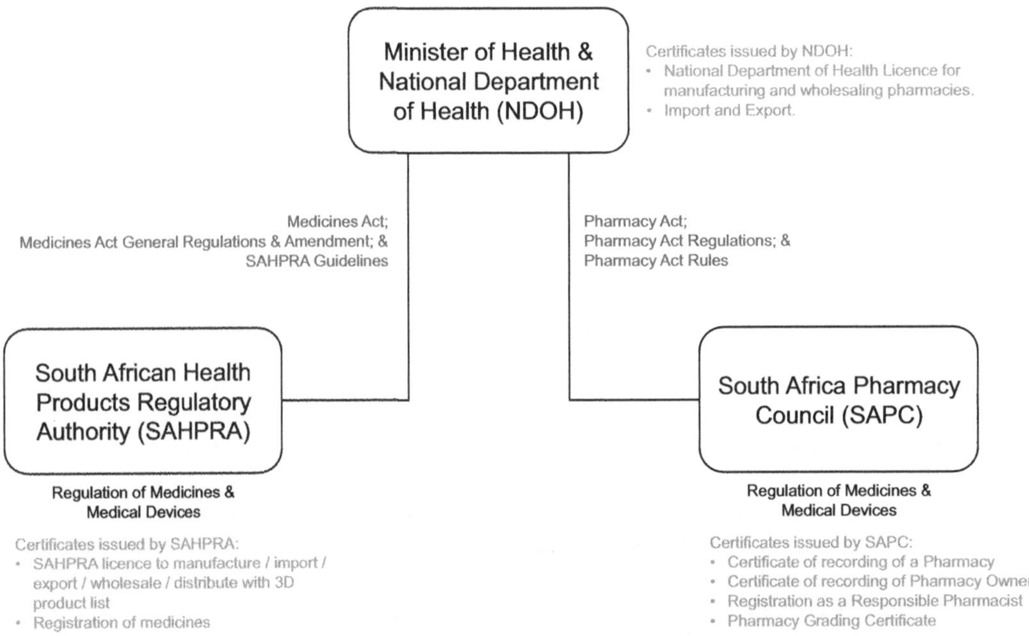

FIGURE 10.1 Structure of regulation of complementary medicines in South Africa.

> Department of Health issues a licence to [Entity] as a ["Manufacturing Pharmacy"] to ["import or export"] "complementary medicines (category D) only"; on condition: provided that the licence holder operates the pharmacy in compliance with the [Licence Number DL02] issued in terms of Section 22c(1)(b) of the Medicines Act. No deviations from the conditions of this licence may take place without prior written approval of SAHPRA. [Entity] is authorized in terms of Section 22A of the Pharmacy Act to own a [manufacturing pharmacy] and licensed to carry on the business of a [manufacturing pharmacy] only located at [location]. ("[ ]" implies could vary)

FIGURE 10.2 Example of wording on license issued by the National Department of Health.

SAHPRA issues a Licence to [import or export] complementary medicines (category D), issued in terms of section 22C(1)(b) of the Medicines Act and regulation 23 of the Medicines Act Regulations made in terms of the same Act, to the Licence Holder indicated herein and licences this party to [manufacture, import or export] Complementary medicines (Category D), inclusive of the activities stipulated herein. Licensed activities – [storage, importing, exporting].

The licence consists of one page in addition to the letter granting and the complete product list and is issued in terms of section 22C(1)(b) on the following conditions: The licence holder and the persons described and names in this licence must"

At all times ensure that all Complementary Medicines (Category D) associated with this licence, irrespective of registration status, comply with the Medicines Act, as amended and the General Regulations made in terms of this Act be inspected at least once within a period of FIVE (5) years for verification of compliance with the minimum requirements as attested to; and pay annual licence retention fees.

The licence needs to be renewed every 5 years.

Verification of appropriate standards and certification thereof will only be undertaken after successful inspection of the premises.

Only products listed or attached to this licence will be permitted association with this licence as Category D medicines and continued rights of sale, provided that an application is submitted for their registration by the prescribed deadlines of the applicable notice issued in terms of section 14 of the Medicines Act.

This licence is issued to the licence holder based on SAHPRA's acceptance of the applicant's attestation of compliance with minimum requirements relevant to Category D medicines at the time of application and the payment of the required licence application and desktop evaluation fees. This licence is not a confirmation of applicable GMP standards, unless presented together with a valid SAHPRA GMP certificate.

"Authority" means the South African Health Products Regulatory Authority established by section 2 of the "Medicines Act". Section 2A of the Medicines Act lists the objects of the Authority which are to provide for the "*monitoring, evaluation, regulation, investigation, inspection, registration and control of medicines, Scheduled substances, clinical trials and medical devices, IVDs and related matters in the public interest.*" Section 2B of the Medicines Act lists the functions of the Authority:

"*(a) ensure the efficient, effective and ethical evaluation or assessment and registration of medicines, medical devices and IVDs that meet defined standards of quality, safety, efficacy and performance, where applicable;*

(b) ensure that the process of evaluating or assessing and registering medicines, medical devices and IVDs is transparent, fair, objective and concluded timeously;

(c) ensure the periodic reevaluation or reassessment and monitoring of medicines, medical devices and IVDs;

(d) ensure that evidence of existing and new adverse events, interactions, information with regard to Postmarketing surveillance and vigilance is being monitored, analysed and acted upon;

(e) ensure that compliance with existing legislation is being promoted and controlled through a process of active inspection and investigation; and

(f) ensure that clinical trial protocols are being assessed according to prescribed ethical and professional criteria and defined standards." ("[ ]" implies could vary)

FIGURE 10.3 Example of wording on a license issued by SAHPRA.

Registration certificate issued by SAPC under the Pharmacy Act. Certificate of recording of a Pharmacy Owner issued in terms of the Pharmacy Act, to certify that [Entity] has been recorded on [date] as the owner of [Pharmacy Number] at [location]. ("[ ]" implies could vary)

FIGURE 10.4 Example of wording on a certificate of recording of a pharmacy owner issued by SAPC.

Registration as a Responsible Pharmacist issued by SAPC under the Pharmacy Act, to certify that [Name with registration number], has been registered as the responsible pharmacist of [Entity] on [date]. ("[ ]" implies could vary)

FIGURE 10.5 Example of wording on a registration as a responsible pharmacist issued by SAPC.

Registration certificate issued by SAPC under the Pharmacy Act. Certificate of recording of a Pharmacy issued in terms of the Pharmacy Act, to certify that [Entity] has been recorded on [date] in the category of a [Manufacturing Pharmacy] and is in possession of [Licence Number] dated [date] to conduct a pharmacy business from [location]. ("[ ]" implies could vary)

FIGURE 10.6 Example of wording issued on a certificate of recording of a pharmacy by SAPC.

Registration certificate issued by SAPC under the Pharmacy Act. Pharmacy grading certificate, to confirm that [Entity] has attained [A-Grade] following an inspection conducted on [date] by SAPC in accordance with the provisions of the Pharmacy Act. This grading shall remain valid until the pharmacy is inspected at a date determined by council or until such certificate is withdrawn. This certificate is valid until [date]. ("[ ]" implies could vary)

FIGURE 10.7 Example of wording issued on a pharmacy grading certificate by SAPC.

Figure 10.5 shows an example of the wording on a registration as a responsible pharmacist issued by SAPC.

Figure 10.6 shows an example of the wording issued on a certificate of recording of a pharmacy by SAPC. An annual fee is payable for the responsible pharmacist and for the pharmacy.

Figure 10.7 shows an example of the wording issued on a pharmacy grading certificate by SAPC.

10.4 REQUIREMENTS TO IMPORT/EXPORT COMPLEMENTARY MEDICINES

This section discusses the requirements for importing, manufacturing, and distributing complementary medicine in South Africa. Section 6(1) and Section 6(2) of the Medicines Act General Regulations outline the rules for importing and transmitting medicines into and through South Africa. Regulation 6(1) specifies the ports of entry that medicines could be imported through. These ports of entry are the main airports and harbors of South Africa. Regulation 6(2) mandates that anyone importing medicines must hold a license, under the Medicines Act, to import medicines. Alternatively, for unregistered medicines, the importer must be authorized by SAHPRA to import the specified unregistered medicines. Regulation 6(3) specifies the requirements for the authorization of importing unregistered medicines. Regulation 7 of the Medicines Act General Regulations addresses the transmission of medicines through South Africa. A medicine storage warehouse should be licensed in terms of Section 22C of the Medicines Act by SAHPRA. The Foodstuffs, Cosmetics, and Disinfectants Act 54 of 1972 controls the sale, manufacture, and import of foodstuffs, cosmetics, and disinfectants. The Customs and Excise Act 91 of 1964 intends to provide for the levying of customs and excise duties, for the prohibition and control of the importation or manufacture of certain goods, and for matters incidental thereto. The importing/exporting entity should be registered with the South African Revenue Service as an importer and/or exporter.

10.4.1 National Department of Health Requirements

In terms of SAHPRA Guideline 7.02, the process starts with an application to the NDoH for a pharmacy premises license. The SAPC and NDoH both review the application, with the SAPC reviewing the application in terms of Good Pharmacy Practice (GPP). After a confirmation of compliance with GPP from SAPC, the applicant may then apply to SAHPRA for the appropriate category D medicine license. SAHPRA evaluates the application in terms of compliance with Good Manufacturing Practices (GMPs), Good Distribution Practices (GDPs), or Good Wholesaling Practices (GWPs) as applicable. SAHPRA also verifies the list of products supplied in the Section 3D product list. Following SAHPRA confirmation to issue the SAHPRA license, SAHPRA then notifies NDoH, which then issues the NDoH license. The applicant then proceeds with SAPC for the recording of the pharmacy and the pharmacy owner and the certificate of registration for the responsible pharmacist.

10.4.2 South African Health Products Regulatory Authority Requirements

There are three types of licenses of category D medicines (SAHPRA Guideline 7.02) for new applications:

- Type DL01—Licence to manufacture, import or export Complementary Medicines (Category D) [manufacturers only]
- Type DL02—Licence to import or export Complementary Medicines (Category D) [holders of certificate of registration]
- Type DL03—Licence to act as a wholesaler of or distribute Complementary Medicines (Category D) [wholesalers or distributors]

SAHPRA's General Information Guideline (December 11, 2023) (SAHPGL-HPA-07) and Guideline 7.02 provide guidelines for the registration process for medicines in South Africa. A primary objective of SAHPRA is to establish the safety, quality, and efficacy of a medicine, and to ensure that all registered medicines are of the required safety, quality, and efficacy. There are administrative requirements for the processing and evaluation of applications. The registration of a medicine in South Africa is governed by the provisions and requirements of the Medicines Act and its regulations and guidelines. The following SAHPRA guidelines are relevant to the compilation of applications for registration of complementary medicines:

- SAHPRA Guideline 7.01: Discipline-Specific Safety and Efficacy
- SAHPRA Guideline 7.04: Health Supplements Safety and Efficacy
- SAHPRA Guideline 7.05: Registration Application ZA-CTD—Quality
- SAHPRA Guideline 7.06: Specified Substances

Adherence to the following practices are required:

- Good Manufacturing Practice
- Good Wholesaling Practice

The SAHPRA General Information Guideline (December 11, 2023) stipulates who an eligible applicant to apply for registration or the holder of the certificate of registration would be. A registered legal entity in South Africa with a responsible pharmacist, that is, a registered pharmacist with SAPC, will qualify as an eligible entity. The responsible pharmacist would be the primary point of contact with SAHPRA. The site master file should be in accordance with SAHPGL-INSP-04 (August 2022) and is the key document that is evaluated. A quality manual and specific standard operating procedures (SOPs) are also required.

The site master file (SAHPGL-INSP-04) (August 2022) provides information on the manufacturer's or wholesaler's standard operating procedures. This is useful in the efficient planning and undertaking of a GMP or a GWP inspection. Requirements and guidelines are provided in SAHPGL-INSP-04 (August 2022) to prepare a site master file. It should entail the following sections:

- C.1. General information
- C.2. Personnel
- C.3. Premises and equipment
- C.4. Documentation
- C.5. Production or wholesaling
- C.6. Quality control
- C.7. Contract manufacture and analysis
- C.8. Distribution, complaints, and product recall
- C.9. Self-inspection

Manufacturing requirements are specified in SAHPGL-HPA–07. Only medicines that are manufactured, packed, and quality-controlled at sites adhering to the current GMP principles prescribed by SAHPRA will be eligible for registration, as per SAHPGL-INS-02 (September 19, 2022).

SAHPGL-INS-02 (September 19, 2022) also states that GMP is a set of principles and procedures to consistently ensure the safety, quality, and efficacy of medicines and related substances and is a participating authority of the Pharmaceutical Inspection Cooperation Scheme (PIC/S). The key objective of PIC/S is to harmonize GMP standards between countries and pharmaceutical inspection authorities. The PIC/S guide to GMP has been adopted by SAHPRA and adjusted for country-specific requirements.

The South African Good Wholesaling Practice for Wholesalers applies to wholesale pharmacies. Wholesalers are responsible for the effective, efficient, and safe handling, storage and distribution of medicines. This includes mitigating risks, such as those related to humidity and temperature, that are not aligned to the required storage and handling conditions and that which may potentially harm the product, in terms of its quality, safety, and efficacy. The provisions of the Pharmacy Act and the Medicines Act apply.

10.4.3 South African Pharmacy Council Requirements

Section 14 of the Pharmacy Act addresses the registration of persons, the recording of licensed premises, and the maintenance of registers.

Section 22A of the Pharmacy Act addresses the ownership of pharmacies. The Minister of Health may establish regulations regarding who can own a pharmacy, the conditions under which ownership is permitted, and the circumstances under which such authority may be revoked.

Section 33 of the Pharmacy Act Regulations addresses the control of pharmaceutical education.

GNR. 553 of April 25, 2003, Regulations Relating to the Ownership and Licensing of Pharmacies, defines different types of pharmacies and related facilities:

> "[C]ommunity pharmacy" means a pharmacy wherein or from which some or all of the services as prescribed in terms of regulation 18 of the Regulations Relating to the Practice of Pharmacy are provided to the general public or any defined group of the general public, but excludes an institutional pharmacy.
>
> "[I]nstitutional pharmacy" means a pharmacy situated in a—
>
> (a) public health facility, wherein or from which some or all of the services as prescribed in terms of regulation 18 of the Regulations Relating to the Practice of Pharmacy are provided to the general public requiring pharmaceutical services, medical or surgical treatment, nursing or other health care from or at that public health facility; or
> (b) private health facility, wherein or from which some or all of the services as prescribed in terms of regulation 18 of the Regulations Relating to the Practice of Pharmacy are provided to persons requiring pharmaceutical services, medical or surgical treatment, nursing or other health care from or at that private health facility.
>
> "[M]anufacturing pharmacy" means a pharmacy wherein or from which some or all of the services as prescribed in terms of regulation 16 of the Regulations Relating to the Practice of Pharmacy are provided and which shall sell medicine only to the wholesale or retail sector or to the State.
>
> "[P]rivate health facility" means any hospital, institution or facility at which provision is made for medical treatment or health care services which is not owned or controlled by the State, and includes facilities such as a clinic, mobile clinic, community health centre, maternity home, or unattached delivery suite, convalescent home, unattached operating theatre and sanatorium but does not include a consulting room, surgery or dispensary of an authorised prescriber.
>
> "[P]ublic health facility" means any hospital, institution or facility at which provision is made for medical treatment or other health care services and includes facilities such as a clinic, mobile clinic, community health centre, maternity home or unattached delivery suite, convalescent home, unattached operating theatre and sanatorium that is owned by the State or organ of the State.
>
> "[W]holesale pharmacy" means a pharmacy wherein or from which some or all the services as prescribed in terms of regulation 17 of the Regulations Relating to the Practice of Pharmacy are provided and which shall sell medicines only to the retail sector or to the State.

Good Pharmacy Practice in South Africa references the following:

- Part 1: Rules relating to the code of conduct for pharmacists and other persons registered in terms of the Pharmacy Act (BN 108 in GG 31534 of October 24, 2008).
- Part 2: Rules relating to good pharmacy practice (BN 129 in GG 27112 of December 17, 2004, as amended, and Chapter 2, paras. 2.13.6 to 2.13.9 and 2.26 to 2.31, inclusively sourced from http://www.pharmcouncil.co.za/G_PublicationsD.asp on November 14, 2016).
- Part 3: Rules relating to the services for which a pharmacist may levy a fee and guidelines for levying such a fee or fees (GenN 432 in GG 40892 of June 6, 2017).
- Part 4: Rules relating to acts or omissions in respect of which the Council may take disciplinary steps (GN R599 in GG 11792 of March 31, 1989, as amended by GN R1549 in GG 13382 of July 5, 1991, GN R339 in GG 13747 of January 31, 1992, and GN R1547 in GG 15067 of August 20, 1993).

10.5 ADVERTISING AND LABELING OF COMPLEMENTARY MEDICINES

10.5.1 Advertising

The advertising of complementary medicines is also regulated by the Medicines Act and General Regulations. Section 19(1) of the Medicines Act states that:

> No person shall **sell** any medicine, unless it complies with the prescribed requirements. Therefore, any person who contravenes provision of this sub-section shall be guilty of an offence.

The definition of *sell* by the Medicines Act

> means sell by wholesale or retail and includes import, offer, **advertise**, keep, expose, transmit, consign, convey or deliver for sale or authorize, direct or allow a sale or prepare or possess for purposes of sale, and barter or exchange or supply or dispose of to any person whether for a consideration or otherwise; and "sale" and "sold" have corresponding meanings.

Selling (implying advertising as well) improperly is considered an offence. Section 29 of the Medicines Act describes the offences for the failure to comply with the provisions of the Act, and Section 30 of the Medicines Act provides for penalties when an offence has been committed. The advertising of complementary medicines, therefore, also undergoes the same level of scrutiny as other registered and scheduled medicines, which can disrupt the sales of complementary medicines.

Regulation 42 of the General Regulations states:

> (1) Medicines which contain a Schedule 0 substance or a substance listed as Schedule 1 may be advertised to the public. . . .
>
> (4) No advertisement for a medicine may contain a statement which deviates from, is in conflict with or goes beyond the evidence submitted in the application for registration of such medicine with regard to its safety, quality or efficacy where such evidence has been accepted by the Authority in respect of such medicine and incorporated into the approved professional information of such medicine.
>
> (5) An advertisement for a medicine shall contain . . .
>
> (c) in the case of a . . .
> (ii) complementary medicine—
> (aa) a statement identifying the discipline of the medicine where relevant;
> (bb) an indication that the medicine must be used in accordance with the applicable complementary discipline and principles where relevant; and
> (cc) if the medicine has not received registration with the Authority the following disclaimer "This unregistered medicine has not been evaluated by the SAHPRA for its quality, safety, or intended use."

Complementary medicines can therefore be advertised to the public subject to the prescribed regulations. Since the roadmap for the processes on registering complementary medicines is not yet in place, all complementary medicines in the South African market would have the disclaimer as in Section 42 (5)(c)(ii)(cc), described earlier, which is: "This unregistered medicine has not been evaluated by the SAHPRA for its quality, safety, or intended use.".

Section 18 prescribes the requirements for labels and advertising. It also refers to the sampling of medicines in 18B:

> (1) No person shall sample any medicine, medical devices or IVD.
> (2) Use of medicine, medical devices or IVDs for exhibition or appraisal purposes shall be as prescribed.
> (3) For the purposes of this section "sample" means the free supply of products, medical devices or IVDs by a device or IVD establishment, manufacturer or wholesaler or its agent to a pharmacist, medical practitioner, dentist, veterinarian, practitioner, nurse or other person

> registered under the Health Professions Act, 1974 (Act No. 56 of 1974), or any professional or person authorized to use the device.

Since no sampling of medicines is allowed in South Africa, it could negatively affect the successful sale of complementary medicines due to:

- Reduced promotional activities through health-care professionals
- Negative impact on awareness among health-care professionals
- Increased reliance on consumer marketing and alternative marketing channels

10.5.2 Labeling

There exist many complementary medicines in the South African market. The large players mainly sell in large South African retail pharmacy chains, while the other players opt for smaller pharmacy chains and other specialty shops, depending on their product category and clientele.

An example of a complementary medicine's packaging artwork compliant to the SAHPRA guidelines and the Medicines Act (Section 10 of the General Regulations) is shown in Figure 10.8 and Figure 10.9. Figure 10.8 shows the immediate packaging or outer packaging of the product, and Figure 10.9 shows the label on the intermediate packaging (like a tube, bottle, etc.).

Along with the product, it is mandatory to have a patient information leaflet (or package insert) which follows the prescribed requirements as stipulated in Section 12 of the General Regulations.

10.6 PURPOSE OF REGULATION OF COMPLEMENTARY MEDICINES

The key purpose of the Medicines Act is to ensure the safety, quality, and efficacy of medicines. The regulations also promote standards, best practices, and international harmonization.

> The quality, safety and efficacy of complementary medicines in South Africa cannot be ascertained unless they have been evaluated and registered by the SAHPRA.
>
> (SAHPRA Guideline 7.02)

Some of the reasons for the regulation of complementary medicines posed by SAHPRA (SAHPRA Guideline 7.02) include that many plants and natural compounds could be poisonous; complementary medicines could interact with other medicines, resulting in side effects or the other medicines reducing or enhancing their effects; and adverse reactions could occur because of complementary medicines. Further concerns are that there are claims made on complementary medicine products without the necessary clinical trials and evidence supporting the claims.

10.7 CONCLUSIONS

SAHPRA published a roadmap and transitional process for the regulation of complementary medicines (SAHPRA Guideline 7.02) in December 2021. However, this was withdrawn due to a Supreme Court judgment (April 11, 2022). It is envisaged that a new roadmap could be published or for complementary medicines to be called for registration as per Section 14 of the Medicines Act, with "call-up notices." In terms of SAHPRA Guideline 7.02 (December 2021), a transitional priority licensing period from February 17, 2020, to February 28, 2022, was provided for the license applications of existing category D medicine manufacturers, importers, exporters, wholesalers, or distributors prior to the publication of new call-up notices issued in terms of Section 14 of the Medicines Act. The process flow (SAHPRA Guideline 7.02) provides for a coordinated approach between NDoH, SAHPRA, and SAPC in issuing new licenses.

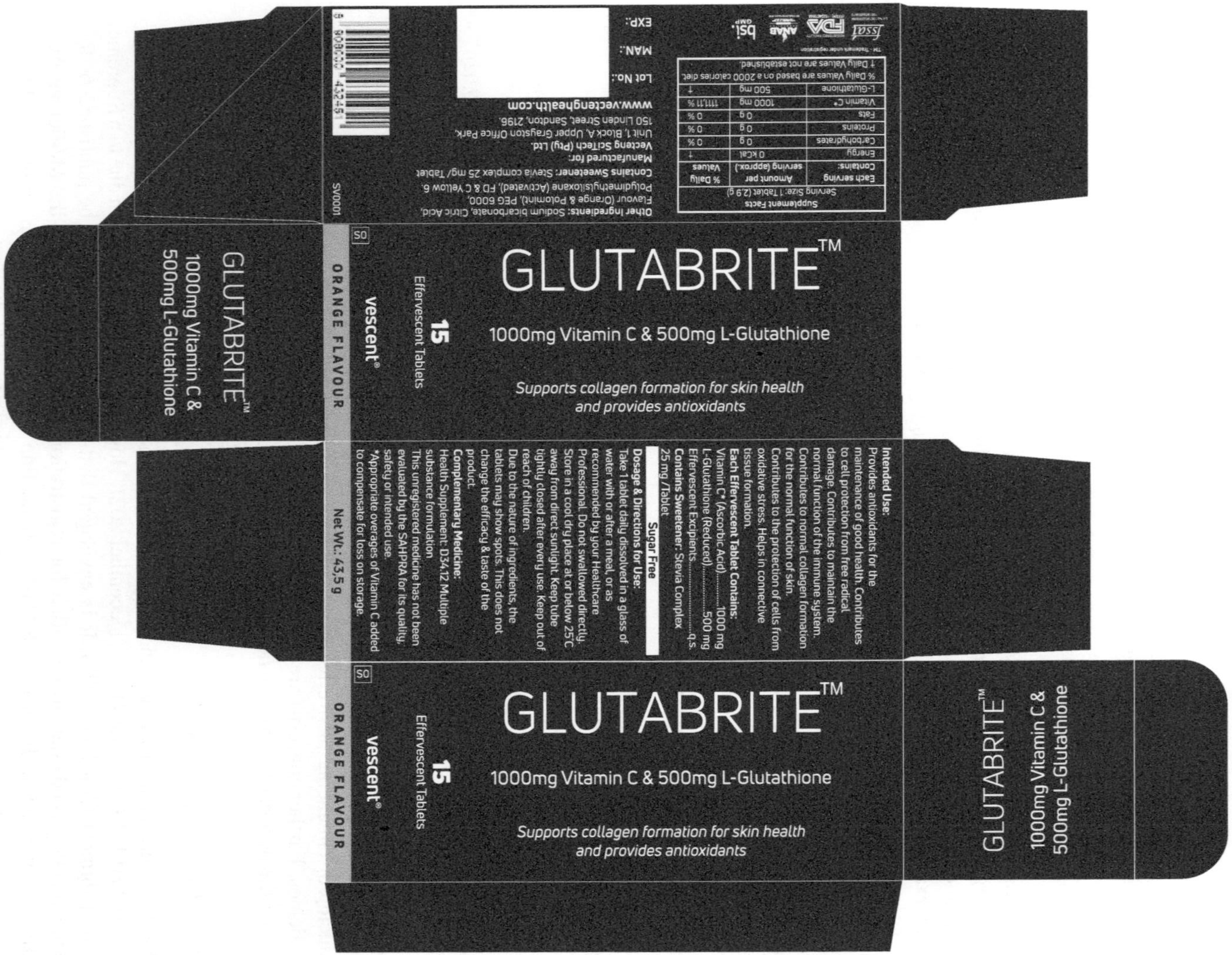

FIGURE 10.8 Example of an immediate packaging design for a complementary medicine compliant to the Medicines Act.

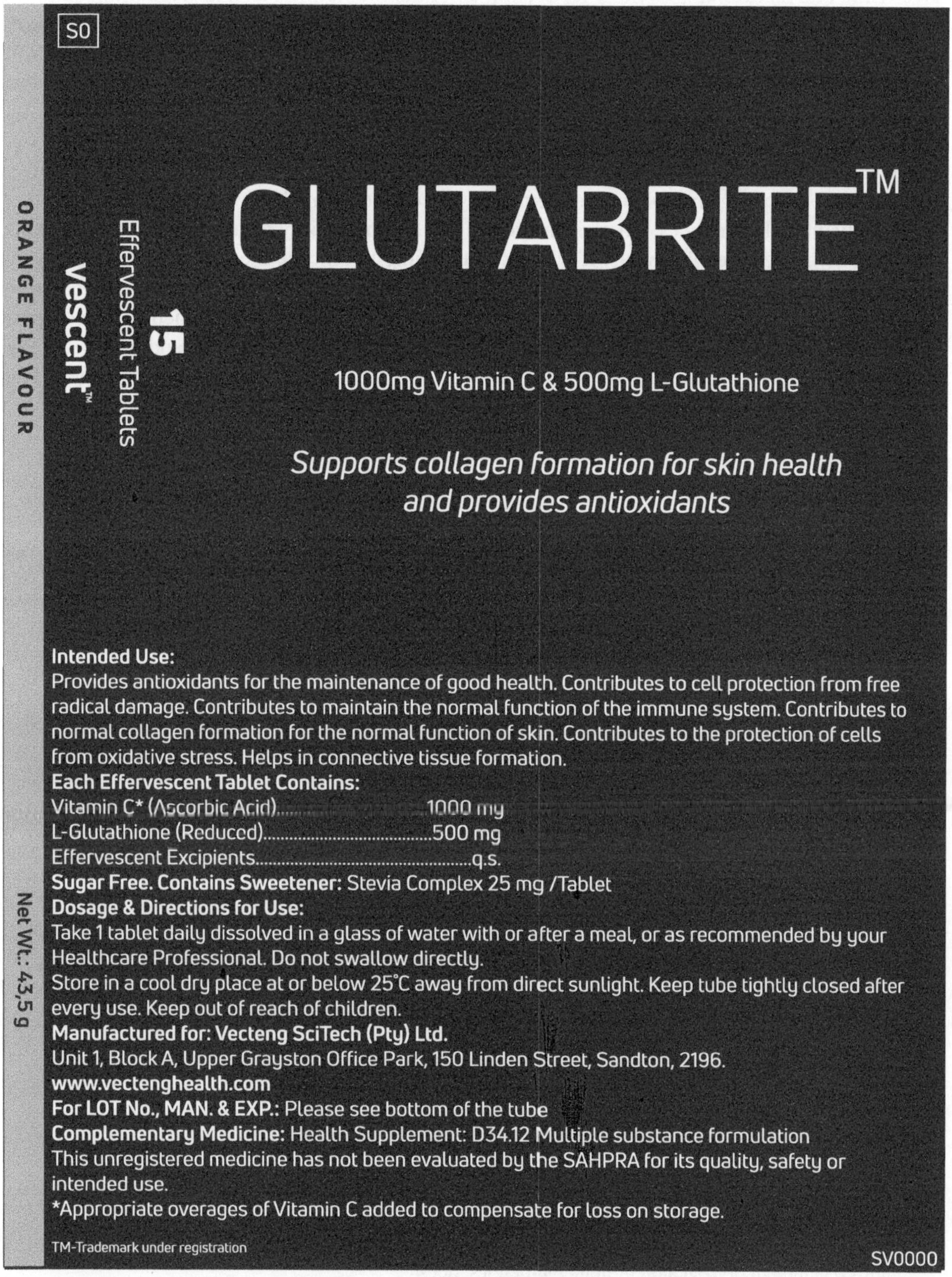

FIGURE 10.9 Example of an intermediate packaging label design for a complementary medicine compliant to the Medicines Act.

In terms of SAHPRA Guideline 7.02, areas for product compliance include product label, professional information and the patient information leaflet (refer to SAHPRA Guideline 7.05), and advertising and marketing (Sections 18, 19, and 20 of the Medicines Act and Regulation 42 of the Medicines Act General Regulations). Guidance on the use of particular expressions in advertising unregistered medicines is published in the SAHPRA Guideline 7.02. SAHPRA Guideline 7.02 describes the process and requirements for the licensing of manufacturers, wholesalers, distributors, importers, and exporters of category D medicines in terms of the Pharmacy Act and the Medicines Act.

It could be that 12 months are provided for before call-up notices are issued. It will be required that applications for registration as complementary medicine be made by the deadlines prescribed by the relevant call-up notices. There is, however, still an uncertainty as to whether the authorities will implement these requirements. It will be challenging to enforce these regulations due to the proliferation, diversity, and large quantities of complementary medicines in the market. It could be argued that regulatory resources should be prioritized to regulate higher-risk scheduled medicines.

Currently, import/export entities should be licensed with the NDoH, SAPC, and SAHPRA, and the products that are imported or exported listed with SAHPRA and sold with the disclaimer "This unregistered medicine has not been evaluated by the SAHPRA for its quality, safety, or intended use."

REFERENCES

Case 256/2021. The supreme court of appeal of South Africa judgment. 2022. https://www.gpwonline.co.za/GPWGazettes.htm

GNR.553 of 25 April 2003: Regulations relating to the ownership and licencing of pharmacies. https://www.sapc.za.org/Publications

Good Pharmacy Practice Manual and Associated SAPC rules. https://www.sapc.za.org/Publications

Government Gazette 866. Foodstuffs, cosmetics and disinfectants act 54 of 1972. https://www.gpwonline.co.za/GPWGazettes.htm

Government Gazette 3530. Customs and excise act 91 of 1964. https://www.gpwonline.co.za/GPWGazettes.htm

Government Gazette 23480. Pharmacy act no 53 of 1974. 3 June 2002. https://www.sapc.za.org/Publications

Government Gazette 40869. Medicines and related substances act no. 101 of 1965. 26 May 2017. https://www.gpwonline.co.za/GPWGazettes.htm

Government Gazette 41064. Medicines and related substances act no. 101 of 1965. General Regulations. 25 August 2017. https://www.gpwonline.co.za/GPWGazettes.htm

Government Gazette 48358. Medicines and related substances act no. 101 of 1965. General Regulations made in terms of the Medicines and related substances act no. 101 of 1965: Amendment. 2023. https://www.gpwonline.co.za/GPWGazettes.htm

SAHPGL-CEM–02. SAHPRA. Guideline for professional information for human medicines (categories A and D). 18 March 2024. https://sahpracm.org.za/catd/cm-home/guidelines_docs/

SAHPGL-CEM–03. SAHPRA. Guideline for patient information leaflet for human medicines (categories A and D). 16 September 2022. https://sahpracm.org.za/catd/cm-home/guidelines_docs/

SAHPGL-CEM-NS–03. SAHPRA. Guideline for proprietary names for medicines. May 2022. https://sahpracm.org.za/catd/cm-home/guidelines_docs/

SAHPGL-CEM-PRE–01. SAHPRA. Guideline on co-packaging of medicines. 17 February 2023. https://sahpracm.org.za/catd/cm-home/guidelines_docs/

SAHGPL-HPA–03. SAHPRA. Guideline on South African eCTD validation criteria. 30 November 2022. https://sahpracm.org.za/catd/cm-home/guidelines_docs/

SAHPGL-HPA–04. SAHPRA. Renewal of registration of human and veterinary medicines. 22 August 2023. https://sahpracm.org.za/catd/cm-home/guidelines_docs/

SAHPGL-HPA–06. SAHPRA. Variations addendum for human and veterinary medicines. 30 September 2022. https://sahpracm.org.za/catd/cm-home/guidelines_docs/

SAHPGL-HPA–07. SAHPRA. General information guideline. 11 December 2023. https://sahpracm.org.za/catd/cm-home/guidelines_docs/

SAHPGL-HPA–08. SAHPRA. Multiple applications guideline. 11 December 2023. https://sahpracm.org.za/catd/cm-home/guidelines_docs/

SAHPGL-INSP–02. SAHPRA. Guideline on good manufacturing practice for medicines. 18 September 2022. https://sahpracm.org.za/catd/cm-home/guidelines_docs/

SAHPGL-INSP–03. SAHPRA. South African good wholesaling practice for wholesalers. July 2016. https://sahpracm.org.za/catd/cm-home/guidelines_docs/

SAHPGL-INSP–04. SAHPRA. Guidelines for preparation of site master file. 18 September 2022. https://sahpracm.org.za/catd/cm-home/guidelines_docs/

SAHPGL-INSP-RC–07. SAHPRA. Guidelines for advertisement of medicines and health products. 6 June 2022. https://sahpracm.org.za/catd/cm-home/guidelines_docs/

SAHPGL-PEM–01. SAHPRA. Availability of medicines for use in a public health emergency. (PHE). August 2023. https://sahpracm.org.za/catd/cm-home/guidelines_docs/

SAHPGL-PEM–02. SAHPRA. Quality and bioequivalence guideline. 23 May 2023. https://sahpracm.org.za/catd/cm-home/guidelines_docs/

SAHPGL-PEM–03. SAHPRA. Guideline for the API Master File (APIMF) procedure. November 2022. https://sahpracm.org.za/catd/cm-home/guidelines_docs/

SAHPRA Guideline 7.02. SAHPRA. Roadmap and transitional process for the regulation of complementary medicines. December 2021. https://sahpracm.org.za/catd/cm-home/guidelines_docs/

SAHPRA Guideline 7.03. SAHPRA. Complementary medicines—use of the ZA-CTD format in the preparation of a registration application. June 2020. https://sahpracm.org.za/catd/cm-home/guidelines_docs/

SAHPRA Guideline 7.04. SAHPRA. Complementary medicines—health supplements safety and efficacy. March 2022. https://sahpracm.org.za/catd/cm-home/guidelines_docs/

SAHPRA Guideline 7.05. SAHPRA. Complementary medicines registration application ZA-CTD—quality. June 2020. https://sahpracm.org.za/catd/cm-home/guidelines_docs/

SAHPRA Guideline 7.06. SAHPRA. Complementary medicines guidance on specified substances. 30 April 2022. https://sahpracm.org.za/catd/cm-home/guidelines_docs/

11 Regulatory Provisions for Food Supplements in India

Maitreyi Zaveri and Bhavita Dhru

11.1 INTRODUCTION

Nutraceuticals are products which are very rich in nutrients and provide lots of nutritional benefits, like the prevention of many diseases. *Nutraceutical* is made up of the word *nutrient*, which is a nutritious food or food component, with *pharmaceutical*, which is a medical drug. They help in improving and maintaining human health, and also, the prevention of some lifestyle diseases, like diabetes, cancers, and osteoporosis, is covered by nutraceuticals.

Today, the market of nutraceuticals is rising, and the request for these products is increasing, since consumers always look for health products to supplement their diet and to enhance their health. People need this for many reasons, such as lifestyle changes and the consumption of food and dietary habits of people. Nutraceuticals are widely available in many forms, and their use varies from country to country. It may range from isolated nutrients to food supplements, processed foods, and plant preparations and may be formulated in oral dosage in the form of pills, capsules, powders, and syrups, and also as dietary supplements, prebiotics, probiotics, and functional food. Different countries have their own regulations and legislative requirements. Some countries follow stringent regulations, whereas in some countries, well-structured and streamlined regulations for nutraceuticals are lacking.

All categories of nutraceuticals/dietary supplements clutch the market.

The terminology and definition of dietary supplements vary throughout the world, with developed countries having regulations in place, followed by some stringent laws and regulatory requirements, whereas in developing countries, the regulations are constantly evolving. [1]

11.1.1 The Food Safety and Standards Authority of India (FSSAI)

In India, FSSAI (Food Safety and Standards Authority of India) regulates all the activities of manufacturing, storage, packaging, import, and sale to ensure the availability of food and food products within the country. FSSAI, which regulates food and food products for special dietary purpose, also regulates many categories of food, such as nutraceuticals, health supplements (HS), food for special dietary use (FSD), and food for special medical purpose (FMP), which includes all the products that contain vitamins, minerals, amino acids, etc. FSSAI was established under the Food Safety and Standards Act (FSS Act) of 2006, which consolidated various acts and orders that up until then had handled food-related issues in various ministries and departments. FSSAI was created to lay down science-based standards for articles of food and to regulate all their aspects (manufacture, storage, distribution, sale, and import), to ensure the availability of safe and wholesome food for human consumption. [2–3]

11.1.2 History of Food Regulation in India [4]

India is the world's largest producer of fruits and vegetables, and by the mid-1990s, the food processing sector laws were framed in a veritable grid of regulations including a multitude of state laws as well as the following national laws.

 DOI: 10.1201/9781003465041-11

TABLE 11.1
History of Food Regulation in India

Export (Quality Control and Inspection) Act	1963
Solvent-Extracted Oil Control Order	1967
The Insecticide Act	1968
Meat Food Products Order	1973
Prevention of Food Adulteration Act (PFA) of 1954 Rules (Ministry of Health and Family Welfare) with Last Amendments	1986
Bureau of Indian Standard Act	1986
Environmental Protection Act	1986
Pollution Control Act	1986
Milk and Milk Products Order 992	
The Infant Milk Substitutes, Feeding Bottles, and Infant Food (Regulation of Production, Supply) Act of 1992 and Rules	1993
Prevention of Food Adulteration Act (PFA) of 1954 Rules (Ministry of Health and Family Welfare) with Last Amendments in 1986	1986
Fruit Industrial Products Act of 1955	1955
Essential Commodities Act of 1995 (Ministry of Food and Consumer Affairs)	1955
Industrial License	
Vegetable Oil Product Control Order of 1998	1998

TABLE 11.2
Timeline of Food Laws in India

Committee appointed by the Central Advisory Board of Health.	1937
Prevention of Food Adulteration Bill passed in 1954, came into effect.	1955
Subject Group on Food and Agro Industries recommended one comprehensive legislation.	1998
Parliamentary Committee on Pesticide Residue emphasized converging all food laws, with a single regulatory body.	2004
Standing Committee of Parliament on Agriculture submitted for an integrated food law.	2005
Law Commission was also asked to give its view.	2005
FSSA Bill passed, came into force in 2007, 2008, and 2009.	2006

FSSAI is mandated by the FSS Act of 2006 to perform the following functions: framing of regulations, laying down the standards and guidelines in relation to articles of food, and specifying the appropriate system to enforce various standards thus notified. [5–6]

11.1.3 The Food Safety and Standards Act of 2006

The Food Safety and Standards Act of 2006 aims to establish a single reference point for all matters relating to food safety and standards by moving from multilevel, multidepartment control to a single line of command. It incorporates the salient provisions of the Prevention of Food Adulteration Act of 1954 and is based on international legislations, instrumentalities, and the Codex Alimentarius Commission. [7]

The Food Safety and Standards Act lays down the mechanisms and guidelines for the accreditation of different bodies engaged in the certification of the food safety management system for food

TABLE 11.3
Food Safety and Standard Act

Name of Act/Order	Year
Vegetable Oil Products (Control) Order	1947
Prevention of Food Adulteration Act	1954
Fruit Production Order	1955
Any other order under the Essential Commodities Act relating to food	1955
Solvent-Extracted Oil, De-Oiled Meat, and Edible Flour (Control)	1967
Meat Food Products Order	1973
Milk and Milk Products Order	1992
Edible Oil Packaging (Regulation) Order	1998

businesses. It also lays down the procedure for the guidelines and accreditation of laboratories and the notification of accredited laboratories.

These accreditation bodies provide scientific advice and technical support to state governments and the central government in matters relating to the framing of policy and rules in areas which have a direct or indirect bearing on food safety and nutrition.

They also collect and organize data regarding food consumption, the incidence and occurrence of biological risks, the presence of contaminants in food, and the residues of various processes, as well as the identification of emerging risks and the introduction of a quick alert system.

Moreover, they create an information network across the country so that the public, consumers, Panchayats, etc. receive fast, trustworthy, and objective information regarding food safety and issues of concern. At the same time, they provide training programs for persons who are intending to get involved in the food business.

Hence, they contribute to the development of international technical standards for food and phytosanitary and sanitary standards.

They always promote general awareness about food safety and food standards.

11.1.4 Required Forms

Form A	Form B	Form C
It is an application for registration or the renewal of a registration.	It is an application for a license/the renewal of a license.	It is a license format.

The FSSAI was started under the Food Safety and Standards Act of 2006. The regulation involves parameters on how these products can be produced/manufactured and marketed, which means sold or distributed or imported by another company. FSSAI defines *dietary supplement* as "[a] product which is taken by mouth and contains a dietary ingredient and/or a new dietary ingredient which is intended to supplement the diet." [8]

Nutraceuticals are products obtained from food and food sources which provide a health advantage by adding to the basic nutritional value in food, which is a fusion of nutrition and pharmaceuticals. DeFelice defined *nutraceutical* as "a food or part of a food that offers medical or health benefits, including the prevention and/or treatment of a disease." It is a more natural way to accomplish therapeutic results with fewer side effects with its nutritious properties and to attain a better quality of life for the people using more of these products. Food plays a pivotal role in maintaining human health, immunity, inflammation regulation, and overall body functioning. Nowadays,

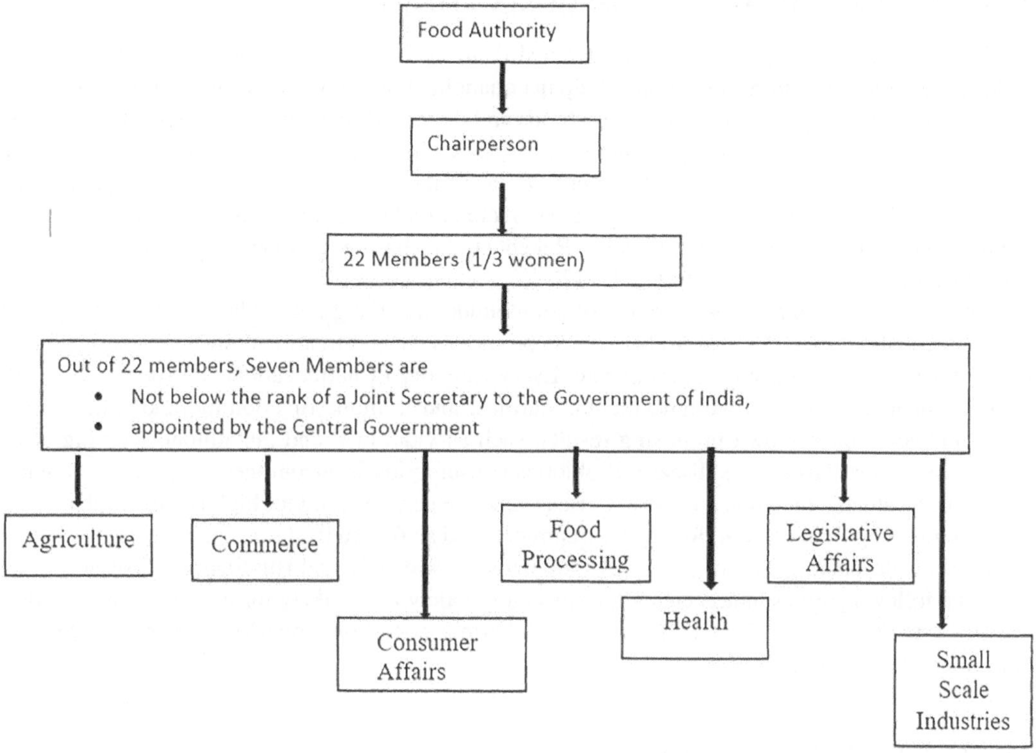

FIGURE 11.1 Food authority.

TABLE 11.4
Indian Registration Process Parameters

Parameter	In India
Defined and known as	Functional food for food supplement
Responsible regulatory authority for the registration of dietary supplements	FSSAI (Food Safety and Standards Authority of India)
Rules and regulations applied	The Food Safety and Standard Regulation
Regulatory agency's website	https://www.fssai.gov.in/
Fees required for registration	Rs 100
Timeline for approval of application	60 days without queries, and 90 days with queries
Validity period	1 to 5 years
Health claims	Allowed ingredients, nutrition claims, functional claims, health maintenance claims, anti-aging claims
Nutritional information on labeling	Nutritional information given in the form of a barcode or a Global Trade Item Number (GTIN)

lifestyle alteration is a major factor for people using more of nutraceutical products, which are mainly products isolated from food and sold in medicinal formulae. Many of them are proven to help in chronic diseases also. Different counties have their own definitions and terms, with various product categories with numerous synonyms used internationally. A formally shared and accepted definition of nutraceuticals is still not available [9, 10].

11.2 NUTRACEUTICALS' ROLE IN HUMAN HEALTH

Worldwide lifestyle changes have influenced a shift in the everyday diet regime. Apart from lifestyle-related diseases, nutraceuticals also help in enhancing body functioning and maintaining daily nutritional needs. Nutrients found in natural foods and their modification into nutraceutical products have been proven to manage and prevent many diseases. Hence, nutraceuticals are used to improve health, delay the aging cycle, inhibit long-term diseases, increase the lifespan, and/or support body structure or function. Nutraceutical usage areas include diabetes, cardiovascular diseases, neurological conditions, arthritis, and gut health. Recent studies have even shown a promising response to nutraceuticals in PCOD and PCOS. [11–12]

Nutraceuticals comprise a wide range of compounds, including antioxidants, different types of phytochemicals, fatty acids, amino acids, and probiotics. In the treatment and prevention of diseases, nutraceuticals play an important role. Every category of nutraceuticals has potential health benefits, such as in human eyesight; mental, cardiac, and immune functioning; and cancer treatment, and some have shown promising results, such as vitamin E and glutathione seeming to be protective against Parkinson's disease. Probiotics consumption is encouraged because of its major role in the treatment and prevention of GI diseases, while garlic is used for high BP and high cholesterol, in lowering lipid content, for cancer treatment, and as free radical scavenger. Oxidative stress causes neurodegenerative diseases, like Alzheimer's, Parkinson's, and Huntington's disease.

In the following areas, nutraceutical products are widely used, likely for diabetes, cancer, allergies, PCOD/PCOS, neurological disorders, ophthalmic conditions, obesity, Parkinson's disease, cardiovascular disease, etc. [13–14]

11.2.1 Categories of Nutraceuticals

There are many evidences and literature reviews available denoting activities which positively affect the human body in the prevention or inhibition of many chronic diseases. They are divided in several categories, according to which they possess activities or therapeutic benefits. [15]

11.2.2 Nutraceuticals' Enhanced Health Benefits

1. **Proactive attitude.** It helps people stay healthy by preventing illnesses.
2. **Natural/plant-based ingredients.** In general, nutraceutical products are prepared by using plants, herbs, fruits, and vegetables which are obtained from natural sources.
3. **Easily available and useful.** They are available in different oral dosages or drinks. It can be easily included in one's daily routine.
4. **Customized as special nutrition.** There are many different products from nutraceuticals, so any person can use supplements that apply to a particular health issue or nutritional need. Hence, people can select products based on their age, lifestyle, or any particular health condition they have.

11.2.3 Nutraceuticals' Drawback

There are no strict regulations in the nutraceutical industry, so that the quality, safety, and effectiveness of nutraceutical products can vary. Some products might make claims that are not substantiated or contain ingredients that are either of low quality or are mixed with unwanted substances.

1. **Deficient scientific evidence.** Many nutraceutical products are not checked by clinical trials/studies to prove their claims. Without this kind of solid evidence, it is difficult for consumers to determine if the products work and how safe they are.

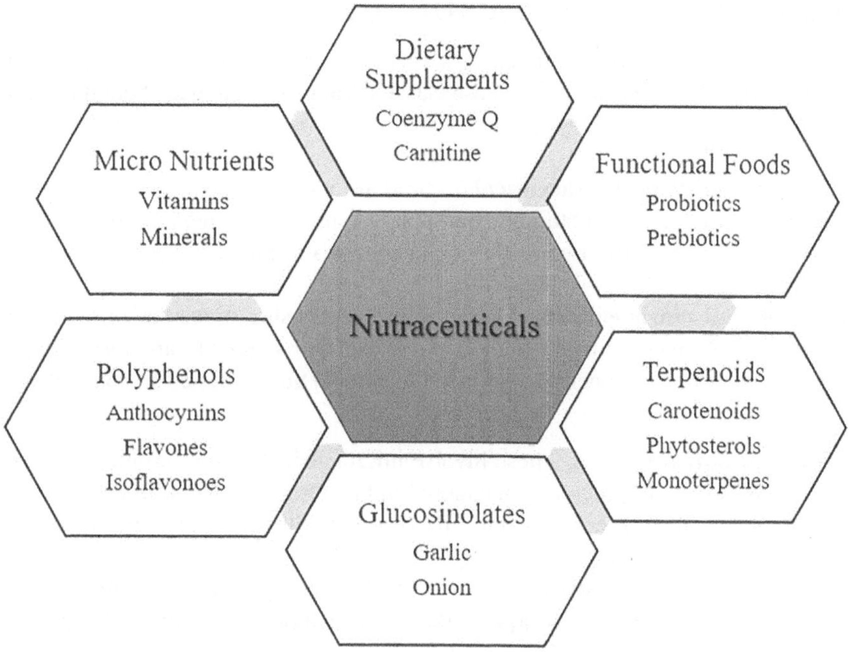

FIGURE 11.2 Categories of nutraceuticals.

2. **Possible interactions and aftereffect reaction.** A bioactive substance interacts with medicines or food supplements, and this interaction can lead to unexpected side effects or diminished effectiveness of either the nutraceutical or the medication.
3. **Financial considerations.** Nutraceuticals often cost more than regular supplements, and this higher price can make them less available for people with lower income or in developing areas. There are also ethical issues with some nutraceutical products, such as misleading claims, exaggerated advertising, or unclear information about ingredients and how they are made.
4. **Bioavailability.** This means that even before they have an opportunity to interact with their intended targets or exert any therapeutic effects, they are quickly removed from the bloodstream and excreted. As a result, their potential benefits are diminished, as they do not stay in the body long enough to make a significant impact.
5. **Effect of placebo outcome.** Sometimes, people might believe that dietary supplements can cure illnesses more effectively than they actually do. This misunderstanding can happen because the body often has its own natural ability to heal itself without extra help from supplements.
6. **Product superiority concerns.** Nutraceuticals that are sold on the international stage frequently boast about their use of organic ingredients and natural chemicals, presenting an appealing image of health and purity. Yet behind these enticing claims lies a potential issue: The protection and effectiveness of these products can be questionable.
7. **Possible interactions.** Nutraceuticals, which are products made from natural sources and marketed for their health benefits, often lack clear information about how they affect existing medical conditions or interact with prescribed medications. Nutraceuticals are meant to complement a balanced diet and a healthy lifestyle, not to replace them. [16]

11.3 CLASSIFICATION OF NUTRACEUTICALS

Nutraceuticals can be classified in various ways, including by their nutrients and properties, activities, mechanisms of action, and sources. These classifications give an idea about the applications of nutraceuticals.

1. **Natural or traditional nutraceuticals.** These are natural products without any changes or modifications. They are obtained from plants, animals, marine environments, or some microorganisms. They are used for basic nutrition. Examples are lycopene, omega-3 fatty acids, etc.
2. **Non-traditional nutraceuticals.** These are prepared using processes or techniques like biotechnology or recombinant technology. These nutraceuticals are formed by adding nutrients or dietary components to improve nutritional quality. They are again divided into two categories:
    - **Fortified nutraceuticals.** These involve micronutrients, essential trace elements, and vitamins. They help increase nutritional value. Orange juice with calcium, vitamins or minerals added in cereals, and milk fortified with cholecalciferol are examples of fortified nutraceuticals.
    - **Recombinant nutraceuticals.** These are obtained by using biotechnology methods or genetic engineering techniques, like the production of yogurt and cheese through enzymatic or fermentation processes. Gold kiwifruit is an example of a nutraceutical.
3. **Potential and established nutraceuticals.** This class of nutraceuticals shows promise of medicinal benefits, but they are only established as medicines after extensive testing and data demonstration.
4. **Herbals.** *Herbals* are botanical products made from fresh or dried parts of plants, like leaves, stems, roots, fruits, seeds, or different extracts. Examples of herbals include garlic, aloe vera, Shatavari, turmeric, and guggul.
5. **Phytochemicals.** These are biologically active phytoconstituents found in plants that provide health benefits, like antioxidant effects, immune boosters, and other health-promoting properties. Examples are flavonoids, phenolic acids, and carotenoids.
6. **Functional foods.** These foods are used to give health benefits. They are used as basic nutrition for increasing antioxidant and anti-inflammatory activities. Examples of functional foods are fermented milk, black chocolate, seeds, fortified juice, dairy products, monocot plants (rice, wheat), soybeans, lentils, citrus fruits, etc.
7. **Dietary supplements.** These are products that provide additional nutrients to the diet, often to address deficiencies or health issues. Examples include folic acid, vitamin A, vitamin D, and iron.
8. **Probiotics.** These are live microbial food ingredients that adhere to specific sites in the gastrointestinal tract, promoting health by eliminating pathogens. Examples are *Lactobacillus acidophilus*, *Bifidobacterium* spp., and *Streptococci.*
9. **Prebiotics.** These are selectively fermented ingredients or fibers which are used to promote the growth of gut bacteria. Examples are oligofructose, galactooligosaccharides, lactulose, inulin, and fructooligosaccharides. [17–18]

11.4 REGISTRATION PROCESS OF NUTRACEUTICALS IN INDIA

Regulation requirements do not mandate clinical trials for old ingredients—that is, a usage history of 15 years or more in India, or 30 years outside, is accepted for registration. Adequate scientific evidence, such as clinical trials and published literature, as support is required for novel ingredients. Nutraceutical products must be registered before entering the Indian market. For nutraceutical product registration in the Indian market, Good Manufacturing Practices are required. In India, the registration process for nutraceuticals must comply with FSSAI regulations. After submission, the application is reviewed, and then it will be approved. After getting the approval, the nutraceutical product will be marketed in India.

11.4.1 Regulatory Overview

The Food Safety and Standards Authority of India (FSSAI) governs the registration process under the Food Safety and Standards Act (FSS Act) of 2006. FSSAI has released guidelines regarding food safety and standards (2011), which cover various aspects of the food industry. [19]

11.4.2 Key Steps in the Registration Process

1. Identify the product category.
2. Before registration, the formulated product must meet safety and quality standards.
3. To get a license, the product must be manufactured in India. Depending on the scale of production, a manufacturing license must be obtained from the relevant Indian licensing authority. After getting a license, it follows that the manufacturing process must comply with the standards.
4. Before applying for registration, a safety/efficacy assessment must be conducted. This covers toxicological studies, clinical trials, etc.
5. For registration, the application must be prepared according to FSSAI requirements, which require the product's composition, manufacturing process, labeling information, and safety assessments.
6. To ensure the product meets the required quality and safety standards, samples must be submitted.
7. Before approval of the registration, FSSAI will review the application, after which FSSAI will grant the registration of the product with specific conditions or limitations, after all requirements are fulfilled.
8. After registration, ensure that the product's labeling complies with FSSAI regulations. [20 21]

11.4.3 List of Documents for Registration

The list of documents includes Form A, a declaration form, Form B, the plan/layout of the unit, a directory list, a machinery list, photo ID and proof of address of proprietor/partner/director(s), signature of authorized person, list of categories to be manufactured, authority letter nominating the responsible person, water analysis report (chemical &and bacteriological), premises possession letter, partnership deed, affidavit of association, recall plan (if applicable), NOCs (where required), which are all required for the registration process.

11.4.4 Challenges in the Registration Process

The registration process for nutraceuticals is not clear in terms of product categorization. Nutraceuticals in tablet or liquid form that contain vitamins and minerals are often classified as drugs, particularly if they are in gelatine capsules, which complicates the registration process. [22]

11.5 LICENSING SYSTEM

FBOs (food business operators) have to register themselves to the FSSAI, and the kind of license or registration they need depends on the annual turnover.

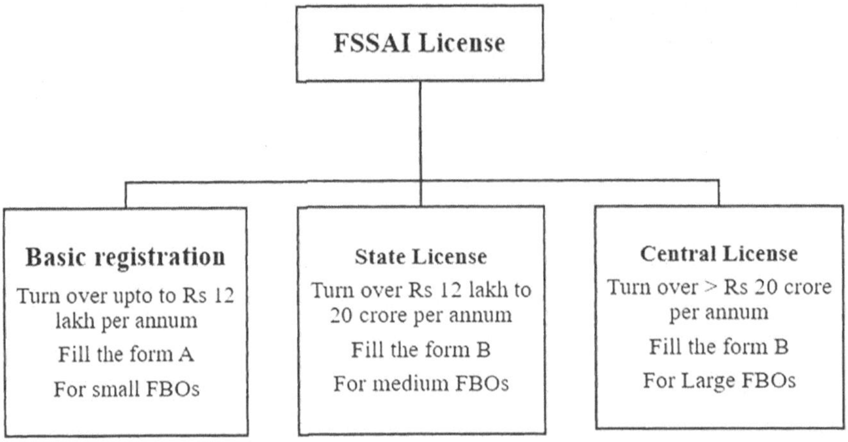

FIGURE 11.3 FSSAI licensing system.

On the basis of annual turnover, the registration process is divided into three categories:

1. **FSSAI basic registration.** FBOs with yearly revenues of less than Rs. 12,00,000 are required to register with the FSSAI. The FSSAI registration Form A must be completed by the applicant in order to get FSSAI basic registration, which is for petty business.
2. **FSSAI state license.** FBOs with annual turnover of more than Rs. 12,00,000 but less than Rs. 20 crores must apply for a state license from the FSSAI. Form B is the FSSAI registration that must be completed by the applicant to get an FSSAI state license, which is for food business operators.
3. **FSSAI central license.** FBOs with annual revenues of more than Rs. 20 crores must obtain an FSSAI central license. To receive an FSSAI central license, the FSSAI registration Form B must be completed. This license is for large food business operators. [21]

11.5.1 Required Documents for FSSAI License/Registration

- Proclamation form
- Authority letter
- Property paper
- Rent agreement paper
- Food safety certificate
- Electricity bill/water bill
- Photo ID proof
- Nominee name

The validity of the FSSAI license is between one and five years. The applicant must apply for a license number for renewal within 30 days of the expiration of the current license. [23]

11.6 LABELING AND PACKAGING REQUIREMENTS

11.6.1 Declaration of Veg and Non-Veg or Vegan Foods

11.6.2 Labeling Requirements

Each and every label of the package must carry information such as the food name, ingredient list, nutritional information, veg or non-veg statue, food additives, complete address, name, FSSAI logo

TABLE 11.5
History of Food Packaging Regulations in India

Sr. No.	Packaging Regulation Detail	Published Year
1	Food safety and standards for the packaging provisions of the packaging and labeling regulations	2011
2	Food safety and standards for packaging regulations	2018
3	Food safety and standards for packaging regulations with respect to food packaging—a new regulation, updated	January 3, 2019
4	Food business operators to comply with all the provisions of these regulations, published in the official gazette	July 1, 2019

TABLE 11.6
Food for Special Dietary Use: Information on the Label

Front of Pack: Following Information on the Label	
Food for special dietary use	It should mention the particular physiological/ health condition/particular usage group.
Whether or not the food is for special dietary use	It is taken under the medical advice of a physician or certified dietician.
Front or Back of Pack	
"This product should not be used by pregnant, nursing, and/or lactating women, children <5 years, adolescents, and the elderly."	Except when medically advised by physician/ certified dietician/nutrition professional.
Product rationale	Product properties, details, characteristics for use of the product.
Nutrient specification	According to normal requirement of nutrient, it becomes reduced, deleted, increased.
"Warning: product is used for the oral, not the parenteral, route."	This product is used for oral consumption only.
Information	Osmolality or osmolarity on acid–base balance.

plus license number, quantity, price, consumer care details, identification number of lot/code/batch, marking date, imported foods label and its country of origin, usage instructions, food allergen information, and food material package (must be sold in retail only).

A color that contrasts with the background of the packaging must be used for the FSSAI logo and license number. The FSSAI logo and license number of the manufacturer or marketer or packer or bottler, if they are different from the brand owner, should be displayed on the label. [24]

11.7 FOOD RECALL PLAN

All food business operators (FBOs) have to submit a detailed recall plan when a recall is being carried out. It is to be submitted as per food recall procedure regulation 7, 2017 to the CEO, FSSAI. A brief step by step procedure and its description are as follow. [25]

TABLE 11.7
Logo That Represents the Type of Food

A brown-colored triangle inside a square symbol is used for non-vegetarian food.	A green-colored filled circle inside a square with green outline symbol is used for vegetarian food.	A green-colored leaf above a green-colored letter *V* is used for vegan food (excluding dairy products or eggs).	The products that are not meant for human consumption are indicated by the symbol of a black *X*.
		VEGAN	

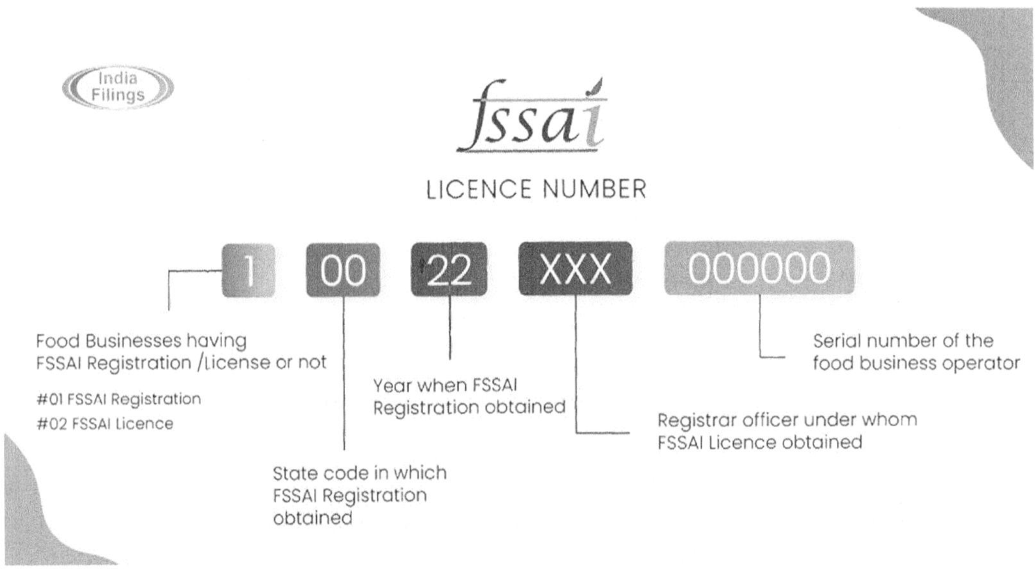

FIGURE 11.4 FSSAI coding of license number.

11.8 HEALTH CLAIMS

Functional food products in Asia tend to be regulated under the conventional food category. In Asian countries, the following group claims are permitted:

1. **Nutrition claims.** Nutrient content claims mean, on the product label, they mentioned the level of certain nutrients, while nutrient comparative claims indicated on the product label describe the nutrient content or energy value as compared to other similar foods.
2. **Structure/function claims.** They include a statement on the label of food products regarding how the product affects the human body.
3. **Risk reduction claims.** In the context of total diet, this regards the reduction in the risk of developing a disease or health-related condition in case of risk reduction claims.

4. **Non-addition claims.** It means any claim that an ingredient or additive has not been added to a food, either directly or indirectly, and the ingredient or additive is one whose presence or addition is permitted in the food and which consumers would normally expect to find in the food.
5. **Conditional claims.** It indicates the word "natural" or "naturally" in the claim statement, where a claim may be made that the food is, by its nature, high or low or free of a specific nutrient, with the name of the nutrient or substance.
6. **Prohibited claims.** The regulations made under the Food Safety and Standards Act of 2006 specify that no claims should be made for the suitability of any food or food product in the prevention, alleviation, treatment, or cure of a disease, disorder, or particular physiological condition.

11.9 NUTRACEUTICALS IN INDIA: OPPORTUNITIES, FUTURE PROSPECTS, AND CHALLENGES

The nutraceutical industry in India is witnessing a significant growth, with numerous opportunities and promising future prospects. Nutraceuticals, encompassing dietary supplements, functional foods, and products derived from traditional medicinal systems, like Ayurveda, are becoming increasingly popular due to the rising health consciousness among consumers.

11.9.1 Opportunities in the Indian Nutraceutical Market

The rising frequency of lifestyle-related diseases/issues is driving consumer interest in nutraceuticals. Opportunities for nutraceuticals in the Indian market are going to be increasing due to the following reasons:

1. Growing health awareness
2. Ageing population
3. Rising middle class
4. Shift toward preventive health care
5. Government support and regulations
6. Ayurveda and traditional knowledge of peoples
7. Innovation and R&D regulations
8. Export potential of industries

11.9.2 Future Prospects of the Nutraceutical Market

1. **Expansion into rural markets.** The potential for growth in rural markets is significant, with increasing health awareness and product availability. A report by Nielsen India highlights the growing demand for health products in rural areas, driven by rising disposable incomes and better access to health-care information.
2. **Digital and e-commerce growth.** The rise of digital platforms and e-commerce is facilitating easier access to nutraceutical products across the country. According to a report by PwC India, e-commerce is expected to be a major driver of growth for the nutraceutical industry, with online sales becoming increasingly important.
3. **Personalized nutrition.** Advances in technology and data analytics are paving the way for personalized nutrition, where nutraceutical products can be tailored to individual health needs. A report by Grand View Research predicts that the global personalized nutrition market will grow significantly, offering new opportunities for the nutraceutical industry.

4. **Partnerships and collaborations.** Collaborations between nutraceutical companies, research institutions, and health-care providers could drive innovation and enhance product offerings. The Indian government is encouraging public–private partnerships (PPP) in the health-care sector, which could extend to nutraceuticals.
5. **Global expansion.** Indian nutraceutical companies have the opportunity to expand globally by adhering to international standards and tapping into growing markets, like the United States, Europe, and Southeast Asia. A report by the Export–Import Bank of India (Exim Bank) highlights the export potential of Indian nutraceuticals.
6. **Sustainability and organic products.** There is a growing global trend toward sustainability and organic products. A report by Euromonitor International suggests that nutraceuticals made from organic and sustainable ingredients are likely to see an increased demand in the coming years.

11.9.3 Challenges

Despite the promising future, the nutraceutical industry in India faces challenges, such as regulatory hurdles, the need for greater consumer awareness, and competition from established global players. However, with strategic initiatives and investments, these challenges can be overcome, making the nutraceutical sector a key growth area in India's economy. [26]

11.10 CONCLUSION

In summary, the nutraceutical industry in India is poised for significant growth, driven by changing consumer behaviors, government support, and the country's rich biodiversity and traditional knowledge. The future prospects are bright, with substantial opportunities both domestically and internationally.

REFERENCES

[1] Food Safety and Standards Authority of India (FSSAI). FSMP introduction in India [Internet]. Available from: https://www.fssai.gov.in/cms/non-specified-food.php. Cited 2024 Jan 15.
[2] Food Safety and Standards Authority of India (FSSAI) [Internet]. Available from: https://www.fssai.gov.in/. Cited 2024 Jan 15.
[3] Schmitz SM, Lopez HL, MacKay D. Nutravigilance: principles and practices to enhance adverse event reporting in the dietary supplement and natural products industry. *Int J Food Sci Nutr.* 2014 Mar;65(2):129–134. https://doi.org/10.3109/09637486.2013.836743. Epub 2013 Oct 11. PMID: 24112316.
[4] Nwosu O, Ubaoji K. Nutraceuticals: history, classification, and market demand. In: *Functional Foods and Nutraceuticals*. 2020. p. 13–22.
[5] Chaudhari S. Nutraceuticals: a review. *World J Pharm Pharm Sci.* 2017;6(8):681–739. https://doi.org/10.20959/wjpps20178–9825
[6] Food Safety and Standards Authority of India (FSSAI). About FSSAI [Internet]. Available from: https://fssai.gov.in/cms/about-fssai.php. Cited 2024 Aug 28.
[7] Food Safety and Standards Authority of India (FSSAI). Food safety and standards act, 2006 [Internet]. Available from: https://fssai.gov.in/cms/act–2006.php. Cited 2024 Aug 28.
[8] Food Safety and Standards Authority of India (FSSAI). Food authority [Internet]. Available from: https://fssai.gov.in/cms/food-authority.php. Cited 2024 Aug 28.
[9] Food Safety and Standards Authority of India (FSSAI). FSMP introduction in India [Internet]. Available from: https://www.fssai.gov.in/cms/non-specified-food.php. Cited 2024 Jan 15.
[10] Food Safety and Standards Authority of India (FSSAI) [Internet]. Available from: https://www.fssai.gov.in/. Cited 2024 Jan 15.
[11] U.S. Food and Drug Administration (FDA). FDA website: Questions and answers [Internet]. Available from: https://www.fda.gov/regulatory-information/search-fda-guidance-documents/guidance-industry-questions-and-answers-regarding-labeling-dietary-supplements-required-dietary. Cited 2023 Mar 4.

[12] U.S. Food and Drug Administration (FDA). CAERS adverse event reporting system [Internet]. Available from: https://www.fda.gov/food/compliance-enforcement-food/cfsan-adverse-event-reporting-system-caers. Cited 2023 Mar 6.

[13] U.S. Food and Drug Administration (FDA). FDA medwatch forms [Internet]. Available from: https://www.fda.gov/safety/medical-product-safety-information/medwatch-forms-fda-safety-reporting. Cited 2023 Mar 6.

[14] Timbo BB, Chirtel SJ, Ihrie J, Oladipo T, Velez-Suarez L, Brewer V, Mozersky R. Dietary supplement adverse event report data from the FDA center for food safety and applied nutrition adverse event reporting system (CAERS), 2004–2013. *Ann Pharmacother.* 2018 May;52(5):431–438. https://doi.org/10.1177/1060028017744316. Epub 2017 Nov 24. PMID: 29171279.

[15] Nori LP, Sowmya P, Madhusree T, Venkateswara Raju K. Nutraceuticals: regulatory landscape and market dynamics in the Indian scenario. *J Young Pharm.* 2023;15(4):595–603.

[16] Bailey RL, Gahche JJ, Miller PE, Thomas PR, Dwyer JT. Why US adults use dietary supplements. *JAMA Intern Med.* 2013 Mar 11;173(5):355–361. https://doi.org/10.1001/jamainternmed.2013.2299. PMID: 23381623.

[17] Lakhanpal G, Gupta N, Bandral JD, Sood M, Bhat A, Singh J. Nutraceuticals classification and their health benefits: a review. *The Pharma Innovation J.* 2022;11(10):1927–1933.

[18] Verma G, Mishra MK. Review on nutraceuticals: classification and its role in various diseases. *Int J Pharm Ther.* 2016;7(4):152–160.

[19] Morgovan C, Ghibu S, Juncan A, Rus L, Butuca A, Vonica Tincu A, Muntean A, Mioara L, Mos F, Gligor F, Olah N-K. Nutrivigilance: A new activity in the field of dietary supplements. *Farmacia.* 2019; 67:537–544. https://doi.org/10.31925/farmacia.2019.3.24.

[20] Tomar S, Musyuni P, Aggarwal G. An overview of regulation for nutraceuticals and concept of personalized nutraceuticals. *J Gen Med.* 2023;19(2):66–74. https://doi.org/10.1177/17411343221150875.

[21] Ramdasi V, Lingala D, Macharam K, Saxena T, Prasanthi D. Nutraceutical regulations in India: a comprehensive review. *Int J Drug Regul Aff.* 2023;11(3):80–86.

[22] Jain PN, Rathod MH, Jain VC, SM V. Current regulatory requirements for registration of nutraceuticals in India and USA. *Int J Drug Regul Aff.* 2018;6(2):22–29. https://doi.org/10.22270/ijdra.v6i2.232.

[23] Food Safety and Standards Authority of India (FSSAI). FSDU Registration requirement [Internet]. Available from: https://www.fssai.gov.in/cms/non-specified-food.php. Cited 2023 Jan 11.

[24] Food Safety and Standards Authority of India (FSSAI). Labelling requirements [Internet]. Available from: https://www.fssai.gov.in/upload/uploadfiles/files/Compendium_Labelling_Display_23_09_2021.pdf. Cited 2024 Feb 12.

[25] Food Safety and Standards Authority of India (FSSAI). Food recall [Internet]. Available from: https://fssai.gov.in/cms/food-recall.php. Cited 2024 Aug 28.

[26] Food Safety and Standards Authority of India (FSSAI). Claims [Internet]. Available from: https://www.fssai.gov.in/upload/uploadfiles/files/Compendium_ Advertising _Claims_Regulations_04_03_2021.pdf. Cited 2023 Feb 19.

12 Navigating Regulatory Requirements

A Comprehensive Guide to Nutraceuticals and Traditional Chinese Medicine

Risha Fillah Fithria and Jinfu Peng

12.1 INTRODUCTION

12.1.1 An Introduction to Nutraceuticals and Traditional Chinese Medicine

In the contemporary global landscape characterized by rapidity and escalating health-related apprehensions, it becomes imperative to delve into alternative methodologies for promoting well-being. Nutraceuticals and traditional Chinese medicine (TCM) have garnered global recognition due to their distinctive therapeutic approaches. This section of the chapter thoroughly examines the intriguing domain of nutraceuticals and TCM, providing valuable perspectives on their historical roots, fundamental principles, and regulatory obligations.

Nutraceuticals, derived from the combination of the terms *nutrition* and *pharmaceuticals*, are commonly known as functional foods or dietary supplements that provide enhanced nutritional and therapeutic advantages. The aforementioned substances encompass various compounds, such as vitamins, minerals, herbal extracts, and bioactive compounds. The definition highlights their dual function in fostering well-being and averting or controlling illnesses [1]. An illustration of the nutritional and therapeutic synergy inherent in nutraceuticals is the well-known cardiovascular benefits of omega-3 fatty acids, commonly found in fish oil supplements [2]. The increasing prevalence of nutraceuticals has led to their integration into the daily health regimens of numerous individuals.

12.2 TRADITIONAL CHINESE MEDICINE (TCM)

TCM is a time-honored medical system with a history spanning more than 2,500 years. The concept is founded on the notion that the human body is an intricate network of interrelated systems, and that preserving equilibrium is vital for overall well-being. TCM comprises diverse therapeutic modalities, such as acupuncture, herbal medicine, dietary therapy, and exercise, with the overarching objective of reinstating equilibrium and vitality within the human body [1]. A concrete instance is the utilization of ginseng (*Panax ginseng*) in TCM, renowned for its adaptogenic characteristics and extensively researched for potential health advantages [3]. TCM formulations frequently embody the delicate equilibrium of yin and yang and the movement of vital energy, known as "qi," which are fundamental principles in traditional Chinese medical philosophy [1].

It is imperative for chemists, students, lecturers, businessmen, industry experts, and individuals engaged in the field to possess a comprehensive comprehension of the regulatory obligations of nutraceuticals and TCM. This comprehensive guide delves into the regulatory frameworks that regulate the manufacturing, labeling, safety, and efficacy of these products, providing detailed insights

DOI: 10.1201/9781003465041-12

into navigating the complex regulatory landscape to ensure compliance with legal and quality standards. Moreover, it analyzes the evolving regulatory environment at both regional and global levels, highlighting the latest advancements and trends, with a focus on quality assurance, safety, and efficacy. This knowledge equips readers to make informed decisions and contribute effectively to the progress of this dynamic field, catering to those looking to broaden their knowledge, students pursuing new opportunities, lecturers educating future professionals, entrepreneurs entering the market, and industry experts staying abreast of regulatory developments. Understanding the origins, principles, and regulatory frameworks associated with these fields is crucial to contributing to the growth and success of this burgeoning industry.

12.2.1 The Significance of Adhering to Regulatory Requirements

Regulatory compliance is crucial in the nutraceuticals and traditional Chinese medicine industry to guarantee product safety, effectiveness, and quality. The primary objective of this section is to emphasize the importance of complying with regulatory obligations for diverse stakeholders, encompassing chemists, students, lecturers, business professionals, industry experts, and individuals engaged in this field.

Within nutraceuticals, where science and nature intersect, the significance of adhering to regulatory standards aligns with the dynamic nature of ongoing research. These products, which contain bioactive compounds derived from natural sources, require careful supervision to ensure they meet their claims. Imagine this regulatory vigilance as a protector of veracity, guaranteeing that nutritional assertions on labels remain robust and steadfast.

Now, redirect your attention toward TCMs, where the ancient knowledge derived from herbal texts intersects with the meticulousness of contemporary scientific knowledge. Regulatory compliance safeguards this intricate equilibrium, guaranteeing the genuineness and security of herbal formulations.

Pharmacists must adhere to regulatory compliance to ensure that they offer their patients products that meet the most stringent criteria for safety and effectiveness. Pharmacists can ensure the delivery of intended therapeutic effects and prevent potential harm or adverse reactions by adhering to the regulations established by the relevant authorities.

Students and lecturers must comprehend the significance of regulatory compliance within the context of their educational and training endeavors. Gaining comprehension of the regulations about nutraceuticals and traditional Chinese medicine provides individuals with the requisite knowledge to make well-informed decisions in their prospective professional endeavors. By establishing a culture of adherence at an early stage, these prospective health-care practitioners will make a valuable contribution to the health-care sector's overall safety and ethical standards.

Business professionals and industry experts must acknowledge that regulatory compliance is not solely a legal requirement but a pivotal element in establishing consumer trust. Adhering to regulations showcases a dedication to ethical business conduct and safeguarding consumer interests. Moreover, strict compliance with regulatory obligations can bolster the standing and trustworthiness of a company or brand, resulting in a rise in market dominance and customer allegiance.

12.3 REGULATORY COMPLIANCE

Regulatory compliance holds a broader significance that transcends the realm of individual businesses and professionals. Ensuring public health and safety is a shared obligation. By adhering to regulatory requirements, the nutraceutical and traditional Chinese medicine industry can effectively uphold the integrity of these products, thereby cultivating trust among consumers and regulatory bodies.

This section will explore the precise regulatory obligations of nutraceuticals and traditional Chinese medicine, encompassing areas such as the registration of products, the development of labels and packaging, the implementation of quality control measures, and the monitoring of post-market activities. This publication aims to offer practical guidance and valuable insights for effectively navigating the intricate terrain of regulatory compliance. By doing so, it seeks to empower readers to make well-informed decisions and actively contribute to the progress and development of the industry.

In summary, ensuring adherence to regulatory requirements is significant in nutraceuticals and traditional Chinese medicine. Implementing quality assurance measures guarantees the safety, effectiveness, and excellence of products; fosters consumer confidence; and upholds the overall integrity of the industry. This section aims to provide chemists, students, lecturers, businessmen, and industry experts with the requisite knowledge and comprehension to effectively navigate the regulatory environment. Through the adoption of regulatory compliance, there is potential for collective progress in the field and the enhancement of the welfare of individuals searching for natural health remedies.

12.3.1 Central Regulatory Authorities and Legislative Measures

Understanding the role of key regulatory bodies and the legislation governing the nutraceutical and traditional Chinese medicine industries is crucial in these sectors' dynamic and constantly changing landscape. Within the expansive territory of China, where the reverberations of age-old knowledge are felt within the realm of contemporary advancements, a committed entity assumes the role of safeguarding TCM and the emerging domain of nutraceuticals. The preservation and the protection of the timeless wisdom of TCM and the potent power of nutraceuticals for future generations are ensured by a force of esteemed regulatory bodies and comprehensive legislation.

The China Food and Drug Administration (CFDA), known as the State Food and Drug Administration (SFDA), previously tasked with overseeing food, drugs, and medical device regulation, also exerted considerable influence in establishing specific criteria and directives for healthy foods and TCM products. In 2018, the oversight of food and pharmaceuticals in China transitioned to the State Administration for Market Regulation (SAMR) (https://www.samr.gov.cn/) and the National Medical Products Administration (NMPA) (https://www.nmpa.gov.cn/index.html). The NMPA holds a critical role in the regulatory landscape, primarily responsible for the regulation and supervision of the safety, efficacy, and quality of drugs (including traditional and ethnic medicines), medical devices, and cosmetics. It operates as the primary authority ensuring the integrity of pharmaceuticals, medical apparatus, and health products throughout the country. Affiliated with SAMR, the Special Food Safety Supervision Department (https://www.samr.gov.cn/tssps/) manages the regulation of health supplements, foods for special medical purposes, and infant formula, among other specialized foods. This department is tasked with analyzing the safety conditions in relevant areas; developing and implementing regulatory measures for the registration, filing, and supervisory management of special foods; and coordinating the investigation and handling of significant legal violations in this sector. Beyond the responsibilities of SAMR and NMPA in the oversight of food and drugs, the National Administration of Traditional Chinese Medicine also participates in the regulation of the cultivation and use of traditional Chinese medicines, including the certification of hospital-based Chinese medicine preparations and Good Agricultural Practice (GAP) bases for herbal materials.

Furthermore, the State Administration for Market Regulation (SAMR) is responsible for the oversight of market competition, the protection of consumer rights, and the supervision of product quality. The enforcement of product safety, labeling, and advertising regulations is paramount, as it guarantees adherence to the standards and guidelines established by the NMPA.

The regulation of nutraceuticals and TCM products in China is governed by several significant legislative and regulatory frameworks. The Food Safety Law sets forth fundamental principles for managing, evaluating, labeling, and inspecting food safety, serving as the basis for regulatory supervision in this domain.

The regulatory framework for registering and approving health food products in China is delineated by the Regulations on Health Food Registration. The preceding regulations delineate the permissible categories of ingredients, the necessary labeling prerequisites, and the scientific substantiation needed to substantiate health assertions, guaranteeing transparency and responsibility in promoting health-oriented food products.

The TCM Law governs the practice, regulation, and development of traditional Chinese medicine in China. The purpose of this is to establish regulatory mechanisms for TCM products, practitioners, and institutions and to safeguard the integrity and authenticity of TCM practices. It defines TCM principles, standards, and practices.

In addition, the manufacturing of health foods, dietary supplements, and TCM products has been subject to implementing Good Manufacturing Practice (GMP) standards. The preceding standards delineate the necessary criteria for properties, machinery, staff, manufacturing procedures, and quality assurance, guaranteeing the product's safety and uniformity.

Establishing and enforcing regulatory frameworks about nutraceuticals and TCM products in China plays a pivotal role in protecting public health and promoting consumer welfare. Regulatory bodies perform inspections, sampling, and testing procedures on products to verify their adherence to safety and quality regulations.

Ensuring rigorous oversight of the marketing, advertising, and labeling of health foods, dietary supplements, and TCM products is imperative to mitigate the dissemination of misleading assertions and deceptive methodologies. The SAMR investigates consumer grievances, product recalls, and breaches of regulatory obligations, ensuring manufacturers and distributors are held responsible for failing to comply [4].

China maintains the integrity of TCM and nutraceuticals by effectively coordinating with regulatory bodies and enacting comprehensive legislation. This ensures that ancient knowledge from the past is seamlessly integrated with modern advancements, creating a future where the pursuit of health and wellness is limitless [5].

12.3.2 China's Regulatory Framework for Nutraceuticals

12.3.2.1 Nutraceuticals: Definition and Classification

The notion of nutraceuticals in China is firmly entrenched within the nation's cultural, historical, and philosophical fabric. These factors have influenced the comprehension and categorization of these contemporary wellness products, endowing them with a distinct essence that profoundly connects with Chinese ethos.

Although the term *nutraceutical* is a recent addition to the vocabulary, its underlying principles have been deeply ingrained in Chinese tradition for a long time. The quest for equilibrium, concord, and comprehensive wellness has been a guiding principle for many generations, and nutraceuticals have arisen as a modern expression of this ancient knowledge.

The Foundation for Innovative Medicine initially introduced the term *nutraceuticals* to encompass a wide range of products, including dietary supplements (such as vitamins and minerals), botanicals (such as ginseng, St.-John's-wort, and ginkgo biloba), functional foods (such as oat, bran, lignin, prebiotics, and polyunsaturated fat products), as well as certain novel foods of transgenic origin (such as lactoferrins), oral vaccines, and the addition of medication to certain food products [6].

In contrast to pharmaceuticals, nutraceuticals encompass many substances derived from food that possess health-enhancing properties. On the other hand, *pharmaceuticals* typically pertain to purified chemicals that exhibit well-defined mechanisms of action toward specific physiological or biochemical targets, primarily to treat diseases [6].

Nutraceuticals, in contrast to functional food, are not considered food itself but as extracts derived from food that can be incorporated into food products to enhance health. A nutraceutical has been shown to possess physiological advantages or safeguard against chronic ailments [7–8].

The initial subgroup is commonly referred to as **functional health foods**. These products are thought to have 1 or more of the 27 acknowledged functions, including improving immunity or controlling metabolism. These products are engineered to provide physiological advantages beyond essential nourishment, acting as powerful companions for optimal physical and mental health.

The subsequent classification is **nutrient supplements** or **dietary supplements**. These products enhance the body's nutritional requirements by offering crucial vitamins, minerals, and other essential nutrients. They function as a contemporary supplement to a well-rounded diet, guaranteeing that the body obtains the necessary nourishment to flourish in a constantly evolving world. The CFDA registration regulations outline the test requirements for dietary food and functional health food [9].

TABLE 12.1
Nutraceuticals' Test Requirements

Test Item	Sub-Category of Nutraceutical
Performing tests for toxicology, hygiene, stability, and active ingredients	Dietary supplements and functional health food
Animal study only	**Functional health food** • Enhancing immune systems • Improving sleep • Alleviating physical fatigue • Enhancing anoxia endurance • Protecting from irradiation hazard • Increasing bone density • Assisting liver protection against injury
Human study only	**Functional health food** • Alleviating eye fatigue • Eliminating acne • Eliminating skin pigmentation • Improving skin's ability to retain moisture • Improving skin oil content function
Animal and human studies	**Functional health food** • Assisting in blood lipid reduction • Assisting in blood sugar reduction • Antioxidative function • Assisting in memory improvement • Alleviating lead excretion • Improving throat function • Assisting in blood pressure reduction • Facilitating milk secretion • Assisting in weight control • Improving child growth • Improving nutritional anemia • Regulating gastrointestinal flora • Facilitating digestion (regularity) • Facilitating bowel movement • Protecting gastric mucosa
Doping test	**Functional health food** • Alleviating physical fatigue • Assisting weight control • Improving child growth

Source: [9].

The categorization of nutraceuticals by the Chinese government enhances regulatory oversight and monitoring of their production, distribution, and consumption. This categorization aims to ensure strict adherence to the utmost quality and safety standards. China has adopted a comprehensive classification system to explore nutraceuticals, providing its citizens with the means to achieve wellness that combines traditional and innovative approaches. This system upholds the principles of balance and harmony that have been important to the nation for centuries [10–11].

12.3.2.2 Regulatory Obligations about Ingredients in Nutraceuticals

In China, there is a long-standing belief that food and medicine originate from a common source. Numerous uncooked herbs, regarded as medicinal, can also be served as everyday food for daily consumption. The Food Sanitation Law (trial) was enacted by the Ministry of Health (MOH) of the Chinese government in 1987. This law prohibits the inclusion of medicine or drugs in food, except for those utilized as raw materials in food preparation or as seasoning. Following this, a compilation of 33 primary resources that possess the potential to serve as both sustenance and therapeutic agents was released. In 2002, the MOH updated the list and expanded it to encompass 87 items [12].

TABLE 12.2
Materials with Dual Applications in Food and Medicine

Clove	Star anise	Sword beans
Fennel seed	Thistle	Chinese yam
Hawthorn fruit	Purslane herb	Black plum
Chinese quince fruit	Hemp seed	*Citrus aurantium*
Solomonseal rhizome	Licorice	Dahurian angelica root
Gingko seed	White hyacinth bean	Hyacinth flower
Arillus fruit, Longan	Foetid Cassia seed	Lily bulb
Nutmeg	Cinnamon bark	Indian gooseberry
Finger citron fruit	Almond seed (bitter, sweet?)	Seabuckthorn fruit
Gordon Euryale seed	*Zanthoxylum schinifolium*	Adzuki bean
Germinated barley	Kelp	Date
Swingle	Bush-cherry seed	Honeysuckle flower
Chinese white olive	Heartleaf houttuynia herb	Ginger root
Dulcis seed	Wolfberry fruit	Gardenia fruit
Spiny amomum fruit	Boat fruited-sterculia seed	Poria
Citron fruit	Aromatic madder	Peach seed
Mulberry leaf	Mulberry fruit	Tangerine peel
Platycodon root	Bitter cardamom	Lotus leaf
Radish seed	Lotus seed	Lophatherum herb
Fermented soybean	Chrysanthemum flower	Chicory grass
Air potato yam	Siberian Solomonseal rhizome	Perilla leaf
Perilla seed	Pueraria root	Black sesame seed
Black pepper	Pagoda tree flower and bud	Dandelion
Grand Torreya seed	Wild jujube and seed	Fresh imperata rhizome
Fresh reed root	Peppermint, menthe	Coix seed
Longstamen onion	Raspberry fruit	Agastache Pogostemon
Zaocys snake	Oyster	Donkey hide gelatin
Chicken gizzard skin or lining	Honey	Viper

Source: L12)

The Food Safety Law was revised in October 2013 by the National Health and Family Planning Commission (NHFPC), and various governmental organizations, including the National Food Safety Risk Assessment Center, the Chinese Center for Disease Control and Prevention, the Chinese Academy of Medical Sciences, the Institute of Medicinal Plant, and others. The revision aimed to provide new information by including details about the specific parts of plants or animal bodies and specialized preparation methods.

In January 2020, a supplementary announcement (No. 8, 2019) was jointly issued by the National Health Commission of the People's Republic of China (NHC) and the State Administration for Market Regulation (SAMR). This announcement introduced six dual-use substances exclusively utilized as spices and condiments. The announcement provided detailed information regarding the Chinese name, Latin name, family name, and edible components of each substance [13], as presented in Table 12.3. In November 2023, a supplementary announcement (No. 9, 2023) [15] (Table 12.4) was jointly issued by NHC and SAMR [14].

12.3.3 The Regulatory Body Responsible for Overseeing the Management of Functional Health Foods

The Chinese government implemented the Food Hygiene Law in 1995, as recorded by the China State Council. The aforementioned legislation established the MOH as the designated governing body tasked with enforcing the law. Furthermore, the responsibility of granting approval for licensing health foods was assigned to the National Centre for Health Inspection and Supervision [15]. There were 104

TABLE 12.3
Six Traditional Substances Historically Served as Both Food and Chinese Medical Materials

No.	Name	Botanical Name/Animal Name	Latin Scientific Name	Designation of Department to Which It Is Affiliated	Part of Plant or Animal	Commentary
1	Angelica	Angelica	*Angelica sinensis* (Oliv.) Diels	Apiaceae	Root	Exclusively as a spice and condiment
2	Kaempferia	Kaempferia	*Kaempferia galanga* L.	Zingiberaceae	Rhizome	
3	Saffron	Crocus	*Crocus sativus* L.	Iridaceae	Stigma	Saffron, commonly called "spices and condiments," is exclusively utilized in culinary contexts.
4	Grass fruit	Grass fruit	*Amomum tsao-ko* Crevost et Lemaire	Zingiberaceae	Fruit	Exclusively as a spice and condiment.
5	Turmeric	Turmeric	*Curcuma Longa* L.	Zingiberaceae	Rhizome	
6	Long pepper	Long pepper	*Piper longum* L.	Piperaceae	Cob	

Source: [13].

TABLE 12.4
Nine Supplementary Substances Traditionally, Both Food and Chinese Medical Materials

Codonopsis pilosula (Franch.) Nannf.	*Cistanche deserticola* Y.C.	*Dendrobium officinale*
Panax quinquefolium L.	*Astragalus membranaceus* (Fisch.)	*Ganoderma lucidum* (Leyss.)
Cornus officinalis Sieb.	*Gastrodia elata*	*Eucommia ulmoides*

Source: [15].

articles in the law. Several aspects, including food standards, inspection procedures, import and export regulations, and handling food safety incidents, are mirrored in the food legislation of other countries. Regulations about healthy foods are addressed explicitly in Articles 22 and 23.

- Food with specific health functions must be submitted to the administrative department of public health under the State Council for examination and approval. Under the State Council, the administrative department of public health will formulate hygiene standards and control measures for its production and marketing (Article 22).
- Food that is claimed to possess specific health benefits should not pose any harm to human health. It is imperative that the information provided in the product description is accurate and that the functions and ingredients of the product align with the information provided in the description, without any inclusion of false information (as stated in Article 23).

The State Food and Drug Administration (SFDA) was given the responsibility of health (functional) food management in 2003. Subsequently, over 4,000 functional foods were authorized, with 10% imported from foreign nations [16]. The year 2009 witnessed the replacement of the Food Hygiene Law with the Food Safety Law[17]. The new document's Article 50 dealt explicitly with Article 23, stating that:

> Food producers are prohibited from incorporating any medicinal substances into food unless such substances are conventionally recognized as both food and Chinese medicine.

The MOH underwent a name change, and its functions were consolidated into a new agency, known as the NHFPC, in the 2013 reform. The SFDA underwent a transfer from the MOH to assume a ministerial status, operating directly under the State Council [18], which holds the highest authority within the government. In addition to this modification, the establishment was rebranded as CFDA. To streamline regulation processes for food and drug safety, the CFDA replaced a large group of overlapping regulators with an entity similar to the US Food and Drug Administration (FDA) [19]. The CFDA is responsible for the comprehensive supervision of food, health food, and cosmetics safety management. It is the competent authority for drug regulation in China [18].

12.3.3.1 The Process of Approval of the Application of Functional Health Foods

According to the "Regulations on the Registration and Filing Management of Health Food," CFDA Order No. 22, revised in 2020, manufacturers or sellers intending to introduce health foods into the Chinese market must submit an application for and acquire a health food registration certificate or filing certificate. The registration process for domestic health foods manufactured in China is handled by the State Administration for Market Regulation (SAMR), while the filing process is conducted by the Provincial Administration for Market Regulation. Both the registration and filing for imported health foods produced abroad must be submitted to SAMR. In the interim, it is required that foreign corporations establish a permanent office in China or designate a Chinese agent to handle the processes of registration, filing, and acquisition of relevant certificates.

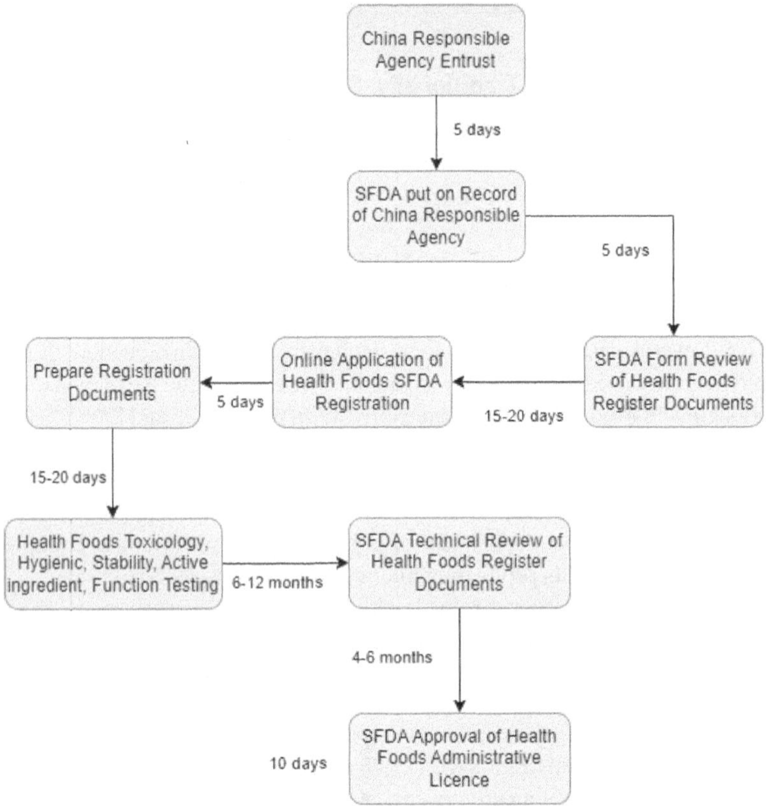

FIGURE 12.1 Registration process of functional health food by SFDA.

Source: [21].

Health food registration/filing can be submitted by manufacturers or companies in China, or alternatively, a local agent can be appointed to facilitate this process. To complete the health food registration/filing process, foreign manufacturers or companies must designate a local Chinese agent. The registration/filing process should be conducted by a local agent with the necessary knowledge and qualifications [20].

The SFDA, now known as the CFDA, was granted the authority to oversee and manage all matters about functional health foods. All food products that purport to possess health-enhancing properties must undergo a comprehensive evaluation and authorization procedure. The preceding components encompass the registration process for every functional food, application security measures, toxicity assessments, functional evaluations (excluding nutritional supplements), stability evaluations, hygiene evaluations, and identification tests for functional ingredients. All tests must adhere to the established protocols conducted by specialized agencies or laboratories that are certified by the MOH and SFDA [21]. The flowchart of the application procedure is depicted in Figure 12.1.

Initially, comprehending a product dossier's underlying objective is vital. It constitutes a comprehensive aggregation of pertinent details about a product, encompassing its constituent elements, production methodology, safety protocols, quality assurance protocols, and more relevant information. The dossier is a thorough record of the product's adherence to regulatory standards and serves as a reference point for regulatory authorities throughout the assessment and authorization procedure.

The process of creating a product dossier encompasses various essential stages. Collecting all necessary information and paperwork about the product is crucial in the first stages. This encompasses information regarding the constituents employed, their origins, and the production procedure. Furthermore, it is imperative to incorporate safety and efficacy data, stability studies, and quality control procedures.

After the necessary data is collected, the subsequent task involves systematically arranging and consolidating the information. This entails the development of a comprehensive table of contents, the accurate labeling of each section, and the assurance of a coherent and succinct overall presentation. It is necessary to adhere to the precise requirements set by regulatory agencies on the format and content of dossiers to ensure a successful submission.

Following the organization of the dossier, it is essential to undertake a comprehensive review and validation process to ascertain its accuracy and comprehensiveness. This stage facilitates the identification of any deficiencies or inconsistencies that require attention before submission. It is recommended to engage with professionals or consultants with expertise in regulatory requirements to obtain direction and assistance throughout this phase.

The dossier has been prepared and is ready for submission to the appropriate regulatory authority. Now it is imperative to adhere to the prescribed submission protocols, encompassing all obligatory forms or supplementary materials. During this step, it is essential to prioritize timeliness and meticulousness to prevent any avoidable delays or rejection of the dossier [20].

To initiate the process, it is necessary to submit the application to the relevant local provincial health administration authorities for a first screening [20]. Once authorized, the application will be forwarded to SFDA for assessment and endorsement. The comprehensive application dossier ought to be structured into five distinct sections:

1. *Administrative data about the enterprise and product.* This should encompass several essential components. These components include a health food registration application form, a copy of the legal registration or certificate of the enterprise or applicant, a testimony affirming the uniqueness and dissimilarity of the product name from any authorized food or drug registration products, a guaranteed document provided by the applicants to ensure the absence of any infringement on other patents, and a copy of the brand registration certificate.
2. *Food-specific information.* This section should include a report on the research and development of the product, a description of the product's formulation or ingredients, the analytical methods used for functional or representative components, a concise description of the manufacturing process and its flowchart, the specifications or standards for the product and raw materials, the quality standards, and the criteria for selecting packing materials.
3. *Laboratory testing data for the product.* This encompasses test reports obtained from testing agencies and any other pertinent documents. The reports typically encompass various components, including a test application form, the requisition provided by testing agencies to the applicant, the toxicology test report, the functional test report, the test reports on illicit substances (specifically for specific purposes, such as alleviating physical fatigue, weight loss, and enhancing growth functions), the test report for functional components and composition, the stability test report, the hygiene test report, and additional test reports that aid in the application process (e.g., identification tests for raw materials or bacterial virulence categories).
4. *Consumer information.* This section includes the product's label and usage instructions, along with three product samples to be included with the application.
5. *Scientific literature evidence.* This should encompass a concise overview and a comprehensive analysis of existing literature, published empirical data in the scientific literature, and any relevant documents from foreign governments (i.e., beneficial for product approval).

Following a comprehensive examination of scientific literature by proficient scientists from many fields, an official certification will be granted to the endorsed product, accompanied by a designated serial number, such as "Foodstuff/Health [year] No. ##" [22]. At this juncture, the commencement of manufacturing may be initiated under the oversight of the local governing body to guarantee adherence to the General Hygiene Regulations and the GMP. Production must also adhere to the approved regulatory procedure to ensure the utilization of appropriate raw materials, adherence to scientifically accepted technical assessment techniques and methods for quality and safety controls, and compliance with standard labeling and advertisement requirements. The responsibility of evaluating sample products in the market to ensure compliance with quality standards will also lie with the local authority [20].

Foreign firms must register with the SFDA and acquire a GMP certificate to sell their functional food items in China. To obtain the certificate in China, it is necessary to obtain the following supplementary materials [23]:

1. A certificate that adheres to the applicable local production quality control standards from the country of origin
2. A certificate for products that have been sold in the producing country for a duration exceeding one year, issued by the certifying agency of the country of origin and the Chinese embassy in the country of origin
3. Instructions and guidelines for usage in the place of origin
4. Applicable product regulations of the manufacturing country or international organizations

All documents must be written in Chinese, ensuring that the content remains consistent with the original version.

The technical assessment procedures and methods, such as standard functional assessment methods, standard toxicological assessment procedures and methods, regulation on nutrient supplements, and standard analytic methods for functional foods, are regularly updated to align with the latest advancements in scientific rigor. This is done to meet the growing demand for safety requirements. The product must be appropriately labeled to mitigate the risk of consumer deception. Furthermore, the language employed in health claims on the labels should be comprehensible, aiding consumers in making well-informed decisions. However, it is crucial to ensure that such language does not deceive customers regarding the therapeutic efficacy of the product, in addition to the product name, total weight, solid weight, ingredients, health function, preferred population for use, methods of consumption, dates of use, storage method, operation, manufacturer's name and address, and health certificate number [20]. To differentiate it from pharmaceuticals, it is recommended to include a "blue hat" symbol (Figure 12.2) in the upper left corner of the label [24].

The registration number structure for domestic health food consists of four Chinese characters, namely, Guo Shi Jian Zhu, followed by the letter *G*, a four-digit year, and a four-digit serial number. The

FIGURE 12.2 A blue hat symbol for health functional food.

Source: © SAMR [24].

registration number structure for imported health food consists of four Chinese characters, namely, Guo Shi Jian Zhu, followed by the letter *J*, a four-digit year, and a four-digit serial number [25].

12.3.4 The Process of Approval of the Application for Dietary Supplements

According to the Food Safety Law of China, all dietary or nutrient supplements are required to obtain a "blue hat" certification, which is identical to the certification for health functional food (Figure 12.2). The efficacy of supplements should align with their advertised claims, and the government possesses the authority to initiate product recall for hazardous items in the marketplace. The registration of dietary or nutrient supplements is not mandatory, but it is necessary to notify them [26]. The flowchart of the application technique is depicted in Figure 12.3 [9].

The notification number format for domestic health food consists of three Chinese characters, namely, Shi Jian Bei, followed by the letter *G*. This is followed by a four-digit year, a two-digit province administrative district code, and a six-digit serial number. The notification number structure for imported health food consists of three Chinese characters, namely, Shi Jian Bei, followed by the letter *J*, a four-digit year, double zero, and a six-digit serial number [25].

The CFDA issued a notice to further stipulate dietary supplement naming on August 27, 2015 [27], to address the issue of misunderstanding arising from the presentation of functions in the names of dietary supplements. This action complied with the supervision requirements of nutritional supplements outlined in the new Food Safety Law. Table 12.5 displays the list of prohibited descriptions for dietary supplements.

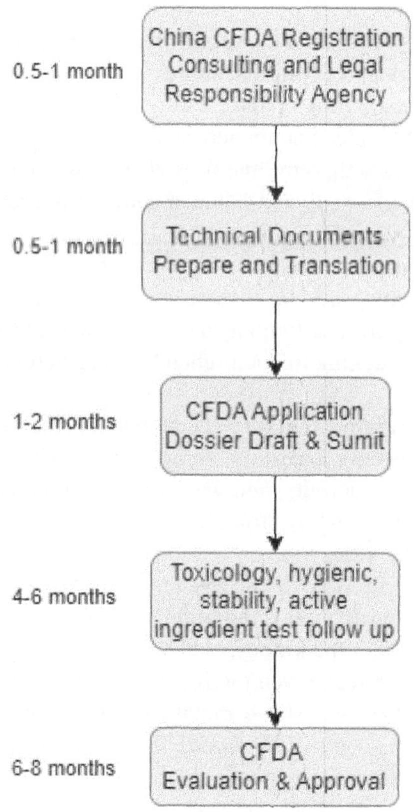

FIGURE 12.3 Registration process of dietary supplements by the SFDA.

Source: [25].

TABLE 12.5
Compilation of Prohibited Dietary Supplements

No.	Type	Illustration of Descriptive Terminology
1	Deceptive	"Ancestral" (祖传), "secret" (秘制), "purified" (精制), and so on
2	Exaggerated	"Special effect" (特效), "treasure" (宝), "magical effect" (神效), and so on
3	Grade	"Best" (最), "champion" (第一), "high-grade" (高级), and so on
4	Prevention and curing of diseases	"Prescription" (处方), "medicine," "cure" (治疗), and so on
5	Personal name	"Hua Tuo" (华佗), "Bian Que" (扁鹊), "Shizhen Li" (李时珍), and so on
6	Region	"China" (中国), "Huaxia" (华夏), and so on
7	Difficulty in being understood	"Nanometer" (纳米), "gene" (基因), "space" (太空), and so on
8	Superstitious	"Sex" (性), "God" (神), "immortal" (仙), and so on
9	Body organ	"Brain" (脑), "eye" (眼), "heart" (心), and so on
10	Misleading	Homophonic words: words which are similar in form

Source: [27].

12.3.5 Framework for Regulating Traditional Chinese Medicine

12.3.5.1 Introduction to TCM

Traditional Chinese medicine (TCM) is a comprehensive medical system established in China for millennia and now receiving global recognition, addressing the prevention, diagnosis, and treatment of diseases. It emphasizes understanding of and treating the human body as a whole, interconnected entity. Key components of TCM include acupuncture, moxibustion (heat therapy), cupping, exercise (such as tai chi and qigong), dietary therapy, and lifestyle management. TCM is based on theoretical concepts like qi (vital energy), yin and yang balance, and the interaction of the human body with its environment.

This section aims to present a comprehensive review of traditional Chinese medicine, encompassing its fundamental concepts and the diverse modalities employed in its application. This knowledge holds significant importance for pharmacists, students, lecturers, business professionals, industry specialists, and others seeking to comprehend the regulatory obligations and intricacies of TCM.

The notion of yin and yang (Figure 12.4) is a fundamental principle in TCM. *Yin* symbolizes the attributes associated with femininity, passivity, and coolness, whereas *yang* symbolizes the attributes associated with masculinity, activity, and warmth. TCM emphasizes maintaining a harmonious equilibrium between yin and yang to promote optimal health. An imbalance can result in the development of illness or disease [28].

Another important principle in TCM is the *five elements hypothesis*, which delineates the five elemental qualities (wood, fire, earth, metal, and water) that characterize the human body and herbs. As an illustration, water symbolizes the kidney, bladder, and ears, whereas fire represents the heart, small intestine, and tongue. Moreover, given the interconnection between the five elements in both the generating circle (wood → fire → earth → metal → water → Wood) and the overcoming circle (wood → earth → water → fire → metal → wood), it follows that all bodily organs are interconnected and susceptible to the influence of specific herbs [30].

The origins of TCM can be traced back to the early Zhou dynasty, which dates back over 3,000 years, or even earlier, to the Shang dynasty (1766–1122 BC). The historical roots of this phenomenon can be traced back to the shamanistic era, during which the prevailing belief was that illness was attributed to ancestral disturbances or demonic influence.

FIGURE 12.4 Yin and yang symbol.

Source: [29].

The emergence of TCM theory may be traced back to the Warring States and Zhou dynasty period (475–221 BC), a time characterized by the convergence of philosophy and science. During this era, there were groundbreaking authors whose concepts redirected the understanding of health and illness from supernatural origins.

Following the collapse of the Qin dynasty, intellectuals endeavored to amass and arrange medical manuscripts originating from the Warring States era. As a result, significant works, such as the *Huangdi Neijing* (*Yellow Emperor's Inner Classic*), were created around 26 BCE. These works provided a comprehensive overview of fundamental concepts in traditional Chinese medicine(TCM), including yin-yang, qi, acupuncture, and herbals [31–32].

Several significant old TCM classics emerged during the subsequent centuries, including the *Shennong Bencaojing* (*Divine Farmer's Classic of Materia Medica*), dating back to approximately 200–250 CE. This renowned text meticulously recorded 365 distinct medicinal ingredients. TCM evolved gradually over centuries and eventually became a widely accepted medical system in China. It incorporates a range of treatments, such as acupuncture, herbal remedies, nutrition therapy, and massage. Nevertheless, the field experienced a fall throughout the 17th century due to the emergence of Western medicine.

In 1929, the Chinese government first prohibited TCM procedures such as acupuncture but later reinstated them after 1949. In the 1950s, research facilities were developed which facilitated the increased incorporation of TCM and Western medicine into contemporary Chinese health care [32–33].

The 11th version of the International Statistical Classification of Diseases and Related Health Problems (ICD) officially incorporated TCM on May 25, 2019. The World Health Assembly, the governing body of the World Health Organization (WHO), has officially classified diseases worldwide in this publication. The official compendium has acknowledged TCM as a conventional health-care system and has formally supported its 129 member governments worldwide. Citing the World Health Organization's website on the subject of ICD-11: The International Classification of Diseases (ICD) serves as the fundamental basis for identifying global health trends and statistics and establishing the international standard for reporting diseases and health problems. ICD-11 was presented to the 72nd World Health Assembly in May 2019. Subsequently, upon receiving approval, member states will commence the utilization of ICD-11 for reporting purposes starting from January 1, 2022 [28].

12.3.5.2 Treatments Using TCM

TCM is a comprehensive and intricate treatment system encompassing several therapeutic techniques to foster healing and well-being. The following modalities enhance bodily strength and facilitate qi and blood circulation: acupuncture, herbal medicine, moxibustion, acupressure, Tui Na massage, cupping, nutrition and lifestyle counsel, meditation, and tai chi and qigong exercises.

The process of **moxibustion** entails the application of heat to specific acupuncture points using a smoldering mugwort plant, commonly referred to as moxa. Moxibustion is a therapeutic technique that enhances blood circulation, mitigates cold and dampness within the body, and facilitates the efficient movement of blood and qi. This approach, which is both safe and non-invasive, can be employed independently, although it is commonly utilized in combination with acupuncture therapy.

Acupressure employs identical acupuncture sites while applying mild yet strong pressure to the skin's surface using the fingers and hands. This technique aims to enhance the flow of qi and blood, thereby promoting the body's inherent healing capabilities. Acupressure is often regarded as a very efficacious approach for addressing tension-related conditions since it harnesses the potency and sensitivity of the human hand to induce muscle relaxation and alleviate tension.

The term **Tui Na massage**, derived from the Chinese medical tradition, refers to a potent body-work known as "pushing grasping." Drawing upon the fundamental principles of Oriental medicine, Tui Na aims to enhance the circulation of qi inside the meridian channels. This therapeutic approach demonstrates notable efficacy in addressing various disorders of the musculoskeletal system, including structural misalignment, orthopedic ailments, and sports-related injuries. Furthermore, it has the potential to address internal ailments.

Cupping is a time-honored method wherein a specialized cup is administered into the skin and secured in position through a vacuum-like suction mechanism. The suction process induces the movement of superficial tissue into the cup, which can either remain stationary or be relocated throughout the body. Cupping facilitates the infusion of fresh blood to the affected region, enhancing circulation. The practice of traditional cupping, also known as fire cupping, involves the application of heat to generate a suction within glass cups. In contemporary society, plastic cups are employed with a compact vacuum pump to generate suction. Small rubber cups that can be manually manipulated without heat have been introduced to enhance convenience and ensure safety.

Dietary treatment and nutritional advice involve choosing certain foods to improve a patient's general health, considering their imbalanced eating habits. TCM posits that individuals possess distinct body constitutions, temperaments, and health problems that undergo transformations as they age. Diet and nutrition can restore the body to equilibrium by eliminating foods that exacerbate imbalances. This idea is highly relevant in contemporary medicine, as a significant proportion of malignancies and other degenerative diseases is attributed to inadequate diet and nutritional insufficiencies.

Agronomic variables, such as the use of toxic pesticides, herbicides, and fungicides in plant growth and cultivation, as well as the commercial farming of fish, poultry, and cattle with GMO feed, added hormones, and antibiotics, contribute to the toxic overload of the environment. These factors were absent in ancient times, when TCM was initially created. However, in the present day of global pollution, they pose a significant and tangible threat to the health of humans, animals, and the environment. TCM recognized the crucial significance of diet and nutrition millennia ago and now considers food therapy as the primary approach to treatment.

Qigong refers to cultivating life force energy by practicing deep breathing techniques and specific kinds of movement, some of which imitate natural animal movements. Medical qigong is a therapeutic approach to enhance, safeguard, and reinstate well-being. The comprehensive advantages of qigong exercises can be attained through proper posture, diaphragmatic breathing, and deliberate mental guidance as part of a daily routine. Qigong is frequently employed to mitigate blood pressure, tension, and anxiety while fostering peaceful sleep and overall well-being. Furthermore, it is believed to alleviate prevalent ageing damage and increase longevity.

Tai chi, classified as a kind of qigong, encompasses a deliberate and controlled regimen of physical exercises grounded in the fundamental concepts of qigong, including breathing, meditation, strength, and balance. Hua Tuo (华佗), a highly esteemed physician who lived from 140 to 208 CE, achieved recognition for his exceptional proficiency in the fields of acupuncture, moxibustion, and medical Da Yian (大雁) exercises. The development of the Wu Qin Xi (exercise of the five animals, 五禽戏) resulted from his astute observation of the animal realm, explicitly focusing on the movements of the tiger, deer, bear, monkey, and crane.

Acupuncture is a therapeutic approach involving inserting sterile needles, similar to human hair, into precise anatomical sites on the skin, referred to as acupuncture points. The extent of needle insertion is contingent upon the specific ailment under consideration and the accurate positioning of the acupuncture point. For instance, shallow insertion is commonly employed in treating diseases that affect the superficial layer, such as the beginning phase of common colds and disorders affecting the skin and muscles. In contrast, treating internal diseases, such as organs and bones, typically necessitates using deeper needle insertion.

TCM has 12 primary meridians, each containing over 700 acupuncture sites evenly distributed on both sides of the body. Additionally, there are two central meridians. The Ren Mai (任脉) is the anterior meridian, also known as the conception vessel, whereas the Du Mai (督脉) is the posterior meridian, also known as the governing vessel.

The regulation of qi flow throughout the energy pathways, known as meridians or channels that interconnect the entire body, is facilitated by acupuncture stimulation. Acupuncture exerts an influence on the endocrine and central nervous systems, including the physiological reactions of both the sympathetic nervous system (yang) and the parasympathetic nervous system (yin), to establish a state of balance and facilitate the body's inherent capacity for healing [28–33].

12.3.5.3 The Role of Herbal Medicine in Traditional Chinese Medicines

Chinese herbal medicine specifically involves the use of natural substances to create formulas that restore balance and enhance health. Chinese herbal remedies mainly contain plant-derived substances. However, specific formulations may incorporate minerals or animal-derived components. The packaging of herbs might vary based on their intended usage, encompassing powders, pastes, creams, or tablets. These substances are selected and combined based on their properties and effects as understood through TCM theories. The theory exploring the mechanisms and application rules of drug nature is known as the theory of drug properties. Its fundamental aspects include the four energies ("qi," cold, hot, warm, and cool), five flavors (sour, bitter, sweet, pungent, and salty), directions of movement (ascending, descending, floating, and sinking), meridian tropism (selective action of a drug on specific parts of the body), toxicity, compatibility, and contraindications. Various plants possess distinct qualities and can harmonize specific bodily systems. When prescribing a particular herb or combination of herbs, the practitioner must consider the patient's yin and yang balance and the components influencing the affected organs to provide an accurate diagnosis.

The idea of qi, translating to "vital life force," is central to TCM. The body is interconnected by the flow of qi, which connects all organs and systems. Qi imbalances result in illness. Furthermore, TCM highlights the dynamic interaction between yin (cold, nourishing energy) and yang (warm, stimulating energy). These energies are harmonized by selecting herbal treatments [28–32].

TCM recognizes the existence of variables referred to as characteristics. The accurate differentiation of symptom patterns is associated with the unique qualities exhibited by each plant in herbal therapy. The aforementioned features pertain to distinct values that include distinctive descriptive attributes. Consequently, they assist the practitioner in identifying the herbal options that exhibit the strongest correlation with a range of symptoms and the individual patient's constitution.

Chinese herbal medicine is categorized based on distinctive attributes, such as flavor, thermal properties, direction, and channel connections. Various qualities possess distinct evaluative criteria, necessitating their consideration during the formulation of herbal medication.

TCM defines several disharmony patterns (syndromes). Every pattern is associated with a distinct set of symptoms and imbalances. As an example:

- Qi deficit is marked by weariness, weakness, and weakened immune system symptoms.
- Blood stasis is characterized by reduced blood flow and the presence of discomfort.
- Damp heat presents as either irritation or an excessive amount of moisture.

Herbal remedies are specifically formulated to target these patterns and reinstate balance. When considering these alternatives, such as selecting the ideal flavor, opting for an incorrect thermal characteristic or an unsuitable meridian may lead to an ineffectual therapeutic intervention. Patients with a heart condition should avoid administering herbs with hot or warm thermal properties. Engaging in such actions not only proves to be futile but also may exacerbate the hot condition. The directional affinity is employed to direct the herb's mechanism of action toward the designated treatment region. Huang Qin (radix Scutellariae) effectively alleviates heat in the upper jiao (chest) region, making it a potential treatment for lung or liver ailments. The plant known as Du Huo (radix Angelicae Pubescentis) exhibits a downward movement and is beneficial for both the therapy and guidance of other herbs in the management of lower-body musculoskeletal disorders [28–33].

TCM seldom uses individual herbs in isolation. Alternatively, practitioners generate personalized formulations by amalgamating many plants. The formulations provided are customized to suit the unique constitution, symptoms, and diagnosis of each individual. The combined utilization of various herbs amplifies therapeutic efficacy. In TCM, the concept of "君臣佐使" (Jun-Chen-Zuo-Shi), or the "monarch, minister, assistant, and courier" formulation strategy, is a fundamental principle for composing herbal remedies. This approach strategically combines different herbs in a prescription to maximize therapeutic efficacy and minimize side effects. Each herb in the formula has a specific role:

- 君 *(Jun): Monarch.* The *monarch* herb is the primary ingredient that targets the main symptoms or the disease itself. It has the greatest effect on the condition being treated.
- 臣 *(Chen): Minister.* The *minister* herb supports the monarch by enhancing its effects or treating secondary symptoms. Sometimes, it can also moderate harsh properties of the monarch herb.
- 佐 *(Zuo): Assistant.* The *assistant* herb can have several roles; it can enhance the effects of the monarch and minister herbs, counteract their possible adverse effects, or directly treat less-prominent symptoms that are not addressed by the monarch or the minister.
- 使 *(Shi): Courier.* The *courier* herb ensures that the effects of the other herbs are directed to the correct part of the body. It can also harmonize and integrate the effects of the entire formula.

This structured approach allows TCM practitioners to create balanced, synergistic herbal formulations tailored to the individual's specific health needs, enhancing both the safety and the effectiveness of the treatment.

Many historical works from influential TCM writers and practitioners contain extensive volumes of herbal formulas, medical applications, and factors related to TCM treatments for personalized protocols. These texts continue to be shared and explored in the present day. The ancestral authors deserve respect, as their wisdom and intuitive understanding have established the foundation for TCM, both in the past and in the present. TCM practitioners express their gratitude for transmitting this cultural legacy and providing their individual contributions to the succeeding generation.

1. The Tang dynasty (Dao), *The Inner Classic of the Yellow Emperor*, dating back to 300 BCE, holds the highest reverence among the ancient writings of TCM. It is thought that this literature was created by combining numerous individual writings into two texts: Suwen (Basic Questions 素问) and Lingshu (Spiritual Pivot 灵枢). It is commonly considered the theoretical basis of TCM and its diagnostic techniques. It encompasses the interplay between natural forces (yin/yang) and the innate human desire to restore and sustain equilibrium and concord.
2. *The Divine Husbandman's Classic of the Materia Medica*, authored by Shen Nong Ben Cao Jing around 200 CE. The manuscript comprehensively analyzes more than 300 herbs, focusing on their thermal properties, taste, toxicity, dose, and various forms. The manuscript and era played a significant role in the continued use of several herbal medicines,

including ma huang (Ephedrae herba) for asthma relief, dang gui (*Radicis Angelica sinensis*) for menstrual management, and huang lian (*Rhizoma copies*) for diarrhea cessation.

3. The treatise titled *Treatise on Cold Pathogenic and Miscellaneous Diseases* was authored by physician Zhang Zhong Jing during the Eastern Han dynasty. It was written by Shang Han Za Bing Lun (150–219 CE). The manuscript primarily addressed therapeutic interventions for epidemic infectious illnesses that induce fevers. Furthermore, the author authored Jin Kui Yao Lüe's (金匱要略) *Essential Prescriptions of the Golden Coffer*, a work that centers on the therapeutic approaches for interior ailments.
4. Yao Xing Ben Cao (600 CE) authored the *Materia Medica of Medicinal Properties*. The manuscript comprehensively analyzed various intricate topics, including combination, reaction, taste, temperature, toxicity, function, primary therapeutic use, processing, and preparation. This text holds significant importance, representing the initial endeavor to integrate herbal empirical facts, clinical application, and classical TCM theory into a unified compendium.
5. The *Materia Medica of Diet Therapy*, authored by Meng Shan in 700 CE, is attributed to Shi Liao Ben Cao (Cao). The significance of nutrition and of using 227 herbs as medicinal agents was underscored.
6. Li Shi Zhen compiled the *Grand Materia Medica* in 1578 CE. This comprehensive compendium has data of about 1,892 compounds, comprising 1,173 derived from plants, 444 derived from animals, and 275 minerals. Furthermore, it suggested 11,000 herbal remedies for particular afflictions.
7. The compilation of Ben Cao Gang Mu Shi Yi (1765 CE) was undertaken by Zhao Xue Min (1765 CE). The present publication offers a comprehensive account of more than 700 natural chemicals documented for the first time.
8. The *Chinese Herbal* (1999 CE) was developed by the Chinese State Administration of TCM with the participation of more than 500 scholars/experts from at least 60 academic and research institutes. The compilation process lasted for ten years. The comprehensive course book comprises ten volumes, encompassing 9,282 pages. It includes 8,534 pictures and 8,980 monographs on Chinese Materia Medica (CMM). A truncated version consisting of two volumes is now accessible. The material examined in this study extends until the year 1994. This documentation is the most comprehensive of CMM until now [28, 34].

Prior to integrating herbal treatments into a patient's health regimen, it is imperative to consult with a certified TCM practitioner for appropriate assistance. TCM practitioners possess a comprehensive understanding of herbal medicine principles, individual constitutions, and patterns of disharmony. Herbal formulae can be customized to meet individual requirements, ensuring maximum effectiveness and safety.

In certain instances, certain herbs have the potential to interact with other drugs or are contraindicated for specific illnesses or populations. For example:

- Ginseng has the potential to impact blood glucose levels and may exhibit interactions with antidiabetic treatments.
- Licorice root can modify potassium levels and may interact with medications such as diuretics.
- Astragalus is generally considered safe; however, it is advisable to refrain from using it during severe illnesses or fever episodes.
- Dang gui (*Angelica sinensis*) should not be consumed while pregnant.

Regarding the drawbacks of herbal remedies, TCM practitioners regularly follow a crucial practice: revealing the patient's prescription history. The diligent methodology guarantees that the health-care professional possesses comprehensive knowledge regarding any concurrent medicinal interventions the patient may receive. The justification for this revelation is rooted in the possible interplay between conventional drugs and natural therapies. Certain herbs are known to contain bioactive chemicals that have the potential to impact drug metabolism, modify absorption rates, or potentially worsen side effects. TCM practitioners can customize treatment recommendations by comprehending the patient's pharmaceutical regimen. In addition, practitioners of TCM consider the holistic health condition of the patient. Determining the most appropriate herbal formula is influenced by various factors, including underlying problems, physiological constitution, and individual sensitivities. The practitioner's instruction is tailored to enhance therapeutic outcomes while reducing dangers, whether addressing qi deficiency, blood stasis, or other patterns of disharmony [31–36].

12.3.5.4 TCM Product Registration and Licensing Requirements

When using TCM strictly as a medication, additional requirements, such as an extensive clinical trial, safety evaluation, and efficacy study, must be met. The following categories apply to TCM drug registration [36]:

1. A pure compound isolated from natural material
2. Recently discovered medicinal natural material
3. TCM raw material substitute
4. Recently discovered medicinal part of TCM herb
5. Effective fraction isolated from natural material
6. TCM product with multiple ingredients
7. TCM injection
8. Changing the route of administration
9. Changing the dose form
10. Changing the producing technology
11. Generic drug

The government agency known as the State Administration of TCM was previously run under the MOH. The organization was tasked with overseeing several aspects of TCM, such as education, clinics, licensing practitioners, scientific research on herbal formulation and quality, and other legal problems related to the practice of TCM [37]. The SFDA has been designated as the sole regulatory organization responsible for supervising all drugs, including both Western and TCM remedies, since 2003. Implementing this measure effectively resolved the existing disparities in standards among provincial government agencies, consolidated the Chinese health-care regulatory system, and enhanced the clarity of pertinent matters. To produce medications in China, foreign pharmaceutical businesses must acquire GMP certificates from the SFDA [4–37].

China has made notable advancements in drug development regulations over the last three decades. This has been achieved by implementing a range of policies to regulate drug manufacture, drug distribution, pharmaceuticals utilized in medical institutions, drug packaging, drug pricing, and advertising, all in alignment with international standards. The SFDA also facilitated research on TCM to foster business growth. The clearance and registration process for TCM in China has become as equally stringent as that for chemical pharmaceuticals and biological goods. The primary criteria for a dossier utilized in the registration process are as follows:

1. Pharmaceutical research materials encompass several aspects, such as the origin and identification of Chinese Materia Medica, cultivation techniques, preparation methods, strip processing, production processes, quality testing data, quality standards, and stability tests.
2. The research materials encompass pharmacological and toxicological aspects, such as pharmacodynamics tests, acute and long-term toxic studies, allergic reactions, hemolytic tests, and studies on potential carcinogenicity.

3. The clinical research materials include clinical trial documentation, ethics committee approval, and testing reports [4–5].

The primary distinction from functional foods lies in the requirement for comprehensive clinical trials and the subsequent approval of the results by SFDA experts. Nevertheless, the manufacturer is responsible for ascertaining whether the product should be classified as functional food or TCM. Therefore, many TCM products could be promoted as functional foods. One of the primary obstacles encountered in utilizing TCM or herbal compounds is the potential toxicity due to activities affected by the increasing economic incentives and inadequate ethical standards among individuals in this sector. In addition to the substandard quality and the absence of uniformity, clients may face significant risks due to variables such as substitution, misidentification of herbs resulting from inadequate understanding, and adulterations with Western pharmaceuticals that have established functions. In conjunction with essential measures, such as the implementation of GAP, GMP, and the application of hazard analysis and critical control point (HACCP) during the production process, the most efficacious approach to ensuring safety is through the monitoring of product performance in the marketplace. This can be achieved by collecting and investigating consumer inquiries, complaints, and adverse reactions. The monitoring of negative occurrences in herbal medications has been implemented in numerous nations, such as the United States, Canada, Australia, South Korea, and European Union countries. China has implemented an adverse drug reaction system to document grievances related to TCM [38–39]. Details regarding the TCM registration process can be found in the Chinese Medicine Regulatory Office [40].

To advertise and promote TCM, obtaining a permit from the Health Sciences Authority in accordance with the Medicines (Medical Advertisements) Regulations is necessary. The clearance of a TCM product listing does not necessarily indicate that the product name and/or its claims will be permitted for advertising. It is incumbent upon the applicant to abstain from utilizing the approval of the TCM product listing as a means of marketing or promoting the product [41].

12.3.5.5 The Nomenclature of Traditional Chinese Medicine

Drug brand names in China adhere to a stringent regulatory framework. As stipulated by the Drug Administration of China, the nomenclature of a pharmaceutical product must exclusively consist of Chinese characters authorized by the Trademark Law. The name should not imply the therapeutic efficacy of the medication, specify the location of treatment, or bear resemblance to the generic name of a drug in terms of sound or structure.

Before translation into Chinese, an alphabetic name is initially verified when assigning names to Western medications. In Chinese, there are two primary name typologies for such drugs:

1. *Pure phonetic translation* refers to using Chinese characters that correspond with the phonetic representation of the drug's original name in the Western language.
2. *Transliteration* in conjunction with Chinese characters provides a significant connotation about the medication or its intended use.

Chinese medication brand names typically possess distinct qualities. These characters are straightforward, easily pronounced, and usually comprise two to four frequently used Chinese characters with fewer strokes. These names also possess favorable implications, often employing characters that symbolize joy, commitment, or tranquility.

Alphabetic brand names also exhibit distinct characteristics. Including prefixes in brand names frequently denotes the specific context in which the drug is intended to be utilized, whilst suffixes can infer the inherent attributes of the drug. The letters *X* and *Z* frequently represent power, velocity, and ingenuity, implying potent therapeutic effectiveness.

Pharmaceutical firms not only have the option to register trademarks for their brand names but are also strongly encouraged to do so for better brand protection. Registering a trademark offers optimal safeguarding for the brand and deters imitators from employing the same or comparable names, thus reducing the likelihood of legal disputes and unauthorized utilization. Registering a trademark confers exclusive rights to the person who registers it and improves the reputation and intangible value of the business.

In China, it is recommended to extend the scope of trademark protection beyond pharmaceuticals and medical services to encompass additional relevant classes, including chemicals, cleaning products, medical instruments, scientific research, and computer telecommunication categories. Obtaining trademarks similar to the primary name can also protect against possible violations.

Generic names are employed instead of brand names in the context of proprietary Chinese medicine (PCM). The variety of names can cause confusion when comparing different medications. The naming criteria for pCm generic names have been released by the China Food and Drug Administration, with a focus on promoting consistency. The names must be limited to a maximum of eight characters and must explicitly indicate the dosage form. It is advised to refrain from using exaggeration, self-praise, and impractical statements. In contrast, generic designations frequently pertain to adverse reactions, primary constituents, or a fusion thereof [42].

12.4 CONCLUSION

In conclusion, "Navigating Regulatory Requirements: A Comprehensive Guide to Nutraceuticals and Traditional Chinese Medicine" emphasizes the significance of maintaining awareness and familiarity with the most recent developments in regulatory affairs. The ever-changing nature of regulatory requirements makes it necessary to maintain constant awareness. Participants in the nutraceuticals and traditional Chinese medicine industries can successfully traverse the ever-changing landscape of rules, guidelines, and policies affecting their respective industries if they routinely study the websites of regulatory authorities, attend industry conferences, and subscribe to publications pertinent to their interests. Adopting a proactive approach to remaining up-to-date increases the likelihood of success in these industries and guarantees compliance with regulations.

REFERENCES

[1] National Medical Products Administration (NMPA). *Regulation of Nutritional Supplements* [Internet]. 2021. Available from: http://www.nmpa.gov.cn/WS04/CL2138/.

[2] Kidd PM. Omega-3 DHA and EPA for cognition, behavior, and mood: Clinical findings and structural-functional synergies with cell membrane phospholipids. *Alternative Medicine Review*. 2012;17(3):237–49.

[3] Choi J, Kim TH, Choi TY, Lee MS. Ginseng for health care: A systematic review of randomized controlled trials in Korean literature. Ferro VA, editor. *PLoS One* [Internet]. 2013 Apr 1;8(4):e59978. Available from: https://dx.plos.org/10.1371/journal.pone.0059978.

[4] Liang Z, Hu H, Li J, Yao D, Wang Y, Ung COL. Advancing the regulation of traditional and complementary medicine products: a comparison of five regulatory systems on traditional medicines with a long history of use. Feng N, editor. *Evidence-Based Complementary and Alternative Medicine* [Internet]. 2021 Oct 27;2021:1–16. Available from: https://www.hindawi.com/journals/ecam/2021/5833945/.

[5] Wang WY, Zhou H, Wang YF, Sang BS, Liu L. Current policies and measures on the development of traditional Chinese medicine in China. *Pharmacological Research* [Internet]. 2021 Jan;163:105187. Available from: https://linkinghub.elsevier.com/retrieve/pii/S104366182031495X.

[6] Santini A, Novellino E. To nutraceuticals and back: Rethinking a concept. *Foods* [Internet]. 2017 Sep 5;6(9):74. Available from: https://www.mdpi.com/2304–8158/6/9/74.

[7] Shahidi F. Nutraceuticals and functional foods: Whole versus processed foods. *Trends in Food Science & Technology* [Internet]. 2009 Sep;20(9):376–87. Available from: https://linkinghub.elsevier.com/retrieve/pii/S092422440800229X.

[8] Vyas N, Kaservani RK, Jain S, Raghuvanshi R, Kasharwani RK, Sharma AK. Nutraceuticals and functional food as future food: A review. *Der Pharmacia Lettre*. 2010;2(1):106–16.

[9] Only F, Animal B. *SFDA Registration Expense and Timetable for Health Food* [Internet]. p. 25–7. Available from: http://www.sfdachina.com/info/121–1.htm.

[10] Yang MS. Regulatory aspects of nutraceuticals: Chinese perspective. In: *Nutraceuticals* [Internet]. Elsevier; 2021. p. 1281–91. Available from: https://linkinghub.elsevier.com/retrieve/pii/B9780128210383000756.

[11] Hee Siang. *Chinese Nutraceutical Industry Background and Future* | Monteloeder [Internet]. 2017. Available from: https://monteloeder.com/chinese-nutraceutical-industry-background-and-future/.

[12] National Health Commission of the People's Republic of China. *Notice of the Ministry of Health on Further Standardizing the Management of Nutritional Supplements Raw Materials* [Internet]. 2002. Available from: http://www.nhc.gov.cn/wjw/gfxwj/201304/e33435ce0d894051b15490aa3219cdc4.shtml [in Chinese].

[13] National Health Commission SA for MR. *Announcement on 6 New Substances Traditionally Both Food and Chinese Medicinal Materials (No. 8 [2019])* [Internet]. 2019. Available from: http://www.nhc.gov.cn/sps/s7885/202001/b941b6138e93414cb08aed926ca3c631.shtml.

[14] National Health Commission, State Administration for Market Regulation. *Announcement on the Addition of 9 New Substances such as Codonopsis Ginseng, Which Are Both Food and Chinese Medicinal Materials According to Tradition.* 2023. Available from: http://www.nhc.gov.cn/sps/s7892/202311/f0d6ef3033b54333a882e3d009ff49bf.shtml.

[15] Ministry of Commerce of the People's Republic of China. *Food Hygiene Law of the People's Republic of China* [Internet]. 2021. Available from: http://english.mofcom.gov.cn/aarticle/lawsdata/chineselaw/200211/20021100050616.html.

[16] Hasler CM. *Regulation of Functional Foods and Nutraceuticals: A Global Perspective*. Wiley-Blackwell; 2007. p. 1–411.

[17] Zhang L. China: Food safety law passed. 2009;2019(17 June). Available from: http://www.loc.gov/law/foreign-news/article/china-food-safety-law-passed/.

[18] Mu X. *Xinhu News*. 2013. *China to Elevate Food, Drug Agency to General Administration.* Available from: http://news.xinhuanet.com/english/china/2013–03/10/c_132221729.htm.

[19] Beckett N, Pountney D. *Consolidation of the China Food and Drug Administration* [Internet]. 2013. p. 20–2. Available from: http://www.lexology.com/library/detail.aspx?g=20a4e134-7414-4f4d-88d5-d7dd36d800a7.

[20] Chemical Inspection and Regulation Service (CIRS). *China Health Food Registration and Filing* [Internet]. 2017. Available from: http://www.cirs-reach.com/news-and-articles/health-food-registration-and-filing.html.

[21] China Food and Drug Administration. *Health Food SFDA Registration Process* [Internet]. 2015. Available from: http://www.sfdachina.com/info/85–1.htm.

[22] Huang G, Lampsley K. *Chinese Health (Functional) Food Regulation*. In: *Regulation in Functional Food and Nutraceuticals: A Global Perspective*. 2005. p. 263–92.

[23] Lennie Tao. *China Health Food Regulation* | ChemLinked [Internet]. Available from: https://food.chemlinked.com/foodpedia/china-health-food-regulation.

[24] Tingmin Koe. *China Proposes "Blue-Hat" Logo to be Printed on All Smallest Individual Packs* [Internet]. 2022. Available from: https://www.nutraingredients-asia.com/Article/2022/03/30/china-proposes-blue-hat-logo-printed-on-smallest-individual-packs.

[25] Amanda Mao. *Unraveling the Mystery—China Health Food Registration and Notification* [Internet]. 2020. Available from: https://www.acuritmedcomms.com/2020/11/13/unravelling-the-mystery-china-health-food-registration-notification/.

[26] Global Regulatory Partner. *Is the Registration of Supplements Required in China?* [Internet]. 2021. Available from: https://globalregulatorypartners.com/registration-of-food-supplements-in-china/.

[27] CIRS. *Dietary Supplement Name with Function Presentation will be Forbidden* [Internet]. 2015. Available from: https://www.cirs-group.com/food/news/Dietary_Supplement_Name_Function_Presentation_Forbidden.html.

[28] Marshall AC. Traditional Chinese medicine and clinical pharmacology. In: *Drug Discovery and Evaluation: Methods in Clinical Pharmacology*, Second Edition [Internet]. Cham: Springer International Publishing; 2020. p. 455–82. Available from: http://link.springer.com/10.1007/978-3-319-68864-0_60.

[29] Wang FL. *Handbook of Traditional Chinese Medicine*. Yin and Yang; 2014. p. 19–30. Available from: https://en.wikipedia.org/wiki/Yin_and_yang.

[30] Chan K. Progress in traditional Chinese medicine. *Trends in Pharmacological Sciences* [Internet]. 1995 Jun;16(6):182–7. Available from: https://linkinghub.elsevier.com/retrieve/pii/S0165614700890197.

[31] Gu S, Pei J. Innovating Chinese herbal medicine: From traditional health practice to scientific drug discovery. *Frontiers in Pharmacology* [Internet]. 2017 Jun 16;8. Available from: http://journal.frontiersin.org/article/10.3389/fphar.2017.00381/full.

[32] Hua Chen J, Jobst K. *Vol. 2, Complementary Therapies in Medicine*. Traditional Chinese Medicine; 1994. p. 138–9. Available from: https://www.nlm.nih.gov/hmd/topics/chinese-traditional/index.html.
[33] Ma D, Wang S, Shi Y, Ni S, Tang M, Xu A. The development of traditional Chinese medicine. *Journal of Traditional Chinese Medical Sciences* [Internet]. 2021 Nov;8:S1–9. Available from: https://linkinghub.elsevier.com/retrieve/pii/S2095754821000909.
[34] BenskyD,ClaveyS,StogerE.*ChineseHerbalMateriaMedica*[Internet].2016.1325p.Availablefrom:https://www.thenile.co.nz/books/dan-bensky/chinese-herbal-medicine-materia-medica/9780939616824.
[35] Hesketh T, Zhu WX. Health in China: Traditional Chinese medicine: One country, two systems. *BMJ* [Internet]. 1997 Jul 12;315(7100):115–7. Available from: https://www.bmj.com/lookup/doi/10.1136/bmj.315.7100.115.
[36] Zhou X, Li CG, Chang D, Bensoussan A. Current status and major challenges to the safety and efficacy presented by Chinese herbal medicine. *Medicines* [Internet]. 2019 Jan 18;6(1):14. Available from: http://www.mdpi.com/2305–6320/6/1/14.
[37] Eisenhart S. Emergo Group. *State Food and Drug Administration (SFDA) Becomes China Food and Drug Administration (CFDA)*. 2013. Available from: https://www.emergogroup.com/blog/2013/03/china-state-food-and-drug-administration-renames-itself-china-food-and-drug.
[38] Shao A. Global market entry regulations for nutraceuticals, functional foods, dietary/food/health supplements. In: *Developing New Functional Food and Nutraceutical Products* [Internet]. Elsevier; 2017. p. 279–90. Available from: https://linkinghub.elsevier.com/retrieve/pii/B9780128027806000158.
[39] Zhang L, Yan J, Liu X, Ye Z, Yang X, Meyboom R, et al. Pharmacovigilance practice and risk control of traditional Chinese medicine drugs in China: Current status and future perspective. *Journal of Ethnopharmacology* [Internet]. 2012 Apr;140(3):519–25. Available from: https://linkinghub.elsevier.com/retrieve/pii/S0378874112001122.
[40] Chinese Medicine Regulatory Office. *Chinese Medicine and Proprietary Chinese Medicines* [Internet]. Available from: https://www.cmro.gov.hk/html/eng/forms/forms_cm_and_pcm_related.html#pcm.
[41] Health Science Authority. *Guidelines on Chinese Proprietary Medicines Product Listing* [Internet]. 2023. Available from: www.hsa.gov.sg › default-source › hprg-tmhs › pl_fsc_guidelines.
[42] Brand Acumen. *Drug Naming in China: Challenges and Importance for Pharmaceutical Manufacturers* [Internet]. 2023. Available from: https://www.linkedin.com/pulse/drug-naming-china-challenges-importance-pharmaceutical.

13 Nutraceutical Regulation in Indonesia

Ayu Shabrina

13.1 INTRODUCTION

According to the 2022 Census Report, Indonesia has a population of over 270 million, making it one of the world's largest archipelagic countries. About 50% of Indonesia's population lives on the island of Java, making this island the most densely populated [1]. The dense population and vast territory, as well as the diverse socio-cultures, have caused variations in the health conditions of Indonesian people, some of whom live in remote areas, borders, and islands.

Population health is influenced by nutritional status. The better a person's nutritional status, the better their ability to grow, develop, and adapt to the environment. The nutritional status of children under 5, which showed 21.6% in 2022, was 21.5% in 2023 [2]. Most malnutrition happens in children between 24 and 35 months old and is concentrated in Central Papua (39.4%) [2]. Additionally, obesity is one of the risks of degenerative diseases which decreases the productive population [3].

Indonesia entered the demographic bonus era in 2012. At that time, the population dependency ratio was 49.6%, meaning that the number of non-productive residents (aged <15 and >65 years) was 49.6% when compared with the productive age group (15–64 years) [4]. The lowest dependency ratio figure will occur in 2028–2031 and end in 2036–2037. It is hoped that this large number of productive residents will increase public awareness on healthy living, especially regarding nutritional needs.

The Indonesian government, in encouraging health improvement, has made various efforts, the main of which are preventive efforts. Some of the preventive efforts that have been carried out include health promotion through community health centers as primary health facility units. The government is attempting to promote public health by utilizing supplements, vitamins, and herbal treatments known as nutraceuticals. The growth in the use of nutraceuticals in Indonesia is currently increasing rapidly, especially after the COVID-19 pandemic. The use of nutraceuticals in Indonesia increased by 314.31% in 2020–2021, during the COVID-19 pandemic, and is currently experiencing a decline again with normal sales [5].

13.2 DISTRIBUTION OF DISEASE AND NUTRITIONAL STATUS IN INDONESIA

Non-communicable diseases, such as stroke, heart disease, diabetes mellitus, and cancer, have been increasing in the past five years. These diseases continue to increase in Indonesia, placing a high-cost burden on the government [6]. The prevalence of hypertension in Indonesia has not significantly decreased, from 34.1% in 2013 to 30.8% in 2023. Diabetes mellitus (DM) increased in the population of >15 years old, from 10.97% in 2013 to 11.7% in 2023. This prevalence is in line with stroke, which increased to 20.2% in 2023. This means that there are still many cases of hypertension and diabetes mellitus that are not detected. Cancer cases based on doctor's diagnosis also increased from 0.6% to 2.2% at all ages, with cases being greater in urban areas compared to rural areas. Cancer also occurs more in women compared to men.

The prevalence of smoking behavior in Indonesia was 58%, and 95% of them were dominated by men [7]. Smoking behavior was dominated by ages 12–24 years old [8]. In Indonesia, 3.3% of

DOI: 10.1201/9781003465041-13

adolescents start drinking at 10 years of age, and the majority of adolescents (38.7%) drink typical alcoholic beverages [9]. Teenagers aged 11–14 years old are likely to consume alcohol 1.5 times higher than do late teens [10]. Insufficient activity was also discovered in 33.3% of the population aged 10 years and above, while 95.5% of the people aged 5 years and over did not consume enough fruits and vegetables [11]. Obesity detected in the population aged 18 years and over is getting higher every year, namely, 10.5%, 14.8%, and 21.8% [12–13].

The 2023 National Survey results show the problem of malnutrition among toddlers and the elderly in Indonesia. Based on this problem, the Indonesian government is trying to increase the perception and use of nutraceuticals, especially in teenagers, to support improved nutrition and decrease metabolic diseases [14]. Based on data in 2023, there was an increase in non-communicable diseases in Indonesia, which can be seen in Figure 13.1.

In several developing countries, the majority of the population still continues to use natural plants, especially to fulfill basic health needs. In Indonesia itself, many medicinal plants are used as basic ingredients for various herbal medicines, medicines, cosmetics, spa ingredients, and raw materials for the food and beverage industry. The tendency for a healthy lifestyle to return to nature has caused people to start switching to using natural ingredients to minimize side effects, such as chemical drugs. Nutraceuticals nowadays have been used to help in treating non-communicable disease, such as hypertension, diabetes, and cancer [15–16].

Indonesia has over 30,000 plant species, 940 of which are nutritional plants (180 species have been employed by the traditional industry), which is a potential market for herbal medicines and phytopharmaceuticals in both Indonesia and Europe [17–18]. In the Research on Medicinal Plants and Herbal Medicine (RISTOJA), which was carried out in 2015, the number of medicinal plants identified was 1,159, consisting of 156 plants that potentially can be developed as nutraceuticals [19–20]. In Indonesia, the development of the use of herbs and nutraceuticals has increased from year to year, especially after COVID-19 [21]. It is known that there has been an increase in the use of nutraceuticals and herbs [22–23]. Currently, the use of herbs and nutraceuticals has returned to normal.

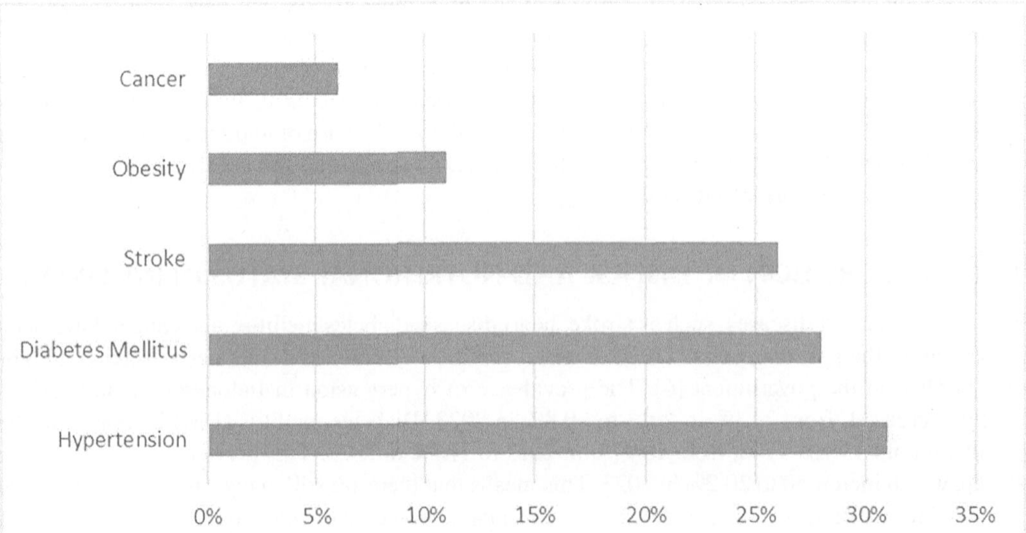

FIGURE 13.1 Graphic presentation of metabolic disorders in Indonesia in 2022.

Source: [14].

13.3 REGULATION OF NUTRACEUTICALS AND TRADITIONAL MEDICINES IN INDONESIA [24]

The Directorate of Traditional Medicines, Health Supplements, and Cosmetics conducts pre-market evaluations to strengthen the regulatory system for the oversight of medicines and foods in the traditional medicines, health supplements, and cosmetics field. These evaluations assess the quality, efficacy/benefits, and safety of products in traditional medicines, health supplements, and cosmetics that are to be marketed. In conducting these assessments, the Directorate of Traditional Medicines, Health Supplements, and Cosmetics also involves experts in the field of traditional medicines, health supplements, and cosmetics, both within and outside the Deputy for the Supervision of Traditional Medicines, Health Supplements, and Cosmetics. The Directorate of Traditional Medicines, Health Supplements, and Cosmetics consists of:

1. The Sub-Directorate of Registration of Products and Advertising for Traditional Medicines and Health Supplements
2. The Sub-Directorate of Cosmetic Registration
3. The Sub-Directorate for the Evaluation of Pre-Clinical/Clinical Tests of Traditional Medicines, Health Supplements, Cosmetics, and Cosmetic Product Information Documents, and the Functional Position Group

The Sub-Directorate of Product Registration and Advertising for Traditional Medicines and Health Supplements is tasked with preparing and implementing policy; developing norms, standards, procedures, and criteria; conducting technical guidance and supervision; and evaluating and reporting on product registration and advertising for traditional medicines and health supplements. In carrying out these duties, the Sub-Directorate of Registration of Products and Advertising for Traditional Medicines and Health Supplements performs the following functions:

1. Materials preparation for policy formulation in the product registration and traditional medicines and health supplements advertising field
2. Materials preparation for policy implementation in the product registration and traditional medicines and health supplements advertising field
3. Materials preparation for the development of norms, standards, procedures, and criteria in the product registration and traditional medicines and health supplements advertising field
4. Materials preparation for providing technical guidance and supervision in the product registration and traditional medicines and health supplements advertising field
5. Execution of monitoring, evaluation, and reporting in the product registration and traditional medicines and health supplements advertising field

The Sub-Directorate of Registration of Products and Advertising for Traditional Medicines and Health Supplements is divided into three sections: Registration of Traditional Medicines, Registration of Health Supplements, and Registration of Advertising for Traditional Medicines and Health Supplements.

The **Traditional Medicine Registration** section is responsible for preparing materials for policy formulation and implementation; developing norms, standards, procedures, and criteria; providing technical guidance and supervision; and monitoring, evaluating, and reporting on the implementation of traditional medicine registration.

The **Health Supplements Registration** section is responsible for preparing materials for policy formulation and implementation; developing norms, standards, procedures, and criteria; providing technical guidance and supervision; and monitoring, evaluating, and reporting on the implementation of health supplement registration.

The **Traditional Medicine and Health Supplements Advertising Registration** section is in charge of preparing materials for policy formulation and implementation; developing norms, standards, procedures, and criteria; providing technical guidance and supervision; and monitoring, evaluating, and reporting on the implementation of traditional medicine and health supplement advertising registration.

13.4 REGISTRATION OF TRADITIONAL MEDICINES AND NUTRACEUTICALS IN INDONESIA [25]

The Ministry of Health, based on Ministerial Decree No. 149/SK/Menkes/IV/1978 concerning medicinal plants, stated that various parts of these plants can be used for different purposes, including:

1. Plants or parts of plants used in traditional medicine or herbal treatments
2. Plants or parts of plants used as precursors for raw pharmaceutical materials
3. Plants or parts of plants that are extracted, with the extracts being used as medicine

Nutraceuticals, also known as health supplements, contain ingredients such as vitamins, minerals, amino acids, and other non-plant materials that can be combined with plant substances to supplement nutritional needs, maintain or improve health functions, and provide nutritional and physiological effects.

Based on the head of the Food and Drug Supervisory Agency's decision Number HK.00.05.4.2411 of 2004 regarding the Basic Provisions for the Classification and Labeling of Natural Material Drugs in Indonesia, Indonesian traditional medicines can be classified into jamu, registered herbal medicines (OHT), and phytomedicines. *Jamu* is used traditionally across generations and only has usage claims that correspond to traditional types of proof (passed down through generations). *OHTs* (registered herbal medicines) are traditional medicines that have been scientifically proven for quality, safety, and efficacy and that use raw materials that meet established standards; typically, pre-clinical trials have been conducted on OHTs. Meanwhile, *phytomedicines* are standardized herbal medicines that have undergone more rigorous scientific validation, including clinical trials. Nutraceuticals are classified within health supplements but can be in the form of food or beverages. In Indonesia, nutraceuticals are commonly referred to as health supplements.

Functional food can bridge the conventional gap between food and medicine, ensuring that the potential benefit of a newly designed functional food is realized. Consuming a nutritious diet is essential for meeting nutritional demands and reducing the risk of disease [26]. However, while functional foods can benefit human health, they must be supplemented by other good lifestyle habits, such as exercise and personal hygiene. The bioactive components found in functional foods, such as dietary fiber, inulin, antioxidants, PUFAs, prebiotics, and probiotics, determine their physiological effects [27].

The general quality requirements for traditional medicine products and health supplements in Indonesia include:

1. Business operators must obtain a good manufacturing practice (GMP) certificate, import certificate, or export certificate within Indonesia.
2. These certificates are required for both supplement raw materials and finished products.
3. The quality standards for health supplement products distributed in Indonesia must comply with the provisions of the *Indonesian Pharmacopoeia* or the *Indonesian Herbal Pharmacopoeia*.
4. Other compendia used as quality standards for supplements in Indonesia include *Materia Medica Indonesia*, the *United States Pharmacopoeia*, the *British Pharmacopoeia*, and/or other scientifically recognized and validated references.

5. Registration is conducted electronically to obtain distribution authorization. The distribution license issued by BPOM (*Badan Pengawas Obat dan Makanan*)/Drug and Food Control of Indonesia is valid throughout Indonesia.
6. Business operators eligible to apply for registration include the pharmaceutical industry, traditional medicine industry (IOT), small traditional medicine business (UKOT), food industry, and importers.
7. The licensing registration process is divided into two categories: new registration and registration variation. *New registration* is used by industries registering health supplement products that do not yet have a distribution license. *Renewal registration* is the health supplement registration for the extension of the distribution license validity. Registration variation is divided into four categories:
    a. The minor variation registration with notification is a type of variation registration for specific aspects that do not affect the safety, efficacy, and/or quality of health supplements and do not change any information in the approved distribution license.
    b. The minor variation registration with approval is a type of variation registration that does not fall under the category of minor variation registration with notification or major variation registration.
    c. Major variation registration is a type of variation registration that affects the administrative aspects, safety, efficacy, and/or quality of health supplements.

The registration process is divided into two stages: pre-registration and registration. The *pre-registration* stage includes the examination and assessment of administrative documents, formula data checking, category determination, and registration fee assessment. The *registration* stage involves the examination and assessment of administrative documents, as well as of safety, efficacy, quality, and labeling.

Industries must ensure that health supplements distributed or imported meet the standards for safety, efficacy, quality, and labeling. The composition of health supplements may include preservatives, colorants, sweeteners, flavorings, emulsifiers, stabilizers, and solvents.

Both active ingredients and additives must meet these minimum requirements. Active ingredients in supplements distributed in Indonesia may derive from natural sources. These natural active ingredients can be in the form of isolates, fractions, or extracts. If in the form of isolates, fractions, or extracts, the solvents used are water and alcohol. Specifically, for alcohol solvents, the active ingredients must meet the solvent residue limits according to Table 13.1.

Additional ingredients permitted for supplements include preservatives, sweeteners, colorings, antioxidants, and flavors. Additional materials must also meet the maximum limits according to Table 13.2

Finished nutraceutical products include powders, effervescents, tablets, capsules, and oral liquids. Supplement tablets can be film-coated, sugar-coated, enteric-coated, lozenges, chewable,

TABLE 13.1

Residue Limits for Solvents Other Than Water for Active Supplement Ingredients Derived from Extracts, Fractions, and Isolates

Solvents	Maximum Limit of Residual Solvent in the Product
Ethanol	1%, or 10,000 ppm
Methanol	0.3%, or 3,000 ppm
n-Hexane	0.029%, or 290 ppm
Ethyl acetate	0.5%, or 5,000 ppm

TABLE 13.2
Additional Ingredients Permitted for Supplements

Type	Material	Synonym	Product Limit (mg/kg of Product)
Natural dye	Carmine	CI Natural Red 4	300
	Carotenes	CI Food Orange 5	600
	Chlorophylls	CI (1975) No. 75810	500
Synthetic dyes	Allura Red AC 2	CI (1975) No. 16035	300
	Fast Green FCF	CI Food Green 3	600
	Iron oxide, red iron	CI Pigment Red 101	7,500
Natural sweetener	Steviol glycosides		2,500 equivalent steviol
Artificial sweeteners	Acesulfame potassium	Acesulfame K	2,000
	Aspartame	APM	5,500
	Saccharin		1,200
Preservative	Methyl paraben	Nipagin M	2,000
	Sodium sorbate		2,000
Antioxidant	α-Tocopherol	D-α-tocopherol	500
	Ascorbic acid	L-ascorbic acid	1,000
	Butylated hydroxyanisole (BHA)	6-(Stearoyloxy)-L-ascorbic acid	400

effervescent, and/or gummy tablets. Capsules may contain powder or granules, or oil-like liquids. Oral liquid supplements can be solutions, emulsions, syrups, and suspensions.

The testing of nutraceutical is crucial for the release of supplement products in Indonesia. Quality parameters for products include organoleptic properties, moisture content, disintegration, dissolution, uniformity of weight, microbial contamination, heavy metal contamination, determination of alcohol content, specific gravity, pH, and the identification and quantification of active ingredients. Testing parameters according to the form of the product can be found in Table 13.3.

Health supplements or nutraceuticals must meet stability tests using temperature and humidity that follow zone IVb. The testing follows applicable international regulations. In addition to stability tests, health supplements must meet labeling criteria: Labels must be printed directly on or attached to the container or packaging, not easily removable or damaged by water, friction, or sunlight. Labels on health supplements in Indonesia must, at a minimum, include:

1. Product name, including generic and/or trade name, with the designation "health supplement"
2. Dosage form
3. Name and address of the industry and/or business operator
4. Name and address of the contract giver and/or contract receiver
5. Name and address of the license giver and/or license receiver
6. Size, content, and net weight; qualitative and quantitative composition; qualitative additives; benefit claims; usage instructions/methods; and contraindications, side effects, and warnings (if any)
7. Marketing authorization number, batch number/production code, expiration date, storage conditions
8. 2D barcode in accordance with legislative regulations
9. Additional information as long as it meets standards and/or requirements for safety, quality, and/or specific material origin according to legislative regulations

TABLE 13.3
Nutraceutical Quality Testing Parameters in Accordance with the Dosage Form

Dosage Form	Organoleptic	Water Content	Disintegration	Dissolution	Weight Uniformity	Microbial Contamination	Heavy Metal Contamination	Specific Gravity and pH	Determination of Alcohol Content	Identified Active Ingredients	Content Determination
Tablet and hard capsule	√	√	√	√	√	√	√			√	√
Soft capsule	√		√		√	√	√			√	√
Granules	√	√				√	√			√	√
Effervescent	√	√				√	√			√	√
Gummy	√	√				√	√			√	√
Oral liquid	√					√	√	√	√	√	√

Pharmaceutical industries and micro-industries are applying for health supplement registration must provide documents including tax identification number (NPWP), GMPs (Good Manufacturing Practices) for nutraceutical or traditional medicine, or approval to produce health supplements in a food production facility, and a power of attorney from company management and the company's notarial deed of establishment. Food industries applying for licensing must have a GMP implementation license, called CPOB in Indonesia, according to the dosage form they wish to register and documents of approval to produce health supplements in a food production facility. To obtain such licensing, industries must have a pharmacist responsible, evidenced by a stamped statement. Health supplements to be exported outside of Indonesia must meet all the aforementioned documents and the documents required by the destination country for export. Health supplements imported from abroad must meet tax documentation and:

1. Name and address of the manufacturer/principal from the country of origin, name of the importer, brand and/or name and size of the product packaging
2. Date issued; validity period of the agency appointment; rights to conduct registration, import, and distribution from the manufacturer/principal of the country of origin
3. Name and signature of the director/leader of the manufacturer/principal from the country of origin
4. Certificate of free sale (CFS), certificate of pharmaceutical product (CPP), or another equivalent document issued by the competent government authority in the country of origin, or another agency designated by the government of the country of origin, with proof, including an appointment letter
5. Additional requirements for functional foods, including a site master file (SMF) from the manufacturer in the country of origin

Importers must complete several documents related to raw materials and testing, including methods and results of raw material and finished product quality testing and results of finished product quality

TABLE 13.4
Length of Time for Evaluating Decisions on Health Supplement Registration Results

Type	Composition/Description	Length of Evaluation Time
Category 1	The single active ingredient is a vitamin or mineral whose safety profile and benefits are known.	15 days
Category 2	Single active ingredients other than vitamins and minerals, or combinations, whose safety and efficacy profiles are already known.	30 days
Category 3	New single active ingredient or combinations; new posology; new claims; new dosage forms; or the safety profile and benefits are not yet known.	50 days
Category 4	New import registration.	50 days
Registration of major variations		7 days
Re-registration	No change.	30 days
Export registration		3 days

testing from an accredited laboratory in Indonesia or a GMP- or GLP-certified (Good Laboratory Practice) industry laboratory in Indonesia. The testing documents contain results of toxicity tests, pharmacodynamic tests, and/or clinical tests.

Pre-clinical/clinical test documents for natural medicine products and health supplements submitted to the *BPOM* will first be evaluated by the *BPOM* evaluator. If necessary, evaluations may involve experts in the field. Evaluations are conducted according to the *BPOM* Chief Regulation No. 21 of 2015 on Clinical Trial Approval Management, and *BPOM* Chief Regulation No. 7 of 2014 on Non-Clinical in vivo Toxicity Test Guidelines. Once the documents are deemed to meet the required provisions, a pre-clinical/clinical test implementation approval letter will be issued, indicating that pre-clinical or clinical tests can be conducted.

Priority services are available under certain conditions, including businesses in the health supplement field that are registered with *BPOM* and have a distribution license number; registration is only used for new domestic products, industries that have never been involved in criminal activities in the drug and food sector, and industries that are administratively compliant and have not received public warnings related to illegal health supplement chemical cases in the last two years and is prioritized for industries that have an adverse event reporting system.

Industries must conduct safety monitoring that includes complaints about adverse events related to health supplements and their other safety aspects while they are on the market. The results of monitoring adverse events and other safety aspects of health supplements must be reported by the licensed industry holder to the *BPOM* through the adverse event monitoring mechanism of health supplements. Management and evaluation of registration of traditional medicines and health supplements are processed using a time-to-response mechanism. The evaluation results can be seen in the following table.

Health supplements submitted through export registration, as referred to in Article 31, must meet the following requirements:

1. They must comply with safety, efficacy, and quality aspects, and be produced by applying Good Manufacturing Practices, as evidenced by a certificate/license of implementation.
2. Health supplement exports are prohibited from being distributed within the territory of Indonesia.

TABLE 13.5
Ingredients Prohibited in Health Supplements in Indonesia

Species	Prohibited Part
Abrus precatorius L.	Seeds
Artemisia spp.	Leaves
Atropa belladonna	All parts
Berberis spp.	Roots, bark, rhizomes
Calotropis gigantea (L) Dryand syn. *Asclepias gigantea* L.	All parts
Cannabis sativa L.	All parts
Catharanthus roseus (L.) G.Don Syn. *Vinca rosea* L.	All parts
Cerbera manghas L. Syn. *Cerbera odollam* Gaertn	Seeds
Chelidonium majus L.	All parts
Chincona spp.	Bark
Datura spp.	All parts
Digitalis spp.	All parts
Ephedra spp.	All parts
Garcinia hanburyi Hook. f. Syn. *Garcinia morella* Gaertn.	All parts
Hyoscyamus niger L.	All parts
Lobelia chinensis Lour.	All parts
Lobelia nicotianifolia Roth ex Schult	All parts
Magnolia officinalis Rehder and E. H. Wilson	All parts
Mahonia spp.	Roots, bark, rhizomes
Melaleuca alternifolia	Leaves, oil (prohibited for oral consumption)
Nicotiana tabacum	Leaves
Papaver spp.	All parts
Punica granatum L.	Bark and root bark
Rauvolfia spp.	All parts
Tinospora crispa (L.) Miers ex Hook.f. and Thoms	Root
Valerian spp.	Root
Veratrum spp.	All parts

Some ingredients that cannot be distributed or are contained in health supplements in Indonesia can be seen in Table 13.5.

Animals that are not permitted as ingredients for health supplements include thus:

1. *Bufo gargarizans* Cantor, *Bufo melanostictus* Schneider, *Bufo vulgaris* Lour (Samsu, Frog Kerok)
2. Glandula parathyreoideae, glandula suprarenalis, glandula thyreoideae, glandula pinealis (pituitary gland), glandula thyreoidea (thymus gland), posterior pituitary, anterior pituitary, ovaries, pancreas, testes, placenta, hormones
3. *Lytta vesicatoria* (*Cantharis*)
4. *Mylabris phalerata* Pall
5. *Mylabris cichorii* Linnaeus
6. Animals that are protected based on Indonesian government regulations

TABLE 13.6
Minimal Information on Labels for Traditional Medicines or Health Supplements

Information	Secondary Packaging	Etiquette/Label/Sachet	Strip/Blisters	Brochure
Product name	√	√	√	√
Health supplements	√	√	–	√
Dosage form	√	√	–	√
Packaging size	√	√	–	√
Composition in qualitative and quantitative aspect	√	√	–	√
Manufacturer's name and address (city and country name)*	√	√	√	√
Importer's name and address (city and country name)*	√	√	–	√
Name and address of licensee/ licensee (city and country name)*	√	√	–	√
Name and address of contract giver/recipient (city and country name)*	√	√	–	√
Registration number	√	√	√	√
Batch number/production code	√	√	√	–
Expired date	√	√	√	–
Claims	√	√	–	√
How to use	√	√	–	√
Side effects, warnings, contraindications, drug interactions (if any)	±	±	–	√
Storage conditions	√	√	–	√
Special information (e.g., relating to the origin of certain ingredients, alcohol content, use of radiation, ingredients derived from GMOs, or genetically modified organisms)	√	√	–	√
Information on sweeteners, colorings, preservatives, and flavors	√	±	–	√

Note:
√ Information must be included.
± Information can be included or "See brochure" can be stated if the packaging cannot contain this information. If no brochure is available, this information must be included.
– Information does not need to be included.
** According to product status.*

Labels on traditional medicine or health supplement packaging designs must contain a minimum of information according to Table 13.6.

Registration of traditional medicines and nutraceuticals in Indonesia must be carried out through the Marketing Permit Agency and is valid for a duration of five years, as long as it does not contradict the terms of the statutory rules. Health supplements whose marketing permit has expired and has not been extended are declared as health supplements that do not have a marketing permit.

TABLE 13.7
Advertising Material Terms and Claims Examples

Product	Advertising Material Regulations	Example Claims
Traditional medicine	• Claims should only indicate relief of symptoms, not treatment, cure, or prevention of diseases. • Can only be included in advertisements using visual media. • Must display approved product packaging (pack shot). • Must include a "Consult your doctor regularly" spot.	• Traditionally used to help relieve mild high blood pressure symptoms. • Traditionally used to help relieve diabetes symptoms. • Helps maintain health in cancer patients.
Health supplements	• Claims must not state that consuming health supplements can replace nutrients from food. • Can only be included in advertisements using visual media. • Must display approved product packaging (pack shot).	• Helps maintain health in people with diabetes. • Helps maintain prostate health. • Helps reduce the risk of cognitive decline in older adults (over 50 years old).

Traditional medicines and health supplements that already have distribution permits can be promoted through advertisements, which are regulated in accordance with the Food and Drug Regulatory Agency Regulation No. 34 of 2022 concerning the supervision of advertising for traditional medicines, quasi-medicines, and health supplements [25]. Advertising and claim provisions can be seen in Table 13.7.

The Food and Drug Supervisory Agency of Indonesia, called *BPOM*, always carries out strict supervision regarding distribution permit regulations. In 2022, *BPOM* obtained several data related to imports and exports of nutraceuticals in Indonesia, which can be seen in Table 1 (Annual Report, POM Directorate) [28].

From 2021 to 2022, the registration for traditional medicine has collected 160 products, with 13 health certificates for traditional medicine, 22 certificates of pharmaceutical products (CPP), 90 certificates of free sales (CFS), and 29 GMP and 6 (to whom it may concern). For quasi-product, there are 101 products, including 44 CFSs, 38 CPPs, 12 health certificates and 7 (to whom it may concern), and 124 for import. For supplements, there are 124 from export, 81 from export and certified as CPP, 180 products as health certificate, 1 (to whom it may concern) and 60 product certified GMP.

13.5 THE SUPERVISION OF THE TRADITIONAL MEDICINE AND NUTRACEUTICAL INDUSTRY IN INDONESIA

Supervision regarding herbal and nutraceutical production in Indonesia is carried out by the *BPOM* regularly. Supervision is carried out regarding the production process and comprehensive risk-based inspections. The assessment graph for traditional medicine and health supplement industrial facilities can be seen in Figure 13.2.

Inspections of 1,284 nutraceutical distributors found that 1,233 distributors met the provisions and 51 facilities did not. Facilities that do not meet these requirements have distributed nutraceuticals without a distribution permit, resulting in product recalls, destruction, strong warnings, and pro justitia.

Based on other assessments at the natural materials extraction center, it is known that all facilities meet GMP requirements. The results of other inspections at small traditional medicine enterprises

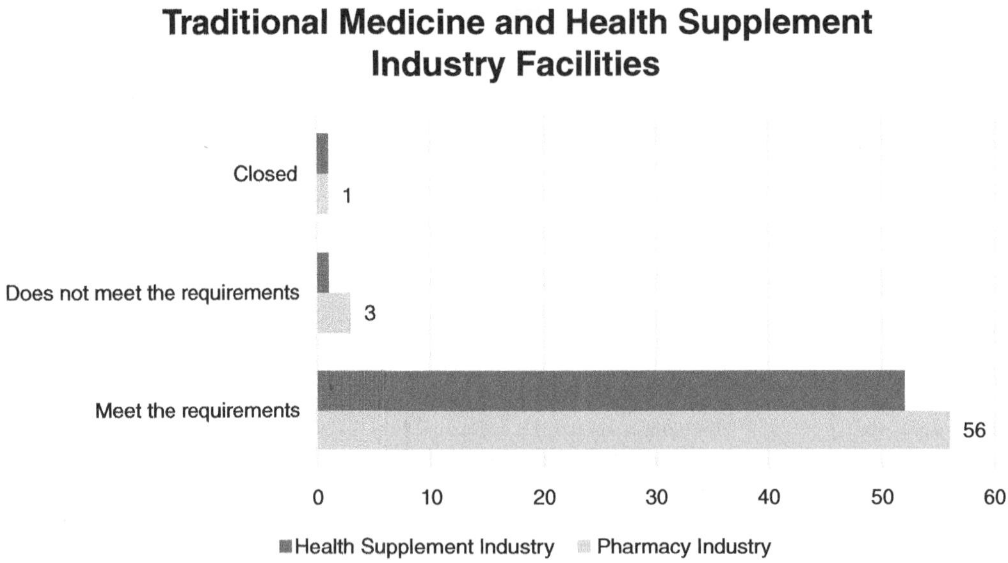

FIGURE 13.2 Graphic presentation of traditional and nutraceutical industry assessment.

(UKOT) in Indonesia, in 278 industries, showed that 234 (84.17%) facilities had complied with GMPs, 16 (5.76%) did not meet the provisions, and 28 (10.07%) facilities were closed [28]. In addition to risk-based assessments, *BPOM* carries out inspections of products whose distribution permits are not registered.

The registration trend for traditional medicines has increased from the 2012 numbers, where 1,953 files received applications, up to 9,747 files in 2020. When compared with 2020, there was an increase of 17% in the receipt of traditional medicine registration application files in 2021. Apart from that, in 2021, there were completed evaluations of 65 pre-clinical test protocols and 10 pre-clinical test result data, as well as 45 clinical test protocols and 13 clinical test results data. All pre-clinical/clinical test protocol assessment files are declared timely because they meet the predetermined timeline.

The registration trend for nutraceuticals has been increasing, where 4,262 files were received in 2021, compared to 2012's 1,029 files. When compared with 2020, there was an increase of 5.47% in the receipt of application files for the registration of health supplements in 2021. *BPOM* is currently facilitating the checking of numbers of product distribution permit via the website http://cekbpom.pom.go.id/.

13.6 HALAL PRODUCTS

Halal is an Arabic term that means "permissible" or "lawful" in English. The phrase relates to what Muslims are permitted to do in their daily lives, notably with food and drink. The inverse of *halal* is *haram*, which is another Arabic term for something forbidden. The Holy Quran determines what is halal (or haram), and Sharia law governs Muslim life. What is halal and haram, like everything else in Muslim law, is founded on the Holy Quran, Sharia laws, and hadith (the teachings of the prophet Muhammad). Muslims are forbidden to consume alcohol (on numerous occasions), goods from carrion (an animal that dies by itself), pork and its derivatives, blood, donkeys, horses, and animals with fangs [29].

The number of Muslim residents in Indonesia was recorded at 241.7 million people as of December 2022, or equivalent to 89.02% of the country's population of 277.75 million people at the end of 2022 [30]. The Indonesian society today, being in the middle-class income category, is of course one of the largest market niches in the country for the development of the national halal industry. The Indonesian government has required products circulating in Indonesia to be halal-certified through the Halal Product Guarantee Law No. 33 of 2014. The products under concern are goods and/or services connected to food, beverages, medications, cosmetics, chemical products, biological products, genetic engineering products, and consumer goods [31].

Law No. 33 of 2014 concerning halal product guarantees (JPH) and Government Regulation No. 31 of 2019 concerning implementing regulations of Law No. 33 of 2014 mandate the supervision of halal product guarantees [32]. The main duties and functions of the POM agency, which intersect with the Halal Product Guarantee Law in terms of post-market supervision, are to supervise product (food) sampling and testing and to monitor conformity between labels that have been approved in the pre-market assessment. The POM agency coordinates with cross-sectors, such as the Halal Product Guarantee Organizing Agency (BPJPH), the Food, Drug, and Cosmetic Review Institute of the Indonesian Ulema Council (LPPOM MUI), the National Standardization Agency (BSN), and experts from the Indonesian Institute of Science and Research (LIPI). The Indonesian government, through Presidential Regulation No. 6 of 2023 concerning halal certification of medicines, biological products, and medical devices, requires that nutraceuticals or health supplements meet halal requirements [33].

Medicines, biological goods, supplements, and medical devices must be halal-certified. The biological products in question are enzymes, monoclonal antibodies, vaccines, and recombinant DNA products. Other products that must have halal certification are food, beverages, and functional food. General provisions for applying for halal certification include halal manufacturing methods (including the provision of materials, processing, storage, and packaging), which must meet the criteria for commitment and responsibility of the material provider, manufacturing process, final product, and monitoring and evaluation.

Materials that must meet halal certification are raw materials, additional materials, packaging, lubricants, grease or sanitizer, and other materials or products that come into direct contact during the production process, either with machines or with the final product, including washing materials.

1. Materials used in making halal health supplements must be halal-certified, except for materials that fall into the category of non-critical materials in accordance with statutory provisions.
2. These should not come from materials that are prohibited according to Islamic law and statutory provisions.
3. They must not be produced from production facilities that are also used to make non-halal products.
4. They must not be mixed with and/or be in contact with substances that are prohibited according to Islamic law and statutory provisions.
5. Animal ingredients and derivative products must come from halal animals slaughtered in accordance with Islamic law, except for halal animals which, according to Islamic law, do not need to be slaughtered.
6. These must fulfill safety and health aspects in accordance with statutory provisions.
7. Alcohol/ethanol materials can be utilized as long as they are not from the wine industry, are not harmful to health, and are not misused.

Manufacturing facilities used to create halal products can be used alongside production facilities used to create items that are not submitted for halal certification, as long as the products do not come from materials that contain banned substances.

Medicinal products, traditional medicines, nutraceuticals, or health supplements that are to be registered for halal certification must comply with *BPOM* regulations and several provisions, namely:

1. Must come from halal ingredients, processed in accordance with Islamic law, using equipment, production facilities, and packaging and storage systems that are not contaminated with non-halal ingredients.

2. Must not use names that refer to something that is haram or that contains pornography and does not have characteristics/sensory profiles that refer to haram products or that have been declared haram based on the fatwa provisions of the Indonesian Ulema Council.
3. Must ensure halal and high-quality goods by proper packaging and labeling, with clear designs, symbols, brands, names, and images.
4. Must be packaged and labeled without violating the principles of Islamic law.
5. Must be identifiable and able to be traced clearly, with guaranteed compliance with halal manufacturing methods.

Nutraceutical products are predominantly in capsule form derived from gelatin. The raw materials for gelatin are generally skin and bones from mammals, such as cows and pigs. In general, the source of gelatin for making capsules is divided into type A gelatin, which comes from the skin of both cows and pigs, and type B gelatin, which comes from the bones of both cows and pigs. Almost 60% of collagen and gelatin used in the world come from pigs, so the requirements for nutraceuticals and health supplements in Indonesia must take this into account. Today, Indonesia has developed hard capsules from glucomannan that are isolated from a local plant called *porang* [34].

Other nutraceutical products that need special attention are vitamins A, E, and C, which are unstable. Vitamin A (B-carotene) is made from fish and mammal liver oil extract, fermented or chemically synthetic. Vitamin E (tocopherol) is made from plants, animals, and chemical synthesis. Vitamin C (ascorbic acid) is made from plants and synthetic chemicals. These vitamins are unstable, so they require additional ingredients that function as stabilizers. Therefore, it is necessary to be wary of the use of coatings and sources of fatty acids as stabilizers.

Other products are in elixir form which contains alcohol and tablets that require lubricants, such as magnesium stearate and stearic acid, which are from fat derivatives. The critical point to which attention needs to be paid in the registration requirements for nutraceuticals and health supplements for elixir products is that those whose halal qualities are doubtful include flavors, colorings, alcohol, and glyceryl, whereas for tablets it is the lubricant component.

13.7 CASES OF HEALTH SUPPLEMENTS CONTAINING NON-HALAL INGREDIENTS

Health supplements have become a lifestyle trend. With massive advertising and claims of being able to increase stamina and maintain body health, health supplements have now become a daily consumption for anyone who wants health and fitness. It must be noted that not all health supplements are halal-certified. It is hoped that consuming food supplements can help people live a better life, so that they can avoid various diseases that require the use of medication. Nowadays, medicines appear in many variants in the treatment of various diseases.

According to the Director General of Pharmaceutical Development and Medical Devices, Ministry of Health, the largest import of pharmaceutical raw materials currently is from China, namely, around 60%. The remaining 25% comes from India, and the rest comes from Europe and America. All exporting countries to Indonesia are non-Muslim countries that are less concerned about the halal status of medicines. Therefore, we need to understand the ingredients in drugs and supplements so that we can avoid drugs and supplements that are not necessarily halal [35].

Herbal medicines and supplements contain plant extracts, mineral ingredients, galenic preparations, or a mixture of these ingredients, which are processed in such a way that they change form into pills or powder without any chemical mixture. Meanwhile, drugs or chemical supplements contain ingredients that are mixed and processed using chemical synthesis, so that compounds with certain pharmacological properties are obtained.

In the process, both herbal and chemical, supplements and medicines involve various main ingredients and other supporting ingredients, one of which is a solvent which functions to dissolve other substances which are generally in solid form without undergoing chemical changes. In the pharmaceutical field, the solvents commonly used are polar (soluble in alcohol and water), semi-polar (less soluble in water and alcohol), and nonpolar (which can dissolve in oil).

13.7.1 Raw Material

Raw materials are the main ingredients for making medicines and supplements. Raw materials can come from animal and vegetable sources. If the raw materials come from pigs and their derivatives, or from animals that are not necessarily halal, then it is clear that the product produced is haram. However, if it comes from halal animals, it must be ensured that it comes from animals slaughtered according to *Sharia law.*

13.7.2 Adjuvant

Materials must also be ensured to be halal, even though their role is not as large as that of raw materials. In synthetic drugs and supplements, critical points in adjuvants need to be considered. For example, tablet coatings may be made with gelatin, which must be ensured to come from halal animals and slaughtered according to Sharia law.

13.7.3 Capsule Shell

Capsule shells are usually made from gelatin. The gelatin capsule technology is chosen by pharmaceutical companies/manufacturers because it is superior in its bioavailability, as well as being easier to modify from a biopharmaceutical perspective. The raw materials for gelatin are skin and bones from mammals, such as cows and pigs. In general, the source of gelatin for making capsules is divided into type A gelatin, which comes from pork skin, and type B gelatin, which comes from cow skin and bones. Muslim customers need to be made more aware of and concerned about halal products, particularly medications, as the world's population continues to rise. This study was done

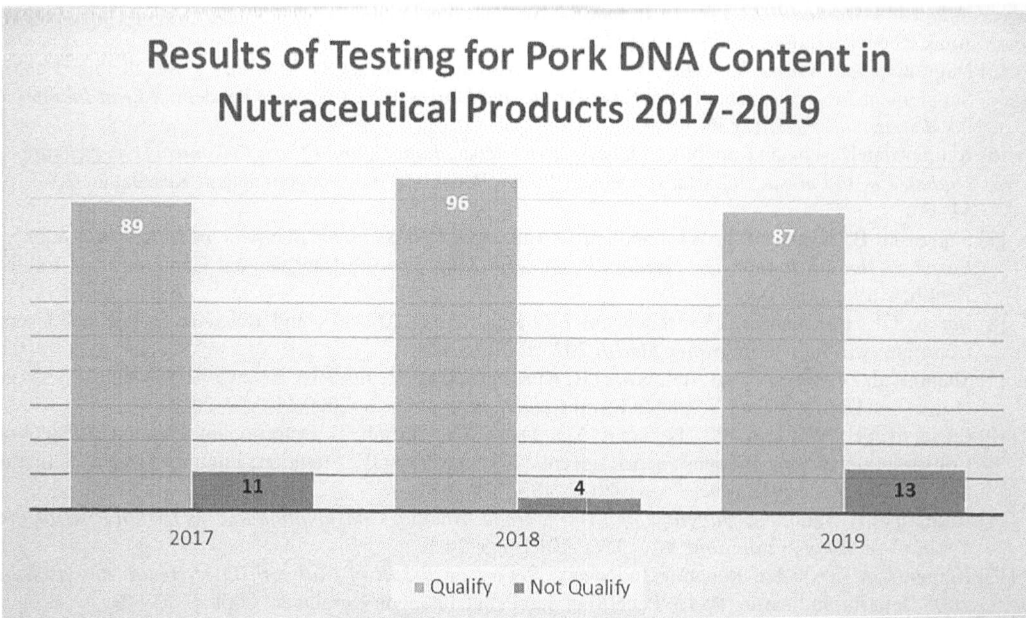

FIGURE 13.3 Result of DNA test from suspect nutraceutical products (37).

to better understand the significance of halal and non-halal concerns in pharmaceutical ingredients, their potential consequences on the Muslim community, and the opportunities and obstacles in the search for substitute chemicals [36].

The POM agency and POM centers throughout Indonesia carry out sampling and laboratory testing of food suspected of containing pork DNA. Samples are taken based on predetermined sampling priorities. The results of testing for pork DNA content in processed food products for 2017–2019 can be seen in Figure 13.3.

In 2018, the Food and Drug Supervisory Agency identified the discovery of supplement products containing non-halal ingredients in circulation. The head of *BPOM* RI said that the finding of pig DNA in medicinal products and multivitamin supplements and joints indicated that there was an inconsistency between pre-market data information and the results of post-market supervision. The results of post-market surveillance tests showed that both products positively contained pork DNA, while the data submitted for and that passed *BPOM* evaluation during product registration (pre-market) used raw material sourced from cows.

These results were followed up by issuing warning letters to business actors to withdraw batches of products detected to contain pork DNA, recommending business actors to export or re-export products to countries that do not require halal provisions and/or destroy the products. For imported products, product imports are stopped for problematic batches and ask business actors to provide guarantees that similar products to be imported meet the requirements. BPJPH currently facilitates tracking of halal products by disbursing certificate numbers on the website https://info.halal.go.id/.

REFERENCES

[1] Badan Pusat Statistik. *Indonesia Statistic Data of 2022* [Internet]. Available from: https://www.bps.go.id/publication/2020/04/29/e9011b3155d45d70823c141f/statistik-indonesia–2020.html

[2] Kementrian Kesehatan Republik Indonesia. *Indonesia Health Survey Report of 2023*. Jakarta, Indonesia: Badan Penelitian dan Pengembangan Kesehatan; 2023. p. 29–55.

[3] Kementrian Kesehatan Republik Indonesia. *Report on Result of National Basic Health Research in 2018*. Jakarta, Indonesia: Badan Penelitian dan Pengembangan Kesehatan; 2018. p. 164–74.

[4] Kementerian Pemberdayaan Perempuan dan Perlindungan Anak Republik Indonesia. *The Profile of Indonesian Women 2023*. Jakarta, Indonesia: Kementerian Pemberdayaan Perempuan dan Perlindungan Anak Republik Indonesia; 2023. p. 124–31.

[5] Nugraheni AP, Purba S, Riani D, Wahyuri, Nurizati. Analysis of Monitoring Results for Health Supplement Advertisements/Promotions Before and During the COVID-19 Pandemic. *Erud Indones J Food Drug Saf.* 2021;1(2):36–43.

[6] Kementrian Kesehatan Republik Indonesia. *Prevalence, Impact, and Efforts to Control Hypertension & Diabetes in Indonesia*. Jakarta, Indonesia: Badan Penelitian dan Pengembangan Kesehatan; 2023. p. 54–60.

[7] Sujatmiko B, Anggaeni T, Wismandanu O, Indraswari N. *Smoking Behavior in Indonesian Society Based on the 5th Indonesian Family Life Survey Data*. The 6th International Conference on Public Health, 2019, p. 110.

[8] Masan GE, Raksanagara AS, Rinawam FR. Knowledge, Attitude, and Behavior in E-Cigar Users Community in Indonesia. *Althea Med J.* 2023;10(3):167–74.

[9] Shaluhiyah Z, Musthofa SB, Indraswari R, Kusumawati A. Health Risk Behaviors: Smoking, Alcohol, Drugs, and Dating among Youths in Rural Central Java. *Kesmas*. 2020;15(1):17–23.

[10] Prasetyo YB, Wardojo SSI, Laksono AD, Dewi YS, Huriah T. Determinant Factors of Alcohol Consumption Among Indonesian Adolescents Through Parents' Attention: Findings from a National Survey. *Florence Nightingale J Nurs*. 2022;30(3):238–44.

[11] Respatiadi H, Tandra S. *Underage and Unrecorded: Alcohol Consumption and its Health Risk for the Youth Case Study in Bandung*. West Java. 2018. p. 5–24.

[12] Kementrian Kesehatan Republik Indonesia. *Report on Result of National Basic Health Research in 2008*. Jakarta, Indonesia: Badan Penelitian dan Pengembangan Kesehatan; 2008. p. 33–40.

[13] Kementrian Kesehatan Republik Indonesia. *Indonesian Health Profile*. Jakarta, Indonesia: Badan Penelitian dan Pengembangan Kesehatan; 2018. p. 35–55.

[14] Amaliah I, David W, Ardiansyah A. Perception of Millennial Generation Toward Functional Food in Indonesia. *J Funct Food Nutraceutical*. 2019;1(1):31–40.
[15] Ferdinand M, Lestari K. Review: Traditional Plants that Function as Antidiabetic Nutraceuticals. *Farmaka*. 2019;17(2):315–23.
[16] Theresa A, Sukohar A, Wardhana MF, Kedokteran F, Lampung U, Farmakologi B, et al. Nutraceuticals in the Management of Obesity. *Nutraceuticals in the Management of Obesity*. 2023;13:472–8.
[17] Dianasari W, Nadjib M. Supervision of Traditional Medicines Containing Undeclared Substance: Analysis of Indonesian FDA Monitoring Data for 2012–2021. *J Indones Heal Policy Adm*. 2022;7(1):196–7.
[18] Embassy of the Republic of Indonesia in Brussels. *Indonesia's Potential Herbal Products Gain in The European Market*. Brussels, Belgium; 2021. p. 9–15.
[19] Kristiana L, Nantabah ZK, Maryani H. Priority Analysis of Potential Medicinal Plants for Mental Emotional Disorders using the Weighted Product (WP) Method: Research Data on Medicinal Plants and Herbal Medicine 2012, 2015, and 2017. *Media Peneliti dan Pengembangan Kesehatan*. 2019;29(3):255–57.
[20] Kementrian Kesehatan Republik Indonesia. *Special Research Exploration of Local Knowledge of Ethnomedicine and Community-Based Medicinal Plants in Indonesia (Osing Ethnicity, East Java Province)*. Jakarta, Indonesia: Center for Research and Development of Medicinal Plants and Traditional Medicine; 2015. p. 21–25.
[21] Gondokesumo ME, Budipramana K, Aini SQ. Study of Jamu as Indonesian Herbal Medicine for COVID-19 Treatment. Proceedings of the 2nd International Conference of Contemporary Science and Clinical Pharmacy. 2021 Sep 9–10; Andalas University, North Sumatra; 2021. p. 40.
[22] Achmadi NK, Setiawan Y, Purwati AA. An Overview Knowledge Level and Use of Herbal Medicines During the COVID-19 Pandemic by Students of Poltekkes Kemenkes Jakarta II Pharmacy Department. *SANITAS*. 2021;12(2):106–14.
[23] Hartanti D, Dhiani BA, Charisma SL, Wahyuningrum R. The Potential Roles of Jamu for COVID-19: A Learn from the Traditional Chinese Medicine. *Pharm Sci Res*. 2020;7(4):12–22.
[24] Badan Pengawasan Obat dan Makanan. *Health Supplement Quality Requirements*. Peraturan BPOM 17 No 17 Tahun 2019; p. 1–5.
[25] Indonesia. *Food and Drug Supervisory Agency Regulation Number 32 of 2022 concerning Criteria and Procedures for Health Supplement Registration*. Jakarta, Indonesia: Kementrian Kesehatan Republik Indonesia; 2022. p. 1–80.
[26] Arozal W, Louisa M, Soetikno V. Selected Indonesian Medicinal Plants for the Management of Metabolic Syndrome: Molecular Basis and Recent Studies. *Front Cardiovasc Med*. 2020;7(May):1–10.
[27] Siahaan EA, Pangestuti R. Functional Foods and Nutraceuticals from the Sea: Prospects and Challenges. *Jurnal Ilmu-Ilmu Perairan, Pesisir dan Perikanan*. 2017;6(3):273–81.
[28] Direktorat Registrasi Obat Tradisional-Suplemen Kesehatan dan Kosmetik BPOM. *Badan Pengawasan Obat dan Makanan (BPOM) Annual Report of 2021*. Jakarta, Indonesia: Badan Pengawasan Obat dan Makanan; 2021. p. 11–15.
[29] Shamsudin SN, Abu Bakar E, Osman S, Mohd Dali NRS. *Muslim Consumer Behavior Towards Halal Nutraceutical Products: A Pilot Study*. 4th International Islamic Business Management Conference, 2022. p. 216–24.
[30] Kementerian Perencanaan Pembangunan Nasional Republik Indonesia dan Badan Penyelenggara Jaminan Produk Halal. *Master Plan Indonesia of Halal Industry in 2023–2029*. Jakarta, Indonesia: KNEKS; 2023. p. 1–12.
[31] Mutmainah N, Hapidin A, Ramdhani MA, Nuryadin BW, Muharir M, Ahyani H. Global Awareness of Halal Products in Indonesia in Era 4.0. *Al-Amwal J Ekon dan Perbank Syari'ah*. 2022;14(1):14.
[32] Indonesia. *Undang-Undang No.33 Tahun 2014 tentang Jaminan Produk Halal*. Jakarta, Indonesia: Menteri Hukum dan Hak Asasi Manusia; 2014. p. 1–7.
[33] Indonesia. *Peraturan Presiden (PERPRES) Nomor 6 Tahun 2023 tentang Sertifikasi Halal Obat, Produk Biologi, dan Alat Kesehatan*. Jakarta, Indonesia: Pemerintah Indonesia; 2023. p. 1–17.
[34] Kurniawan F, Ulfin I, Harmami H, Juwono H. *Indonesian Halal Hard Capsule*. Proceeding of Dosen Muda Indonesia Nasional Seminar. Semarang, Indonesia; 2019 Sep 19–20. p. 59–67.
[35] Batubara C, Harahap I. Halal Industry Development Strategies: Muslims' Responses and Sharia Compliance in Indonesia. *J Indones Islam*. 2022;16(1):103–32.
[36] Herdiana Y, Sofian FF, Shamsuddin S, Rusdiana T. Towards Halal Pharmaceutical: Exploring Alternatives to Animal-Based Ingredients. *Heliyon*. 2024;10(1):e23624.
[37] Badan Pengawasan Obat dan Makanan. *Annual Report of 2019 Badan Pengawas Obat dan Makanan (BPOM)*. Vol. 4. Jakarta, Indonesia: Badan Pengawas Obat dan Makanan Republik Indonesia; 2022.

14 Complementary Medicines
Australian Regulatory Requirements, Market Trends, and Challenges in the Australian Market

Aditi Kirtane

14.1 INTRODUCTION

Australian adults are spending significant amounts on such products as compared to the prescription medicines. The main reasons for this increased use are that these are easily available, are more affordable, do not require prescription, and are usually more appealing. The other aspects to consider are increased awareness about health and well-being, the ageing population, consumer preference of using organic and natural products, abundance of choice, and innovative product range. As per the Complementary Medicines Australia (CMA) report [1], more than 70% of the Australian population uses complementary medicines. The complementary medicines industry reached $5.69 billion in revenue in 2020 due to the high demand for these products, with an estimated steady growth of 7% year on year, as per Austrade estimates.

The complementary medicine sector has evolved over 30 year and has become a world-class industry in Australia. It has strengthened the area of research, employment, manufacturing, and exports. Australian complementary medicines are in high demand, and the sector has achieved significant growth during COVID and in post-pandemic times. Australian brands are globally esteemed and respected, with China sourcing more complementary medicines from Australia than from nearly any other country in the world. Therefore, regulating complementary medicines and enforcing these laws are crucial for sustaining the current growth trend and meeting both market and safety standards.

14.2 REGULATORY AFFAIRS RELATED TO COMPLEMENTARY MEDICINES IN AUSTRALIA

The Australian Regulatory Guidelines for Complementary Medicines (ARGCM) offer essential information for manufacturers, sponsors, health-care professionals, and the general public regarding the regulation of complementary medicines in Australia. Part A of ARGCM version 7.0 outlines the general guidance on these regulations, presenting an overview of the regulatory framework and providing direction for sponsors, health-care professionals, and the general public (ARGCM Part A: General Guidance on Complementary Medicine Regulation in Australia). [2]

14.2.1 Complementary Medicines: What They Are

In Australia, medicinal products containing ingredients such as certain herbs, vitamins, minerals, nutritional supplements, homeopathic medicines, and aromatherapy products are classified as "complementary medicines." These products are regulated by the Therapeutic Goods Administration (TGA) under the Therapeutic Goods Act of 1989 (the Act) and the accompanying Therapeutic Goods Regulations of 1990 (the Regulations).

 DOI: 10.1201/9781003465041-14

Part 1(2) of the Regulations provides the following definitions:

Complementary medicine means a therapeutic good consisting wholly or principally of 1 or more designated active ingredients, each of which has a clearly established identity and a traditional use.

Designated active ingredients, for a complementary medicine, means an active ingredient, or a kind of active ingredient, mentioned in Schedule 14 (to the Regulations).

Schedule 14 to the Regulations provides a list of designated active ingredients:

Designated Active Ingredients

1. An amino acid
2. Charcoal
3. A choline salt
4. An essential oil
5. Plant or herbal material (or a synthetically produced substitute for material of that kind), including plant fibers, enzymes, algae, fungi, cellulose and derivatives of cellulose, and chlorophyll
6. A homoeopathic preparation
7. A microorganism, whole or extracted, except a vaccine
8. A mineral, including a mineral salt and a naturally occurring mineral
9. A mucopolysaccharide
10. Non-human animal material (or a synthetically produced substitute for material of that kind), including dried material, bone and cartilage, fats and oils, and other extracts or concentrates
11. A lipid, including an essential fatty acid or phospholipid
12. A substance produced by or obtained from bees, including royal jelly, bee pollen, and propolis
13. A sugar, polysaccharide or carbohydrate
14. A vitamin or provitamin

14.2.2 Regulation of Complementary Medicines in Australia

14.2.2.1 Australian Register of Therapeutic Goods (ARTG)

Unless exempt, any therapeutic product that makes therapeutic claims must be registered on the Australian Register of Therapeutic Goods (ARTG) before it can be legally brought into, shipped from, produced, or distributed for use in Australia.

14.2.2.2 Registered and Listed Complementary Medicines

Within the regulatory framework, complementary medicines are categorized as either registered or listed on the ARTG, depending on their ingredients and the indications made for the medicine. Most complementary medicines are listed; however, some are registered. [3, 4]

14.2.2.2.1 Listed Medicines

AUST L–listed medicines are deemed to have a lower risk profile, as they may only use low-risk ingredients and low-level indications from pre-approved lists. These medicines are not individually assessed by the TGA before being released to the market; rather, they are automatically entered into the ARTG upon completion of their application. Listed medicines display an AUST L number on their labels. The listing process facilitates quick market access for these complementary medicines. When submitting a listing application, a sponsor legally certifies under Section 26A of the Act that their medicine complies with all relevant legislative requirements concerning safety, quality, and efficacy.

Since listed medicines are not individually assessed, several mechanisms are employed to ensure the safety and quality of both the ingredients used and the final listed products.

Most ingredients that can be included in listed medicines are those which were used in therapeutic goods supplied in Australia before the Act came into effect in 1991. Since then, all new active and excipient ingredients have been subject to a safety assessment by the TGA. If someone wants to use an active or excipient ingredient that is not currently approved for use in listed medicines, the substance must undergo evaluation by the TGA before it can be used.

14.2.2.2.2 Assessed Listed Medicines

Introduced by the TGA in 2018 as a sub-category of listed medicines, AUST L(A)–assessed listed medicines carry an AUST L(A) number. These medicines may only use low-risk ingredients approved for listed medicines but are allowed to have at least one intermediate indication, which is a level above those permitted for standard AUST L–listed medicines, and may also include lower-level indications. Assessed listed medicines are entered in the ARTG following self-certification by the applicant regarding the product's safety and quality, combined with a TGA pre-market assessment of the efficacy evidence supporting the proposed indications. The key distinction between AUST L(A)–assessed listed medicines and standard listed medicines is that the assessed pathway allows for higher-level indications due to a pre-market evaluation by the TGA of robust scientific evidence, including clinical trials, to support the product's efficacy. Intermediate-level indications might include the prevention or alleviation of non-serious conditions, such as preventing cold sores, or the management of more serious conditions, like "relieving symptoms" associated with rheumatoid arthritis.

14.2.2.2.3 Registered Medicines

AUST R–registered complementary medicines are deemed to be at a higher risk compared to listed medicines due to the nature of their ingredients or the indications made. Unlike listed medicines, all registered medicines undergo a comprehensive assessment by the TGA for quality, safety, and efficacy before they are permitted to enter the Australian market. Registered medicines have an AUST R number on the medicine's label.

According to the Act, medicines must be registered on the Australian Register of Therapeutic Goods (ARTG), where they

- do not solely comprise ingredients permitted for use in listed medicines; or
- contain an ingredient or ingredient component that is subject to the conditions of a Schedule or relevant appendix to the Poisons Standard; or
- are required to be sterile; or
- have indications that are not indications permitted for use in listed medicines.

Before being approved for inclusion on the ARTG, registered medicines undergo rigorous evaluation by the TGA to ensure they meet the required standards for quality, safety, and efficacy.

14.2.3 Scheduling of Registered Complementary Medicines

Registered complementary medicines may be subject to the conditions of a schedule (not Schedules 4, 8, and 9) or an appendix of the Poisons Standard. For example:

Schedule 2: "Pharmacy Medicine"
Schedule 3: "Pharmacist Medicine"

Before submitting an application for registration, it is crucial to consider any applicable scheduling requirements. The decision to place a medicine in a schedule considers factors such as toxicity, intended use, potential for abuse, safety, and the necessity of the substance.

14.2.4 Complementary Medicine Presentation

Section 3(1) of the Act provides the following definition:

"Presentation" in relation to therapeutic goods, means the way in which the goods are presented for supply, and includes matters relating to the name of the goods, the labelling and packaging of the goods and any advertising or other informational material associated with the goods

The overall "presentation" of a medicine should include things such as (but is not limited to):

- medicine name
- medicine label
- packaging
- promotional/advertising material
- graphics
- package inserts
- dosage form
- indications

14.2.5 Complementary Medicine Labels

A product's "label" encompasses both the label affixed to the container (such as a bottle, tube, sachet, or blister pack) and the primary packaging (like a carton). It is essential to ensure that the product label and any accompanying printed information (such as a package insert) comply with all relevant legislation, including advertising requirements, before the product can be supplied in Australia.

According to the Act:

Specific documents relating to medicine labelling requirements include:

- The Therapeutic goods labelling Order as current and in force Part 5–1 (Advertising and generic information) of the *Therapeutic Goods Act 1989*
- The Therapeutic Goods Advertising Code
- The Therapeutic Goods Regulations 1990
- The Therapeutic Goods (Permissible Ingredients) Determination
- The Medicines Advisory Statements Specification 2016
- The Poisons Standard (the SUSMP) (note: Australian states and territories vary in the way they adopt the Poisons Standard)
- The TGA approved terminology for medicines

Complementary medicines must also adhere to any general and specific conditions applicable to registered or listed therapeutic goods under Section 28 of the Act.

14.2.6 Advertising of Complementary Medicines

The goal of regulating therapeutic goods advertising in Australia is to ensure that the marketing and promotion of therapeutic goods, including complementary medicines, are carried out in a way that upholds product quality and safety, is socially responsible, and avoids misleading or deceiving consumers.

Advertisements for complementary medicines are regulated through a mix of co-regulatory and self-regulatory mechanisms established under the Act and the Regulations. This oversight involves the TGA, the Therapeutic Goods Advertising Code Council (TGACC), the Australian Self-Medication Industry (ASMI), and Complementary Medicines Australia (CMA).

To ensure compliance with public benefit standards, advertisements for complementary medicines aimed at consumers through designated media must receive prior approval from a delegate of the secretary of the Department of Health.

Australian-based websites that promote the use or supply of therapeutic products are regarded as advertisements and must adhere to all provisions of the Therapeutic Goods Advertising Code.

14.2.7 Good Manufacturing Practices (GMPs)

In Australia, Good Manufacturing Practice (GMP) guidelines for complementary medicine are set by Therapeutic Good Administration (TGA). The GMP clearance is a non-statutory process used to confirm that overseas manufacturing sites adhere to Good Manufacturing Practice (GMP) principles for products supplied to Australia. Introduced to lessen the regulatory burden on the industry while ensuring that manufacturing and quality control practices are satisfactory, the TGA can issue the GMP clearance to sponsors of medicines or active pharmaceutical ingredients (APIs) manufactured abroad. This clearance is granted when the Australian sponsor provides acceptable evidence demonstrating that the overseas manufacturer complies with GMP principles or equivalent standards.

Sponsors must obtain GMP clearance for overseas manufacturers of their registered or listed products to comply with Sections 25(1)(g), 26(1)(g), 26A(3), and 26(AB)(4) of the Therapeutic Goods Act of 1989. This requirement supports the Act's primary objective of ensuring the safety, quality, efficacy, and timely supply of therapeutic goods for Australian consumers.

Key aspects of GMP required for complementary medicines are:

1. Facility and equipment
2. Personnel
3. Quality control
4. Documentation and recordkeeping
5. Raw material
6. Finished product testing
7. Stability testing
8. Complaints and recalls

14.2.8 TGA Post-Market Regulatory Activity of Complementary Medicines

The TGA ensures the ongoing safety, quality, and efficacy of therapeutic goods in the market through vigilance activities. It employs a risk management approach to regulate these products.

The types of post-market regulatory activities include the following.

14.2.8.1 Adverse Events Reporting for Complementary Medicines

The TGA receives reports of adverse events from various sources, including consumers, healthcare professionals, the pharmaceutical industry, international medicines regulators, and medical or scientific experts on TGA advisory committees.

14.2.8.2 Inspection of Manufacturers

The TGA regularly inspects manufacturers of complementary medicines using a *risk-based approach* to ensure adherence to *GMP standards*. Overseas manufacturers supplying complementary medicines to Australia must also meet these high standards. If a manufacturer is approved by a recognized international regulator, it may not be necessary for TGA inspectors to conduct a site inspection.

14.2.8.3 Sampling of Medicines in the Marketplace for Testing

The TGA conducts a laboratory testing program that supports desk-based compliance reviews of listed complementary medicines, evaluations of registered medicines, and other post-market regulatory activities. This program prioritizes testing for therapeutic goods deemed to have higher risk, while also providing for responsive testing in response to marketplace issues, such as adverse events and complaints about specific medicines.

14.2.8.4 Compliance Reviews of Listed Complementary Medicines

The regulatory process for listed complementary medicines enables early market access for low-risk products. To support this early access, a thorough risk-based system for post-market monitoring is employed. The TGA reviews a portion of listed complementary medicines to ensure they comply with regulatory requirements. These reviews may include thus:

- **Random reviews.** A certain number of newly listed medicines are selected at random by computer.
- **Targeted reviews.** Specific listed medicines are reviewed if they are suspected of potential non-compliance issues.

14.2.8.5 Product Recalls of Therapeutic Goods

A *product recall* involves removing therapeutic goods from the Australian market due to concerns about their quality, efficacy, or safety. A recall is mandated whenever there is a risk to public safety caused by product non-compliance. Recalls can be initiated for issues such as labeling or packaging errors, contamination problems, or an unexpected increase in side effects.

Overall, the regulation of complementary medicines in Australia aims to strike a balance between ensuring consumer access to these products and protecting public health. It is important for companies involved in the manufacture, supply, or advertising of the complementary medicines in Australia to familiarize themselves with these regulations to ensure compliance.

14.3 RECENT TRENDS IN COMPLEMENTARY MEDICINES IN AUSTRALIA

In recent years, especially post-pandemic, lots of trends have changed with respect to the complementary medicines market. Along with consumer focus, even the market dynamics has changed. Along with increased focus on self–health care, consumers are also looking for ease of access to complementary medicines, leading to increased online shopping. The industry outlook and government policies have significantly influenced the future of complementary medicines in Australia. Some of the recent trends are as follows.

14.3.1 Increased Demand in Australia

The Australian population has been getting more focused on their health and well-being over the years. Along with the importance of physical fitness and nutrition, people seem to use complementary medicines to achieve the desired health benefits, with a focus on preventative health care. The other important factor is the growing ageing population in Australia. As a result, there is a growing demand for prescription medicines, and people are increasingly turning to complementary medicines to manage the symptoms of chronic diseases. Almost 70–75% of Australians have used at least some type of complementary medicine in the last year. Complementary medicines seem more natural and less toxic, so people feel more comfortable using it as compared to prescription medicines. One study suggested that a substantial segment of the Australian population reported using supplements, with 34% of men and 47% of women. It also found that supplement users were more likely to be female, older, and more highly educated. They also demonstrate healthier lifestyle behaviors and generally have a superior health status compared to non-users of supplements.

Supplement use is more prevalent among individuals who engage in healthier lifestyle behaviors, such as increased physical activity, non-smoking, and higher consumption of fruits and vegetables while consuming less alcohol. Several studies indicate that non-users of supplements often have less-healthy diets, characterized by higher fat intake, lower fiber consumption, and reduced fruit intake. [5–7]

Previous studies have demonstrated that individuals with higher socio-economic status are often more health-conscious, which may lead them to take more dietary supplements. [5]

14.3.2 Increased Export Potential

According to the CMA Industry Audit and Trends 2020 report, Australia is the leading exporter of nutrition and health food to China, with exports exceeding $1 billion. [1–8] Australian complementary medicines are recognized for their quality, safety, and efficacy. In Australia, complementary medicines are highly regulated. These products are manufactured in compliance with the pharmaceutical standards under the principles of Good Manufacturing Practice (GMP) standards. These products need to meet the strict quality guidelines set by the TGA. Thus, the export potential and consumer confidence for the Australian-made complementary medicines have grown every year, and the major market is Asia, especially China.

14.3.3 Regulatory Changes

In 2016, the Australian government endorsed several recommendations from the Review of Medicines and Medical Devices Regulation (MMDR), aimed at enhancing consumer access to therapeutic goods and reducing unnecessary regulatory burdens for industry while ensuring safety in Australia. In response to these recommendations, the TGA has introduced reforms to simplify the regulation of complementary medicines. These reforms include establishing a new approval process for listed complementary medicines and improving the availability of information to aid consumer decision-making.

14.3.4 Emphasis on Growth of Australian Manufacturing and Exports

In October 2020, the federal government released **Modern Manufacturing Strategy** (MMS), in which medical products were outlined as one of the National Manufacturing Priorities (NMP) [9–10]. The primary objective is to forge a partnership between the government, the industry, and the science and research community to cultivate a manufacturing environment that will not only generate positive economic outcomes but also meet the growing global demand for complementary medicines. This initiative aims to position Australia as a leading high-quality and sustainable manufacturing nation, contributing to a robust, modern, and resilient economy for all Australians. Additionally, it will foster long-term confidence in Australian-made complementary medicines. The Asian market presents significant opportunities due to its high demand for health food products, with projections indicating a 10% annual growth rate until 2025. Consequently, there is a strong focus on collaborating with medicines regulators to apply key manufacturing principles to complementary medicines, potentially reducing costs and enhancing the sector's capacity for future growth. Additionally, there is a heightened emphasis on government support programs designed to help Australian exporters expand their businesses in emerging and growth markets, including providing advice, building capacity, and expediting export opportunities.

14.3.5 Integration with Conventional Medicine

With a growing ageing population, there is an increased burden on the primary health-care sector in Australia. The focus is slowly growing on the preventative health-care management strategies. The best public health approach is to consume a diet rich in nutritious foods that satisfies all macro- and micronutrient needs. However, due to the fast lifestyle, changing dietary habits, economic conditions, and increased health awareness, more and more people are inclined toward the use of complementary medicine. Due to the complex nature of the prescribed medicines, which are manufactured from a certain chemical substance, people relate with it as a toxic substance and thus want to use something natural. The overall cost of prescription medicines has gone up over the years. Thus, there is an increased demand to integrate the use of complementary medicines and conventional medicines. A sensible initial step toward integration is to develop guidelines for the effective incorporation of complementary medicines into primary care, backed by relevant research and clinical experience. It is also important to train health-care professionals such as physicians and pharmacists so that they can advise on the proper usage of complementary medicines as per the appropriate requirement per patient. The quality use of medicines involves asking patients about their use of complementary medicines and participating in informed conversations about potential interactions between complementary medicines and prescription drugs.

Recent trends in complementary medicines in Australia show that there is a greater shift in consumer preferences, regulatory changes, and industrial development. This industry continues to grow in the broader economy and is forecasted to grow by 22% (CAGR, compound annual growth rate) over the next ten years in Australia, with a focus on quality, safety, and efficacy, as well as a growing interest in holistic approaches to health and wellness [11].

14.4 MARKETING AND OTHER CHALLENGES FOR COMPLEMENTARY MEDICINES IN AUSTRALIA

Even though the demand and usage of complementary medicines is growing steadily, the industry is still establishing its roots in Australia. Marketing complementary medicines in Australia presents several challenges. Seven out of ten people in Australia use complementary medicines; however, how they purchase their medicines has changed since the COVID-19 pandemic. With multiple brands and variety of product range available, there is a need for developing a holistic system with better global regulatory compliance and safety profiles. The number of Australians using complementary medicines or seeking advice from complementary therapists is increasing. However, this rise brings with it various safety and legal concerns. [12]

Some of the key challenges are as follows.

- **Regulatory Compliance:** Under Australian law, all complementary medicines are evaluated for the safety and quality of their ingredients, though efficacy is not always assessed. Only those complementary medicines classified as "high risk" undergo efficacy evaluation. [12] Additionally, complementary therapists are regulated by state governments in Australia, resulting in varying laws across different states. Australian complementary medicines regulations are very stringent and costly, but the aim is to ensure product safety, efficacy, and quality. This also helps provide a good reputation of Australian products in the global market. However, these processes also create a financial imbalance for local businesses due to delaying marketing times and reducing the competitiveness [13]. Thus, there is a need for establishing global regulatory guidelines to streamline the approval processes and ensuring the uniformity in product quality and availability.
- **Consumer Purchase Choice:** In the past, Australian consumers purchased complementary medicines mainly from pharmacies or outlets that provided health advice. However, since the pandemic, this trend has changed, with more choices available in the supermarkets and on the online market forums. The online chatbots are assisting consumers with personalized health advise. The following chart shows the recent trend in the sales channels. [14]
- **Product Export Challenges:** The complementary market is growing within Australia and overseas. Australian complementary medicines are in high demand in Asian countries, especially in China and India. However, achieving success in China necessitates

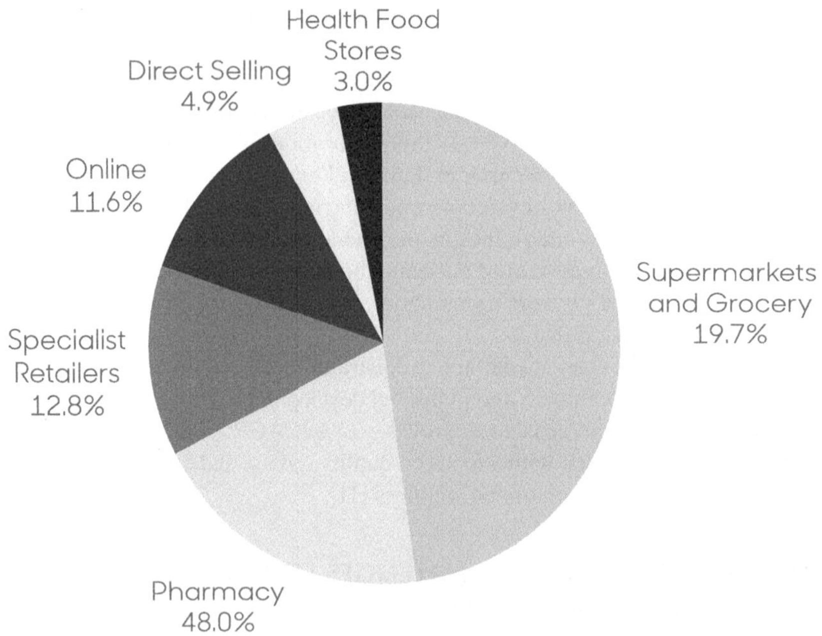

FIGURE 14.1 Australian complementary medicine sales by channel, 2020.

a deep understanding of the local regulatory environments. This also calls for a systematic approach to develop an appropriate route to market strategies, including cross-border e-commerce, B2C direct selling, and online pharmaceutical channels [14]. There is also a need to create a skilled workforce and industrial infrastructure to be able to cater to the market demand based on projected global industry CAGR of between 19% and 26.5% from 2021 to 2030.

- **Regulation of Complementary Medicines Practitioner in Australia:** Under the Health Practitioner Regulation National Law, practitioners of complementary medicines are now recognized as registered health-care professionals across all Australian states and territories. This measure ensures consistent enforcement of clinical practice regulations throughout Australia. These practitioners are highly trained and follow high professional standards and techniques; however, with a changing choice of purchase, these standards need to evolve in alignment with current regulations. Most complementary therapists are associated with professional organizations, which often require members to uphold specific standards of care. However, membership is generally voluntary and not legally mandated. [12] While there are statutory regulations governing complementary medicines clinical practice in Australia, these regulations need to be regularly reviewed to keep pace with the evolving profession and consumer behavior. [15]
- **Consumer Confidence:** Complementary medicines are a product of choice for a lot of consumers due to the ease of availability, cost-effectiveness, etc. However, usage of these medicines also comes with the risk of adverse reaction with other medications and/or existing health conditions. A significant factor undermining confidence in the industry is publication bias. False claims and inappropriate marketing lead to a loss of confidence in the use of complementary medicines. There have been instances where Australian-based and international companies have tried to obfuscate the negative results related to complementary medicines. Therefore, it is essential to focus on evidence-based research, reduce publication bias, protect intellectual property, and distinguish between false claims and legitimate listed or registered complementary therapies and medicines. Proper monitoring

and regulation of these practices will, in the long term, enhance both consumer and clinician confidence in the complementary medicines industry. [16]

- **Workforce:** The Australian industry is currently facing acute workforce shortage. Hence, there is a need for a skilled workforce in areas such as manufacturing, research and development, labor, GMP compliance, etc. This leads to increased operating costs and affects the industry's competitive edge. To address these issues, there is a need to invest in the training and development and channelizing the workforce to achieve optimum industrial operations.
- **Manufacturing Costs and Supply Chain Issues:** Overall establishment cost, higher wage rates, and limited government support affect manufacturing costs in Australia. This puts Australian manufacturers under pressure as compared to their overseas competitors. Unfortunately, this can lead to relocation of manufacturing outside Australia and thereby to a potential compromise with product quality [13].

14.5 CONCLUSION

Looking past the pandemic, Australia's complementary medicine sector is in a fast-growing phase. The stringent regulations and GMP specifications ensure the quality, efficacy, and safety of the product, which are listed or registered in Australia. As a result, consumer confidence is rising, and the demand for complementary medicines is increasing, both in Australia and in international markets. On the other hand, consumers also want to access the products so that these can be made available conveniently, to meet their changing lifestyles, via practitioner, in store (pharmacy, grocery, or health food store), and/or through online channels. Sustainability and traceability are high on the agenda, with increasing awareness of the needs for recyclable, reusable, or compostable packaging. They also want to buy from companies that blend profit with purpose. Complementary medicines companies that engage with consumers, earn trust, and demonstrate value will be embraced by consumers [14].

Nutraceuticals/complementary medicines are regulated differently across the USA, Europe, and Australia. Variations in regulation, challenges in technical analysis, and the rise of global internet sales have increased public health risks and economic burdens. Although these medicines show potential benefits, a comprehensive and trustworthy regulatory system is necessary to fully realize their benefits. However, there is a pressing need to enhance awareness and provide clear usage guidelines for both consumers and health-care practitioners. It is crucial for manufacturing companies to adhere to Good Manufacturing Practices, for regulatory authorities to maintain vigilant monitoring, and for regular market research and risk assessments to be conducted across all countries. The complementary medicine industry in Australia is still emerging, and establishing an effective national and international regulatory system could ensure safer options for consumers while helping the industry build a robust safety profile and brand value.

Navigating through these challenges requires a strategic and holistic approach to marketing that considers the regulatory requirements, consumer preferences, and industry trends. Companies that effectively tackle these challenges will be well-positioned to thrive in both the competitive Australian market and the global complementary medicine sector.

REFERENCES

[1] Complementary Medicines Australia, Industry Audit & Trends 2020

[2] Australian regulatory guidelines for complementary medicines ARGCM Version 7.2, February 2018

[3] Indication requirements for non-prescription medicines | Therapeutic Goods Administration (TGA) https://www.tga.gov.au/how-we-regulate/supply-therapeutic-good/supply-non-prescription-medicine/non-prescription-medicine-requirements-and-standards/indication-requirements-non-prescription-medicines

[4] Understand the non-prescription medicine pathways. Understand the non-prescription medicine pathways | Therapeutic Goods Administration (TGA)

[5] Burnett AJ, Livingstone KM, Woods JL, McNaughton SA. Dietary supplement use among Australian adults: findings from the 2011–2012 National Nutrition and Physical Activity Survey. *Nutrients*. 2017;9(11):1248. https://doi.org/10.3390/nu9111248.

[6] Yu X, Smith W, Webb K, Mitchell P, Leeder S. Prevalence and predictors of dietary supplement use in an older Australian population. *Aust. J. Nutr. Diet.* 1999;56:69–75.

[7] Kofoed CLF, Christensen J, Dragsted LO, Tjønneland A, Roswall1 N. Determinants of dietary supplement use—Healthy individuals use dietary supplements. *Br. J. Nutr.* 2015;113:1993–2000.

[8] Complementary Medicines Australia Pre-Budget Submission 2021–22

[9] Complementary Medicines Australia Pre-Budget Submission 2022–23

[10] The Australian Government's Modern Manufacturing Strategy. https://www.australiaminerals.gov.au/__data/assets/pdf_file/0004/106447/make-it-happen-modern-manufacturing-strategy.pdf

[11] Australia's Complementary Medicines Industry Audit 2019. https://www.cmaustralia.org.au/resources/Documents/CMA-INDUSTRY_AUDIT-V4.pdf

[12] Complementary therapies—safety and legal issues. Complementary therapies—safety and legal issues—Better Health Channel

[13] Securing the future of complementary medicines manufacturing in Australia. A strategic business case. https://doi.org/10.26183/kj2q-ee25. https://westernsydney.edu.au/cmmanufacturing

[14] Gillian Fish. Growing but challenged Australia's supplement brands face a different set of tests in the second year of COVID, the 6 AM Agency. Gill Fish_NBJ Global Issue Sept 2021 Final.pdf (cmaustralia.org.au)

[15] Bhuyan DJ, Dissanayake IH, Jaye K, Chang D. Traditional and complementary medicine in Australia: Clinical practice, research, education, and regulation. Int J Ayurveda Res. 2022 Jan–Jun;3(1):16–29. doi: 10.4103/ijar.ijar_4_22

[16] Sarris J. Current challenges in appraising complementary medicine evidence. *Med J Aust* 2012;196(5):310–311. https://doi.org/10.5694/mja11.10751. Published online: 19 March 2012; Current challenges in appraising complementary medicine evidence. *Med J Aust* (mja.com.au)

15 A Comprehensive Guide to South Korea's Food Regulation

Sahil Sorathiya and Hardik Bhatt

15.1 INTRODUCTION

Korea has a complex legal framework and agency-based food safety regulating structure with the main objective of safeguarding public health by ensuring that food is safe for customers. The National Assembly is Korea's main legislative body, but the president and government agencies also have the authority to promulgate orders and decrees to effectively carry out laws as per the Constitution. Despite the absence of an explicit constitutional provision regarding food, the government is required to safeguard the health of its population, leading to the creation of laws such as the Framework Act on Food Safety and the Food Sanitation Act.

Korea's food safety system is mostly based on two laws:

1. **The 1962 Food Sanitation Act.** This regulation ensures food safety and sanitation to preserve public health and governs the production, distribution, and oversight of potentially dangerous food items.
2. **Food Safety Framework Act of 2008.** This legislation controls food imports, monitors the food business, and prevents unsafe food from entering Korea, with provisions for business authorization, preventive measures, and post-fact disciplinary rules. Various government organizations, primarily the Ministry of Agriculture, Food, and Rural Affairs (MAFRA) and the Ministry of Food and Drug Safety (MFDS), enforce food legislation, leading to occasional overlaps or gaps in regulations. [1]

15.2 SCOPE OF THE REGULATORY BODY

The objectives of the law are:

- **Consumer protection.** Preventing the distribution of harmful food products to safeguard public health.
- **Food safety.** Establishing strict standards and regulations for food production, processing, and distribution.
- **Transparency.** Promoting open communication between the government, food industry, and consumers to build trust and confidence in the food supply.
- **Global alignment.** Keeping pace with international food safety standards and practices to facilitate trade and ensure consistency.
- **Traceability.** Implementing systems to track food products throughout the supply chain for rapid response in case of food safety incidents.

To achieve its goals, Korean food law includes the following:

- **Strict regulations.** Food businesses must follow detailed rules concerning hygiene, labeling, and the use of food additives.
- **Regular inspections.** Government authorities perform frequent inspections to ensure compliance.

DOI: 10.1201/9781003465041-15

- **Penalties for violations.** Businesses that violate food safety laws face severe punishments.
- **Consumer education.** Promoting consumer awareness of food safety issues and their rights.
- **International cooperation.** Aligning Korean food standards with global practices to facilitate trade.

South Korea's interest in health-boosting foods began in the late 1990s. Before that, products claiming nutritional benefits were simply called "health supplements," similar to dietary supplements in the United States. As South Korea's economy grew and living standards improved, people became more health-conscious. These supplements, promising various health benefits, became popular. However, this popularity also brought problems. Many companies made exaggerated or false claims about their products, causing public concern. To protect consumers and maintain public health, there was a clear need for regulations on these supplements' ingredients, labeling, and advertising. Consumers and the supplement industry itself called for stricter rules. In response, the South Korean government developed the Health Functional Food Act, which became law in 2003. This law changed the name of these products to "health functional foods" and established a new regulatory framework to oversee their safety, effectiveness, and labeling.

The rising popularity of health supplements led to some issues as many companies made exaggerated or false claims about their products, causing public concern. To protect consumers and public health, it became necessary to establish regulations on these supplements' ingredients, labeling, and advertising. Both consumers and the supplement industry itself called for stricter rules. In response, the South Korean government introduced the Health Functional Food Act in 2003. This law renamed these products as "health functional foods" and created a new regulatory framework to ensure their safety, effectiveness, and proper labeling. [2]

15.3 DISTINGUISHING BETWEEN GENERAL FOOD, HEALTH FOOD, AND HEALTH FUNCTIONAL FOOD

15.3.1 General Food

Nutritional function. General foods provide essential nutrients necessary for sustaining life and health. This includes vitamins, minerals, proteins, and other nutrients that contribute to the body's overall functioning.

Sensory and symbolic function. These foods offer sensory experiences, such as taste, smell, and visual appeal, enriching the eating experience beyond just nutritional needs.

Bioregulatory function. Beyond basic nutrition, general foods can influence overall health by supporting bodily functions and processes.

15.3.2 Health Food

Definition. Health food refers to products that are commonly perceived to offer health benefits. However, this category is broader and not always regulated in the same way as health functional foods (HFFs).

Distinction from HFFs. While health foods are generally recognized for their beneficial effects on health, they may not adhere to specific regulatory standards or labeling requirements.

15.3.3 Health Functional Food

Definition. Health functional foods are specifically designed and regulated to provide health benefits by supporting or improving physiological functions. They are manufactured according to strict guidelines and are recognized through official certification.

Regulation and labeling. To be classified as HFFs, the product must be produced in compliance with detailed regulations and is typically labeled with a certification mark or the phrase "Health Functional Food." This distinguishes them from general health foods and natural foods, which do not have the same regulatory requirements.

15.3.4 Differences from Medicines

Health functional food. These products are not intended to treat or prevent diseases in the way medicines do. Instead, their purpose is to support overall health by maintaining normal bodily functions or enhancing physiological processes.

Medicines. Medications are specifically designed to diagnose, treat, prevent, or manage diseases, often with targeted therapeutic claims.

15.3.5 Functional Ingredients

Notified ingredients. Listed in the Health Functional Food Codex, these ingredients meet established standards and do not require additional certification. Examples include vitamins, minerals, and dietary fiber.

Individually recognized ingredients. These ingredients are not included in the Codex but have been individually evaluated and approved by the Ministry of Food and Drug Safety. They require the submission of detailed data and are subject to stringent evaluation before approval. Over 200 such ingredients exist, and some may eventually be included in the Codex if specific criteria are met.

15.3.6 Manufacturing and Labeling

Manufacturing standards. Health functional foods must adhere to prescribed manufacturing standards and use approved functional ingredients along with other compliant raw materials.

Functional labeling. Products are labeled to indicate their specific health benefits, such as "supporting intestinal health," "improving blood cholesterol," "enhancing blood circulation," "regulating blood pressure," "reducing body fat," "controlling blood sugar levels," "boosting immunity," and "supporting joint and cartilage health."

15.3.7 Types of Functional Benefits

Nutrient function. Involves essential nutrients that support the body's growth, development, and maintenance. This includes vitamins for various physiological processes, minerals for bone health, proteins, and dietary fibers.

Disease risk reduction function. Focuses on reducing the risk of specific diseases or health conditions through dietary intake. For example, calcium and vitamin D are associated with a reduced risk of osteoporosis, Xylitol is approved for reducing the risk of tooth decay.

Bioactive function. Contributes to health by influencing the body's normal function or biological activity in beneficial ways.

15.3.8 Functional Marking

Functional foods are products fortified with specific ingredients to deliver health benefits beyond basic nutrition. To be labeled as functional, these foods must contain sufficient quantities of active components to produce the claimed effects and adhere to established quality standards. Common health claims associated with functional foods include supporting intestinal health, improving blood cholesterol and circulation, managing blood pressure, reducing body fat, controlling post-meal blood sugar spikes, enhancing immunity, and promoting joint and cartilage health. [3]

15.4 CATEGORIES OF HEALTH FUNCTIONAL FOOD

Health Functional Foods are of two types: generic and product-specific. The standardization, safety, and efficacy of a new active component are reviewed by the Korean Food and Drug Administration to admit blessing as a product-specific HFF. Conforming to transnational principles and securing public health require constant advancement of the Health Functional Food Act. [4]

1. **Generic HFFs.** Generic HFFs are foods that contain common, well-known health-boosting ingredients, such as vitamins, minerals, and essential fatty acids. These ingredients have clear benefits for our bodies, such as helping us grow, develop, and function normally. The amount of these nutrients in a generic HFF is carefully controlled to ensure they are safe and effective. Examples of generic HFFs include:
   - Vitamins and minerals
   - Essential amino acids and proteins
   - Dietary fiber
   - Essential fatty acids
   - Green tea extract
   - Soy protein
   - Red yeast rice
   - Fructooligosaccharide
   - Phytosterol

   For vitamins and minerals, companies can claim health benefits based on the nutrient's role in the body, such as growth, development, or normal functions. However, the amount used must be within specific limits to prevent side effects. Other generic HFFs, like dietary fiber, have more general health claims, such as aiding digestion or weight management.
2. **Product-specific HFFs.** If a food company wants to use a new ingredient in a health food product that is not already on the generic HFF list, it needs to go through a strict approval process. They must provide the government with lots of scientific evidence to prove that the ingredient is safe and effective. The government then carefully examines the ingredients and the product to make sure it meets high standards for safety, quality, and effectiveness. This process takes about four months. Once approved, the product can be sold as a product-specific HFF. The KFDA reviews the application, checking things like:
   - The ingredient's origin and properties
   - How much of the active ingredient is in the product
   - How the product is made
   - How the amount of the active ingredient is measured
   - The product's stability and purity
   - Scientific proof that the product is safe and works as claimed

HFFs are special foods designed to be good for human health beyond just basic nutrition. They are made with ingredients that have been carefully studied to make sure they are safe and beneficial. To protect consumers, the Korean government has strict rules for HFFs. All ingredients must pass rigorous tests to prove they are safe and effective. Plus, HFFs are made in special factories with high standards (GMP-certified). The government also closely watches how HFFs are advertised and what claims are made about them. They want to make sure consumers get accurate information. Every HFF that passes these checks gets a special mark to show it is approved. Approved HFF products are marked with the official "health functional food" label, indicating they meet all regulatory requirements.

There are two categories of functional ingredients approved for use in HFFs in South Korea:

1. **Notified functional ingredients.** These are ingredients that have been publicly notified in the codes of HFFs.

2. **Individually recognized ingredients.** Business operators can apply for individual recognition of functional ingredients or specific HFFs by submitting data on safety, functionality, and other requirements to the Ministry of Food and Drug Safety (MFDS). Once the MFDS grants recognition, the business can use the individually recognized ingredient(s) in their products.

15.4.1 Recognition Process

To obtain individual recognition for a functional ingredient, business operators are required to submit specific data as outlined in Articles 12 and 13 of the "Regulation on Approval of Functional Ingredients and Standard/Specification for Health Functional Food." This data undergoes a thorough review process by the MFDS, which assesses it against established standards and specifications for safety and functionality. The review includes consultation with the Health Functional Food Deliberation Committee. Once an ingredient successfully completes this review, it is officially recognized as a functional ingredient. However, its use is limited to the business operator who received the recognition.

Addition of a new functional ingredient. If a functional ingredient has been officially recognized for at least six years and has been used in at least 50 different products, it can be added to the Code. This means it becomes a standard ingredient that food companies can use without requiring special approval.

Updating information about an existing ingredient. If there is a change to an already-recognized ingredient, such as a new health claim, a different recommended daily intake, or an altered production method, it can be added to the Code after three years. However, the company that originally obtained the ingredient approval cannot be the entity requesting the change. [5]

The Ministry of Food and Drug Safety (MFDS) in South Korea has implemented a comprehensive food safety management system to ensure the provision of safer food to consumers. This system is applied throughout three key stages: manufacturing, distribution, and consumption.

1. **Manufacturing stage**
   a. **Reporting requirements.** Anyone planning to manufacture or process foods or food additives must submit a manufacturing report to the local government before or within seven days of starting production.
   b. **Self-quality inspection system.** Manufacturers and processors must conduct regular inspections to ensure food safety before distribution and to verify compliance with relevant standards and regulations.
   c. **HACCP System.** The Hazard Analysis Critical Control Point (HACCP) system is a scientific, preventive management approach designed to ensure food safety by identifying, evaluating, and managing potential hazards from raw materials through production and distribution. The HACCP system includes:
      i. **Hazard analysis (HA).** Identifying all possible biological, chemical, and physical hazards.
      ii. **Critical control points (CCP).** Managing specific points in the process to prevent, eliminate, or reduce hazards to acceptable levels.

The Ministry of Food and Drug Safety (MFDS) mandated the application of HACCP through the 2002 Food Sanitation Act. Certain items, such as processed fish products, frozen foods, ice cream, non-pasteurized beverages, and retort food products, have been required to apply HACCP standards. The list has expanded to include snacks, candies, breads, chocolates, fish sausages, beverages, instant foods, noodles, and special-purpose foods, with mandatory HACCP implementation phased in from 2014 to 2020 based on business size and annual sales.

2. **Distribution stage**
   a. **Collection and inspection.** To prevent food safety issues and ensure sanitary conditions, food products distributed for sale are regularly collected and tested. Products that fail to meet safety standards are promptly recalled or disposed of.
   b. **Food traceability.** The MFDS operates a food traceability system to track and manage food hazards. This system allows for effective recalls, cause analysis, and accurate consumer information. The enforcement of food traceability has been phased in, especially for businesses dealing with infant and baby foods or HFFs with significant annual sales. The system includes regular investigations and assessments.
   c. **Hazardous food sales prevention system.** Approximately 120,000 food products are inspected annually to ensure safety. Defective products identified through these inspections are reported to the MFDS in real time. The MFDS then prevents these products from reaching consumers by sending alerts to checkout counters in convenience stores, supermarkets, and other retail outlets. Initially implemented as a pilot project in Lotte Mart in 2009, the system has since been expanded to cover all major food retailers nationwide.
3. **Consumption stage**
   a. **Monitoring false and exaggerated advertising.** The MFDS monitors all advertising media, including the internet, broadcasts, newspapers, and magazines, to prevent consumer deception through false or exaggerated claims about food products' therapeutic effects or misleading comparisons to medicines.
   b. **Consumer Food Sanitation Guard System.** To enhance food sanitation monitoring, the Consumer Food Sanitation Guard System encourages consumer participation in monitoring food service businesses. This system complements the efforts of public officers and promotes greater transparency in food sanitation activities.
   c. **Management of food standards and specifications.** Regulations establish legal standards for managing harmful elements in food products. These standards ensure that food manufacturing, processing, and preservation methods comply with safety requirements, preventing risks to human health. The standards cover chemical, biological, and physical elements that could negatively impact health if consumed through food.

This multi-stage approach by the Ministry of Food and Drug Safety (MFDS) helps maintain high standards of food safety from production to consumption. This ensures that consumers have access to safe and reliable food products. [6]

15.5 FOOD SUPPLEMENT CLAIMS

In 2004, only specific types of food supplements, such as pills or liquids, could make health claims. However, the rules changed in 2008, when food companies lobbied for more lenient regulations. As a result, almost any processed food could now claim to be healthy, even if it was not good for human health. This change allowed many more food products to make health claims. *Health functional foods* are defined as "food manufactured or processed with ingredients or components that possess functionality useful for the human body." [7]

All new ingredients that are not generic HFFs undergo efficacy evaluation before being marketed. This evaluation is based on their ingredients or physiologically active components to ensure the accuracy of health claims. Authorization is granted on a product-by-product basis by issuing a certificate without regulatory amendments. The HFFA does not define what constitutes "substantiation" for a claim made for an HFF. Instead, it gives the Korean Food and Drug Administration (KFDA) the exclusive authority to evaluate the safety and efficacy of HFFs before their introduction to the market. It also places the responsibility on manufacturers or distributors to provide all evidence backing up the claims of their products, by developing substantiation for claims or relying on existing information.

Korean food laws have clear rules about what kind of health claims food companies can make. There are three main types of claims a food company can make:

1. **Nutrient claims.** These describe how a specific nutrient in the food helps the body grow, develop, or function normally. To make these claims, the food must contain enough of that nutrient to meet daily needs. For instance, a claim might say that calcium is good for strong bones.
2. **General health claims.** These suggest that the food can help improve overall health or well-being in some way. For instance, a label might say that eating a certain yogurt can help with digestion.
3. **Disease risk reduction claims.** These claims link eating the food to a lower chance of getting a specific disease. However, the law is very careful about these claims and does not allow companies to say that food can prevent or cure a disease. [4]

15.6 FOOD ADDITIVES

Food additives are substances added to food products during manufacturing, processing, cooking, or preservation to enhance qualities, such as taste, color, and shelf life. These additives may also include substances used for sterilizing equipment and packaging materials, which might indirectly affect the food.

Under Article 7 of the Food Sanitation Act, the Minister of Food and Drug Safety sets and announces the standards for food additives to ensure public health. Additives must meet these standards to be manufactured, imported, or used. Those failing to meet standards cannot be used in food products.

The standards aim to protect public health by ensuring that food additives are safe and used appropriately. This includes setting guidelines for their manufacturing, processing, usage, preservation, and disposal.

15.6.1 Types and Functions of Food Additives

Food additives are categorized based on their functions.

TABLE 15.1
Food Additives and Their Function

Food Additives	Function	Example
Sweeteners	Enhance the sweetness of food	Acesulfame potassium
Anti-caking agents	Prevent clumping of food particles	Silicon dioxide
Defoamers	Reduce or prevent foam formation	Silicon resin
Gum base	Provide viscosity and elasticity in chewing gum	Rosin
Flour improvers	Improve the baking quality or color of flour or dough	Chlorine
Color-developing agents	Stabilize or enhance food color	Sodium nitrite
Preservatives	Extend shelf life by inhibiting microbial growth	Sorbic acid, benzoic acid
Propellants	Release food from containers	Carbon dioxide
Acidity regulators	Adjust the pH level of food	Citric acid
Antioxidants	Prevent deterioration due to oxidation	Disodium
Fungicides	Eliminate microorganisms on food surfaces	Sodium hypochlorite

(Continued)

TABLE 15.1 ***(Continued)***
Food Additives and Their Function

Food Additives	Function	Example
Humectants	Retain moisture and prevent drying	Polydextrose
Stabilizers	Maintain uniform dispersion of ingredients	Cellulose
Filtration agents	Remove impurities	Diatomaceous earth
Nutritional enhancers	Restore or enhance nutrients	Vitamin C
Emulsifiers	Mix immiscible substances	Lecithin, liquid paraffin
Release agents	Prevent food from sticking to containers	Liquid
Coagulants	Bind or firm food ingredients	Magnesium chloride
Processing agents	Act as catalysts or aids in food processing	Nickel
Gel-forming agents	Create gels for texture	Gelatine
Thickeners	Increase food viscosity	Pectin
Coloring agents	Add or restore color	Grape peel pigment
Scale preventatives	Prevent scale formation in equipment	Cheonggukjang
Extraction solvents	Dissolve or extract components	Isopropyl alcohol
Fillers	Protect food from spoilage by injecting gases into packaging	Nitrogen
Leavening agents	Release gases to make dough rise	Sodium bicarbonate
Bleaching agents	Remove color from food	Anhydrous sulfurous acid
Surface treatment agents	Smooth or tidy food surfaces	Talc
Coating agents	Create a protective or glossy film on food	Beeswax
Flavor enhancers	Improve food flavor	L-monosodium glutamate
Spices	Add unique flavors or enhance existing flavors	Vanillin
Enzymes	Facilitate specific biochemical reactions	Alpha-amylase

15.6.2 Application for Additive Designation or Revision

To apply for the designation or revision of food additives, one must submit detailed documentation, including the additive's origin, manufacturing method, safety data, and technical necessity. The process ensures that only additives meeting safety and quality standards are approved. [8]

15.7 NOTIFICATION AND REGISTRATION PROCESS

15.7.1 Health Functional Food Functional Ingredient Certification Application Procedure

To certify a functional ingredient for use in HFFs, businesses must submit a comprehensive application to the Ministry of Food and Drug Safety. The process involves several stages and requires detailed documentation to demonstrate the safety and efficacy of the ingredient.

Eligibility for the certification is available for functional ingredients that have not yet been recognized. These ingredients are used by manufacturers and importers of HFFs, provided they submit adequate safety and functionality data for evaluation.

The regulation of the recognition process is governed by the "Regulations on the Recognition of Health Functional Food Functional Ingredients and Standards" (Ministry of Food and Drug Safety Notice No. 2021–66, dated July 29, 2021).

15.7.1.1 Required Documents

1. **Application form.** Completed application form for functional ingredient approval, including both electronic and physical copies.
2. Two copies of all documentation, including one copy of the ingredient's details. A CD containing the submitted materials. Samples of the product or prototype and the functional ingredient (or indicator component) in its standard form. Test reports from a recognized domestic health functional food inspection agency.

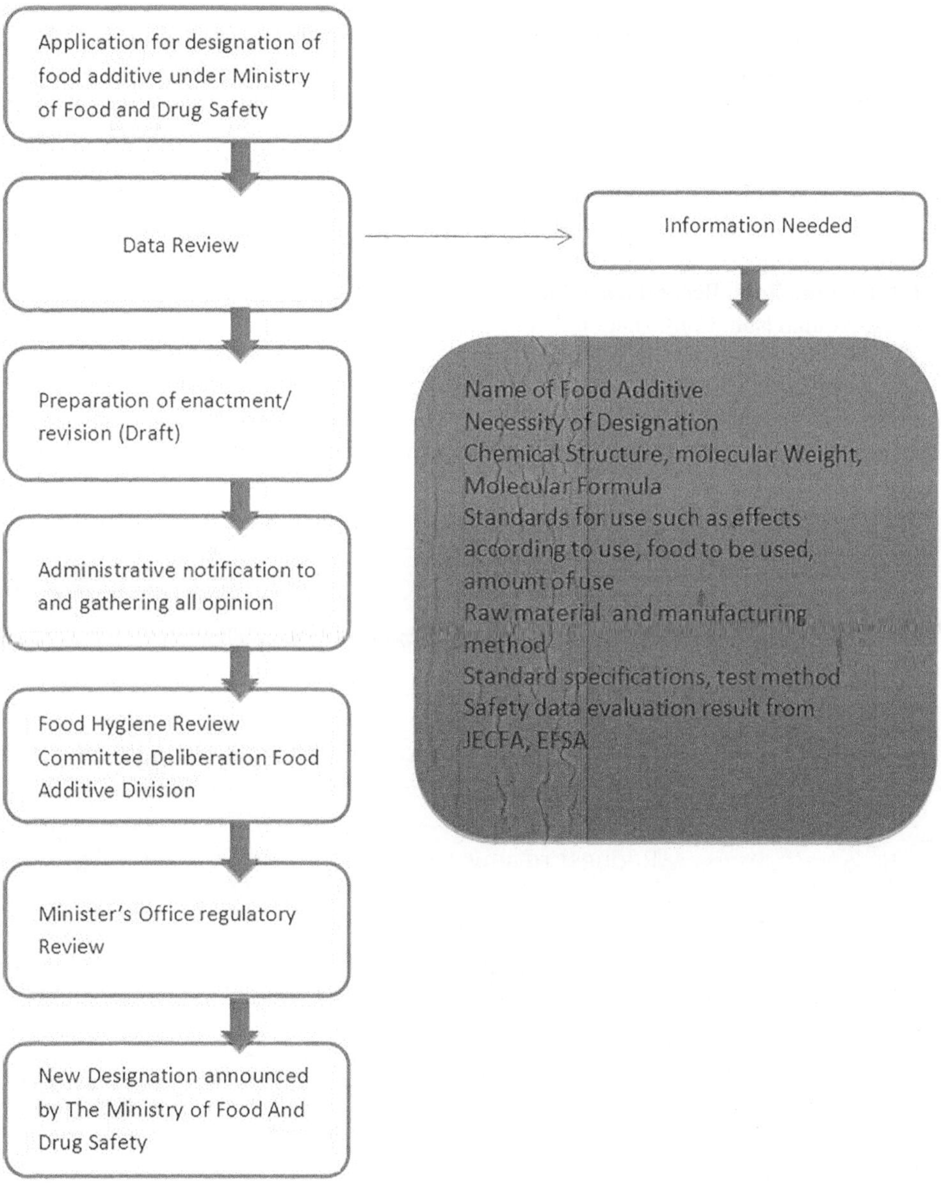

FIGURE 15.1 Application for additive designation of food additives.

Content and Requirements of Submission Materials:

1. **Summary of submission**. A complete summary of the submission materials.
2. **Origin and development**. Details of when and where the ingredient was developed, including scientific names and specific usage sites if it is a natural product.
3. **Recognition and permits**. Information on domestic and international approvals and standards.
4. **Usage status**. Records of the ingredient's use in other countries, including distribution and consumption data.
5. **Manufacturing method**. Detailed process description, including safety and functional evaluations.
6. **Characteristics data**. Information on the ingredient's properties and the functional components standardization.
7. **Specifications and test methods**. Details on the acceptable limits and test methods for functional and hazardous substances.
8. **Safety data**. Comprehensive data on the ingredient's safety, including historical usage records and scientific research.
9. **Functional content**. Evidence of the ingredient's health benefits, supported by trials and studies.
10. **Intake guidelines**. Recommended intake levels and precautions.
11. **Drug comparison**. Confirmation that the ingredient is not similar to a drug. [9]

15.7.1.2 Application Process

Applications can be submitted electronically or via the Office of Customer Support at the Ministry of Food and Drug Safety. The processing department is the Nutritional Function Research Team, which handles the evaluation.

FIGURE 15.2 Functional ingredient certification application procedure.

15.7.1.3 Fees

New Submissions: 1,900,000 won
Additions/Changes: 800,000 won

15.7.1.4 Processing Time

New Ingredients: Approximately 120 days
Functional Additions: Approximately 60 days

15.8 LICENSING AND SALE

15.8.1 Health Functional Food Manufacturing License Application Procedure

The licensing process adheres to Article 5 of the Act on Health Functional Foods and Article 3 of its Enforcement Regulations.

1. **Specialized manufacturing.** Includes facilities dedicated solely to producing HFFs. Applies to businesses engaged specifically in the manufacture and sale of these products.
2. **Venture manufacturing businesses.** Refers to ventures that contract with specialized manufacturers to produce HFFs as per Article 2 of the Act on Special Measures for the Promotion of Venture Companies.

15.8.1.1 Required Documentation

1. **For Specialized Manufacturing**
   Business license application. Complete form for HFF manufacturing.
   Product and method description. Detailed description of the products to be manufactured and the manufacturing methods.
   Facility layout. Diagram of manufacturing facilities and a list of key machinery and equipment.
   Quality manager appointment. Notification of the appointed quality manager.
   Education certificate. Proof of education for the business owner, if pre-requisite training has been completed.
   Water quality report. Required if non-tap groundwater is used in manufacturing or cleaning, issued by a recognized inspection agency.
2. **For Venture Manufacturing Businesses**
   Business license application. Same as previous.
   Technical data. Information on functional ingredients used in the HFF.
   Manufacturing manual. Details on the product types and manufacturing processes.
   Quality manager appointment. Notification of the appointed quality manager.
   Consignment contract. Agreement with a manufacturing company handling production.
   Sales training certificate. Proof of prior training, if applicable.

15.8.1.2 Application Submission

Submission Points: Regional Food and Drug Administration (Civil Complaint Office)
Fees: 50,000 won (or 45,000 won for electronic submissions, as per the Electronic Government Act)
Processing Time: Standard applications, 14 days; venture manufacturing, 10 days

15.8.1.3 Conditions Where Permission Is Denied

Recent license revocation. Applications may be denied if the business license was revoked less than six months prior for similar activities.

Incomplete revocation period. Individuals or representatives with recent business license revocation may be barred from starting similar businesses.

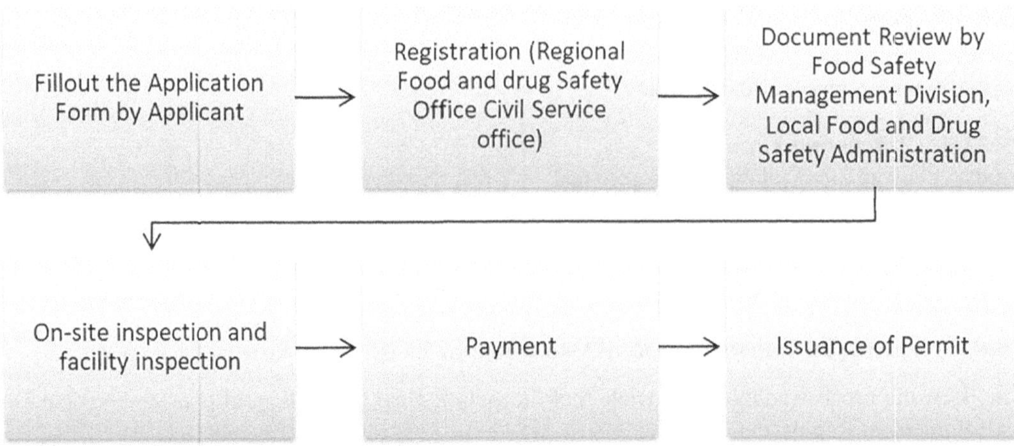

FIGURE 15.3 Manufacturing license application procedure.

TABLE 15.2
Changes to Permits of Manufacturing License Application

	Permission for Changes	Notification of Changes
Change type	• Includes changes in the business location	• Includes changes in business ownership, facility details, or consignment arrangements
Regulations	Governed by Article 5, paragraph 1 of the Health Functional Food Act and related enforcement decrees	Governed by Article 5, paragraph 2 of the Health Functional Food Act
Required documents	• Application for change and business license • Updated facility layout and equipment list • Water quality report, if applicable	• Change notification form and business license • Details of changes to facilities • Consignment contract for venture manufacturers
Submission points	• Regional Food and Drug Administration (Civil Complaint Office)	• Regional Food and Drug Administration (Civil Complaint Office)
Fees	• 30,000 won (income stamp)	• 30,000 won (income stamp)
Processing time	• Standard changes: 14 days • Venture manufacturing: 10 days	• 7 days

Bankruptcy. Those who are bankrupt or under guardianship cannot apply.

Premature resumption. Businesses under suspension cannot resume similar operations before the suspension period ends.

Ongoing suspension. Entities with unresolved business suspensions cannot engage in similar businesses. [10]

15.8.2 Procedure for Reporting Health Functional Food Sales Businesses

The reporting process is governed by Article 6 of the Act on Health Functional Foods and Article 5 of the Enforcement Regulations for the Act on Health Functional Foods. Reports should be submitted to the sanitation department at the city, county, or ward office.

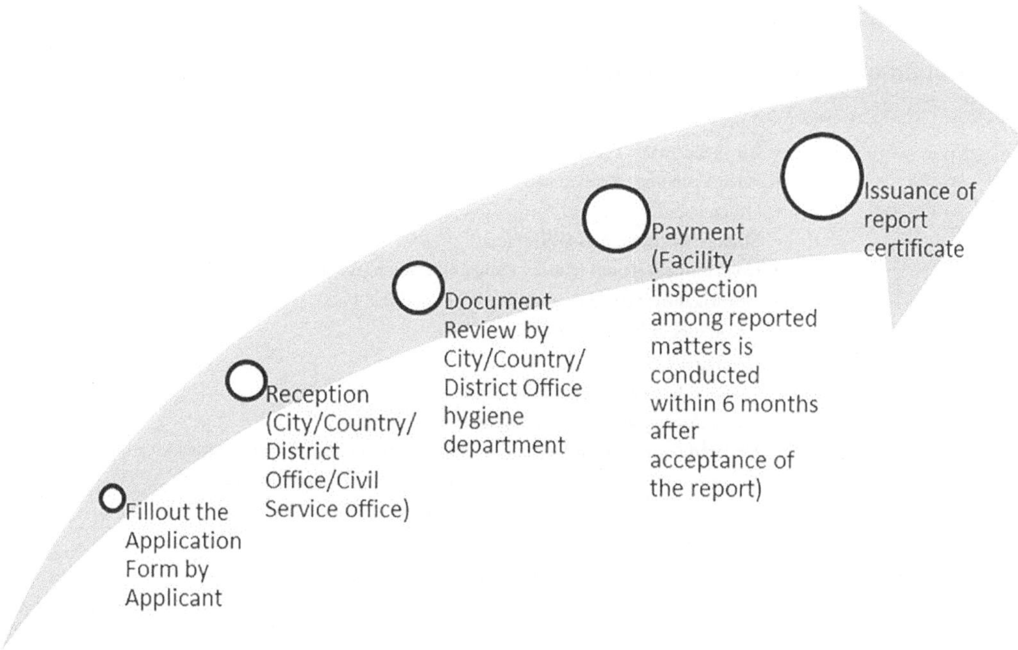

FIGURE 15.4 Procedure for reporting health functional food sales businesses.

15.8.2.1 Reporting Categories

1. **General sales.** Encompasses the sale of HFFs through various methods, including in-person sales, door-to-door, multi-level marketing, telephone solicitation, e-commerce, and mail-order, as per Article 2 of the Consumer Protection in Electronic Commerce Act.
2. **Specialized distribution.** Involves the distribution and sale of HFFs that are manufactured under the business's own brand.

15.8.2.2 Required Documentation

1. **For General Sales**
   - **Business declaration form.** Completed form declaring the sales activity.
   - **Sales training certificate.** Proof of prior training, if applicable.
   - **Lease agreement.** For rented storage facilities, if applicable.
   - **Consignment contract.** For businesses distributing HFFs, if applicable.
   - **Cable or ferry business license.** Required if sales occur at cable docks or ferry sites, as per the Cable and Ferry Business Act.
   - **Facility operation contract.** Required if sales are conducted at military or welfare facilities, as outlined in relevant acts.
   - **State property use permission.** Required if using state-owned property, such as railway stations.
   - **Urban rail facility contract.** Required if sales occur at urban rail facilities, in accordance with the Urban Railway Act.
2. **For Specialized Distribution**
   - **Business declaration form.** Same as previous.
   - **Consignment production contract.** Agreement with the manufacturing company.
   - **Sales training certificate.** If applicable.

TABLE 15.3
Notification of Changes to Sell Business

	Notification of Changes
Change type	• Business owner's name (or representative's name if a corporation) • Business name or office name • Office location • Storage facility location (if rented) • Consignment manufacturing company's name (for distribution businesses)
Regulations	• Governed by Article 6 of the Health Functional Food Act and its corresponding Enforcement Regulations
Required documents	• Change notification form • Updated business registration certificate • New lease agreement (if changing storage location) • Updated consignment production contract (for changes in sales office location)
Submission points	• City, county, or district office responsible for business reporting
Fees	• Location change: 26,500 won • Other changes: 9,300 won
Processing time	• Immediate (generally processed on the same day)

Fees: Receipt fee of 28,000 won
Processing Time: Standard processing time of 3 days

15.8.2.3 Conditions for Reporting Denial

- **Business closure orders.** Reporting may be denied if applying to continue the same type of business within six months after a closure order, except if the entire facility has been demolished.
- **Recent business closure.** If the applicant (or representative in case of a corporation) seeks to engage in a business similar to the one previously closed.
- **Incapacity or bankruptcy.** Individuals who are guardians, are bankrupt, or have not been reinstated cannot report. [11]

15.9 FOOD LABELING REGULATIONS

15.9.1 Food Labeling Requirements

The Ministry of Food and Drug Safety (MFDS) mandates detailed labeling for food products to ensure consumers receive accurate and essential information. According to the Act on the Labeling and Advertising of Food, labels must include:

- **Product identification.** Labels must specify the product name, type, and identity and principal location of the business responsible for its production or distribution.
- **Dates.** Labels are required to show both the manufacturing date and the expiration or quality retention date.
- **Ingredients and quantity.** A complete list of ingredients, along with the net contents (by weight or volume), must be provided.
- **Nutritional information.** Nutritional details, such as calories, fat, carbohydrates, sugars, and sodium, must be included.
- **Storage and safety instructions.** Guidelines for proper storage and handling, as well as any safety warnings, must be clearly stated.

15.9.2 Functions of Food Labeling

Food labels serve several identical functions:

- **Basic product details.** Provides essential information about the product, including its name, type, and business details.
- **Health and safety information.** Offers guidance on safe storage and handling, as well as detailed nutritional information.
- **Marketing and claims.** Includes any claims related to the product's health benefits or specific attributes, such as low fat or high fiber.

15.9.3 Types of Food Labeling

1. **Food equipment labeling.** To prevent consumer injuries from unsafe equipment, the MFDS requires that food-related apparatuses be clearly labeled. This includes items like tableware, scissors, and disposable gloves. The label must indicate that the equipment is food-safe, either through direct imprinting or by using stickers, if necessary. This labeling requirement was phased in over several years, starting with metal in 2015 and concluding with all material types by 2018.
2. **Allergen labeling.** For foods containing common allergens—such as eggs, milk, peanuts, soybeans, and others—labels must clearly list these allergens. If allergens are present due to cross-contamination from shared manufacturing processes, labels must include a warning about potential allergen presence. Additionally, allergens must be prominently highlighted in a different background color on the label to ensure clear visibility.
3. **Substantiation of claims.** As of July 30, 2019, new regulations require that all claims made in food labeling and advertising be substantiated with objective evidence. This measure aims to protect consumers from misleading claims and ensure that any statements about the product's benefits are backed by factual information.
4. **Nutrition labeling.** Mandatory nutrition labeling applies to various processed foods, including snacks, dairy products, and beverages. Labels must provide information on essential nutrients, such as calories, fats, sugars, and proteins. Special provisions are in place for foods targeted at children, requiring detailed nutritional information for items sold by large franchise businesses. Voluntary nutrition labeling is also implemented by some restaurants and food service establishments, including family restaurants and food courts. This allows consumers to make informed choices even when dining out.
5. **GMO labeling.** The MFDS oversees the labeling of genetically modified organisms (GMOs). During import, products labeled as GMOs are checked for compliance with labeling standards. For businesses using GMO raw materials, the MFDS reviews the necessary documentation, including identity preservation certificates and inspection reports, and conducts inspections to ensure proper labeling. Since 2013, the MFDS has been responsible for GMO labeling, consolidating management and oversight under one regulatory framework.
6. **Health functional food labeling.** Health functional foods must adhere to specific labeling standards to ensure quality and transparency. Labels must include the product's name, raw materials, expiration date, and functional claims. Additionally, HFFs are required to provide intake instructions and disclaimers clarifying that the products are not intended for disease treatment. Claims about health benefits must be verified through a voluntary review process to ensure accuracy. This comprehensive labeling framework aims to safeguard consumer health, ensure transparency, and promote informed decision-making by providing clear and accurate product information. [12]

15.10 ROUTE TO MARKET FOR HEALTH FUNCTIONAL FOOD AND NOVEL FOOD IN SOUTH KOREA

South Korea has a well-defined regulatory framework for HFFs and novel foods, overseen primarily by the Ministry of Food and Drug Safety (MFDS). The process involves several key steps, from product development to market entry.

15.10.1 Health Functional Foods

In South Korea, the regulatory pathway for introducing HFFs and novel foods to the market is meticulously overseen by the Ministry of Food and Drug Safety (MFDS). For HFFs, which are products that provide health benefits beyond basic nutrition, such as dietary supplements and vitamins, the process begins with a pre-market notification. Companies must determine if their products fall under the category of HFFs or traditional foods and then submit a detailed application to the MFDS. This application must include comprehensive information about the product's ingredients, manufacturing processes, and any health claims being made. To support these claims, companies need to provide robust evidence, such as clinical trials or scientific research data. The MFDS evaluates the application to ensure it meets safety and efficacy standards. Upon approval, a notification number is issued, which must be displayed on the product label. The product must also comply with strict labeling and packaging requirements, and ongoing post-market surveillance ensures continued compliance and safety.

15.10.1.1 Regulatory Process

1. **Pre-Market Notification**
    - **Product classification.** Determine if the product falls under HFFs or traditional foods. Health functional foods require pre-market notification.
    - **Documentation.** Submit a detailed application to the MFDS, including product ingredients, manufacturing processes, and claims.
    - **Safety and efficacy data.** Provide evidence supporting health claims, including clinical trials or scientific research.
2. **Approval and Registration**
    - **Evaluation.** The MFDS reviews the application for compliance with safety and efficacy standards. This may include a review of clinical data or literature.
    - **Approval.** Upon successful evaluation, the MFDS grants approval and issues a notification number that must be displayed on the product label.
3. **Labeling and Packaging**
    - **Regulations.** Adhere to specific labeling requirements, including health claims, ingredient list, nutritional information, and usage instructions.
    - **Packaging:** Ensure that packaging complies with South Korean standards and includes required information.
4. **Post-Market Surveillance**
    - **Monitoring.** The MFDS monitors market products for safety and compliance. Companies must report adverse events or safety issues.
    - **Quality control.** Regular checks and updates to maintain product standards.

15.10.2 Novel Foods

Novel foods are foods or food ingredients that do not have a history of safe consumption within South Korea. This category includes new, innovative, or significantly modified food products. For novel foods—foods or food ingredients that do not have a history of safe consumption within South Korea—the process is more rigorous. Companies must submit a comprehensive dossier to the MFDS detailing the novel food's safety, composition, and intended use. This submission includes safety data, such as toxicological studies and information on the food's consumption history in other countries, if applicable. The MFDS conducts an extensive scientific review, often involving expert panels or consultations with external experts, to assess the novel food's safety and suitability. If the food meets all safety criteria, the MFDS grants approval and issues a certification, allowing the product to be marketed with the appropriate certification number. The product must be labeled according to regulatory requirements, clearly indicating its novel status and any specific conditions for safe use. The advertising of novel foods must also follow stringent guidelines to ensure that claims are accurate and not misleading. Both categories of products are subject to post-market monitoring, with companies required to report any adverse effects or compliance issues to the MFDS.

15.10.2.1 Regulatory Process

1. **Pre-Market Approval**
   - **Product submission.** Submit a comprehensive dossier to the MFDS detailing the novel food's safety, composition, and intended use.
   - **Safety assessment.** Provide data on the safety of the novel food, including toxicological studies, consumption history, and potential risks.
2. **Scientific Review**
   - **Evaluation.** The MFDS conducts a scientific evaluation, often involving expert committees or panels, to assess the safety and suitability of the novel food.
   - **Consultation.** Possible additional consultations with scientific experts or regulatory bodies.
3. **Approval and Certification**
   - **Decision.** If the novel food meets safety standards, the MFDS grants the approval and issues a certification.
   - **Market authorization.** The product can then be marketed with the appropriate certification number.
4. **Labeling and Marketing**
   - **Compliance.** Ensure that labeling meets regulatory requirements, including clear disclosure of the novel nature of the product and any specific conditions for safe use.
   - **Advertising.** Follow guidelines for advertising novel foods, ensuring that claims are substantiated and not misleading.
5. **Post-Market Monitoring**
   - **Surveillance.** Monitor the product in the market for any adverse effects or compliance issues.
   - **Reporting.** Regularly report any issues or safety concerns to the MFDS.

By understanding and adhering to these regulatory pathways, companies can successfully navigate the South Korean market for HFFs and novel foods.

15.11 IMPORTED FOOD SAFETY MANAGEMENT IN KOREA

The Food Safety Management System is established under the Special Act on Imported Food Safety Control. Its goal is to promote safe and healthy food imports by managing and improving food quality and ensuring transparency in food safety information.

15.11.1 Aspects of Food Safety Control

1. **Pre-Importation Safety Measures**
   - **On-site inspections.** Foreign food facilities are inspected to ensure they meet safety standards.

- **Sanitation assessments.** Livestock products and special sanitary control items like fish heads and offal are evaluated.
- **Sanitary control.** Measures are put in place for imported fishery products.

2. **Customs Clearance**
   - **Border inspections.** Includes document reviews, field inspections, close inspections, and random sampling.
   - **Inspection orders.** Specific orders for inspections are issued when necessary.
3. **Distribution Safety**
   - **Product collection and inspection.** Imported food products are collected and inspected in the domestic supply chain.
   - **Overseas direct purchases.** Inspections of food products purchased directly from abroad are conducted.

The Special Act on Imported Food Safety Control governs overall imported food safety practices. Notices by MFDS include specifics on on-site inspections, sanitation requirements, and customs clearance procedures.

Regulations include the following:

1. **Food product standards.** Standards for food, food additives, and HFFs. Approval processes for functional ingredients in health foods. Specifications for food utensils, containers, and packaging.
2. **Import declaration requirements.** Include processed products, HFFs, agricultural products, livestock products, fishery products, and related utensils, containers, and packaging.
3. **Food categories.**
   - **Food.** All kinds, except those used as medicine.
   - **Food additives.** Substances added to foods or used in their processing.
   - **Food utensils, containers, and packages.** Items that come into direct contact with food.
   - **Health functional foods.** Products designed to supplement nutrients or provide health benefits.
   - **Livestock products.** Include meat, raw milk, eggshells, and processed meat and dairy products.

15.11.2 Proactive Safety Control Measures for Imported Food

To ensure the safety of imported food, proactive measures are taken, particularly through registration and on-site inspections for foreign food facilities. The process is broken down as follows.

15.11.2.1 Registration of Foreign Food Facilities

Foreign food facilities must be registered with the Ministry of Food and Drug Safety (MFDS) before importing products into Korea. Businesses that bring food into the country or operate food production facilities outside of the country need to register. Registration is done by submitting an online application through the Imported Food Information Maru website (impfood.mfds.go.kr). This process typically takes three days. Documents required for registration are a letter of consent for registration and an official government document proving business registration (in English).

15.11.2.2 Registration Process for Foreign Establishments

Foreign establishments must apply for registration with MFDS via the government of the exporting country before importing livestock products. Documents required for registration are as follows:

- Application for registration or changes in registration.
- Checklist showing regular inspection and supervision by the exporting country's government.

- Business approval or registration documents from the exporting country.
- If the establishment has its own standards equivalent to Korea's standards, provide a summary of the business plan and a process flowchart with critical control points.
- For other establishments, provide a summary of their sanitary control standards and a process flowchart.

On-site inspection for foreign food facilities. If a foreign facility has a history of non-compliance, or if there are safety concerns, the MFDS requires the conduct of an on-site inspection.

Notification. The facility must respond to the MFDS within 15 days after receiving notification about the inspection. If there is no response, or if the government of the exporting country refuses the inspection, the MFDS may suspend imports from that facility.

15.11.2.3 Import Sanitation Assessment

To ensure the sanitary control of livestock products and food subject to special sanitation control. For the assessment, the sanitation questionnaire is sent to the exporting country for review and response by conducting overseas on-site inspection to get approval. Consult with the exporting country on requirements and health certificates. Finally, register the foreign food facilities and establishments.

Products from countries whose sanitary control systems are deemed equivalent to Korea's are allowed for import, provided the facilities are registered and meet the required standards.

15.11.2.4 Registration as a "Good Foreign Food Facility"

Facilities that actively inspect and manage their sanitation are registered as "good foreign food facilities" by the MFDS. Registration details include the facility's name, address, operator, exported products, and other relevant information.

15.11.3 Customs Clearance Safety Control Measures

To ensure the safety of imported food, several key activities are undertaken during the customs clearance process.

15.11.3.1 Import Declaration and Inspection

When a business intends to import food or related products for sale or use, they must file an import declaration with the Minister of Food and Drug Safety (MFDS). This can be done directly by the business or through a proxy.

Initial import declaration. For the first-time import of a food product, a close inspection is mandatory. This involves thus:

- **Document review.** Checking all relevant paperwork.
- **Field inspection.** Inspecting the physical product.
- **Close inspection.** Detailed examination of the product.
- **Random sampling.** Taking samples for further testing.

Ongoing imports. For subsequent imports of the same product from the same company, clearance is granted only if previous inspections showed compliance.

Advanced technologies. An intelligent imported food management system uses statistical analysis of hazard information, including country of origin, product type, and history of hazardous substances, to enhance inspection efficiency. Machine learning helps in selecting which facilities or products need further inspection, whether on-site, during customs, or in the market.

Responsibility. Importers of products with a history of non-compliance or high hazard levels are required to undergo inspection by a designated inspection agency before filing their import declaration. This measure is intended to enhance importer accountability and ensure safety.

15.11.3.2 Safety Control in Distribution

Collection and inspection. Foods frequently consumed, those with a history of non-compliance or those flagged for potential safety issues are collected and inspected.

Instruction and supervision. Importers with a history of administrative issues receive guidance and oversight to prevent illegal practices. If products are found non-compliant or recalled, they are promptly discarded to ensure market safety.

Safety control for overseas direct purchases online. With the rise in direct purchases from overseas through the internet and social media, the MFDS focuses on inspecting these products to ensure they meet safety standards.

Public warnings. Consumers are advised against purchasing such products online through public notices. If harmful substances are detected in directly purchased products, MFDS requests Korea Customs Service to withhold these products and publish a list of affected products on the "FoodSafetyKorea" website. [13]

15.12 REGISTRATION AND CONTROL STANDARDS FOR GOOD IMPORTERS

This notice establishes the procedures for registering and overseeing "good importers" of food products under South Korean regulations. These standards apply to facilities and establishments outside Korea that produce or process imported food, food additives, HFFs, and livestock products.

15.12.1 Definitions

1. **Imported food.** Includes foods, food additives, apparatus, and packaging from abroad, HFFs, and livestock products.
2. **Foreign food facility.** An overseas facility where non-livestock imported food is produced or processed.
3. **Foreign food establishment.** An overseas establishment that handles livestock products.
4. **Good importers.** Foreign food facilities or establishments that meet sanitary control standards and are registered according to the Act.
5. **Applicant.** A person applying to register as a good importer.
6. **On-site inspection.** An inspection conducted at a foreign food facility to verify sanitary control status.

15.12.2 Scope of Application

The notice applies to facilities producing or processing imported food or persons importing or selling such food. It covers thus:

1. Foods and additives as per the Food Sanitation Act.
2. Apparatus and packaging as per the Food Sanitation Act.
3. Health functional foods as per the Health Functional Foods Act.
4. Livestock products (excluding certain items like meat and raw milk).

Standards for assessing sanitary conditions are outlined in the Ministry of Food and Drug Safety's guidelines on inspecting foreign food facilities and establishments.

1. **Registration Procedures**
   - **Application.** Applicants must inspect foreign facilities or establishments, prepare a report with specific details, and submit it to the Minister of Food and Drug Safety. An inspection by a designated institution is also permissible.
   - **Supplementary data.** The Minister may request additional data if initial submissions are insufficient, with a deadline of 60 days.
   - **Rejection of application.** Applications can be rejected if supplementary data is not provided on time.
   - **Exemptions.** Facilities inspected recently may be exempt from on-site inspections during registration.
   - **Re-registration.** When re-registering, if the type of products and facilities remains the same, some documents may be waived, and on-site inspections may be exempted.

2. **Registration Management**
   - **Additional registrations.** For adding new foods or facilities, similar documentation is required.
   - **Suspension of products.** If a product is no longer imported, the registration must be withdrawn.
   - **Revoked registration.** If registration is revoked or the business closes, the certificate must be returned.

3. **Renewal of Registration**
   - **Application.** Renewal applications must be submitted 60 days before the current registration expires. If there are no changes, fewer documents may be needed.
   - **Inspection exemptions.** Previous inspections or compliance history might exempt facilities from new inspections.
   - **Document verification.** Verified documents from exporting countries can lead to exemption from on-site inspections.

4. **Matters to Be Observed**
   - **Annual inspections.** Good importers must inspect foreign facilities at least once a year.
   - **Joint inspections.** If multiple importers register the same facility, they may jointly conduct inspections.
   - **Production suspension.** If production is suspended for over a year, inspections can be postponed.
   - **Checklist and reporting.** Inspections must follow specific checklists, and reports must be submitted in Korean or English.

5. **Follow-Up Management**
   - **Tracking compliance.** The Minister will monitor compliance with standards.
   - **Non-compliance.** If issues arise with imported food or facilities, the Minister may conduct inspections.
   - **Suspension of measures.** Non-compliance or issues in inspections may lead to a suspension of preferential measures.

6. **Preferential Measures**
   - **Inspection exemptions.** Certain imports may be excluded from random sampling.
   - **Publication.** Names of good importers may be published.
   - **Labeling.** Special labels may be used on packaging.
   - **Rapid inspections.** Support for quicker import inspections may be provided.

7. **Review of Regulations**
    - The minister will review the validity of this notice every three years and make necessary improvements. [14]

15.13 PROMOTING HEALTHY DIETS AND SAFE FOOD CONSUMPTION

TABLE 15.4
Encouraging Healthy Diets through Reduced Sugar and Sodium Intake

Sodium reduction efforts	Excessive sodium intake is linked to chronic health issues, like hypertension and strokes. The World Health Organization (WHO) emphasizes reducing sodium consumption for better health. In Korea, the average sodium intake in 2014 was 3,890 mg per day, a 20% decrease from 4,878 mg per day in 2010. Despite this improvement, it remains nearly double the recommended limit of 2,000 mg per day. To address this, the Ministry of Food and Drug Safety (MFDS) aims to lower daily sodium intake to 3,500 mg by 2020, following the initial reduction target of 3,900 mg by 2017.
Raising public awareness	Efforts to boost public awareness about reducing sodium intake include ongoing campaigns, public service announcements, contests, and educational programs in collaboration with various government ministries and local authorities. Initiatives like the "Samsam (Low-Sodium) Food Service" and low-sodium meal service weeks are also part of these efforts.
Creating a low-sodium food environment	With the rise in processed foods and dining out, strategies include creating sodium reduction guides for high-sodium foods and promoting low-sodium meal services. The MFDS has introduced a "Samsam (Low-Sodium) Food Service" system to ensure that at least one daily meal contains less than 1,300 mg of sodium and has designated restaurants that meet these standards.
Sugar intake reduction initiatives	The Korea National Health and Nutrition Examination Survey found that sugar intake increased from 59.6 g in 2007 to 72.1 g in 2013, with a significant portion coming from processed foods. This trend is concerning, especially among younger age groups, where sugar intake exceeds WHO recommendations. To combat this, projects aimed at reducing sugar intake are being implemented.

TABLE 15.5
Ensuring Food Safety and Preventing Food Poisoning

Food poisoning prevention	The rise in dining out has increased food poisoning cases, which now affect 72% of those eating out. In response, the government has intensified its efforts to prevent food poisoning through targeted inspections, guidance, and preemptive measures. A cooperative approach involves multiple ministries and local governments to manage food safety and investigate outbreaks.
Food poisoning prediction and monitoring	The "Food Poisoning Prediction Map" on the MFDS website helps predict potential food poisoning risks by analyzing data from various sources, including meteorological data and health records. This map aims to enhance preventive measures.
Contamination monitoring	To address food contamination, the Korea Integrated Pathogen Information Network (KIPIN) has been established to track pathogens throughout the food supply chain, improving response and traceability.

TABLE 15.6
Reinforcing Safety Measures for Children's Food

Legislative and safety measures	In 2007, the MFDS introduced "Comprehensive Safety Measures for Children's Food," followed by the enactment of the Special Act on Safety Management of Children's Dietary Life in 2008. This legislation designates "green food zones" around schools to ensure that food businesses within these areas adhere to high sanitation standards and avoid high-calorie, low-nutrition, and high-caffeine foods.
Promoting healthy food choices	The MFDS prohibits the sale and advertisement of high-calorie, low-nutrition foods in schools and restricts advertising aimed at children. The "quality authentication system" ensures that safe and nutritious foods are available for children, with ongoing education and promotional activities to support healthy eating habits.
Support for safe meal services	The MFDS, in collaboration with local governments, has set up the "Center for Children's Food Service Management." This center provides expert support to ensure the safety and nutritional quality of food served in children's facilities, recognizing the importance of balanced nutrition during critical developmental stages.

Source: [15].

15.14 REFERENCE AMOUNTS CUSTOMARILY CONSUMED PER EATING OCCASION

15.14.1 Definition

The reference amounts customarily consumed per eating occasion, or reference amounts, represent the standard values used to define typical serving sizes for various foods. These values are outlined under the "Labeling Standards of Foods, Etc." and are established in accordance with Articles 4 and 5 of the Act on the Labeling and Advertising of Food, as well as Articles 5(3) and 6(4) of its Enforcement Rule. These reference amounts are based on typical single-serving intakes observed in consumers aged 3 and older, supplemented by market research data. They are essential for determining packaging sizes, nutritional labeling of processed foods, assessing the caloric and nutrient content of popular children's foods, and setting criteria for sodium content comparisons.

15.14.2 Purpose

The primary aim of these reference amounts is to provide clear serving size guidelines that help consumers interpret nutritional information more effectively. By standardizing the reference amounts for each type of food, consumers can better assess the nutritional content of food products, making it easier to make healthier dietary choices. This initiative also supports improved understanding of food labels and contributes to more informed consumption habits.

15.14.3 Pathways for Setting Reference Amounts

1. **Data analysis.** The process begins with analyzing recent data from the Korea National Health and Nutrition Examination Survey (2016–2019). This analysis helps identify areas where reference amounts need updating, or establishing, based on current consumption patterns.
2. **Market survey.** Next, a market survey is conducted to review existing product packaging and serving sizes. This helps assess whether the current reference amounts are accurate or need adjustments based on real-world product offerings.

TABLE 15.7
List of Nutrients with Their Daily Intake Range and Functions

Nutrient	Primary Functions	Daily Intake Range
Vitamin A	Supports vision, skin health, and cell growth	210–1,000 μg retinol equivalents
Beta-carotene	Supports vision, skin health, and cell growth	0.42–7 mg for vision and skin; 1.26+ mg for cell growth
Vitamin D	Aids in calcium and phosphorus absorption; bone health, reduces osteoporosis risk	1.5–10 μg
Vitamin E	Protects cells from oxidative damage	3.3–400 mg α-TE
Vitamin K	Essential for blood clotting and bone health	21–1,000 μg
Vitamin B1	Supports carbohydrate and energy metabolism	0.36–100 mg
Vitamin B2	Involved in energy production	0.42–40 mg
Niacin	Essential for energy production	Nicotinic acid: 4.5–23 mg; nicotinamide: 4.5–670 mg
Pantothenic acid	Supports fat, carbohydrate, protein metabolism and energy production	1.5–200 mg
Vitamin B6	Essential for protein and amino acid utilization; maintains homocysteine levels	0.45–67 mg
Folic acid	Supports cell and blood production, fetal neural development, and homocysteine levels	120–400 μg
Vitamin B12	Crucial for normal folate metabolism	0.72–2,000 μg
Biotin	Involved in fat, carbohydrate, and protein metabolism and energy production	9–900 μg
Vitamin C	Supports connective tissue, iron absorption, and cell protection	30–1,000 mg
Calcium	Essential for bone and tooth formation, nerve and muscle function, blood clotting, and osteoporosis prevention	210–800 mg
Magnesium	Supports energy use and maintains nerve and muscle function	94.5–250 mg
Iron	Crucial for oxygen transport, blood production, and energy production	3.6–15 mg
Zinc	Supports immune function and cell division	2.55–12 mg
Copper	Aids in iron transport, use, and cell protection	0.24–7.0 mg
Selenium	Protects cells from oxidative damage	16.5–135 μg
Iodine	Essential for thyroid hormone synthesis, energy production, and neural development	45–150 μg
Manganese	Supports bone formation, energy use, and cell protection	0.9–3.5 mg
Molybdenum	Required for enzyme activity	7.5–230 μg
Potassium	Maintains water and electrolyte balance	1.05–3.7 g
Chromium	—	0.015–9 mg
Dietary fiber	Dietary fiber supplement	More than 5 g
Protein	Builds body tissues, enzymes, hormones, antibodies; supports fluid balance, energy, glucose, and lipid synthesis	More than 12 g
Essential fatty acids	Supplementation	Linoleic acid: more than 4 g; linolenic acid: more than 0.6 g

Source: [16].

3. **Feedback incorporation.** Feedback from industry professionals and consumers is gathered to refine the reference amounts. This includes evaluating demands for adjustments and understanding how well current reference amounts are utilized and understood by the public.
4. **Guideline review.** Finally, the reference amounts are reviewed in the context of established dietary guidelines, such as the Dietary Guidelines for Koreans. This step ensures that the reference amounts align with current nutritional standards and social norms, with improvements made through expert consultations.

15.15 GMP FOR HEALTH FUNCTIONAL FOODS

The Korean Food and Drug Administration (KFDA) has issued guidelines for Good Manufacturing Practices (GMPs) in HFFs to ensure product quality, consumer safety, and public health. These guidelines outline procedures for application, facility standards, quality assurance, team formation, self-regulation, sanitation, quality control, and product preservation. While not yet mandatory, GMP compliance will be required for venture and toll manufacturing companies starting February 2006, aiming to improve manufacturing processes and product stability.

15.15.1 Definition of GMP

GMP, which stands for Good Manufacturing Practice, outlines the standards and protocols necessary to ensure the safe and high-quality production of HFFs. These standards cover all stages of the manufacturing process, from the initial procurement of ingredients to the final stages of production, packaging, and distribution. GMP also encompasses the requirements for workplace structure and facilities

15.15.2 Regulatory Framework

GMP regulations for HFFs are guided by Article 22 of the Act on Health Functional Foods and the Good Health Functional Food Manufacturing Standards, as stipulated in the Ministry of Food and Drug Safety Notice No. 2022–31, dated April 20, 2022.

15.15.3 Objectives of GMP

- **Error reduction.** Minimize human errors during production.
- **Pollution prevention.** Prevent contamination and ensure product quality remains consistent.
- **Quality assurance.** Establish a robust system for maintaining high standards throughout the manufacturing process. [17]

15.16 EVALUATING AND MANAGING THE SAFETY OF HEALTH FUNCTIONAL FOODS

15.16.1 Evaluation of Safety

The safety of HFFs is rigorously assessed by the Ministry of Food and Drug Safety before these products reach consumers. This procedure includes significant steps:

1. **Scientific review.** The Ministry conducts a thorough scientific review to ensure that the HFF does not pose any specific risks or adverse effects when consumed as directed. This review includes an analysis of all existing safety data and considers potential interactions

with medications and effects on vulnerable populations, such as children, pregnant women, and lactating women.

2. **Assessment of safety data.** The safety evaluation involves reviewing historical data on the HFF, including:
   - **Manufacturing methods.** Ensuring that the production processes adhere to safety standards.
   - **Intake level.** Assessing whether different levels of intake impact safety.
   - **Biological effects.** Evaluating the biological utility and effects of the ingredients.
   - **Toxicity testing.** Conducting toxicity tests if safety cannot be established through existing data.
3. **Proof of safety.** If existing evidence does not suffice to confirm safety, additional toxicity tests may be required. These tests are designed to verify that the HFF is safe when consumed according to recommended guidelines.

15.16.2 Management of Safety

Once a HFF has been approved, its safety is managed through several oversight measures:

1. **Standard inspections.** Approved HFFs undergo regular inspections to ensure compliance with established standards. These inspections verify thus:
   - **Functional ingredients.** The product contains the stated amount of functional ingredients.
   - **Hygiene standards.** The product meets hygiene criteria, including limits on heavy metals (such as lead, mercury, arsenic, cadmium), pesticide residues, and microbial contaminants (like *E. coli*).
2. **Imported products.** Health functional foods imported into the country are subject to rigorous inspections at the point of entry. This is crucial for detecting illegal pharmaceutical ingredients that may be present in these products.
3. **Ongoing monitoring.** The Ministry conducts collection inspections and random checks of products in circulation to manage and mitigate any potential safety issues that arise post-market.

15.16.3 Managing Risks of Unauthorized Claims

Illegal additives. There have been instances where general foods, not officially recognized as HFFs, have been found to contain illegal pharmaceutical ingredients under the guise of offering health benefits. Consumers should be cautious, especially with imported dietary supplements or products lacking proper labeling in Korean.

Internet purchases. Special attention should be given to products purchased online, particularly those not labeled with Korean regulatory markings. These products may not adhere to local safety standards and could pose risks to health.

The safety of HFFs is ensured through a comprehensive evaluation process by regulatory authorities, ongoing inspections, and monitoring. Consumers should remain vigilant about the authenticity and regulatory compliance of HFFs, especially when purchasing from less-regulated sources.

15.16.4 Managing Adverse Events in Health Functional Foods

Even with rigorous safety evaluations conducted before the distribution of HFFs, adverse reactions may still occur. These reactions can result from misuse, contamination, or individual physiological responses. Until the exact cause is scientifically identified, such incidents are termed *adverse events* rather than *side effects. Adverse events* are essentially reports of subjective symptoms experienced by users. The management of adverse events follows a structured process divided into three main phases:

1. **Adverse event collection.** Systematic gathering of information on adverse events. The Ministry of Food and Drug Safety actively collects reports of adverse events from a variety of sources, including consumers, business owners, and medical experts. If a patient experiences symptoms potentially related to a HFF, the responsible health-care provider may report the case directly to the Ministry. Alternatively, patients can report adverse events to the designated Health Functional Food Adverse Event Report Center.
2. **Adverse event analysis.** Scientific analysis and classification of reported adverse events. The collected data is entered into a computerized integrated database for comprehensive analysis. Adverse events are categorized based on their severity, from mild to severe. Cases are further classified according to WHO criteria: death, life-threatening conditions, hospitalization requirements, disability or loss of function, birth defects.

   The analysis involves evaluating various factors, such as:
   - The frequency of adverse events in relation to product sales
   - The presence of reporting biases for specific products
   - The amount of product consumed and any misuse by users

   For serious cases, additional steps are taken to obtain detailed information from reporters, including their health history, current medications, and medical treatment details. Consultation with medical records and clinical specialists helps in a thorough scientific evaluation of the adverse events.
3. **Dissemination of results.** Efficiently communicate findings from the adverse event analysis to the public. Results are shared through various channels, including:
   - The Ministry of Food and Drug Safety's Food Safety Information Portal
   - The Integrated Civil Complaint Reporting Center
   - Safety information publications and reports

The management of adverse events involves collecting detailed reports, analyzing them scientifically, and disseminating the results to ensure consumer safety. This process helps identify and address potential issues associated with HFFs, contributing to ongoing public health protection.

REFERENCES

[1] second fao/who global forum of food safety regulators. Bangkok, Thailand, 12–14 October 2004 Food Control and Food Safety System in Korea. https://www.fao.org/4/ae186e/ae186e00.htm

[2] Jong-Soo K. Regulatory framework of Health Functional food in Korea. https://www.koreascience.or.kr/article/CFKO200424717918087.pdf. 2004.

[3] 식품안전나라 (Food safety Nation). https://www.foodsafetykorea.go.kr:443/portal/board/boardDetail.do?menu_no=3811&menu_grp=MENU_NEW01&bbs_no=bbs464.

[4] Kim JY, Kim DB, Lee HJ. Regulations on health/functional foods in Korea. *Toxicology.* 2006 Apr 1;221(1):112–8. https://doi.org/10.1016/j.tox.2006.01.016. https://www.sciencedirect.com/science/article/abs/pii/S0300483X06000448?via%3Dihub

[5] Health Functional Food Code. https://www.mfds.go.kr/eng/brd/m_15/down.do?brd_id=eng0001&seq=70011&data_tp=A&file_seq=1. 2021.

[6] 식품의약품안전처. Ministry of Food and Drug Safety>Our Works>Food>Food Safety | Ministry of Food and Drug Safety. https://www.mfds.go.kr/eng/wpge/m_9/de0110011001.do.

[7] Kim JY, Kim SJ, Jeong S. *Regulations on health/functional foods in Korea.* Elsevier eBooks. 2019, p. 497–507. https://www.sciencedirect.com/science/article/abs/pii/B9780128164679000319?via%3Dihub.

[8] 식품안전나라 (Food safety Nation) Health Functional Food Additive. https://www.foodsafetykorea.go.kr:443/portal/board/boardDetail.do?menu_no=3725&menu_grp=MENU_NEW01&bbs_no=bbs430.

[9] 식품안전나라Health Functional Food Functional Ingredient Certification Application Procedure. https://www.foodsafetykorea.go.kr:443/portal/board/boardDetail.do?menu_no=2664&menu_grp=MENU_NEW01&bbs_no=bbs053
[10] 식품안전나라 Health Functional Food Manufacturing License Application Procedure. https://www.foodsafetykorea.go.kr:443/portal/board/boardDetail.do?menu_no=2664&menu_grp=MENU_NEW01&bbs_no=bbs053
[11] 식품안전나라 Procedure for Reporting Health Functional Food Sales Businesses. https://www.foodsafetykorea.go.kr:443/portal/board/boardDetail.do?menu_no=2664&menu_grp=MENU_NEW01&bbs_no=bbs053
[12] 식품의약품안전처. Ministry of Food and Drug Safety>Our Works>Food>Food labeling System | Ministry of Food and Drug Safety. https://www.mfds.go.kr/eng/wpge/m_14/de0110051001.do
[13] 식품의약품안전처. Ministry of Food and Drug Safety>Our Works>Food>Imported food Safety | Ministry of Food and Drug Safety. https://www.mfds.go.kr/eng/wpge/m_11/de0110021001.do
[14] Registration and control Standards of good importers. https://www.mfds.go.kr/eng/brd/m_15/down.do?brd_id=eng0001&seq=72443&data_tp=A&file_seq=1
[15] 식품의약품안전처. Ministry of Food and Drug Safety>Our Works>Food>Promoting Healthy diets and Safe food consumption | Ministry of Food and Drug Safety. https://www.mfds.go.kr/eng/wpge/m_13/de0110041001.do
[16] Information>Publications | 영문 홈페이지 Reference amounts customarily consumed per eating occasion. https://www.nifds.go.kr/file/Reference%20amounts%20customarily%20consumed%20per%20eating%20occasion%20in%20photos.pdf
[17] GMP for Health Functional Foods. https://www.foodsafetykorea.go.kr:443/portal/board/boardDetail.do?menu_no=2664&menu_grp=MENU_NEW01&bbs_no=bbs053

16 Navigating the Maze of Regulatory Guidelines for Nutraceuticals in the Gulf Countries

Nikunjana Patel, Tasneem Rangwala, Niyati Acharya, and Sanjeev Acharya

16.1 INTRODUCTION [1]

Nutraceuticals, coined from the fusion of *nutrition* and *pharmaceuticals*, represent a class of products integrating the advantageous properties of both food and medicine. These substances are essential for maintaining and regulating regular physiological processes, which enhance general health in people. The rapid expansion of the nutraceutical market is propelled by shifts in population demographics and evolving health-conscious trends. Within this market, nutraceutical products encompass a diverse range of food categories, including polyunsaturated fatty acids, dietary fiber, prebiotics, antioxidants, probiotics, natural foods, and herbal foods. They are crucial allies in the fight against common health problems like osteoporosis, diabetes, arthritis, cancer, heart disease, and high cholesterol level. The emergence of nutraceuticals signifies a paradigm shift in medicine and health care, catalyzing the transformation of the food industry into a research-driven sector.

In recent years, a transformative shift in dietary health paradigms has emerged, placing greater emphasis on the positive aspects of nutrition. This shift reflects the evolving lifestyles of individuals, characterized by altered dietary habits. The pervasive consumption of junk food has surged, resulting in numerous health issues stemming from inadequate nutrition. Notably, obesity has attained global recognition as a pressing concern. Furthermore, in many developing countries, cardiac disease continues to be the primary cause of death, with cancer, osteoporosis, and arthritis coming in close second.

In response to the shortcomings of conventional medical approaches, characterized by their high costs and heavy reliance on technology for disease treatment, consumers are increasingly turning toward complementary or alternative solutions. Nutraceuticals have gained traction in this regard, offering appealing alternatives amid the bureaucratic complexities of managed health care. The ancient adage "Let food be thy medicine and medicine be thy food," attributed to Hippocrates over two millennia ago, resonates deeply in contemporary contexts. This age-old understanding is embodied in the emerging class of natural products known as nutraceuticals, which blur the lines between food and medicine. [2]

Regulation stands as a cornerstone of the nutraceutical industry, serving a multitude of critical purposes that are indispensable for both consumers and manufacturers alike. Here is a more detailed exploration of why regulation is paramount for the nutraceutical sector:

- **Consumer safety assurance.** Nutraceuticals occupy a unique space where food and medicine converge, making it imperative to guarantee their safety for consumption. Regulatory frameworks establish stringent standards governing every aspect of production, from

DOI: 10.1201/9781003465041-16

ingredient sourcing to manufacturing processes, ensuring that products meet strict safety criteria. By adhering to these regulations, manufacturers can mitigate potential health risks associated with the consumption of nutraceuticals, safeguarding consumer welfare.

- **Maintaining product integrity and quality control.** Nutraceuticals are valued for their purported health benefits, necessitating consistent product quality across all batches. Regulations set forth comprehensive guidelines covering sourcing, manufacturing, storage, and packaging practices, thereby ensuring that nutraceutical products maintain their integrity from production to consumption. This meticulous oversight minimizes the likelihood of contamination, adulteration, or distribution of substandard products, preserving the efficacy and safety of nutraceuticals on the market.
- **Substantiating efficacy claims.** Nutraceuticals often tout various health claims, ranging from boosting immunity to promoting cardiovascular health. Regulatory bodies mandate that these claims be substantiated by robust scientific evidence, requiring manufacturers to conduct thorough research and testing to support their assertions. Regulations that enforce this requirement stop deceptive or fraudulent advertising, allowing customers to make educated decisions about the products they select.
- **Ensuring accurate labeling and transparency.** Clear and accurate labeling is essential for empowering consumers to make educated choices about the products they purchase. Regulatory standards mandate that nutraceutical labels provide comprehensive information regarding ingredients, nutritional content, recommended dosage, and potential allergens. This transparency fosters consumer trust and confidence in product safety and efficacy, promoting informed decision-making.
- **Establishing dosage and formulation standards.** Nutraceuticals must adhere to standardized dosage levels and ingredient formulations to mitigate the risk of adverse effects from overconsumption. Regulatory agencies set forth guidelines specifying acceptable dosage ranges and ingredient combinations, thereby safeguarding consumer health and minimizing potential risks associated with improper usage.
- **Ensuring batch-to-batch consistency.** Consistency in product composition is crucial to delivering reliable efficacy and safety to consumers. Regulatory oversight mandates that nutraceutical manufacturers adhere to strict protocols to maintain batch-to-batch consistency. By enforcing these standards, regulations ensure that consumers receive consistent quality and performance from nutraceutical products throughout their shelf life.
- **Monitoring and addressing adverse events.** Regulatory bodies play a pivotal role in monitoring and addressing adverse events associated with nutraceutical consumption. By maintaining surveillance systems and promptly investigating reported incidents, regulatory agencies can identify emerging safety concerns and take appropriate actions, such as product recalls or formulation adjustments, to mitigate risks and protect public health.
- **Facilitating international trade and compliance.** Nutraceuticals are traded globally, necessitating adherence to international regulatory standards. By harmonizing regulations and ensuring compliance with global norms, regulatory frameworks facilitate the smooth movement of nutraceutical products across borders. This harmonization fosters confidence in the safety and quality of nutraceuticals among consumers worldwide, promoting international trade and market growth.
- **Instilling consumer confidence and trust.** Ultimately, a robust regulation instills confidence in consumers by assuring them that nutraceutical products undergo rigorous scrutiny and adhere to stringent safety and quality standards. This confidence is essential for fostering trust in the nutraceutical industry, empowering consumers to embrace these products as integral components of their health and wellness regimens.

In essence, regulation serves as essential to the nutraceutical industry, ensuring consumer safety, maintaining product integrity, and fostering transparency and trust in the marketplace. By upholding these standards, regulatory frameworks play a pivotal role in advancing public health and promoting the responsible growth of the nutraceutical sector.

16.2 WORLDWIDE REGULATORY FRAMEWORKS FOR NUTRACEUTICALS [3]

Regulatory frameworks for nutraceuticals exhibit significant variation worldwide, with each country or region implementing specific rules tailored to address local needs while upholding global standards of safety and efficacy. Here is a comprehensive overview of nutraceutical regulations across various regions.

16.2.1 United States

- The Dietary Supplement Health and Education Act (DSHEA) of 1994 regulates nutraceuticals in the United States, bringing them under the jurisdiction of the Food and Drug Administration (FDA).
- Manufacturers bear the responsibility of ensuring product safety, while the FDA maintains post-market surveillance mechanisms to monitor adverse events.

16.2.2 European Union

- The European Food Safety Authority (EFSA) keeps an eye on health claims for food products like nutraceuticals. They make sure these claims are backed by science, so you can trust what you see on labels.
- To promote openness and consumer confidence, the European Commission oversees the use of nutrition and health claims under the Nutrition and Health Claims Regulation.

16.2.3 Canada

- Health Canada oversees nutraceuticals classified as natural health products (NHPs), requiring pre-market assessments to demonstrate safety, efficacy, and quality.
- Stringent regulations aim to protect public health while fostering innovation in the nutraceutical industry.

16.2.4 India

- The Food Safety and Standards Authority of India (FSSAI) regulates nutraceuticals under specific regulations covering health supplements, nutraceuticals, and other functional foods.
- Nutraceutical manufacturers must adhere to stringent quality standards and undergo regulatory scrutiny to ensure consumer safety.

16.2.5 China

- The China Food and Drug Administration (CFDA) regulates health food, a category that encompasses nutraceuticals, requiring rigorous registration processes for approval.
- Regulatory oversight aims to address safety concerns and combat the proliferation of substandard or counterfeit nutraceutical products.

16.2.6 Japan

- The Ministry of Health, Labour, and Welfare (MHLW) oversees nutraceuticals classified as foods for special dietary uses (FOSHU), ensuring compliance with strict criteria for health claims and safety.
- Japan's regulatory framework emphasizes consumer protection and the promotion of public health through evidence-based regulation.

16.2.7 New Zealand and Australia

- In Australia, the Therapeutic Goods Administration (TGA) regulates nutraceuticals, classified as complementary medicines, requiring pre-market assessments to assess safety, quality, and efficacy.
- Stringent regulatory requirements uphold product integrity and ensure consumer confidence in nutraceuticals.

16.2.8 Brazil

The National Health Surveillance Agency (ANVISA) in Brazil oversees the regulation of nutraceuticals, known as "functional foods," imposing standards to safeguard public health and promote responsible marketing practices.

16.2.9 Middle East

Regulatory frameworks for nutraceuticals in the Middle East vary across countries, with each nation having its own regulatory authorities and frameworks:

- **United Arab Emirates (UAE).** The Emirates Authority for Standardization and Metrology (ESMA) sets standards for nutraceuticals, ensuring compliance with quality and safety requirements.
- **Saudi Arabia.** The Saudi Food and Drug Authority (SFDA) regulates nutraceuticals, enforcing standards to protect public health and ensure product safety.
- **Qatar.** The Qatar General Organization for Standardization (QS) and the Qatar Food and Nutrition Authority collaborate to establish regulations governing nutraceuticals, prioritizing consumer safety and product quality.
- **Oman.** The Public Authority for Consumer Protection (PACP) oversees consumer protection, ensuring nutraceuticals meet safety and quality standards.
- **Kuwait.** The Public Authority for Food and Nutrition (PAFN) regulates nutraceuticals, emphasizing compliance with safety and quality standards to safeguard consumer health.
- **Bahrain.** The Ministry of Health in Bahrain regulates nutraceuticals, ensuring adherence to safety and quality standards for consumer protection.

The regulatory frameworks for nutraceuticals vary globally, but they share common goals of promoting consumer safety, ensuring product quality, and fostering innovation within the industry. Navigating these regulations requires careful attention to local requirements and collaboration with regulatory authorities to ensure compliance and to maintain consumer trust.

16.3 HEALTH AUTHORITIES/MINISTRIES OF HEALTH: NUTRACEUTICAL MARKET REGULATORS [4]

Companies planning to sell dietary supplements internationally must first understand the regulations and classification systems established by each country's ministry of health (MoH) and health authority (HA). This knowledge is essential for navigating the complex regulatory requirements and ensuring compliance with relevant regulations.

A comprehensive understanding of product classification is crucial, as dietary supplements may be categorized differently across regions. Common classifications include dietary supplements, food supplements, or health supplements, each subject to distinct regulatory frameworks.

Following is a comparative table (Tables 16.1–4) outlining the product classification and respective health authorities across the region-wise groups of countries.

Furthermore, each market in the GCC and APAC areas has distinct procedures for licensing foreign or domestic items within its borders. While some markets follow the regulatory processes of globally governed countries, others create strong domestic regulations adapted to their own requirements.

In the APAC region, most nations need product registration with their respective health authorities before marketing may begin. Compliance-based markets, such as those of Singapore and Hong Kong, provide greater flexibility, allowing items to launch provided they meet local requirements. Thailand, Malaysia, Indonesia, the Philippines, Vietnam, and Taiwan, on the other hand, demand full product registration with their national health authorities. Markets such as Japan, South Korea, and China have severe standards that require extensive paperwork and testing details based on the product's formulation and claims.

Similarly, in Middle Eastern countries, companies must ensure compliance with country-specific regulations and GCC Standardization Organization (GSO) standards before introducing products to the market. Product registration typically involves collaboration with a designated local agent to navigate the regulatory landscape effectively. Similar procedures apply to the African and LATAM markets, where adherence to regional and national regulations is essential for market entry.

Furthermore, each market has its own rules for legalizing foreign/local items within its borders. Certain markets follow the lead of regulated countries while creating robust internal regulations.

TABLE 16.1
Nutraceutical Product Classification and Health Authorities across the Association of Southeast Asian Nations (ASEAN)

Country	Classification of Nutrition Product	Health Authority
Brunei	Health supplements	Ministry of Health
Cambodia	Health supplements	Ministry of Health
Indonesia	Food supplements	National Agency for Drug and Food Control (NADFC), Republic of Indonesia Badan Pengawas Obat Dan Makanan (BPOM), Republik Indonesia
Malaysia	Health supplements	National Pharmaceutical Regulatory Agency (NPRA)
Philippines	Food supplements	Food and Drug Administration (FDA) of the Philippines
Singapore	Health supplements	Health Sciences Authority (HSA)
Thailand	Health supplements	Thailand Food and Drug Administration (TFDA)
Vietnam	Dietary supplements	Ministry of Health (MoH) and Drug Administration of Vietnam (DAV)

TABLE 16.2
Nutraceutical Product Classification and Health Authorities across Middle East Countries

Country	Classification of Nutrition Product	Health Authority
Bahrain	Health supplements	National Health Regulatory Authority (NHRA)
Egypt	Health supplements	Egyptian Drug Authority (EDA)/ Egyptian MoH*
Israel	Health supplements	MoH*
Jordan	Health supplements	MoH*
Kuwait	Health supplements	MoH*
Lebanon	Health supplements	MoH*
Oman	Food supplements	MoH*
Qatar	Health supplements	National Health Authority (NHA)
Saudi Arabia	Food supplements	Saudi Food and Drug Authority
Turkey	Health supplements	MoH*
United Arab Emirates (UAE)	Health supplements	MoH*

**MoH:* Ministry of Health.

TABLE 16.3
Nutraceutical Product Classification and Health Authorities across Africa

Country	Classification of Nutrition Product	Health Authority
South Africa	Complementary medicine	South African Health Products Regulatory Authority (SAHPRA)
Nigeria	Dietary/health supplements	National Agency for Food and Drug Administration and Control (NAFDAC)
SADC	Health supplements	MoH*
Kenya	Health supplements	MoH*

**MoH:* Ministry of Health.

TABLE 16.4
Nutraceutical Product Classification and Health Authorities across Latin American Countries (LATAM)

Country	Classification of Nutrition Product	Health Authority
Brazil	Dietary/health supplements	The National Health Surveillance Agency (Agência Nacional de Vigilância Sanitária) (ANVISA)
Peru	Dietary/health supplements	The General Directorate of Medicines, Supplies, and Drugs (Dirección General de Medicamentos, Insumos y Droga) (DIGEMID)
Colombia	Health supplements	Ministry of Health (MoH)
Mexico	Food supplements	The Federal Committee for Protection from Sanitary Risks (Comisión Federal para la Protección contra Riesgos Sanitarios) (COFEPRIS)
Argentina	Health supplements	Administración Nacional de Medicamentos, Alimentos y Tecnología Médica (National Administration of Drugs, Food, and Medical Technology) (ANMAT)

16.4 NUTRACEUTICAL MARKET: DEMAND GOES GLOBAL

The global demand for dietary supplements has surged notably amid the COVID-19 pandemic, as people worldwide have become increasingly health-conscious. This heightened awareness has led to significant growth in the dietary supplements market, attracting greater investment interest. Concurrently, regulatory organizations have responded by adopting tough standards to ensure the safety and efficacy of these products, demonstrating the industry's commitment to consumer health.

Geographically, the global nutraceutical market is segmented into regions such as the Middle East/North Africa (MENA), Asia-Pacific (APAC), and Africa and Latin America (LATAM), all experiencing remarkable growth in demand for dietary supplements. Here is a closer look at the market data from these regions:

- The Asia-Pacific (APAC) nutraceuticals market is expected to increase rapidly over the next five years, with a compound annual growth rate (CAGR) of 7.5%. According to reports, by 2033, the market might be worth $175,487.1 million, up from $90.482 million in 2023. [5, 6]
- The market for dietary supplements in the Middle East and Africa is also expected to witness a robust growth, with a projected 8.35% CAGR from 2022 to 2027, as per market intelligence. [7]
- Africa's nutraceutical market, projected at USD 11.10 billion in 2024, is expected to grow to USD 17.31 billion by 2029, representing a notable 9.28% CAGR over the forecast period 2024–2029. [8]
- In Latin America (LATAM), the dietary supplements market is on a steady upward trajectory, with a projected CAGR of 7.96% from 2023 to 2028. The market valuation is estimated to rise from USD 13.25 billion in 2022 to USD 19.43 billion by 2028. [9]

For regulating nutraceuticals within the Middle East, the GCC—the Gulf Cooperation Council—plays an important role. The GCC has established primary laws governing various aspects, including standards and regulations related to nutraceuticals, as outlined on its official website.

The global nutraceutical market is experiencing a significant growth fueled by increased consumer health awareness. As the industry continues to expand, stakeholders must prioritize adherence to regulations to ensure product safety and efficacy in meeting consumer needs across diverse regional markets.

16.5 ABOUT THE GULF COOPERATION COUNCIL (GCC) [10]

Establishment and membership. The Gulf Cooperation Council (GCC) was founded on May 25, 1981, with the participation of six Arab countries: Bahrain, Kuwait, Oman, Qatar, Saudi Arabia, and the United Arab Emirates (UAE).

Objectives. The GCC aims to promote economic, political, and security cooperation among its member states. It seeks to achieve unity through coordination, integration, and interconnection across various fields, including trade, investment, defense, and cultural exchange.

Geographical coverage. The GCC covers a total area of 2,672,700 km^2 in the Arabian or Persian Gulf region of the Middle East.

Official language and cultural integration. Arabic serves as the official language of the GCC, fostering deepened cultural ties and mutual understanding among member states.

16.5.1 Key Objectives Outlined in the GCC Charter

- Coordination and integration across all fields to achieve unity
- Strengthening relations and cooperation among the peoples of member states in various domains

- Formulation of similar regulations in administrative, economic, and financial affairs
- Promotion of scientific and technological advancement in industries such as manufacturing, agriculture, and mining
- Establishment of joint ventures and encouragement of private sector cooperation for the benefit of member states' populations

Participation in regional initiatives. The GCC, along with Yemen, is part of the Greater Arab Free Trade Area (GAFTA), promoting economic integration and trade facilitation among Arab nations.

Recent developments and expansion. Morocco and Jordan have submitted applications to join the GCC, reflecting the organization's continued relevance and appeal as a platform for regional cooperation and development.

Economic significance. GCC member states boast some of the world's fastest-growing economies, driven by revenues from oil and natural gas, as well as significant investments in infrastructure and development projects.

Resilience and growth. Despite economic challenges in the past, the GCC economies have demonstrated resilience and are currently experiencing a renewed period of rapid growth and development.

The GCC is a political and economic coalition of six Arab countries in the Arabian Gulf, sometimes known as the Persian Gulf, in the Middle East. The member states of the GCC are the United Arab Emirates, Bahrain, Oman, Kuwait, Saudi Arabia, and Qatar.

The GCC was formed on May 25, 1981, with the purpose of encouraging economic, political, and security cooperation among its member countries. The organization facilitates collaboration in various areas, including trade, investment, defense, and cultural exchange. The GCC having Arabic as an official language has played an important role in the development and integration of Gulf countries, promoting regional stability and collaboration. Currently, it covers an area of 2,672,700 km^2.

The objectives outlined in the GCC Charter serve as guiding principles for the Gulf Cooperation Council (GCC) member states, aimed at fostering cooperation and integration across various domains associated with nutraceuticals, food, and other sectors. Here is an elaboration on these objectives:

- **Coordination, integration, and interconnection.** The primary goal of the GCC is to coordinate, integrate, and establish interconnections between member states across all sectors. This aims to achieve unity among the member states, fostering collaboration and synergy in their endeavors.
- **Strengthening relations and cooperation.** The GCC seeks to deepen and strengthen relations, links, and areas of cooperation among the peoples of the member states. This entails enhancing mutual understanding, cultural exchange, and cooperation in various fields to promote solidarity and unity among member states.
- **Formulating similar regulations.** The GCC aims to formulate similar regulations across various fields, including economic and financial affairs, commerce, customs, communications, education, culture, social and health affairs, information, tourism, and legislative and administrative affairs. This harmonization of regulations facilitates smoother interactions and trade between member states.
- **Stimulating scientific and technological progress.** The GCC endeavors to stimulate scientific and technological progress in key sectors, such as industry, mining, agriculture, water, and animal resources. This includes fostering innovation, research, and development to enhance competitiveness and sustainability in these sectors.

- **Establishing joint ventures and encouraging cooperation.** The GCC promotes the establishment of joint ventures and encourages cooperation by the private sector for the benefit of the peoples of the member states. This aims to leverage resources, expertise, and investments to support economic growth, job creation, and prosperity across the region.

Regarding the Greater Arab Free Trade Area (GAFTA), while Yemen and the GCC nations are part of this trade agreement, the impact on the GCC's framework is expected to be limited. The GCC prioritizes its integration efforts, aiming for deeper collaboration and unity among member states.

Furthermore, recent applications from Morocco and Jordan to join the GCC indicate a growing interest in the council's objectives and regional cooperation. The GCC, which boasts among the world's fastest-growing economies, has seen tremendous growth due to increased oil and natural gas income, as well as vigorous investment and development activities. Despite recent economic downturns, most GCC economies have rebounded and are experiencing rapid growth once again, underscoring the resilience and dynamism of the region.

16.5.2 The Gulf Cooperation Council (GCC) Organizational Structure

The GCC's elaborate organizational structure, which includes the Ministerial Council, Supreme Council, and Secretariat General, demonstrates the commitment to achieving unity, coordination, and collaboration among member countries in pursuit of common goals in the economic, social, and cultural sectors. Each component plays a distinct yet interrelated role in advancing the interests and objectives of the GCC, reflecting the collective efforts toward regional development and integration.

16.5.2.1 Supreme Council

- The GCC's highest authority; consists of the heads of state from all six member countries.
- Convenes once a year in ordinary session, with provisions for emergency sessions upon the request of any two member states.
- Rotates the chairmanship among member states, ensuring equitable representation.
- Responsibilities include setting overall policy directives for the GCC and endorsing recommendations from the Ministerial Council or Secretariat General.
- Resolutions are adopted through majority voting, reflecting consensus among member states.

16.5.2.2 Ministerial Council

- Comprised of the foreign ministers of all six member countries.
- Meets every three months in an ordinary session, with the flexibility for emergency sessions as deemed necessary.
- Tasked with formulating policies and recommendations aimed at fostering cooperation and coordination across economic, social, and cultural domains among member states.
- Serves as a key forum for dialogue and decision-making on matters of regional significance, ensuring alignment with the overarching goals of the GCC.

16.5.2.3 Secretariat General

- Headquartered in Riyadh, Saudi Arabia; serves as the administrative arm of the GCC.
- Responsible for generating reports, conducting studies, and managing budgets on behalf of the organization.
- Develops rules and regulations to ensure Supreme and Ministerial Council decisions are implemented effectively.
- Assists member states in executing GCC initiatives and programs, ensuring effective coordination and cooperation.

- Secretary General Dr. Abdullatif bin Rashid AlZayani appointed for a three-year term by the Supreme Council based on the Ministerial Council's nomination.

16.6 ABOUT THE GCC STANDARDIZATION ORGANIZATION (GSO)

The Gulf Cooperation Council (GCC) Standardization Organization (GSO) is a regional entity dedicated to advancing standardization efforts among its member states. Established in May 2004, the GSO operates under the umbrella of the GCC, which includes six member nations: Saudi Arabia, the United Arab Emirates, Bahrain, Qatar, Oman, and Kuwait. Subsequently, Yemen joined as a member in January 2010.

GSO's major goal is to coordinate and unify standardization operations within the GCC area. GSO works together with member states' national standardized agencies to enable the growth of the manufacturing and service sectors, encourage intra-regional trade, protect consumers, and strengthen environmental and public health standards.

A key objective of GSO is to bolster the Gulf economy by fostering the growth of industries and enhancing the competitiveness of Gulf-based products and services. By mitigating technical barriers to trade, GSO endeavors to stimulate economic integration and cooperation among member states.

GSO's initiatives are in line with the broader objectives of the GCC, including the establishment of a customs union and a common market. Through its endeavors, GSO contributes to strengthening economic bonds within the region and fostering the collective prosperity of its member states.

For further insights into GSO and its undertakings, interested parties are encouraged to explore the official website of the Gulf Cooperation Council.

Vision of GSO: To be the trusted regional center of excellence in standardization activities and the first enabler for the Gulf common market.

Mission of GSO: To achieve integration and interdependence between the member states of the Gulf Cooperation Council in the areas of organizing and unifying standardization activities and following up on their implementation to ensure the safety of products, efficiency of services, and ease of transfer in the Gulf common market, in addition to consumer protection, to improve the quality of life in GCC member states and strengthen their economies to compete in global markets.

The Gulf Cooperation Council Standardization Organization (GSO) upholds several core values that guide its operations and interactions with stakeholders. Here is an elaboration on the values of GSO:

Meeting beneficiaries' expectations. GSO places a paramount focus on meeting the needs and surpassing the expectations of its member states. By understanding the diverse requirements and priorities of member states, GSO endeavors to tailor its strategies and initiatives to effectively address their needs, thereby fostering trust and satisfaction among its beneficiaries.

Professionalism and excellence. GSO is committed to maintaining the highest standards of professionalism and excellence in all aspects of its operations. This entails a dedication to continuous improvement, adherence to best practices, and a relentless pursuit of excellence in delivering services and support to its members. By upholding professionalism, GSO strives to instill confidence and credibility in its endeavors.

Transparency. Transparency is a cornerstone of GSO's operations, characterized by effective and clear communication about all aspects of its business. GSO believes in open and honest communication with its stakeholders, including member states, partners, and the public. By fostering transparency, GSO aims to build trust, enhance accountability, and promote informed decision-making.

Integration with partners. GSO recognizes the importance of collaboration and integration with its partners to achieve common goals and sustain relations. By actively engaging with stakeholders, including national standardization bodies, international organizations, and industry stakeholders, GSO seeks to leverage collective expertise, resources, and synergies to drive positive outcomes and advance standardization efforts in the region. Through collaboration and integration, GSO endeavors to foster stronger relationships and promote mutual understanding and cooperation among all stakeholders involved.

16.6.1 Member Countries of the GCC Standardization Organization (GSO)

16.6.1.1 The United Arab Emirates (UAE)

The United Arab Emirates (UAE) saw the establishment of the Ministry of Industry and Advanced Technology (MoIAT) in July 2020, marking a significant milestone in the country's industrial development. MoIAT's primary objective is to revitalize the UAE's industrial sector and bolster its contribution to the national GDP. To achieve this goal, MoIAT focuses on several key initiatives and strategies:

- **Development of laws and policies.** MoIAT is tasked with developing comprehensive laws and policies aimed at creating a conducive environment for industrial growth. These regulations are designed to establish a world-class industrial development framework that attracts foreign direct investment (FDI) and top-tier talent.
- **Enhancement of in-country value (ICV).** MoIAT places a strong emphasis on enhancing in-country value (ICV) by encouraging local manufacturing and production. By promoting the production of goods within the UAE, MoIAT aims to diversify the economy, reduce dependency on imports, and create new opportunities for local businesses.
- **Promotion of entrepreneurship.** MoIAT actively promotes entrepreneurship and innovation within the UAE. By providing support and resources to aspiring entrepreneurs, MoIAT aims to cultivate a vibrant startup ecosystem that drives economic growth and fosters innovation across various industries.
- **Job creation and talent development.** MoIAT prioritizes job creation and talent development as key components of its industrial development strategy. Through various training programs and initiatives, MoIAT seeks to upskill local talent and equip them with the necessary skills to thrive in the rapidly evolving industrial landscape.
- **Boosting exports of "made in the Emirates" products.** MoIAT is committed to boosting the export of locally manufactured goods under the banner of "made in the Emirates." By promoting the export of high-quality UAE products, MoIAT aims to enhance the country's reputation as a global hub for manufacturing excellence.
- **Acceleration of industrial development.** MoIAT accelerates industrial development by spearheading initiatives, such as the establishment of industrial zones and the development of advanced technology capabilities. These efforts are aimed at creating a diversified, knowledge-based economy that is resilient and self-sufficient.

MoIAT has a major role in driving the UAE's industrial transformation, fostering innovation, and positioning the country as a global leader in industrial excellence and sustainability.

16.6.1.2 Kingdom of Bahrain

In the Kingdom of Bahrain, the Standards and Metrology Directorate (BSMD) within the Ministry of Industry and Commerce serves as the national standardization body. Established in 1988 under

Legislative Decree No. 16 of 1985, BSMD plays a pivotal role in setting and maintaining quality standards across various industries in Bahrain.

BSMD is actively involved in ensuring compliance with national and international standards to enhance the quality and safety of products and services within the kingdom. As a member of several regional and international organizations, including the Arab Industrial Development and Mining Organization (AIDMO) and the Arab Center for Standardization and Metrology, BSMD collaborates closely with counterparts in the region to harmonize standards and promote trade facilitation.

Additionally, BSMD is a founding member of the Gulf Standardization Organization (GSO), which aims to unify standardization efforts among Gulf Cooperation Council (GCC) member states. Through its participation in the GSO, BSMD contributes to the development of common standards and regulations that promote intra-GCC trade and enhance consumer protection.

BSMD's responsibilities encompass a wide range of activities, including:

- Developing and implementing national standards and regulations to ensure product quality and safety
- Conducting testing, certification, and inspection activities to verify compliance with established standards
- Providing technical assistance and training programs to industry stakeholders to enhance their understanding of standardization requirements
- Representing Bahrain in regional and international standardization forums to contribute to the development of global best practices

BSMD aims majorly to promote economic development, consumer protection, and international competitiveness by establishing and enforcing quality standards in the Kingdom of Bahrain.

16.6.1.3 Kingdom of Saudi Arabia

In the Kingdom of Saudi Arabia, the Saudi Standards, Metrology, and Quality Organization (SASO) plays an important role in ensuring standardization, metrology, as well as quality across various sectors. Established by Royal Decree No. M/10 dated March 3, 1992H (April 17, 1972G), SASO operates as an independent entity with its own budget and legal personality.

SASO is administered by a board of directors, chaired by the Minister of Commerce and Industry, which includes representatives from various ministries and corporate sectors involved in standardization in the Kingdom. This board is responsible for formulating the general policies and strategies of SASO, ensuring alignment with the national priorities and objectives.

The organization's mandate encompasses several key functions:

- **Standardization.** SASO develops and adopts national standards across various sectors to ensure product quality, safety, and reliability. These standards serve as benchmarks for manufacturers, importers, and consumers, facilitating trade and protecting public health and safety.
- **Metrology.** SASO regulates metrological activities in the Kingdom, including measurement standards, calibration, and testing. By maintaining accurate and reliable measurement standards, SASO promotes fairness in trade and supports industrial and scientific endeavors.
- **Quality assurance.** SASO implements quality assurance programs and initiatives to enhance the quality and competitiveness of Saudi products and services. Through certification, inspection, and accreditation services, SASO helps businesses meet international quality requirements and gain access to global markets.
- **Consumer protection.** SASO works to safeguard consumer interests by enforcing product safety and quality regulations. Through market surveillance and enforcement activities,

SASO ensures that products comply with established standards and are safe for consumption or use by consumers.

- **International collaboration.** SASO actively participates in regional and international standardization bodies and organizations to harmonize standards, facilitate trade, and promote technical cooperation. By collaborating with global partners, SASO contributes to the development of international standards and strengthens Saudi Arabia's position in the global marketplace.

SASO focuses on various aspects, like in advancing standardization, metrology, and quality assurance efforts in the Kingdom of Saudi Arabia, contributing to economic development, consumer protection, and international competitiveness.

16.6.1.4 Sultanate of Oman

In the Sultanate of Oman, the Directorate General for Standards and Metrology (DGSM) is the national standards body under the Ministry of Commerce and Industry, having been founded by Royal Decrees No. 39/76 and No. 1/78, respectively. DGSM is responsible for standardization and metrology activities within the country.

DGSM develops and implements national standards across various industries to ensure product quality, safety, and reliability, also supervising the National Enquiry Point and Information Centre (NEPIC), providing stakeholders with information on standards and regulations.

Key responsibilities of DGSM include developing standards, maintaining measurement standards, conducting conformity assessment activities, providing capacity-building programs, and collaborating with international standardization bodies.

16.6.1.5 Qatar State

In 2002, the Qatar General Organization for Standardization (QS) was established in Qatar to serve as the national standards body in Qatar, which is currently affiliated with the Ministry of Environment. QS is responsible for developing and publishing Qatari product and service standards, which are voluntary as well as mandatory technical regulations. These standards and regulations are developed in collaboration with relevant stakeholders from industry, academia, and regulatory authorities.

QS majorly has its significant focus on ensuring the high quality as well as safety of commodities, products, and materials, aligning with emerging trends and developments. By establishing and enforcing standards, QS aims to promote consumer protection, facilitate trade, and support the overall development of Qatar's economy.

16.6.1.6 Kuwait State

In Kuwait, the establishment of the Standards and Industrial Services Affairs (KOWSMD) dates back to 1967, when it was initially formed as a unit within the Ministry of Commerce and Industry. Over time, its significance grew, leading to its formal recognition as a department following the enactment of Law No. 128 for standardization in 1977.

In a significant development, in 1996, the department underwent a structural transformation, transitioning to the jurisdiction of the Public Authority for Industry. This transition marked a pivotal moment, elevating the department into a sector encompassing standards and industrial services, thereby expanding its scope and influence within Kuwait's industrial landscape.

Comprising two distinct departments, KOWSMD's organizational structure reflects its multifaceted responsibilities. One department is dedicated to standards and metrology, focusing on the development and enforcement of quality standards across various sectors. The other department specializes in quality laboratories, ensuring the efficacy and accuracy of testing procedures to maintain high-quality standards.

By overseeing standards and industrial services, KOWSMD is facilitating in regulating and enhancing Kuwait's industrial sector. Through its efforts in promoting adherence to quality and safety measures, KOWSMD contributes to fostering a conducive environment for economic growth and development, thus bolstering Kuwait's industrial competitiveness on both regional and global scales.

16.6.1.7 Republic of Yemen

The Yemen Standardization, Metrology, and Quality Control Organization (YSMO) is the major body in Yemen in charge of standards and metrology, quality control, and the regulation of jewelry and precious metals. Established by Law No. (44) of 1999 and Republican Decree No. (52) of 2000, YSMO operates with a legal mandate aimed at safeguarding the health and safety of consumers, as well as the environment, while also supporting the national economy.

As Yemen's sole regulatory body in these domains, YSMO plays a pivotal role in upholding and enforcing quality standards across various sectors. By setting and maintaining stringent standards, YSMO ensures that products meet established criteria for safety, reliability, and performance. Moreover, the organization oversees metrological measurements to guarantee accuracy and precision in industrial processes and trade.

In addition to its regulatory functions, YSMO is committed to fostering economic development by promoting compliance with international standards and facilitating trade. By establishing a framework that prioritizes quality and safety, YSMO contributes to enhancing consumer confidence, protecting public health, and advancing the competitiveness of Yemen's economy on both domestic and international fronts.

16.6.2 Organizational Structure of GSO

16.6.2.1 GSO President

The GCC Standardization Organization (GSO) operates under the leadership of a president, who is responsible for overseeing the organization's functions and strategic direction. The president is appointed for a three-year term that can be renewed for another term, ensuring continuity and stability in the organization's leadership

Currently, His Excellency Mr. Saud Al-Khusaibi holds the esteemed position of GSO president. Assuming office in April 2018, Mr. Al-Khusaibi brings a wealth of experience and expertise to his role. With a distinguished career in standards and metrology, he has demonstrated a strong commitment to advancing standardization efforts in the Gulf region.

As president, Mr. Al-Khusaibi plays a multifaceted role in steering the GSO toward achieving its objectives. He leads initiatives aimed at harmonizing standards across member states, facilitating trade, and promoting consumer safety. His leadership ensures that the GSO remains at the forefront of standardization efforts, adapting to evolving challenges and opportunities in the global marketplace.

Under Mr. Al-Khusaibi's guidance, the GSO has intensified its efforts to address emerging issues in various sectors, including industry, agriculture, and health care. By fostering collaboration among member states and international partners, he promotes the adoption of best practices and innovative approaches to standardization.

Mr. Al-Khusaibi's tenure as GSO president has been marked by notable achievements and milestones. Through strategic partnerships and stakeholder engagement, he has enhanced the organization's effectiveness in promoting quality assurance and regulatory compliance. His leadership underscores the GSO's commitment to promoting economic growth, enhancing competitiveness, and safeguarding public health and the environment across the GCC region.

As the GSO president, Mr. Al-Khusaibi continues to advocate for the harmonization of standards and the adoption of international best practices, ensuring that the Gulf region remains a leader in standardization and quality assurance. His visionary leadership and dedication to excellence position the GSO as a trusted regional authority in standardization and metrology.

16.6.3 International Cooperation

The GCC Standardization Organization (GSO) is dedicated to fostering strong and enduring partnerships with stakeholders worldwide, aimed at sharing expertise, transferring international knowledge, and promoting best practices in standardization and related activities. To achieve these goals, the GSO has signed several memorandums of understanding (MoUs) and technical cooperation agreements with standards agencies at the national, regional, and worldwide levels.

These memorandums of understanding and agreements serve as critical foundations for collaboration, information exchange, and alignment and harmonization activities. These agreements improve coordination and collaboration in standardization activities by integrating the GSO's and its member states' work programs to those of relevant partners.

Furthermore, the partnerships formed through MoUs and agreements allow the GSO to use technical cooperation as a supplement to foster alignment with international standards-setting organizations, such as the International Organization for Standardization (ISO), the International Electrotechnical Commission (IEC), and the International Telecommunication Union (ITU).

Through technical cooperation initiatives, the GSO and its partners work together to address common challenges, share resources, and build capacity in standardization activities.

International cooperation forms a cornerstone of the GSO's efforts to advance standardization practices and contribute to global harmonization efforts. By collaborating with a diverse network of stakeholders, the GSO aims to enhance the quality, safety, and interoperability of products and services across the Gulf region and beyond, ultimately promoting sustainable development and facilitating trade in the global marketplace.

16.6.4 Standards Development

16.6.4.1 Standards and Technical Regulations Development Objectives

- **Enhancing competitiveness.** The primary aim is to enhance the competitiveness of products and services by aligning them with the latest production methods and technology across various industries.
- **Removing trade barriers.** Another significant goal is to remove technical trade barriers, resulting in increased trade between GSO member states and foreign business partners.
- **Consumer protection.** Standards are created to safeguard customers by defining acceptable and internationally accepted norms of quality for goods, products, and processes, so ensuring their safety and dependability.
- **Harmonization and common language.** Standardization efforts aim to harmonize concepts and technical applications among member states, fostering a common language among Gulf manufacturers and facilitating smoother business operations.
- **Process of developing standards and technical regulations.** The development process involves the adoption of international standards, such as those from the ISO (International Organization for Standardization), IEC (International Electrotechnical Commission), OIML (International Organization of Legal Metrology), and CODEX (Codex Alimentarius Commission), as GSO standards to cater to the diverse needs of industrial, commercial, and service sectors within the member states. This process ensures that GSO standards are aligned with global best practices and tailored to meet the specific requirements of the Gulf region.

16.7 MARKET OVERVIEW FOR GCC

Nutraceuticals, derived from food sources with both nutritional and medicinal properties, play a crucial role in enhancing overall health and well-being and are gaining significant traction across Gulf Cooperation Council (GCC) countries. These products offer physiological benefits and are instrumental in preventing chronic diseases, supporting mental health, and promoting longevity. As of 2022, the global nutraceutical industry has been on the rise, driven by heightened consumer awareness of health and wellness, demographic shifts, and lifestyle changes, among other factors.

In the Gulf region, including countries like Saudi Arabia, the United Arab Emirates (UAE), Qatar, Kuwait, Bahrain, and Oman, the nutraceutical market has garnered interest as part of an effort to foster healthier lifestyles and combat lifestyle-related diseases. Several factors contribute to the growth of the nutraceutical industry in these countries:

- **Increasing health awareness.** Consumers in the GCC region are becoming increasingly health-conscious, actively seeking products that promote overall well-being and address specific health concerns.
- **Government initiatives.** Some Gulf countries have initiated health and wellness programs, with governments actively supporting the development and promotion of nutraceutical products to improve public health outcomes.
- **Rising disposable income.** Economic prosperity in the region has led to higher disposable incomes among consumers, enabling greater spending on health and wellness products, including nutraceuticals.
- **Urbanization and lifestyle changes.** Urbanization and evolving lifestyles have prompted a greater emphasis on preventive health care, driving up the demand for nutraceuticals as individuals seek to maintain optimal health and vitality.
- **Regulatory support.** Regulatory bodies in select Gulf countries are implementing or exploring regulations to ensure the safety, efficacy, and quality of nutraceutical products, instilling consumer confidence and facilitating market growth.

TABLE 16.5
List of GSO Guidelines Related to Food and Food Supplements

Guideline No.	Title	Ref
GSO CAC/GL 55:2015	Guidelines for Vitamins and Minerals Food Supplements	[15]
GSO 839	Food Packages, Part 1: General Requirements	[16]
GSO 9:2022	Labeling of Prepackaged Foodstuffs	[17]
GSO 2333	Health and Nutrition Claims	[18]
GSO 382	Maximum Limits of Pesticide Residues in Agricultural and Food Products, Part 1	[19]
GSO 2500	Additives Permitted for Use in Food Stuff	[20]
GSO 2233	Requirements of Nutritional Labeling	[21]
GSO 193	Contaminants and Toxins in Food and Feed	[22]
GSO 2055–1	Halal Food, Part 1: General Requirements	[23]
GSO 1863	Food Packages, Part 2: Plastic Package—General Requirements	[24]
GSO 168	Requirements of Storage Facilities for Dry and Canned Foodstuff	[25]
GSO 2210	Ginseng Products	[26]
GSO 2097	Royal Jelly	[27]
GSO 147	Honey	[28]

The nutraceutical industry in GCC countries is poised for continued expansion, fueled by a combination of growing consumer awareness, government support, economic prosperity, changing lifestyles, and regulatory advancements. This presents significant opportunities for industry players to capitalize on the burgeoning demand for health-enhancing products in the region.

16.7.1 Market Insights [11–13]

The GCC nutraceuticals market is expected to develop at a CAGR of approximately 9% during the forecast period 2022–2027. The market is rising as people become more conscious of the importance of maintaining a healthy lifestyle, resulting in a higher consumer demand for functional foods and supplements, such as nutraceuticals. In addition, an active lifestyle and modern-day issues like obesity and diabetes due to improved lifestyle-related eating habits are also demonstrating an upsurge in the demand for nutraceuticals across the GCC.

In the same view, increasing the utilization of these elements in the diet to address and treat cardiovascular disorders and malnutrition is another prominent aspect stimulating the industry. As a result, it creates chances for major businesses to expand their production and distribution capacities in order to meet the growing demand across the GCC.

Besides, skyrocketing health-care costs and the mounting geriatric population shall also demonstrate a higher inclination toward functional food and supplements like nutraceuticals to fulfill the unmet nutrient needs in the body. Such changes in the consumer's attitude toward these foods to attain health and wellness benefits, like improved bowel movements, cleansing of internal organs, higher retention and passage of nutrients into cells, enhanced blood clotting, and other functions, are positively influencing the overall market growth across GCC.

Moreover, as people are increasingly consuming herbal medicines to mitigate the use of conventional drugs, the demand for nutraceuticals is set to rise and boost market expansion and also generate lucrative opportunities for the players through 2027.

Market size and growth. The nutraceutical market is witnessing significant expansion, with a compound annual growth rate (CAGR) projected to be around 7.1% over the forecast period 2023–2030 in major GCC countries, like Saudi Arabia, Kuwait, UAE, and Egypt. The market size is expected to reach USD 13,973.57 million in the year 2022, up from USD 15.5 billion in 2033.

Product segmentation. The nutraceutical market in GCC countries encompasses a wide range of products, including dietary supplements, functional foods and beverages, herbal and botanical extracts, vitamins and minerals, and more. Each segment caters to specific consumer preferences and health needs.

Key players. The market is characterized by the presence of both domestic and international players, including leading pharmaceutical companies, like Pfizer Inc, Bayer AG, Nestlé SA, Danone, and BASF SA; food and beverage manufacturers, like Vitabiotics Ltd., PepsiCo Inc., and General Mills; and specialized nutraceutical firms. Key players are focusing on product innovation, strategic partnerships, and expansion initiatives to gain a competitive edge in the market.

Consumer preferences. Consumers in GCC countries are increasingly prioritizing health and wellness, driving demand for nutraceutical products that offer functional benefits, such as immune support, digestive health, weight management, and cognitive enhancement. There is a growing preference for natural and organic ingredients, as well as products free from artificial additives and preservatives.

Distribution channels. Nutraceutical products are distributed through various channels in the GCC region, including pharmacies, specialty health stores, supermarkets and hypermarkets, online retailers, and direct sales channels. The accessibility of these products across different channels enhances consumer convenience and market penetration.

Regulatory environment. Regulatory authorities in GCC countries are actively working to establish and enforce regulations governing the manufacturing, labeling, and marketing of nutraceutical products. Compliance with stringent quality standards and regulatory requirements is essential for market entry and consumer trust.

Market opportunities. The GCC nutraceutical market offers lucrative opportunities for industry participants, driven by favorable demographic trends, rising health consciousness, increasing disposable incomes, and supportive government initiatives aimed at promoting public health and wellness. Key growth areas include personalized nutrition, functional beverages, fortified foods, and nutraceutical ingredients.

In summary, the GCC nutraceutical industry is poised for continued growth, driven by evolving consumer preferences, regulatory developments, and strategic investments by market players. With a growing emphasis on preventive health care and holistic well-being, the demand for innovative and high-quality nutraceutical products is expected to surge across GCC countries in the coming years.

16.8 GCC AND GSO REGULATORY REQUIREMENTS FOR FOOD SUPPLEMENTS REGISTRATION (14)

Food supplements registration in the Gulf Cooperation Council (GCC) region, overseen by both GCC and the Gulf Standardization Organization (GSO), involves adherence to stringent regulatory requirements to ensure product safety and compliance with established standards. Here is an overview of the regulatory framework governing food supplements registration:

- **Legislative authority.** GCC member states, including Kuwait, Qatar, Bahrain, Oman, the Kingdom of Saudi Arabia (KSA), and the United Arab Emirates (UAE), have the authority to develop and implement national food legislation. Additionally, regulations established by the GSO contribute to harmonizing standards across member countries.
- **GSO technical regulations.** In August 2016, the GSO introduced new technical regulations concerning food safety, covering a wide range of aspects, such as food products, packaging materials, and animal feed. These regulations are formulated to align with international standards, including Codex Alimentarius, ISO 22000, EFSA, and US FDA guidelines, while also considering local religious, cultural, and climatic factors.
- **Adoption of standards.** GSO standards are often based on international regulations but are modified to suit the specific requirements of GCC countries. Existing standards from member states are reviewed, updated, and adopted by the GSO after ensuring alignment with Codex Alimentarius and ISO standards.
- **Health control measures.** All imported food supplements undergo rigorous health inspections at entry point to ensure compliance with GCC technical food safety regulations. These regulations cover various aspects, such as labeling requirements for prepackaged foodstuffs, expiration dates, nutrition labeling, and more.
- **Compliance assurance.** Manufacturers and distributors of food supplements must ensure full compliance with GCC and GSO regulations throughout the registration process. This includes providing detailed documentation, conducting safety assessments, and adhering to labeling and packaging requirements.
- **Quality assurance.** Quality control measures are integral to the registration process, with authorities conducting inspections and audits to verify the safety and efficacy of food supplements. Non-compliant products may be subject to penalties or removal from the market to safeguard public health.
- **Continuous monitoring.** Regulatory authorities in GCC countries maintain vigilant oversight of the food supplements market, conducting regular inspections and surveillance activities to detect and address any instances of non-compliance or safety concerns.

In summary, the regulatory framework for food supplements registration in GCC and GSO countries prioritizes consumer safety and product quality. By adhering to established standards and undergoing thorough evaluation processes, manufacturers can ensure the lawful distribution of safe and effective food supplements within the region.

16.9 GSO FOOD SUPPLEMENT GUIDELINE (GSO 55:2020) [29]

16.9.1 Scope of the Guideline

The GSO Food Supplement Guideline applies to food supplement products intended to enrich specific nutritional elements in the regular diet. These supplements encompass a wide range of ingredients, including but not limited to vitamins, minerals, fatty acids, amino acids, enzymes, prebiotics and probiotics, collagen, dietary fibers, melatonin, propolis, pollen, herbs, and food herbal extracts. The guideline sets standards for the formulation, labeling, and safety of these products within the Gulf region.

16.9.1.1 Definition

Food supplements are products designed to complement the normal diet by providing additional nutrients or physiological benefits. These supplements are typically concentrated and consumed in specific doses. They may contain various components, such as vitamins, minerals, fatty acids, amino acids, enzymes, prebiotics and probiotics, collagen, dietary fibers, melatonin, propolis, pollen, herbs, or food herbal extracts.

16.9.2 General Requirements

Compliance with halal standards. All components of food supplements must adhere to the requirements outlined in the technical regulations of GSO 2055–1, "Halal Food—Part 1: General Requirements." This ensures that the production and sourcing of ingredients are in accordance with halal principles.

Permitted ingredients. The guideline specifies a list of permitted vitamins and minerals (Table 16.6). These ingredients must be included in the formulations as indicated in Tables 16.7a, 16.7b, and 16.7c. This ensures that only approved substances are used in food supplements, enhancing consumer safety and regulatory compliance.

Additionally, the guideline may cover aspects such as quality control measures, packaging requirements, and labeling standards to provide comprehensive guidance for the manufacturing, distribution, and sale of food supplement products in the Gulf region.

The quantity of vitamins and minerals in the food supplements are within the limits mentioned in Table 16.7.

- The levels of vitamins and minerals in food supplements must adhere to the specified limits outlined in Tables 16.7a, 16.7b, 16.7c.
- Any health claims featured on product labels must meet the requirements outlined in GSO 2333, "Health and Nutrition Claims."
- Additives utilized in these supplements must align with the regulations detailed in GSO 2500, "Additives Permitted for Use in Foodstuffs."
- Food supplements should not contain any ingredients intended for non-food or therapeutic purposes.
- Pesticide residues within the product must not surpass the maximum limits set forth in GSO 382, "Maximum Limits of Pesticide Residues in Agricultural and Food Products—Part 1."

TABLE 16.6
Vitamins and Mineral Forms That Are Used in Dietary Supplements

A. Vitamins

1. Vitamin A
   a. Retinol
   b. Retinyl acetate
   c. Retinyl palmitate
   d. Beta-carotene
2. Vitamin E
   a. D-alpha-tocopherol
   b. DL-alpha-tocopherol
   c. D-alpha-tocopheryl acetate
   d. DL-alpha-tocopheryl acetate
   e. D-alpha-tocopheryl acid succinate
3. Vitamin D
   a. Cholecalciferol
   b. Ergocalciferol
4. Vitamin K
   a. Phylloquinone (phytomenadione)
5. Vitamin B1
   a. Thiamin hydrochloride
   b. Thiamin mononitrate
6. Vitamin B2
   a. Riboflavin
   b. Riboflavin 5'-phosphate, sodium
7. Niacin
   a. Nicotinic acid
   b. Nicotinamide
8. Pantothenic acid
   a. D-pantothenate, calcium
   b. D-pantothenate, sodium
   c. Dexpanthenol
9. Vitamin B6
   a. Pyridoxine hydrochloride
   b. Pyridoxine 5'-phosphate
10. Folic acid
    a. Pteroylmonoglutamic acid
11. Vitamin B12
    a. Cyanocobalamin
    b. Hydroxocobalamin
12. Biotin
    a. D-biotin
13. Vitamin C
    a. L-ascorbic acid
    b. Sodium L-ascorbate
    c. Calcium L-ascorbate
    d. Potassium L-ascorbate
    e. L-ascorbyl 6-palmitate

B. Minerals

Calcium carbonate
Calcium chloride
Calcium salts of citric acid
Calcium gluconate
Calcium glycerophosphate
Calcium lactate
Calcium salts of orthophosphoric acid
Calcium hydroxide
Calcium oxide
Magnesium acetate
Magnesium carbonate
Magnesium chloride
Magnesium salts of citric acid
Magnesium gluconate
Ammonium molybdate (molybdenum (vi))
Sodium molybdate (molybdenum (vi))
Ferric sodium diphosphate
Ferrous lactate
Ferrous sulphate
Ferric diphosphate (ferric pyrophosphate)
Ferric saccharate
Elemental iron (carbonyl + electrolytic + hydrogen reduced)
Cupric carbonate
Cupric citrate
Cupric gluconate
Magnesium lactate
Magnesium hydroxide
Magnesium oxide
Magnesium sulphate
Ferrous carbonate
Ferrous citrate
Ferric ammonium citrate
Ferrous gluconate
Ferrous fumarate
Zinc oxide
Zinc carbonate
Zinc sulphate
Manganese carbonate
Manganese chloride
Manganese citrate
Manganese gluconate
Manganese glycerophosphate
Manganese sulphate
Sodium bicarbonate
Sodium carbonate
Sodium chloride
Sodium citrate
Sodium gluconate
Sodium lactate
Sodium hydroxide

TABLE 16.6 *(Continued)*
Vitamins and Mineral Forms That Are Used in Dietary Supplements

Cupric sulphate	Sodium salts of orthophosphoric acid
Copper lysine complex	Potassium bicarbonate
Sodium iodide	Potassium carbonate
Sodium iodate	Potassium chloride
Potassium iodide	Potassium citrate
Potassium iodate	Potassium gluconate
Zinc acetate	Potassium glycerophosphate
Zinc chloride	Potassium lactate
Zinc citrate	Potassium hydroxide
Zinc gluconate	Potassium salts of orthophosphoric acid
Zinc lactate	Sodium selenate
Potassium fluoride	Sodium hydrogen selenite
Sodium fluoride	Sodium selenite
Magnesium glycerophosphate	Chromium (iii) chloride
Magnesium salts of orthophosphoric acid	Chromium (iii) sulphate

TABLE 16.7A
Limits of Minerals in Dietary Supplements

		Age Categories					
		Adults	Adolescents		Children		Infants
		19 Years and Older	14–18 Years	9–13 Years	4–8 Years	1–3 Years	0–12 Months
Calcium	Min	65	65	65	65	65	—
(mg/day)	Max	1,500	1,500	1,500	1,500	1,500	—
Cobalt	Min	0.006	0.006	0.004	0.004	0.004	—
(µg/day)	Max	44	44	44	44	44	—
Chromium	Min	2.2	—	—	—	—	—
(µg/day)	Max	500	—	—	—	—	—
Copper	Min	65	65	35	35	35	—
(µg/day)	Max	8,000	6,500	4,000	2,500	700	—
Iodine	Min	14	14	6	6	6	—
(µg/day)	Max	800	800	400	200	133	—
Iron	Min	1.4	1.4	0.6	0.6	0.6	0.6
(mg/day)	Max	45	45	40	40	40	40
Magnesium	Min	20	20	12	12	12	—
(mg/day)	Max	500	350	350	110	65	—
Manganese	Min	0.13	—	—	—	—	—
(mg/day)	Max	9	—	—	—	—	—
Molybdenum	Min	2.5	—	—	—	—	—
(µg/day)	Max	2,000	—	—	—	—	—
Selenium	Min	3.5	—	—	—	—	—
(µg/day)	Max	200	—	—	—	—	—
Silicon	Min	>0	—	—	—	—	—
(mg/day)	Max	84	—	—	—	—	—
Phosphorus	Min	62	62	62	62	62	—
(mg/day)	Max	2,000	2,000	2,000	2,000	2,000	—

(Continued)

TABLE 16.7A *(Continued)*
Limits of Minerals in Dietary Supplements

		Age Categories					
		Adults	Adolescents		Children		Infants
		19 Years and Older	14–18 Years	9–13 Years	4–8 Years	1–3 Years	0–12 Months
Zinc (from non—source zinc picolinate) (mg/day)	Min	0.7	0.7	0.4	0.4	0.4	0.2
	Max	50	34	23	12	7	2
Zinc (from source zinc picolinate) (mg/day)	Min	0.7	—	—	—	—	—
	Max	25	—	—	—	—	—
Potassium (mg/day)	Min	>0	—	—	—	—	—
	Max	200	—	—	—	—	—
Sodium (mg/day)	Min	—	—	—	—	—	—
	Max	—	—	—	—	—	—
Fluoride (mg/day)	Min	—	—	—	—	—	—
	Max	—	—	—	—	—	—

TABLE 16.7B
Limits of Vitamins Other Than Vitamin A in Dietary Supplements

		Age Categories					
		Adults	Adolescents		Children		Infants
		19 Years and Older	14–18 Years	9–13 Years	4–8 Years	1–3 Years	0–12 Months
Biotin (μg/day)	Min	1.8	1.8	1	1	1	—
	Max	500	500	500	500	500	—
Folic acid (μg/day)	Min	30	30	15	15	15	—
	Max	1,000	800	600	400	300	—
Niacin/ niacinamide (mg/day)	Min	1.0	1.0	0.6	0.6	0.6	—
	Max	500	30	20	15	10	—
Pantothenic acid (mg/day)	Min	0.4	0.4	0.2	0.2	0.2	—
	Max	500	500	500	500	500	—
Riboflavin B2 (mg/day)	Min	0.08	0.08	0.04	0.04	0.04	—
	Max	100	100	100	100	100	—
Thiamin B1 (mg/day)	Min	0.07	0.07	0.04	0.04	0.04	—
	Max	100	100	100	100	100	—
Vitamin B6 (mg/day)	Min	0.10	0.10	0.05	0.05	0.05	—
	Max	100	80	60	40	30	—
Vitamin B12 (μg/day)	Min	0.14	0.14	0.09	0.09	0.09	—
	Max	1,000	1,000	1,000	1,000	1,000	—

TABLE 16.7B ***(Continued)***
Limits of Vitamins Other Than Vitamin A in Dietary Supplements

		Age Categories					
		Adults	Adolescents		Children		Infants
		19 Years and Older	14–18 Years	9–13 Years	4–8 Years	1–3 Years	0–12 Months
Vitamin C (mg/day)	Min	6.0	6.0	2.2	2.2	2.2	—
	Max	2,000	1,800	1,200	650	400	—
Vitamin D (µg/day)	Min	1.0	1.0	0.8	0.8	0.8	0.5
	Max	25	25	25	25	25	25
Vitamin E (mg AT/day)	Min	1.0	1.0	0.6	0.6	0.6	—
	Max (DL-alpha-tocopherol)	500	400	300	150	100	—
	Max (D-alpha-tocopherol)	1,000	800	600	300	200	
Vitamin K1, vitamin K2, and vitamin K1 + K2 (mg/day)	Min	6	6	3	3	3	—
	Max	120	75	60	55	30	—

TABLE 16.7C
Limits of Vitamin A in Dietary Supplements

Vitamin A (µg RAE/Day)

All-Trans Retinyl Palmitate (Max)	All-Trans Retinal Acetate (Max)	All-Trans Retinol (Max)	(Min)		
600	600	600	30	0–12 months	Infants
600	600	600	30	1–3 years	Children
900	900	900	30	4–8 years	
1,700	1,700	1,700	30	9–13 years	Adolescence
2,800	2,800	2,800	65	14–18 years	
3,022	3,000	3,003	65	19 years and older	Adult

- Contaminants and toxins present in the product must not exceed the maximum limits specified in GSO 193, "Contaminants and Toxins in Food and Feed."
- The product must be devoid of any pathogenic microbes and mycotoxins.
- Competent authorities retain the right to request additional documentation or requirements to validate the safety of the product or its ingredients.
- Products must comply with the variance limits detailed in Table 16.9.
- Ingredients utilized in food supplements must have scientific backing to demonstrate their safety.

- If the product contains elevated levels of zinc, it must be fortified with an appropriate amount of copper.
- Food supplements tailored for pregnant women must contain a minimum of 400 μg of folic acid per day.
- The combined maximum dosage of cobalt and vitamin B12 should not exceed 1,000 μg of vitamin B12 per day.
- If the product contains vitamin E, the combined total of DL-alpha-tocopherol (synthetic form) and D-alpha-tocopherol (natural form) must not exceed 1,000 mg of alpha-tocopherol from all sources, with a maximum of 1,500 IU per day of D-alpha-tocopherol and 1,100 IU per day of DL-alpha-tocopherol.

Ingredients

The ingredients which can be added to food supplements are as follows:

- Vitamins
- Minerals
- Fatty acids
- Amino acids
- Algae, such as spirulina
- Prebiotics and probiotics
- Plants or herbs and their extracts
- Melatonin
- Coenzyme Q10 (CoQ10)
- Bovine colostrum
- Dietary fibers

Collagen

Food supplements containing collagen must have clear labeling indicating its presence, including phrases like "collagen" and "food grade." Furthermore, the label should specify the collagen's source (such as animal-derived), its form (e.g., decomposed collagen or collagen peptide), and the concentration of collagen per serving as listed on the nutritional label.

Enzymes

Food supplements may include various enzymes, like amylase and protease, sourced from organisms like *Aspergillus flavus oryzae* or *Aspergillus niger*, bromelain, ficin, invertase, papain, pectinase, pepsin, rennet, protein-coagulating enzymes, lactase, and lipase. These enzymes serve various functions and contribute to the supplement's effectiveness.

Other Permitted Ingredients

Ingredients not explicitly mentioned can be added to food supplements after a thorough safety evaluation by the competent authority. This ensures that only safe and suitable components are included in the supplements, enhancing their overall quality and safety for consumption.

Ginseng

Ginseng is a common ingredient in food supplements known for its potential health benefits. Its inclusion should be noted on the label to inform consumers about its presence and potential effects.

Royal Jelly, Propolis, and Pollen

These natural bee products are sometimes included in food supplements. Special precautions should be provided on the label for vulnerable groups, such as infants, pregnant or lactating women, and individuals allergic to bee products. This helps consumers make informed choices and ensures their safety when using the supplements.

Packaging

In accordance with the guidelines specified in GSO 839 and GSO 1863, the packaging of food supplements must meet certain criteria to ensure consumer safety and regulatory compliance. Here are some detailed considerations:

- **Material safety.** Packaging materials must be chosen carefully to ensure they are safe for storing food supplements. This includes materials that are free from harmful chemicals or substances that could contaminate the product.
- **Pharmaceutical form.** Food supplements should not be presented in a form that resembles pharmaceutical products, such as capsules, tablets, or pills. This is to prevent confusion among consumers and to ensure that food supplements are not mistaken for medication.
- **Allowed forms.** Food supplements can be presented in a variety of forms to cater to different preferences and needs. These forms may include chewable tablets, mouth or throat sprays, liquids, powders, lozenges, ampoules, drops, or effervescent tablets. Each form has its own advantages and may be suitable for different types of supplements or consumer preferences.
- **Packaging requirements.** The packaging of food supplements should provide clear and accurate information about the product, including its ingredients, dosage instructions, and any warnings or precautions. Packaging should also be designed to protect the product from contamination and to ensure its stability during storage and transportation.
- **Compliance with regulations.** All packaging materials and designs must comply with the regulations set forth by GSO, as outlined in standards such as GSO 839 and GSO 1863. These regulations are designed to ensure the safety, quality, and integrity of food supplements sold in the GCC market.

By adhering to these guidelines, manufacturers can ensure that their food supplements are safely packaged and compliant with regulatory requirements, thereby instilling confidence in consumers and promoting the responsible marketing and distribution of these products.

Transport and Storage

Transportation and storage are critical aspects of maintaining the quality and safety of food supplements. Here are detailed guidelines to ensure proper handling:

- **Temperature control.** Food supplements must be transported and stored at appropriate temperatures to prevent spoilage and preserve efficacy. It is recommended to keep storage temperatures below 25°C. Some supplements requiring refrigeration should be stored between −1.5°C and 10°C to maintain quality.
- **Contamination prevention.** During transportation and storage, food supplements should be shielded from contamination by implementing robust packaging and handling protocols. Protection against moisture, sunlight, pests, and other contaminants is essential to uphold product integrity.
- **Handling procedures.** Careful handling of food supplements is paramount to avoid damage to packaging and product quality. Employing proper handling techniques minimizes the risk of spills, leaks, or breakage that could compromise the product.
- **Regulatory compliance.** Adherence to regulations established by relevant authorities, including GSO standards, is imperative. Compliance ensures that transportation and storage practices meet stringent quality and safety requirements.

Following these guidelines guarantees that food supplements are transported and stored in a manner that upholds their quality, efficacy, and safety, meeting consumer expectations and regulatory standards alike.

Labeling

Labeling requirements for food supplements are crucial to provide consumers with accurate and necessary information about the product. Here is a comprehensive breakdown of the detailed requirements, ensuring compliance with GSO regulations:

- **Language.** All information on the product label should be prominently displayed in Arabic, with the option to include additional languages alongside Arabic.
- **Key information.** The label must prominently feature details such as the product name, brand name, and quantity or weight of the supplement.
- **Ingredient list.** A comprehensive list of ingredients used in the food supplement must be provided on the label, including all components, additives, and any other substances present in the product.
- **Nutritional details.** The label should include a nutrition facts panel outlining the nutritional content per serving size, covering information on calories, macronutrients (such as proteins, carbohydrates, and fats), as well as vitamins and minerals.
- **Usage instructions.** Clear instructions on how to use the supplement should be included on the label, including dosage recommendations, frequency of consumption, and any special instructions.
- **Health claims.** Any health claims or benefits associated with the supplement must be scientifically supported and comply with GSO regulations on health and nutrition claims.
- **Storage guidelines.** Information on proper storage conditions to maintain the supplement's quality and efficacy should be provided on the label, including temperature requirements or precautions against exposure to light or moisture.
- **Expiry date.** The label must display the expiry date or "best before" date of the product, ensuring consumers are informed about its shelf life for safety and quality assurance.
- **Product name.** Adhering to the guidelines set forth by the Gulf Standardization Organization (GSO) and GSO 9 regarding the labeling of prepackaged foodstuffs, the product should be identified as a "food supplement," followed by the main active ingredient, such as vitamins or minerals, and any optional ingredients it may contain. If the supplement contains more than three active ingredients, including vitamins or minerals, the label should include one of the following phrases to denote its composition:
    - "Multivitamin" should be used if the supplement contains more than three vitamins.
    - "Multi-mineral" should be used if the supplement contains more than three minerals.
    - "Multivitamin and multi-mineral" should be used if the supplement contains a combination of three or more vitamins and minerals.

By accurately labeling the product in accordance with these guidelines, manufacturers ensure that consumers have clear and informative packaging that assists them in making informed choices about their dietary supplementation.

Nutritional Facts

Understanding the nutritional facts of food supplements is vital for consumers to make informed decisions about their dietary choices. Here is a comprehensive breakdown of the key labeling requirements to ensure compliance with GSO regulations:

- **Nutritional information.** The label must detail the nutritional content per 100 g or mL, as well as per serving size, covering carbohydrates, protein, fat, saturated fat, trans fat, and sodium content. Additionally, active ingredient quantities should be specified according to Table 16.6.

- **Serving size.** Clearly indicate the serving size in grams or milliliters, especially for unique forms like chewable tablets or effervescent tablets, by specifying the number of units per serving.
- **Energy content.** Declare the energy content per 100 g or mL and per serving in calories to provide consumers with an understanding of the product's calorie intake.
- **Ingredient declaration.** Provide a comprehensive list of all ingredients used in production to ensure transparency and inform consumers of any potential allergens or additives.
- **Daily value percentages.** Express nutrient values as a percentage of the daily recommended intake to help consumers understand the nutritional significance of the product.
- **Allergen declaration.** Ensure conformity with allergen regulations if the product contains any allergenic ingredients to safeguard consumers with allergies.
- **Target group and preparation method.** Clearly state the target demographic and recommended preparation method, if applicable, to guide consumers in usage.
- **Recommended daily dose.** Specify the recommended daily dose for consumption, tailored to the target group, to promote safe and effective use.
- **Phrases and warnings.** Include essential phrases and warnings, such as usage under medical supervision and precautions for children, to enhance consumer safety.

The following phrases must be written as applicable for the respective product categories:

- "Used under medical supervision."
- "Keep out of reach of children."
- "Do not exceed the recommended daily dose."
- "Food supplements should not be used as a substitute for a varied diet."

The necessary warnings shall be mentioned according to the nature and components of the product, as mentioned in Table 16.8.

Net weight and expiry dates. Provide information on the product's net weight, production date, and expiration date to ensure consumers are aware of product freshness and shelf life.

Storage conditions. Offer guidance on proper storage conditions to maintain product quality and safety, both before and after opening the package.

Contrast and legibility. Ensure adequate contrast between text and background for easy readability, enhancing consumer comprehension.

Specific use or purpose statements. Include specific use statements for products with high doses of certain nutrients, such as niacin or magnesium, to guide consumers in appropriate consumption. Products containing more than 35 mg of niacin, niacinamide, or a combination of both per day require a specific use or purpose statement. Some of the examples for which specific use or purpose statement is required include:

- Products with more than 35 mg of iron per day
- Products providing over 350 mg of magnesium per day
- Products offering more than 40 mg of zinc per day

Precaution statement. Warn consumers about potential risks, such as hypervitaminosis A, associated with high doses of specific nutrients, like vitamin A and beta-carotene. By meticulously following these comprehensive labeling requirements, manufacturers can equip consumers with transparent and accurate information, empowering them to make well-informed decisions regarding their dietary supplement selections.

TABLE 16.8
Warnings and Cautions Related to the Consumption of Types of Dietary and Associated Daily Doses

Ingredient	Daily Dose	Caution
Chromium sourced from chromium picolinate	≥200 µg	Consult a health-care provider/doctor if you have a kidney disorder and/or diabetes.
Chromium sourced from chromium picolinate	All doses	Consult a health-care provider/doctor for use beyond 6 months.
Chromium sourced from chromium picolinate	All doses	Do not use this product if you are pregnant or breastfeeding.
Iron	When the package contains >250 mg of iron	Keep out of reach of children. There is enough iron in this package to harm a child.
Iron	>35 mg	Some people may experience constipation, diarrhea, and/or vomiting.
	All doses	Stop use if hypersensitivity occurs.
Manganese	>5 mg	Consult a health-care provider/doctor prior to use if you have a liver disorder.
Selenium	>70 µg	Consult a health-care provider/doctor prior to use if you have a non-melanoma skin cancer.
Vanadium	All doses	Consult a health-care provider/doctor prior to use if you are pregnant or breastfeeding
Vitamin E	≥180 mg AT	Consult a health-care provider/doctor prior to use if you have cancer.
	≥268 mg AT	Consult a health-care provider/doctor prior to use if you have cardiovascular disease or diabetes.
	≥360 mg AT	Consult a health-care provider/doctor prior to use if you are taking blood thinners.
Vitamin K (K1 and/or K2)	All doses	Consult a health-care provider/-doctor prior to use if you are taking blood thinners.
Zinc sourced from zinc picolinate	All doses	Do not use this product if you are pregnant or breastfeeding.
Zinc sourced from zinc picolinate	All doses	Consult a health-care provider/doctor for use beyond 3 months.
Zinc	Infants 0–12 months: ≤2 mg Toddlers 1–3 years: 5–7 mg Children 4–8 years: 8–12 mg Adolescents 9–13 years: 16–23 mg Adolescents 14–18 years: 25–34 mg Adults 19 years and older: 31–50 mg	Zinc supplementation can cause copper deficiency. If you are unsure, whether you are taking enough copper, consult a health-care practitioner prior to use.
Potassium	≥100 mg	Do not use this product with other potassium-containing supplements or with potassium-containing salt substitutes.
Magnesium	>350 mg	Some people experience diarrhea.

TABLE 16.9
The Percentage of Variance Limits for Vitamins and Minerals Added to Dietary Supplements

Percentage of Variances Allowed for Vitamins and Minerals in Dietary Supplements	The Nutritional Component
20–50+%	Vitamins
20–45+%	Minerals

16.10 HALAL PRODUCTS, PART 1: GENERAL REQUIREMENTS FOR HALAL FOOD GSO 2055–1 (GSO 05/FDS/2055–1:2014) [23]

16.10.1 Scope

This standard serves to establish the fundamental requirements for halal food across every stage of the halal food chain. It encompasses processes ranging from the initial receiving of ingredients to the final stages of distribution and halal food service. Compliance with these standards ensures the integrity and adherence to halal principles throughout the entire supply chain.

16.10.2 Terms and Definitions

- **Islamic Sharia.** This term refers to the comprehensive code of conduct and moral principles derived from the teachings of Islam, including its beliefs, sentiments, and prescribed actions, as revealed through Prophet Muhammad (PBUH).
- **Halal food.** Halal food encompasses a wide range of food and beverages that are permissible for consumption according to Islamic dietary laws. This includes items that are allowed to be eaten, drunk, injected, or inhaled. Halal food must meet specific criteria outlined in Islamic guidelines to be considered lawful.
- **Tazkeya (halal slaughtering to drain blood).** Tazkeya is the process of halal slaughtering, which is conducted in accordance with the requirements set forth in Islamic law. This procedure ensures that blood is drained from the animal's body following specific guidelines outlined in the standard (GSO 993: Animal Slaughtering Requirements According to Islamic Rules).
- **Halal certificate.** A halal certificate serves as an official documentation confirming that a product, service, or facility complies with halal requirements prescribed by Islamic Sharia. These certificates are issued for various entities, including halal Tazkeya facilities, farms, slaughterhouses, food processing plants, and products containing meat, poultry, or their derivatives.
- **Genetically modified foods (GMF).** Genetically modified foods are products or by-products derived from genetically modified organisms (GMOs). These foods have undergone genetic modifications through biotechnological processes, involving the transfer of genes from one organism to another. Such modifications alter the DNA of the food, resulting in genetically engineered products. It is important to note that the permissibility of consuming genetically modified foods may vary within Islamic dietary laws and require careful consideration.
- **Food additives.** *Food additives* encompass any substance that is not typically consumed directly as food and is not commonly used as a food component, irrespective of its nutritional value. These substances serve various technological purposes during food manufacturing, processing, packaging, transportation, or storage. Whether added for organic reasons or otherwise, they may directly or indirectly influence food properties or their sub-products.
- **Aquatic animals.** Aquatic animals are organisms inhabiting water environments and unable to survive outside of them. This category encompasses a wide array of species, including fish, shellfish, amphibians, and marine mammals. These organisms depend entirely on aquatic habitats for their sustenance and livelihood.

16.10.3 Halal Certification Bodies

Halal certification bodies are entities tasked with issuing halal certificates, accredited by recognized accreditation authorities. These certificates serve as official confirmation that products, services, or facilities adhere to halal requirements outlined by Islamic Sharia. Ensuring integrity and authenticity in the market, these bodies play a critical role in certifying halal products and services.

16.10.4 Accreditation Authorities of Halal Certification Bodies

Accreditation authorities are official agencies authorized to accredit and oversee halal certification bodies. Their role involves regulating and standardizing the accreditation process, ensuring that certification bodies adhere to established guidelines. By upholding credibility and reliability in issuing halal certificates, these authorities contribute to maintaining the integrity of halal products and services.

16.10.5 General Requirements for Halal Food

- **Adherence to Islamic rules.** All food products, derivatives, and extracts must strictly adhere to Islamic rules governing what is permissible and prohibited, as outlined in Table 16.9
- **Comprehensive compliance.** Throughout the entire food chain, from receipt to service, procedures, derived from Islamic rules, must be diligently followed in the preparation, handling, packaging, transportation, distribution, storage, and display of halal food products.
- **Use of halal ingredients.** The ingredients used in halal food production, including additives and raw materials, must be free from any non-halal components. This requirement should be supported by officially legalized documents detailing the components, including those of packaging materials.
- **Safety standards.** Halal food items must not contain any toxic substances or hazardous pollutants that could pose a risk to consumer health.
- **Prevention of Najasah contamination.** Halal food products must be devoid of any contamination by Najasah (impurity) that is deemed forbidden by Islamic laws.
- **Separation of halal and non-halal products.** Throughout the entire food supply chain, strict segregation must be maintained between halal and non-halal products to ensure clear differentiation and prevent any mixing or contamination.
- **Verification procedures.** Official regulatory authorities are empowered to verify the compliance of products with halal requirements and to take appropriate actions as per the local legislation.
- **Hygienic practices for meat.** In the production of meat and its related products, a strict adherence to health requirements outlined in GSO/CAC/RCP 58: Code of hygienic practice for meat is mandatory.
- **Animal slaughtering requirements.** The process of animal slaughtering (Tazkeya) must strictly adhere to Islamic Rules as specified in GSO 993: "Animal Slaughtering requirements According to Islamic Rules.
- **Food hygiene principles.** Halal food production must adhere to the general health requirements specified in GSO 1694, "General principles of food hygiene."
- **Cleanliness of equipment.** All equipment, tools, production lines, and associated materials used in the preparation of halal food must be thoroughly cleaned and free from any non-halal materials.
- **Cleaning procedures.** When transitioning equipment from non-halal to halal food production, rigorous cleaning procedures must be followed to eliminate any traces of non-halal products. This transition should not be repeated on a regular basis.
- **Use of halal-certified cleaning materials.** Cleaning agents, greases, oils, or fats used in the maintenance of machinery or equipment in contact with halal foods must not contain any non-halal components or materials.

16.10.6 Packaging and Wrapping Materials

- **Exclusion of non-halal materials.** It is imperative that packaging and wrapping materials used for halal food products are completely devoid of any non-halal components or substances.
- **Preventing contamination.** Throughout the manufacturing process, these materials must not come into contact with equipment contaminated by non-halal substances. They should be physically segregated from any non-halal materials to prevent contamination.
- **Safety assurance.** Packaging materials should adhere to strict safety standards and must not contain any substances that pose a risk to human health. Measures should be in place to ensure that hazardous materials are not incorporated into packaging intended for halal food products.

16.10.7 Halal Food Retail, Handling, and Service Outlets

All sale, distribution, and provision facilities of halal products shall meet the following requirements:

- **Exclusive handling of halal products.** All facilities involved in the sale, distribution, and provision of food must strictly deal only with halal products that adhere to the provisions outlined in the halal food standard.
- **Segregation of equipment.** Tools, equipment, and utensils utilized during the handling, service, and sale of halal food items must be completely segregated from those used for non-halal food. This ensures the prevention of cross-contamination.
- **Transition to halal production.** In cases where a facility typically deals with non-halal products but intends to transition to halal production, a thorough cleaning process must be undertaken in accordance with Islamic regulations before commencing halal operations.
- **Exclusion of non-halal products.** It is strictly prohibited to serve or offer non-halal food and drinks within the same premises designated for halal products. This ensures the integrity and compliance of the establishment with halal food standards.

16.10.8 Storage, Display, and Transportation

Segregation and labeling. Halal food items must be stored, displayed, and transported separately from non-halal products. They should be clearly categorized and labeled as "halal" to prevent any mixing or confusion.

Use of suitable transportation. During transportation, appropriate means should be employed to ensure that halal products remain segregated from non-halal items. This prevents any inadvertent mixing and maintains the integrity of the halal food chain.

16.10.9 Labeling

- **Product name and type.** The label must clearly state the name and type of the product for easy identification.
- **Ingredient list.** An exhaustive list of ingredients should be provided, ensuring complete adherence to Islamic dietary rules. The origin of each ingredient must be explicitly disclosed.
- **Declaration of fats, oils, and meat derivatives.** Products containing fats, oils, or meat derivatives, such as gelatin or rennet, must specify their sources on the label.
- **Declaration of GMO ingredients.** Food items containing genetically modified organisms (GMOs) must clearly indicate so on the label.
- **Halal slogan affixation.** If the intention is to include the "halal" slogan on the label, the criteria outlined in the standard GSO 2055–2: "Halal" must be followed.

TABLE 16.10
Classification of Non-Halal Animals

Species	Non-Halal
Animal	According to Islamic rules, there are specific criteria for animals that may not be slaughtered. These include animals slaughtered by individuals not adhering to Islamic beliefs, such as polytheists, irreligious people, seculars, atheists, magis, apostates, and other disbelievers, excluding Christians and Jews. Additionally, dead animals, suffocated animals, fatally beaten animals (Mawqouza), falling animals (Motaradiah), butted animals (Nateehah), and animals eaten by beasts of prey are also prohibited from being slaughtered. Furthermore, animals on whose slaughtering the name of Allah is not mentioned, those slaughtered on idols, and those slaughtered in a manner not in accordance with the approach of Allah are deemed unfit. Finally, animals contaminated with prohibited animals of harm, if consumed, are also included in the list of prohibited animals for slaughter according to Islamic rules.
	Dogs, cats, pigs, domestic donkeys, mules, elephants, monkeys, foxes, and other similar animals are among the prohibited species for consumption under Islamic rules.
	Predators like lions, cheetahs, bears, and other fanged beasts of prey are prohibited for consumption under Islamic rules, with the exception of hyenas.
	Owls, kites, peregrines, sparrows, hawks, falcons, and eagles, including birds of prey with sharp claws.
	Mice, rats, squirrels, moles, bats, hedgehogs, martins, centipedes, and reptiles, including serpents, snakes, monitors, lizards, geckos, chameleons, and Coleoptera (with exceptions, like dab lizards and jerboas).
	Scorpions, ants, bees, woodpeckers, hoopoes, and other insects, worms, and animals forbidden to be killed by Islamic dietary laws are not permitted for consumption, with exceptions made for locusts and any unavoidable bee parts found in honey.
	Animals fed with unclean items are prohibited, unless they are confined and fed with permissible feeding according to Islamic rules for at least three days. Additionally, odious creatures, such as shells, larvae, and similar animals, are not allowed for consumption.
Aquatic	Toxicant or harmful species are prohibited for consumption unless toxicant or harmful substances are effectively removed during the preparation process.
Plantation	Plants, along with their products and derivatives, containing toxicants and/or harmful or narcotic substances, are restricted for consumption, unless these harmful components are effectively removed during the preparation process.
Beverages	Beverages that are intoxicating or contain alcohols, narcotics, calming agents, or any toxic or harmful substances are prohibited for consumption.
Genetically modified food	Foods that are produced through genetic modification from a prohibited species or a combination of species where one of them is prohibited are not permitted for consumption.
Blood and blood derivatives from a human or animal origin	Blood and all its derivatives, as well as any bodily fluids from either human or animal sources, such as vomit and urine, are strictly prohibited for consumption.
Food additives	Any food additives, products, or their derivatives containing toxicants, harmful substances, or calming agents are forbidden, especially if they are derived from non-halal materials.
Enzymes	Enzymes sourced from non-halal origins are strictly prohibited.
Microorganisms	Microorganisms, including germs, fungi, and harmful yeast, produced in non-halal environments or using non-halal ingredients, are prohibited.

- **Adherence to slaughter requirements.** Labels for slaughtered animals and their products must comply with the guidelines specified in the standard GSO 993: "Animal Slaughtering Requirements According to Islamic Rules."
- **Declaration of food additives.** Any food additives used must be declared as per the requirements outlined in the reference standard for tools, equipment, and utensils.

- **Identification of scaled fish.** Fish with scales, shrimp, and fish eggs from scaled fish should be labeled as "scaled fish." Other aquatic animals and their by-products should be clearly labeled as "non-scaled fish," as applicable.

16.11 CONCLUSION

The nutraceutical market in the GCC region is experiencing a significant surge, driven by a convergence of factors including heightened health awareness, governmental backing, and shifting consumer preferences toward healthier lifestyles. This presents a highly promising opportunity for Indian manufacturers to penetrate and thrive in this burgeoning market.

Indian manufacturers hold a distinct advantage owing to their well-established presence in the nutraceutical industry, coupled with advanced formulation capabilities and a wide-ranging product portfolio. Leveraging these strengths, Indian companies can effectively address the evolving demands of consumers in the GCC countries and tap into the escalating demand for nutraceutical products.

Furthermore, strict adherence to the stringent regulatory standards set forth by the Gulf Standardization Organization (GSO) is imperative for Indian manufacturers seeking to establish credibility and trust within the market. By ensuring full compliance with GSO regulations, Indian firms can underscore their unwavering commitment to quality and safety, thus bolstering their competitive edge in the GCC market.

To capitalize on the lucrative opportunities offered by the GCC market, Indian manufacturers must adopt a multifaceted approach encompassing innovative product development, strategic partnerships, and targeted marketing initiatives. By aligning their offerings with the specific preferences and requirements of consumers in the region, Indian companies can carve out a formidable presence and position themselves as key players in the rapidly expanding nutraceutical landscape of the Gulf countries.

REFERENCES

[1] Adelaja AO, Schilling BJ. Nutraceuticals: blurring the line between food and drugs in the twenty-first century. *Choices*. 1999;14(4):35–9.

[2] Domínguez Díaz L, Fernández-Ruiz V, Cámara M. The frontier between nutrition and pharma: The international regulatory framework of functional foods, food supplements and nutraceuticals. *Critical Reviews in Food Science and Nutrition*. 2020 May 30;60(10):1738–46.

[3] Komala MG, Ong SG, Qadri MU, Elshafie LM, Pollock CA, Saad S. Investigating the regulatory process, safety, efficacy and product transparency for nutraceuticals in the USA, Europe and Australia. *Foods*. 2023 Jan 16;12(2):427.

[4] Meher B. Nutraceutical Regulations: A country-by-country review of global regulatory requirements. *Nutritional Outlook Nutritional Outlook*. 2022;25(4).

[5] APAC Nutraceutical Market Size & Share Analysis—Growth Trends & Forecasts (2024 –2029). https://www.mordorintelligence.com/industry-reports/asia-pacific-nutraceutical-market.

[6] Asia Pacific Nutraceuticals Market Outlook (2023 to 2033). https://www.futuremarketinsights.com/reports/asia-pacific-nutraceuticals-market.

[7] Middle-East and Africa Dietary Supplement Market Size & Share Analysis—Growth Trends &Forecasts(2024–2029).https://www.mordorintelligence.com/industry-reports/middle-east-and-africa-dietary-supplement-market.

[8] Middle East and Africa Nutraceuticals Market Research Report—Segmented by Type, Application and Country (KSA, UAE, Israel, rest of GCC countries, South Africa, Ethiopia, Kenya, Egypt, Sudan and Rest of MEA)—Analysis on Size, Share, Trends, COVID-19 Impact & Growth Forecast (2024 to 2029). https://www.marketdataforecast.com/market-reports/middle-east-and-africa-nutraceuticals-market.html

[9] GCC Nutraceuticals Market Research Report: Forecast (2022–2027). https://www.marknteladvisors.com/research-library/gcc-nutraceuticals-market.

[10] MENA Nutraceuticals Market: Current Analysis and Forecast (2022–2030). https://univdatos.com/report/mena-nutraceuticals-market.

[11] Africa Nutraceutical Market Size. https://www.mordorintelligence.com/industry-reports/africa-nutraceutical-market/market-size.html.
[12] Latin America Nutraceuticals Market Research Report—Segmented by Type, Application and Country (Brazil, Mexico, Argentina, Chile and Rest of Latin America)—Analysis on Size, Share, Trends, COVID-19 Impact & Growth Forecast (2024 to 2029). https://www.marketdataforecast.com/market-reports/latin-america-nutraceuticals-market.html
[13] About GCC: The charter. https://www.gcc-sg.org/en-us/AboutGCC/Pages/Primarylaw.aspx.html
[14] Food Supplements Registration GCC, GSO—Overview. https://foodsupplements.freyrsolutions.com/food-supplements-registration-gcc-gso.
[15] GSO CAC/GL 55:2015: Guidelines for Vitamins and Minerals Food Supplements. https://www.gso.org.sa/store/standards/GSO:704988/GSO%20CAC-GL%2055:2015.
[16] GSO 839:2021: Food Packages—Part 1: General Requirements. https://www.gso.org.sa/store/standards/GSO:780890/GSO%20839:2021
[17] GSO 9:2022: Labeling Of Prepackaged Food Stuffs. https://www.gso.org.sa/store/standards/GSO:810825/GSO%209:2022
[18] GSO 2233:2021: Gulf Technical Regulation, Requirements of Nutritional Labeling. https://www.gso.org.sa/store/standards/GSO:781113/GSO%202233:2021
[19] GSO 382:2021: Gulf Technical Regulation, Maximum Limits of Pesticide Residues in Agricultural And Food Products. https://www.gso.org.sa/store/standards/GSO:781069/GSO%20382:2021.
[20] GSO 2500:2022/Amd1:2023: Amendment to Gulf Technical Regulation, Additives Permitted For Use In Food Stuffs. https://www.gso.org.sa/store/standards/GSO:817002/GSO%202500:2022-Amd1:2023.
[21] GSO 2233:2021: Gulf Technical Regulation, Requirements of Nutritional Labeling. https://www.gso.org.sa/store/standards/GSO:781113/GSO%202233:2021.
[22] GSO 193:2021: Gulf Technical Regulation, Contaminants and Toxins in Food and Feed. https://www.gso.org.sa/store/standards/GSO:781074/GSO%20193:2021
[23] GSO 2055–1:2015: Gulf Technical Regulation, HALAL FOOD—Part 1: General Requirements. https://www.gso.org.sa/store/standards/GSO:693304/GSO%202055–1:2015.
[24] GSO 1863:2021: Gulf Technical Regulation, Food Packages—Part 2: Plastic Package—General Requirements. https://www.gso.org.sa/store/standards/GSO:780896/GSO%201863:2021
[25] GSO 168:2022: Gulf Technical Regulation, Requirements of Storage Facilities for Dry and Canned Foodstuffs. https://www.gso.org.sa/store/standards/GSO:807794/GSO%20168:2022
[26] GSO 2210:2021: Gulf Technical Regulation, Ginseng Products. https://www.gso.org.sa/store/standards/GSO:780728.
[27] GSO 2097:2010: Gulf Technical Regulation: Royall Jelly. https://www.gso.org.sa/store/standards/GSO:532040/file/1120/preview
[28] GSO 2500:2022, Additives Permitted for Use in Food Stuffs. https://www.gso.org.sa/store/standards/GSO:811154/GSO%20147:2022
[29] GCC STANDARDIZATION ORGANIZATION (GSO), Draft Standard DS: GSO 55/2020, Food Supplements. https://members.wto.org/crnattachments/2020/SPS/BHR/20_6778_00_e.pdf

17 Global Future Trends Related to Nutraceuticals

Aabir Pramanik, Sonali Maheshwari, Niyati Acharya, and Yashwant Pathak

17.1 INTRODUCTION

The term *nutraceutical* combines *nutrients* and *pharmaceuticals*, categorizing a consumable product that neither completely comes under the umbrella of food medicines, falling on the midline of food and medicine. As per Stephen DeFelice, who coined the term *nutraceutical*, they refer to "food or part of a food that provides medical or health benefits, including the prevention and/or treatment of a disease." [1]

The nutraceutical industry in India has been witnessing rapid growth, driven by increasing health consciousness and changing consumer lifestyles. This sector encompasses functional foods and beverages as well as dietary supplements, which are sought after for their health benefits, including improved immune health, bone health, brain health, mental wellness, and overall physical wellness. Interest in nutraceuticals has surged globally in recent years, fueled by an increasing awareness of their potential health benefits beyond basic nutrition. This trend is marked by a rising demand for dietary supplements, functional foods, and fortified beverages that claim to boost health, prevent chronic diseases, and enhance overall well-being. The COVID-19 pandemic has further accelerated this trend, as consumers seek preventive measures to bolster their immune systems and maintain health amidst unprecedented health challenges. Academic research supports this trend, highlighting the efficacy of various nutraceuticals in managing conditions such as cardiovascular diseases, metabolic disorders, and inflammatory responses. This burgeoning interest is reflected in the expanding market, which is projected to continue its robust growth, fueled by advancements in biotechnology, increased consumer education, and a shift toward preventive health care. As nutraceuticals become an integral part of modern health strategies, ongoing scientific research and rigorous clinical trials are critical to evaluate their benefits and ensure their safe and effective use.

The market is expected to experience significant growth, starting from $382.4 billion in 2024 and reaching approximately $765.7 billion by 2033. This trend highlights the increasing consumer demand for health-enhancing products and the growing recognition of nutraceuticals' potential benefits. Such robust growth is anticipated to be driven by advancements in biotechnology, greater consumer education on preventive health care, and a continued focus on health and wellness in the post-pandemic era. This trend underscores the importance of ongoing research and innovation within the nutraceutical industry to meet the evolving needs and preferences of consumers globally. This chapter shall present a comprehensive analysis of the evolving landscape of nutraceuticals, examining global trends, consumer behaviors, product categories, and emerging applications in health and performance enhancement. Several emerging trends in nutraceuticals and their effects on the central nervous system (CNS), on immune system modulation, on sports nutrition, and even on their potential applications in space nutrition are elucidated.

DOI: 10.1201/9781003465041-17

17.2 TRENDS IN NUTRACEUTICALS INDUSTRY

17.2.1 The Comprehensive Approach to Health: Embracing Holistic Nutrition for Overall Wellness

In recent years, holistic nutrition has gained significant traction, especially in light of the global COVID-19 pandemic. The holistic approach to health and wellness emphasizes the interconnectedness of the body, mind, and environment. Unlike conventional nutrition, which often focuses on individual nutrients and their immediate effects on health, holistic nutrition takes a broader perspective. It considers the overall lifestyle, emotional well-being, and environmental factors contributing to one's health. This holistic approach is gaining traction as individuals look for more natural and sustainable methods to enhance their health and well-being. The COVID-19 pandemic has emphasized the need for strong health systems and resilient immune responses. Consequently, there has been a surge in interest in preventive health practices such as holistic nutrition. This movement is fueled by the understanding that true health encompasses more than just the absence of illness; it is a state of complete physical, mental, and social well-being. [2] Holistic nutrition advocates for the consumption of whole, unprocessed foods that are rich in essential nutrients. It also emphasizes the importance of mental health, physical activity, and a supportive environment. This integrative avenue addresses the root causes of health issues rather than merely targeting to treat the symptoms, leading to long-term health benefits.

17.2.1.1 Benefits of Holistic Nutrition

Holistic nutrition offers a wide range of health benefits, beginning with enhanced immune function. By providing essential nutrients through whole foods rich in vitamins and minerals, the defenses of the body against infections and diseases are fortified, ensuring optimal immune function. In addition to it, a diet abundant in fiber from fruits, vegetables, and whole grains improves digestive health, preventing gut issues, such as constipation and irritable bowel syndrome. The inclusion of fermented foods further supports a healthy gut microbiome. Mental health also sees significant improvement through nutrient-dense foods and stress-reducing practices, which lower the risk of anxiety, depression, and other mental health disorders. This comprehensive approach addresses both physical and emotional well-being, fostering a balanced and fulfilling life. Sustainable weight management is another key benefit, as holistic nutrition emphasizes whole foods and mindful eating, promoting balanced diets and sustainable lifestyle changes over quick fixes or fad diets. By focusing on a healthy lifestyle and nutrient-rich food items, holistic nutrition reduces the risk of chronic diseases, addressing their root causes and offering long-term health benefits. [3–4]

17.2.2 Rising Demand for Clean Label Nutrition

The demand for clean label nutrition has seen a significant rise in recent years, driven by increasing consumer awareness and a growing preference for transparency and natural ingredients in food products. Clean label nutrition emphasizes the importance of minimally processed foods with natural ingredients which are free from artificial additives, preservatives, and genetically modified organisms (GMOs). Clean label nutrition involves the use of simple, recognizable ingredients in food products, ensuring that consumers can easily understand what they are consuming. The core principles of clean label nutrition include transparency, simplicity, and the use of natural ingredients. This movement is a response to the growing mistrust among consumers toward overly processed foods with long ingredient lists filled with unpronounceable chemicals and additives.

17.2.2.1 Key Drivers of Demand for Clean Label Nutrition: Consumer Awareness and Education

1. **Consumer awareness and education.** With the advent of the internet and social media, consumers are more informed than ever about the potential health risks associated with

artificial additives and processed foods. This increased awareness has led to a demand for products that are perceived as safer and healthier.

2. **Health and wellness trends.** The global health and wellness trend has significantly influenced the demand for clean label products. Consumers are increasingly seeking foods that contribute to their overall health and well-being. Clean label products, which are free from synthetic additives and preservatives, are viewed as healthier options that align with these wellness goals.
3. **Regulatory and certification bodies.** Certifications from reputable organizations such as the Clean Label Project, Non-GMO Project, and various organic certifying bodies have bolstered consumer trust in clean label products. These certifications provide an added layer of assurance that the products meet specific standards of purity and safety.
4. **Transparency and trust.** The demand for transparency in food labeling is a significant driver of the clean label movement. Consumers want to know where their food comes from, how it is made, and what it contains. Brands that offer full transparency and disclose detailed ingredient information are more likely to gain consumer trust and loyalty.

17.2.2.2 Benefits of Clean Label Nutrition

The benefits of clean label nutrition are multifaceted, encompassing health, environmental, and market advantages. Health-wise, clean label products typically contain fewer artificial ingredients, reducing the risk of issues like allergies, sensitivities, and chronic conditions, while being richer in essential nutrients from whole and natural sources, thus promoting overall health. Environmentally, clean label nutrition supports sustainability through the use of natural and organic ingredients, which often involve more sustainable farming practices, such as reduced pesticide use, less soil degradation, and improved water conservation. Market-wise, the rapidly growing demand for clean label products presents significant opportunities for food manufacturers and retailers to innovate and expand their offerings, enabling them to stand out in a competitive market and build a loyal customer base.

17.2.2.3 Challenges

1. **Cost and accessibility.** One of the main challenges associated with clean label products is their higher cost compared to conventional products. Natural and organic ingredients are often more expensive to produce and source. This can make clean label products less accessible to price-sensitive consumers.
2. **Maintaining product quality and shelf life.** Removing artificial preservatives and additives can impact the shelf life and stability of products. Manufacturers need to find natural alternatives that can maintain product quality without compromising safety or shelf life. This requires significant research and development investment.
3. **Regulatory compliance.** Navigating the complex landscape of food regulations and certifications can be challenging for manufacturers. Ensuring that products meet clean label standards while complying with local and international regulations requires meticulous attention to detail and substantial documentation.

17.2.2.4 Future Prospective

The demand for clean label nutrition is set to rise as consumers are becoming extremely health-conscious and environmentally aware. Innovations in food technology and ingredient sourcing will likely make these products more affordable and accessible. Increased regulatory support and clearer guidelines will help standardize claims and boost consumer confidence. Brands that prioritize transparency, sustainable practices, and natural ingredient sourcing will thrive in this evolving market, which holds the potential to significantly impact global food production and consumption patterns.

Clean label nutrition represents a significant shift toward more transparent, natural, and health-conscious food choices. Driven by consumer demand for simplicity and authenticity, the clean label movement is reshaping the nutraceutical and food industries. Despite challenges related to cost, product quality, and regulatory compliance, the benefits of clean label nutrition for health, the environment, and market growth are substantial. As this trend continues to evolve, it offers exciting opportunities for innovation and consumer engagement in the pursuit of better health and wellness. [3]

17.2.3 Enhancing Immunity through Ayurveda

The ancient Indian system of medicine, Ayurveda, has been practiced for thousands of years and continues to be highly relevant today. It emphasizes the balance of the body, the mind, and the spirit to achieve optimal health. One of the key areas where Ayurveda has made significant contributions is in the enhancement of the immune system. In an era where immunity is of paramount importance, especially during global health challenges like the COVID-19 pandemic, Ayurveda offers valuable insights and remedies for bolstering immune defenses naturally and sustainably.

In Ayurveda, immunity, or "Ojas," is considered the essence of all bodily tissues and is responsible for overall health, vitality, and longevity. Ojas is seen as the ultimate product of proper digestion and metabolism, linking physical health with mental and spiritual well-being. According to Ayurvedic philosophy, a robust immune system is achieved through the balance of three fundamental energies, or doshas: Vata, Pitta, and Kapha. When these doshas are in harmony, the body's natural defenses are strong, and it can resist diseases effectively.

17.2.3.1 Ayurvedic Principles for Strengthening Immunity

1. **Balanced diet (Ahara).** A balanced diet is crucial for maintaining Ojas. Ayurveda recommends consuming fresh, organic, and seasonal foods that are appropriate for one's dosha type. Foods such as fruits, vegetables, whole grains, nuts, and seeds, rich in essential nutrients, antioxidants, and immune-boosting properties, are emphasized. Specific herbs and spices, like turmeric, ginger, garlic, and black pepper, are recognized for their immune-enhancing properties.
2. **Herbs and supplements (Rasayana).** Ayurveda includes a category of rejuvenating herbs known as Rasayanas, which are believed to enhance longevity, vitality, and immunity. Some of the most commonly used Rasayanas include:
   a. **Ashwagandha.** Has adaptogenic properties; helps the body cope with stress and strengthens the immune system.
   b. **Guduchi (Tinospora cordifolia).** Acts as an immunomodulator, enhancing the body's resistance to infections.
   c. **Amla (Indian gooseberry).** Boosts immunity and detoxifies the body as it is a rich source of vitamin C and antioxidants.
   d. **Tulsi (holy basil).** Possesses antimicrobial, anti-inflammatory, and immune-boosting properties.
3. **Daily routine (Dinacharya).** Establishing a consistent daily routine is fundamental in Ayurveda for maintaining health and immunity. Dinacharya involves practices such as waking up early, oral hygiene, physical exercise, meditation, and proper sleep. Regular detoxification practices, like oil pulling and nasal cleansing (Neti), are also recommended to remove toxins and support the immune system.
4. **Mind–body practices.** Yoga and meditation are integral parts of Ayurvedic principles, promoting physical strength, mental clarity, and emotional balance. Practices like pranayama (breathing exercises), asanas (yoga postures), and dhyana (meditation) enhance the flow of prana (life energy), reduce stress, and support immune function. Regular practice of these techniques is known to improve resilience against diseases.
5. **Seasonal and lifestyle adjustments (Ritucharya).** Ayurveda emphasizes the importance of aligning one's lifestyle and diet with seasonal changes. Ritucharya involves specific guidelines for different seasons to balance the doshas and support the immune system. For example, consuming warm, nourishing foods during winter and light, cooling foods during summer helps maintain balance and prevent seasonal illnesses.

17.2.4 The Impact of Clean Nutrition on Mental Wellness: Understanding the Link between Diet and Mental Health

Clean nutrition emphasizes the consumption of minimally processed foods that are free from artificial additives, preservatives, and other synthetic ingredients. While much of the focus has been on physical health benefits, the impact of clean nutrition on mental wellness has not gone unnoticed. It supports cognitive function and emotional balance and played a crucial role in preventing and managing mental health disorders. The brain, like any other organ, requires a range of nutrients to function optimally. The foods we consume provide the building blocks for neurotransmitters, the chemical messengers that influence mood, sleep, and cognitive function. A diet lacking in essential nutrients can lead to imbalances in these neurotransmitters, contributing to mental health issues, such as depression, anxiety, and cognitive decline.

17.2.4.1 Essential Nutrients for Optimal Mental Wellness

Omega-3 fatty acids, found in fatty fish, flaxseeds, and walnuts, play a vital role in maintaining brain cell structure and aiding communication between neurons. Studies have shown that omega-3 supplementation can reduce symptoms of depression and anxiety, underscoring their importance in mental wellness. B vitamins, such as B6, B12, and folate, are crucial for the production of neurotransmitters, like serotonin and dopamine. Deficiencies in these vitamins increase the risk of depression and cognitive decline. Foods rich in B vitamins include leafy greens, legumes, eggs, and whole grains. Antioxidants, including vitamins C and E and polyphenols found in fruits, vegetables, nuts, and seeds, protect the brain from oxidative stress and inflammation, which are linked to cognitive decline and mood disorders. A diet high in antioxidants supports overall brain health and emotional well-being. Magnesium, found in nuts, seeds, leafy greens, and whole grains, is involved in numerous biochemical reactions in the brain and helps regulate mood, reducing symptoms of anxiety and depression. Adequate magnesium intake is crucial for maintaining mental wellness. Amino acids, the building blocks of proteins, are essential for the synthesis of neurotransmitters. For example, tryptophan is a precursor to serotonin, while tyrosine is a precursor to dopamine. Lean meats, fish, dairy, and plant-based proteins are excellent sources of amino acids.

17.2.4.2 How Clean Nutrition Supports Mental Wellness

Chronic inflammation has been linked to the development of various mental health disorders, such as depression and anxiety; thus, managing inflammation is critical. Emphasizing foods such as fruits, vegetables, nuts, seeds, and fatty fish in one's diet can help reduce inflammation. Keeping blood sugar levels in check also plays a significant role in maintaining mood and cognitive function. Diets high in refined sugars and carbohydrates can cause blood sugar spikes and crashes, leading to irritability and mood swings. Clean nutrition, which prioritizes whole grains, lean proteins, and healthy fats, helps stabilize blood sugar levels and promotes emotional stability. Additionally, supporting gut health is vital for catering to the needs of the communication system between the gut and the brain. A healthy gut microbiome is critical for mental health as it affects neurotransmitter production and the immune response. Lastly, avoiding harmful additives is important because artificial additives, preservatives, and food colorings in processed foods have been linked to behavioral issues and mental health disorders. Clean nutrition, by eliminating these harmful substances, reduces the risk of negative mental health outcomes.

17.2.4.3 Practical Strategies for Adopting Clean Nutrition to Enhance Mental Wellness

Whole foods should be prioritized as the foundation of meals, including fruits, vegetables, whole grains, lean proteins, nuts, and seeds, which provide essential nutrients crucial for brain health and emotional stability. Healthy fats, such as avocados, nuts, seeds, and fatty fish, should be incorporated to support brain function and help reduce inflammation. Omega-3 supplements can be beneficial if enough fatty fish is not regularly consumed. Adequate intake of high-quality proteins from sources like lean meats, fish, dairy, eggs, and plant-based proteins should be ensured, as they supply the necessary amino acids for neurotransmitter production vital for cognitive function. Antioxidant-rich foods, including a variety of colorful fruits and vegetables, such as berries, leafy greens, and nuts, should be focused on to help protect brain health and support overall well-being. Additionally, gut health should be supported with fiber-rich foods and fermented options like yogurt, kefir, sauerkraut, and kimchi, as a balanced gut microbiome is essential for mental health and overall digestive wellness. Lastly, processed foods, refined sugars, and artificial additives should be limited, opting instead for natural sweeteners, like honey or maple syrup, in moderation and choosing whole-food snacks over processed alternatives to maintain stable mood and energy levels throughout the day.

Clean nutrition plays a pivotal role in supporting mental wellness by providing the essential nutrients required for optimal brain function and emotional balance. By prioritizing whole, minimally processed foods and avoiding harmful additives, individuals can significantly enhance their mental health. As research continues to uncover the intricate connections between diet and mental well-being, adopting clean nutrition practices offers a powerful and natural approach to achieving and maintaining mental wellness. [3]

17.2.5 The Shift toward Convenience: How Modern Consumers and Millennials Are Shaping the Nutraceutical Market

The modern consumer landscape is marked by an increasing preference for convenience in all aspects of life, and the nutraceutical market is no exception. Nutraceuticals, which encompass functional foods, dietary supplements, and beverages designed to offer health benefits, are evolving to meet the demands of a fast-paced lifestyle. Among the various demographics, Millennials, in particular, are driving this shift toward convenient nutraceutical forms. This trend is reshaping the industry, prompting innovations in product formulation, packaging, and delivery methods.

Today's consumers are more health-conscious and informed than ever before. They seek products that not only support their health and wellness goals but also fit seamlessly into their busy schedules. The demand for convenience is a reflection of broader societal changes, including increased work hours, the rise of dual-income households, and the pervasive influence of technology which prioritizes efficiency and immediacy.

17.2.5.1 Forms of Convenient Nutraceuticals

1. **Gummies.** Gummies have emerged as a popular nutraceutical form, particularly among younger consumers. They offer a palatable and enjoyable way to consume vitamins and supplements, eliminating the need for swallowing pills. Gummies are often fortified with essential nutrients, such as vitamins C, D, B-complex, and omega-3 fatty acids. Their appealing taste and ease of consumption make them a preferred choice for both children and adults.
2. **Effervescent powders.** Effervescent powders provide a convenient solution for those who prefer not to take traditional capsules or tablets. These powders dissolve quickly in water, creating a fizzy, flavored drink that delivers nutrients effectively. Effervescent formulations are popular for delivering vitamins, minerals, and electrolytes, often marketed for their enhanced absorption and quick onset of action.
3. **Protein bars.** Protein bars are a staple in the diet of health-conscious individuals, athletes, and busy professionals. These bars offer a portable and nutritious snack option that supports muscle recovery, energy levels, and overall health. They are available in various flavors and formulations, catering to specific dietary preferences, such as gluten-free, vegan, and high-protein options.
4. **Ready-to-eat products.** Ready-to-eat nutraceutical products, such as meal replacement shakes, fortified snacks, and functional beverages, provide on-the-go nutrition. These products are designed to deliver balanced nutrients quickly and conveniently, making them

ideal for those with hectic lifestyles. Ready-to-eat options are often enriched with vitamins, minerals, proteins, and other functional ingredients that promote health and wellness.

17.2.5.2 The Millennial Influence

Millennials, defined as individuals born between 1981 and 1996, are a significant force in the nutraceutical market. This generation values convenience, health, and sustainability, influencing their purchasing decisions. Millennials are digital natives, comfortable with technology, and often seek products that offer instant gratification and align with their fast-paced lives.

1. **Health and wellness priorities.** Millennials prioritize health and wellness, often seeking nutraceuticals that support their active lifestyles, mental well-being, and long-term health goals. They are well-informed consumers, researching products extensively before making purchases. This demographic is drawn to products that offer clear health benefits, backed by scientific evidence.
2. **Convenience and accessibility.** For Millennials, convenience is a key factor in product selection. They prefer nutraceuticals that are easy to incorporate into their daily routines, whether it is a quick protein bar after a workout or an effervescent vitamin drink in the morning. The convenience of these products allows them to maintain their health without disrupting their busy schedules.
3. **Innovative and sustainable packaging.** Sustainability is another important consideration for Millennials. They favor products with eco-friendly packaging and those produced by companies with strong environmental and social responsibility practices. Innovative packaging solutions that reduce waste and enhance convenience, such as single-serve sachets and biodegradable containers, are particularly appealing to this demographic.

17.2.5.3 Market Implications

The rising demand for convenient nutraceuticals is driving significant changes in the industry. Manufacturers are investing in research and development to create innovative products that meet consumer expectations for convenience, taste, and efficacy. This trend is also leading to the expansion of distribution channels, including online platforms, which offer consumers easy access to a wide range of nutraceutical products.

1. **Product innovation.** The nutraceutical industry is witnessing a surge in product innovation, with new formats and delivery systems being developed to enhance convenience. Companies are exploring novel ingredients, improved formulations, and advanced manufacturing technologies to create products that cater to the needs of modern consumers.
2. **Marketing and branding.** Effective marketing strategies are essential to capture the attention of convenience-seeking consumers. Brands are leveraging social media, influencer partnerships, and targeted advertising to reach Millennials and other tech-savvy demographics. Clear messaging that highlights the health benefits, convenience, and sustainability of nutraceutical products is crucial for attracting and retaining customers.
3. **Regulatory considerations.** As the nutraceutical market evolves, regulatory bodies are also adapting to ensure product safety and efficacy. Manufacturers must navigate complex regulatory landscapes to bring new products to market, adhering to guidelines that govern ingredient sourcing, labeling, and health claims. Compliance with these regulations is essential for maintaining consumer trust and avoiding legal issues.

The rising consumer demand for convenience in the nutraceutical market is a reflection of broader lifestyle changes and evolving health priorities. Millennials, in particular, are driving this trend, seeking products that offer both health benefits and ease of use. As the industry continues to innovate and adapt, the focus on convenience will remain a key driver of growth and transformation. By prioritizing product development, sustainable practices, and effective marketing strategies, nutraceutical companies can successfully meet the needs of modern consumers and thrive in a competitive market. [3]

17.2.6 The Rise of Plant-Based Meat and Milk: Health, Sustainability, and Market Dynamics

The global food industry is witnessing a significant shift as consumers increasingly turn toward plant-based alternatives to traditional animal products. This rise in plant-based meat and milk products is driven by a combination of health, environmental, ethical, and technological factors. As awareness of the impacts of animal agriculture grows, so does the demand for sustainable and health-conscious alternatives. Plant-based meat and milk products are at the forefront of this transformation, offering consumers new ways to enjoy their favorite foods while aligning with their values and dietary preferences.

One of the primary drivers behind the rise of plant-based food is the perceived health benefit associated with these products. Traditional meat and dairy products are often high in saturated fats and cholesterol, which can contribute to various health problems, such as heart disease, obesity, and certain cancers. In contrast, plant-based alternatives are typically lower in saturated fats and free of cholesterol, thus a healthier option for many consumers.

17.2.7 Environmental Impact

The environmental benefits of plant-based meat and milk are significant and play a crucial role in their rising popularity. Animal agriculture is a major contributor to greenhouse gas emissions, deforestation, water use, and pollution. Shifting to plant-based alternatives can substantially reduce the environmental footprint of food production.

1. **Reduction in greenhouse gas emissions.** Livestock farming is responsible for a substantial portion of global methane emissions, a potent greenhouse gas. Plant-based foods generally produce fewer emissions throughout their life cycle. For example, producing a kilogram of plant-based meat emits a fraction of greenhouse gases compared to producing a kilogram of beef.
2. **Water and land use.** Animal agriculture is water-intensive, requiring vast amounts of water for raising animals and growing feed crops. Plant-based alternatives require significantly less water. Similarly, the land required to produce plant-based foods is much smaller compared to the land needed for grazing and growing animal feed. This reduction in land use helps preserve natural habitats and biodiversity.
3. **Resource efficiency.** Producing plant-based foods is more resource-efficient. The energy conversion from plant to meat is much less efficient compared to direct consumption of plant-based foods. This efficiency translates into lower resource use and environmental impact.

17.2.7.1 Ethical Considerations

Ethical concerns about animal welfare are another significant factor driving the shift toward plant-based meat and milk. Many consumers are increasingly aware of the conditions in which animals are raised and slaughtered in industrial farming operations. Choosing plant-based products allows them to align their dietary choices with their ethical values.

1. **Animal welfare.** Industrial farming practices often involve confined spaces, lack of natural behaviors, and inhumane treatment. By opting for plant-based alternatives, consumers

can avoid contributing to these practices and support more humane and sustainable food systems.

2. **Sustainability and ethics.** The ethical considerations extend beyond animal welfare to include broader issues of sustainability and fair trade. Consumers are increasingly interested in the sourcing and production practices behind their food, favoring companies that prioritize ethical and sustainable practices.

17.2.7.2 Technological Advancements

Advancements in food technology have played a crucial role in the development and popularity of plant-based meat and milk. Innovations have enabled the creation of products that closely mimic the taste, texture, and nutritional profile of their animal-based counterparts. Recent advancements have revolutionized the landscape of plant-based foods in several key areas. Firstly, companies such as Beyond Meat and Impossible Foods have spearheaded the innovation of plant-based meats that closely mimic the taste and texture of real meat. These products utilize advanced processing methods and ingredients like heme (as seen in Impossible Burgers) to achieve authentic meat-like flavors and textures. Emerging technologies, like cellular agriculture and fermentation, are pushing the boundaries further, promising even more realistic meat substitutes from plant sources. Secondly, there has been significant progress in the development of plant-based milks. Options such as almond, soy, oat, rice, and hemp milk have seen substantial improvements in taste, texture, and nutritional profiles due to technological advancements. These developments have enhanced the appeal of plant-based milks across a wider consumer base, reflecting growing demand for sustainable and health-conscious alternatives to traditional dairy products.

17.2.7.3 Market Growth and Consumer Trends

The market for plant-based meat and milk has grown rapidly, driven by changing consumer preferences and increased availability of high-quality products. This growth is reflected in both the retail sector and the food service industry.

1. **Retail expansion.** Supermarkets and grocery stores are dedicating more shelf space to plant-based products, reflecting their popularity. Brands are continually expanding their product lines to include a variety of flavors and formulations, catering to diverse consumer tastes.
2. **Food service adoption.** Restaurants, cafés, and fast-food chains are incorporating plant-based options into their menus. High-profile partnerships, such as those between Impossible Foods or Beyond Meat and major fast-food chains, have brought plant-based foods into the mainstream.
3. **Consumer demographics.** The appeal of plant-based products extends beyond vegetarians and vegans, to include flexitarians—those who primarily eat plant-based but occasionally consume meat. This broader consumer base is driving the sustained growth of the plant-based market.

The rise of plant-based meat and milk is a multifaceted trend driven by health benefits, environmental sustainability, ethical considerations, and technological advancements. As consumer awareness continues to grow, the demand for plant-based alternatives is likely to increase, further transforming the food industry. By offering nutritionally rich, environmentally friendly, and ethically produced options, plant-based products are well-positioned to meet the evolving needs of modern consumers and contribute to a more sustainable future. Nutraceuticals, categorized according to their intended use and the specific names under which they are marketed, encompass a wide range of products. Detailed definitions and examples of these categories are provided in **Table 17.1**. [3]

TABLE 17.1
Types of Nutraceuticals, Their Intended Use, and Their Trade Name

Sr. No.	Category	Definition	Examples
1	Herbal supplements	Supplements derived from herbs, plants, standardized extracts, or phytochemicals aiding in the promotion of health, and also in the prevention of certain chronic conditions, are known as herbal supplements or herbal botanicals.	Milk thistle, used in liver health and detoxification, contains silymarin, which acts as a potent antioxidant.
2	Dietary supplements	Supplements consumed via oral route, aiding in supplementing the diet, that is, providing additional nutrients that may be lacking or insufficient in one's diet, of an individual are known as dietary supplements.	Multivitamins, calcium supplements.
3	Fortified foods	Food products in which nutrients, such as vitamins, minerals, and other micronutrients, have been added to enhance the nutritional value and compensate for dietary lack are known as fortified foods.	Fortified cereals with added vitamins and minerals, like folic acid, niacin, riboflavin, and thiamine.
4	Recombinant foods	Genetically modified organisms (GMOs) made using biotechnology techniques to enhance specific traits, such as nutritional content, resistance to pests or diseases, or shelf life, are known as recombinant foods.	Genetically modified corn contains genes that confer resistance to certain pests or herbicides. Golden Rice is a genetically modified rice variety which contains beta-carotene.
5	Nutraceutical enzymes	Enzymes derived from natural sources, such as plants, animals, or microorganisms, to promote overall health by aiding in digestion or metabolic processes by supplementing the body's enzyme production are known as nutraceutical enzymes.	Amylase helps convert carbohydrates into sugars. Bromelain has anti-inflammatory properties and may help with digestion.
6	Probiotics	Live microorganisms, or "good" bacteria, that are administered in required quantities to confer health benefit and can restore the balance of beneficial bacteria in the gut are known as probiotics.	*Lactobacillus casei* found in dairy products such as yogurt may help improve digestion. *Lactobacillus plantarum*, found in fermented foods like kimchi, may help improve gut health and reduce inflammation.
7	Prebiotics	Non-digestible fibers which act as food for probiotics, helping the "good" microbes proliferate and grow by producing short-chain fatty acids that provide various health benefits, are known as prebiotics.	Inulin is used as a prebiotic supplement to support gut health and promote the growth of beneficial bacteria in the colon. Fructooligosaccharides are short-chain carbohydrates found naturally in fruits and vegetables, such as bananas and tomatoes, that are often used as prebiotic supplements to support gut health.
8	Marine nutraceuticals	Supplements derived from marine sources, such as fish, algae, or shellfish, and that contain bioactives, such as antioxidants, omega-3 fatty acids, vitamins, minerals, peptides, etc., known to provide health benefits are known as marine nutraceuticals.	Astaxanthin, a carotenoid antioxidant found in certain types of microalgae and seafood, such as shrimp, crab, lobster, and salmon, acts as a potent antioxidant. Eicosapentaenoic acid and docosahexaenoic acid, omega-3 fatty acids found in fatty fish and certain types of algae, are used in brain health.

9	Cosmeceuticals	Cosmetic products that contain bioactive compounds that help in providing medicinal benefits, such as improving the appearance and health of the skin, hair, and nails, are known as cosmeceuticals. They are a combinatory product of cosmetics and pharmaceuticals that helps in targeting specific concerns, such as ageing, hyperpigmentation, acne, and dryness.	Retinoids, including retinol, are known for their anti-ageing benefits. Hyaluronic acid acts as a humectant, aiding in the retention of moisture in the skin.
10	Farmaceuticals	Food products derived from agricultural sources using biotechnology, which aids in providing health benefits beyond basic nutrition, are known as farmaceuticals. They are aggregate products of "farm" and "pharmaceuticals."	Allicin, a bioactive compound found in onion, acts as an antimicrobial and antioxidant agent. Berberine, a bioactive compound found in various plants, such as barberry, acts as an antimicrobial and anti-inflammatory agent.
11	Functional foods	Food products that provide additional health benefits beyond basic nutrition are referred to as functional foods. They often overlap with nutraceuticals, medical foods, probiotics, designer foods, pharmafoods, and vitafoods.	Oily fish are rich in omega-3 fatty acids, which help support heart health, brain function, and the reduction of inflammation. Berries, high in antioxidants, such as anthocyanins and flavonoids, provide anti-inflammatory properties.
12	Medicinal foods	Food products that contain bioactive compounds, such as antioxidants, vitamins, minerals, and phytochemicals, and have specific health benefits beyond basic nutrition and are traditionally used in various cultures for their therapeutic properties are known as medicinal foods.	Turmeric contains curcumin, a bioactive compound with potent anti-inflammatory and antioxidant properties. Green tea, rich in catechins, acts as a potent antioxidant.
13	Nutricosmetics	A category of nutritional supplements and foods formulated to improve beauty, particularly the health and appearance of skin, hair, and nails, by providing essential nutrients, vitamins, minerals, and bioactive compounds.	Collagen supplements promote skin elasticity, reduce wrinkles, and support overall skin health. Biotin supplements strengthen hair and nails, supporting skin health.
14	Nutrigenetics	It is a nutritional science branch that studies the relationship between the genetic makeup of an individual and their response to specific nutrients and diets. It focuses on how genetic variations influence the way a person absorbs, metabolizes, and utilizes nutrients, ultimately affecting their health and the risk of developing certain diseases.	1. Gene: MTHFR (methylenetetrahydrofolate reductase) Variation: Mutations in the MTHFR gene can reduce the efficiency of folate metabolism, affecting homocysteine levels. Nutrigenetic Response: Individuals with MTHFR mutations may benefit from a higher intake of folate-rich foods or methylated folate supplements. 2. Gene: HLA-DQ2 and HLA-DQ8 Variation: Presence of these gene variants is associated with an increased risk of celiac disease, an autoimmune disorder triggered by gluten. Nutrigenetic Response: Individuals with these variants may need to follow a strict gluten-free diet.

17.3 PRODUCT DEVELOPMENT OF PROBIOTICS–PREBIOTICS FOR THE CENTRAL NERVOUS SYSTEM

Nutraceuticals, derived from "nutrition" and "pharmaceutical," refer to products that provide health benefits beyond basic nutrition. The central nervous system (CNS) benefits from a balanced gut microbiota, influenced by probiotics and prebiotics. These nutraceuticals, derived from natural sources, support mental and cognitive health by modulating the gut–brain axis. Probiotics are live microorganisms that confer health benefits on the host when administered in adequate amounts. Common probiotic strains include *Lactobacillus* and *Bifidobacterium* species. Prebiotics are non-digestible food ingredients that beneficially affect the host by selectively stimulating the growth and/or activity of specific bacteria in the colon. Examples include inulin, fructooligosaccharides (FOS), and galactooligosaccharides (GOS) **(Figure 17.2)**. [5–6] This section discusses various probiotics and prebiotics, their mechanisms, their functions, the specific marketed products, the challenges in product development, and the implications for CNS health.

17.3.1 PROBIOTICS FOR CNS HEALTH

Very few species are used as probiotics for improving CNS health that work through the brain-gut axis as shown in **Figure 17.1**. Here are some key probiotic strains used in nutraceuticals:

17.3.1.1 *Lactobacillus rhamnosus GG*

L. rhamnosus is a type of bacteria found in human intestines. It belongs to the genus Lactobacillus, a type of bacteria that produces the enzyme lactase. This enzyme breaks down the sugar lactose—which is found in dairy—into lactic acid. [7] This strain is known for its ability to modulate the gut microbiota, enhance gut barrier function, and produce neurotransmitters, such as GABA, which is crucial for regulating mood and anxiety. The daily dose is 1 capsule containing at least 10 billion live bacteria. [8] Probiotic, Sacctik-GG, LR 2G, Culturelle, and SuperFlora GG are widely available, contain *Lactobacillus rhamnosus* GG and are marketed for overall digestive health and immune support, with implications for reducing stress and anxiety.

17.3.1.2 *Bifidobacterium longum*

Bifidobacterium longum is a Gram-positive, catalase-negative rod-shaped bacterium residing in the human gastrointestinal tract, classified under the genus Bifidobacterium. [9] This bacterium is instrumental in synthesizing B vitamins, including folate. Moreover, it facilitates the production of short-chain fatty acids (SCFAs), such as butyrate, which contribute to several physiological functions. These functions include enhancing gut lining integrity, regulating blood sugar levels, improving lipid metabolism, modulating the immune system and inflammatory responses, and reducing blood pressure and oxidative stress. The neuroprotective properties of butyrate further underscore the importance of *Bifidobacterium longum* in maintaining central nervous system health. [10] Longum, Alflorex, Epigenetics, Probiotic-Pro BB536, and Align are popular probiotic supplements containing *Bifidobacterium longum* and is marketed for maintaining digestive balance, which in turn supports mental health.

17.3.1.3 *Lactobacillus Helveticus R0052 and Bifidobacterium Longum R0175*

A probiotic formulation containing *Lactobacillus helveticus* R0052 and *Bifidobacterium longum* R0175 can mitigate brain function abnormalities caused by chronic stress by reducing the HPA axis response. [11] This combination also studied for its psychotropic-like properties, showing promise in reducing anxiety and depression. [12] Probid and Probio'Stick contain these strains, marketed for reducing stress and improving mood.

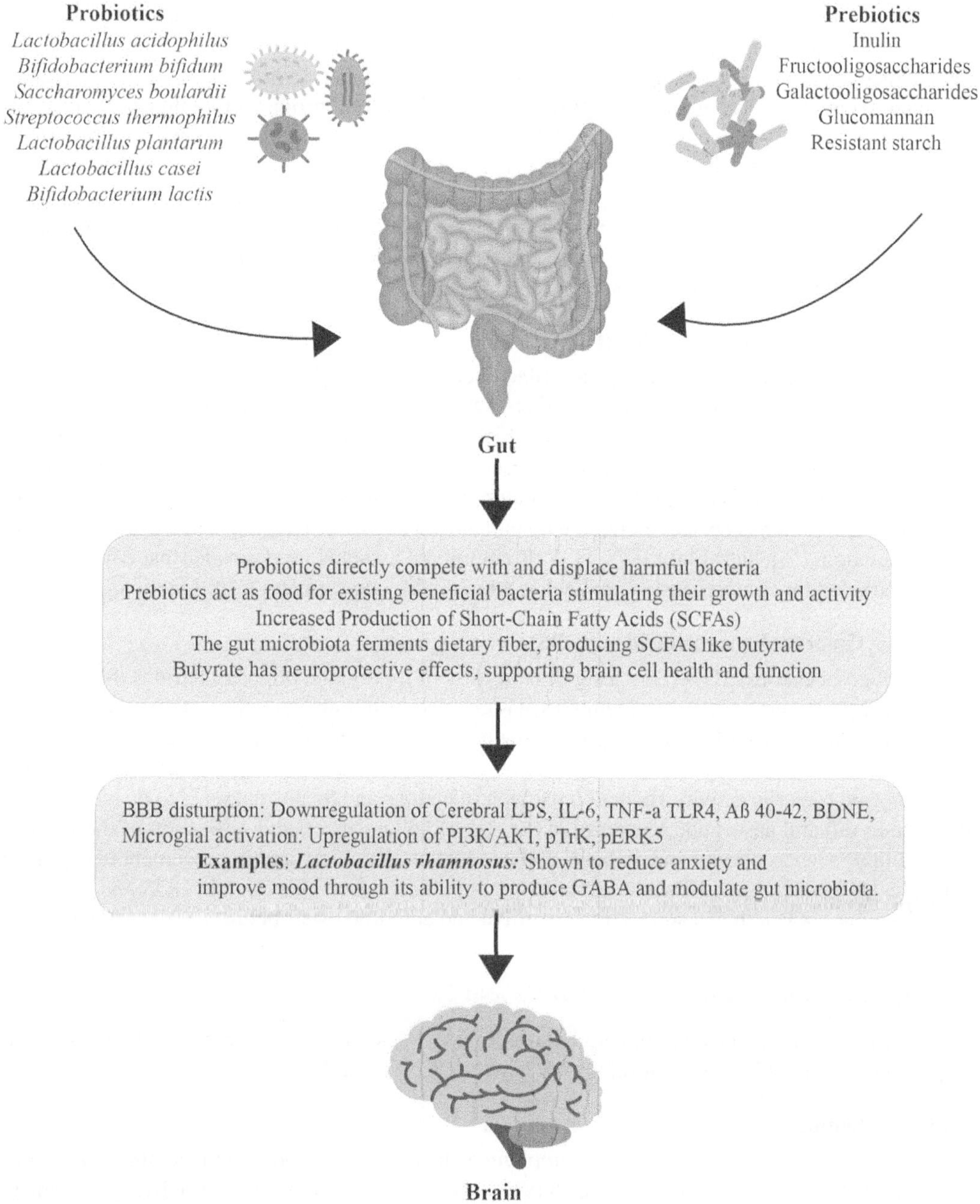

FIGURE 17.1 Action of probiotics and prebiotics via the gut–brain axis.

17.3.2 Prebiotics for CNS Health

Prebiotics work via the immune-modulating pathway or increase neurotransmitter production. Here are some key prebiotics used in nutraceuticals.

17.3.2.1 Inulin

Inulin serves as a prebiotic, fostering the proliferation of beneficial bacteria in the gut. As a fermentable fiber, inulin is metabolized by bacteria in the large intestine, supporting a healthy microbiota. Structurally, inulin is an oligosaccharide known as a fructan, comprising chains of fructose molecules linked together. This prebiotic is naturally present in the roots of various foods, including whole wheat, onions, garlic, and Jerusalem artichokes. Inulin is frequently extracted from chicory root and incorporated into various food products to enhance their nutritional value. Inulin supplements are available in multiple forms, including powders, chewables, and capsules. Typically, these supplements provide 2 to 3 g of fiber per serving. It promotes the growth of beneficial gut bacteria, such as *Bifidobacteria* and *Lactobacilli*. It helps in the production of SCFAs, which have anti-inflammatory and neuroprotective effects. [13] Inulin, Prebiotic Inulin FOS, FiberChoice contains inulin and is marketed for improving digestive health and promoting a balanced gut microbiota.

17.3.2.2 Fructooligosaccharides (FOS)

Fructooligosaccharides (FOS) are nondigestible fibers composed of short fructose chains. They are a type of carbohydrate called oligosaccharides. FOS occur naturally in many plants, which include blue agave, yacon root, garlic, onion leeks, chicory root, Jerusalem artichokes, Asparagus, and bananas. Fructooligosaccharides are subtly sweet and low-calorie. [14] FOS is classified as generally recognized as safe (GRAS). FOS stimulates the growth of beneficial bacteria and enhances SCFA production. It has been associated with enhanced gut barrier integrity and a decrease in neuroinflammation. [15] NutraFlora FOS, FOS, Prebiotin are a widely used powder that contains FOS and is marketed for comprehensive gut health, which indirectly supports CNS function.

17.3.2.3 Galactooligosaccharides (GOS)

Galactooligosaccharides (GOS) are a significant class of food-grade oligosaccharides that are nondigestible by humans. Acting as prebiotics, GOS selectively enhances the activity of beneficial gut microbiota, particularly promoting the growth of microorganisms such as *Bifidobacterium* and *Lactobacillus*. The primary benefits of GOS consumption include a bifidogenic effect, suppression of pathogenic bacterial activity, reduction in colon infections and the formation of toxic metabolites, increase in mineral absorption, decrease in the risk of colon cancer, and mitigation of the severity of conditions such as obesity and diabetes and a reduction in anxiety and improvement in cognitive function. [16] GosPrebio, LiftMode, Bimuno are popular products that contain GOS and marketed for enhancing gut health and supporting cognitive and emotional well-being.

17.3.3 Synbiotics: Combining Probiotics and Prebiotics

Synbiotics are formulations that integrate both probiotics and prebiotics to amplify their combined impact on gut microbiota and central nervous system (CNS) health.

17.3.3.1 Symprove

Symprove is a liquid-based probiotic supplement designed to support gut health and overall well-being. Unlike many probiotics, Symprove uses a unique water-based delivery system to ensure the survival of live bacteria through the digestive tract. It is formulated to improve the gut microbiome and positively influence the gut–brain axis, enhancing central nervous system (CNS) health. Symprove contains four strains of live bacteria: *Lactobacillus rhamnosus* for gut lining adhesion and immune modulation, *Lactobacillus plantarum* for maintaining gut permeability and antimicrobial production, *Lactobacillus acidophilus* for food breakdown and nutrient absorption, and *Enterococcus faecium* for balancing gut microbiota and enhancing gut barrier function. The delivery system protects these bacteria from the stomach acid, allowing them to reach the intestines alive, where they modulate immune function, enhance gut barrier integrity, and produce neuroprotective short-chain fatty acids (SCFAs). Symprove has been shown to alleviate irritable bowel syndrome (IBS) symptoms by balancing gut microbiota and reducing inflammation. [17] [18]

17.3.3.2 VSL#3

VSL#3 is a high-potency probiotic supplement formulated to support gut health and overall well-being. It features a multi-strain composition with a high concentration of live bacterial cultures, designed to restore and maintain a balanced gut microbiota, crucial for digestive health, and that of the central nervous system (CNS) as well via the gut–brain axis. VSL#3 contains eight strains of live, freeze-dried lactic acid bacteria and bifidobacteria, including *Lactobacillus paracasei, Lactobacillus plantarum, Lactobacillus acidophilus, Lactobacillus delbrueckii* subsp. *bulgaricus, Bifidobacterium longum, Bifidobacterium breve, Bifidobacterium infantis,* and *Streptococcus thermophilus.* These strains work synergistically to restore gut microbiota balance, modulate immune function, enhance gut barrier integrity, and produce beneficial metabolites, like short-chain fatty acids (SCFAs). VSL#3 has been shown to manage gastrointestinal disorders, such as IBS, IBD, and pouchitis; alleviate symptoms of anxiety and depression; and support immune health. Available in capsules, sachets, and packets, the dosage varies, depending on individual needs, with common usage being one to two doses per day with food. [19–21] VSL#3 is marketed for managing digestive disorders and supporting a healthy gut–brain axis, which benefits CNS function.

17.4 IMMUNE SYSTEM BOOSTERS: THE LATEST TRENDS IN NUTRACEUTICALS

These products, derived from food sources with extra health benefits in addition to the basic nutritional value found in foods, are poised at the intersection of food and medicine. The immune system, a complex network of cells and proteins that defend the body against infection, is a critical focus of nutraceutical research. This chapter explores the current trends in nutraceuticals concerning their effects on the immune system, highlighting recent advancements, popular ingredients, and future directions.

In the wake of global health challenges, particularly the COVID-19 pandemic, there has been an unprecedented surge in the demand for immune-boosting nutraceuticals. Consumers are increasingly seeking products that can enhance their body's natural defense mechanisms. This trend is reflected in the growing market for supplements containing vitamins, minerals, and herbal extracts, known for their immunomodulatory properties.

17.4.1 Key Ingredients and Their Mechanisms

Several nutraceutical ingredients have been extensively studied for their effects on the immune system. Here are some of the most prominent ones **(Figure 17.2)**. [22–26]

17.4.2 Innovations in Delivery Systems

The efficacy of nutraceuticals is often influenced by their bioavailability, which is the degree and rate at which a substance is absorbed into the bloodstream. Recent innovations in delivery systems are making nutraceuticals more effective:

1. **Liposomal encapsulation.** This technology involves encapsulating nutrients in liposomes, which are tiny vesicles that can enhance the absorption and bioavailability of the ingredients. [27]
2. **Nanoemulsions.** These are fine oil-in-water dispersions that improve the solubility and bioavailability of poorly water-soluble nutraceuticals, making them more effective. [28]
3. **Sustained-release formulations.** These formulations allow for the gradual release of nutrients over time, ensuring a steady supply to the body and enhancing their efficacy. [29]

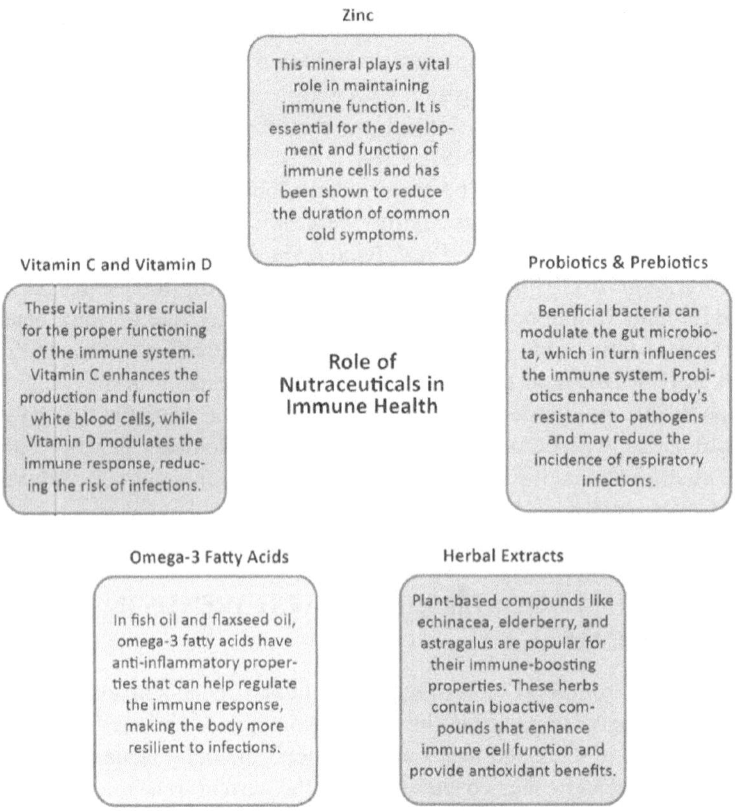

FIGURE 17.2 Role of nutraceuticals in boosting immunity: development of nutraceuticals for space and some potential measures to curb them.

Source: [49,50].

17.4.3 Clinical Evidence and Research Trends

The scientific community continues to explore the potential of nutraceuticals in modulating the immune system. Recent studies have provided compelling evidence supporting their use:

1. **Randomized controlled trials (RCTs).** Numerous RCTs have demonstrated the efficacy of various nutraceuticals in enhancing immune function and reducing the severity of infections. [30]
2. **Analyses and systematic reviews.** These comprehensive reviews of multiple studies provide strong evidence for the immune-boosting effects of specific nutraceuticals, such as vitamin D and probiotics. [31]
3. **Mechanistic studies.** Research at the molecular level is uncovering the specific pathways through which nutraceuticals exert their effects on the immune system, paving the way for more targeted and effective interventions. [32]

17.5 NUTRACEUTICAL TRENDS TRANSFORMING SPORTS NUTRITION

These products, derived from food sources with additional health benefits, beyond basic nutrition, are increasingly being recognized for their potential to enhance athletic performance, recovery, and overall health. This chapter delves into the current trends in nutraceuticals within the realm of sports nutrition, highlighting popular ingredients, recent advancements, and future directions. The global sports nutrition market has experienced robust growth, driven by a rising awareness of health and fitness, the increasing prevalence of sports and physical activities, and the desire for improved athletic performance. Nutraceuticals are at the forefront of this growth, offering athletes and fitness enthusiasts a variety of supplements designed to enhance performance, endurance, and recovery.

17.5.1 Key Nutraceutical Ingredients and Their Benefits

Several nutraceutical ingredients have been identified for their benefits in sports nutrition. Protein supplements, including whey, casein, and plant-based proteins, are key to muscle regeneration and growth, while branched-chain amino acids (BCAAs) and essential amino acids (EAAs) help reduce muscle soreness and improve recovery. [33] Creatine, one of the most researched and effective supplements, enhances strength, power, and lean muscle mass, making it particularly beneficial for high-intensity, short-duration activities, such as weightlifting and sprinting. [34] Beta-alanine, an amino acid, increases muscle carnosine levels, buffering acid in muscles and improving endurance while reducing fatigue during high-intensity exercise. [35] Omega-3 fatty acids, known for their anti-inflammatory properties, can help reduce muscle soreness and inflammation, thus enhancing recovery and overall performance. [36] Caffeine, a well-known stimulant, enhances endurance, focus, and alertness, making it a popular ingredient in pre-workout supplements. [37] Finally, electrolytes are essential for maintaining hydration and electrolyte balance; supplements containing sodium, potassium, magnesium, and calcium are vital for athletes, particularly in endurance sports. [38]

17.5.2 Clinical Evidence and Research Trends

Researchers are actively exploring the potential of nutraceuticals in sports nutrition. Recent studies have provided robust evidence supporting their use:

17.5.2.1 Randomized Controlled Trials (RCTs)

Numerous RCTs have demonstrated the efficacy of various nutraceuticals in enhancing athletic performance and recovery. [39]

17.5.2.2 Meta-Analyses and Systematic Reviews

These comprehensive reviews of multiple studies provide strong evidence for the benefits of specific nutraceuticals, such as creatine and beta-alanine. [35]

17.5.2.3 Mechanistic studies

Research at the molecular level is uncovering the specific pathways through which nutraceuticals exert their effects, paving the way for more targeted and effective supplements. [39]

17.5.3 Future Directions

The future of nutraceuticals in sports nutrition is promising, with several emerging trends. Personalized nutrition is becoming increasingly viable due to advances in genomics and metabolomics, allowing for nutraceutical supplementation tailored to an individual's genetic makeup and metabolic profile. There is also a growing emphasis on sustainability and natural sourcing, driven by consumer demand for environmentally friendly and clean label products. Additionally, combination formulas that include multiple performance-enhancing ingredients are gaining popularity, offering synergistic benefits and improved efficacy. Nutraceuticals are increasingly recognized for their potential to enhance athletic performance and recovery. With ongoing research and innovation, these products are becoming more effective and widely accessible. As the demand for sports nutrition continues to rise, nutraceuticals are set to play a crucial role in helping athletes achieve their performance goals and maintain overall health. [40–41] Nutraceuticals are increasingly recognized for their potential to enhance athletic performance and recovery. With ongoing research and innovation, these products are becoming more effective and widely accessible. As the demand for sports nutrition continues to rise, nutraceuticals are set to play a crucial role in helping athletes achieve their performance goals and maintain overall health.

17.6 ENHANCING SPACE NUTRITION WITH CUTTING-EDGE NUTRACEUTICALS

Space nutraceuticals refer to dietary supplements and functional foods designed specifically for use in space to enhance the health and performance of astronauts. These nutraceuticals aim to address the unique challenges posed by the space environment, such as microgravity, radiation exposure, and the psychological stress of long-duration missions. Several significant limitations in the current space nutrition systems drive the need for innovations. The prevalence of processed foods compared to fresh foods is a significant issue. Astronauts mainly rely on processed and packaged meals because fresh produce can only be consumed in the early stages of a mission due to limited storage duration and high expenses. For a theoretical three-year Mars mission with a six-member crew, the projected daily energy requirement is 12 MJ (megajoules) per person, translating to about 22 t of water-containing food. Even with complete water recycling and partial food dehydration, transport costs remain prohibitively high, around 20,000 euros per kilogram. Another limitation is the lack of quality for resource-intensive refrigerated and frozen food. [42–43]

Space missions face strict constraints on water and electricity, critical resources that are scarce and valuable due to the weight limitations of space shuttles. Developing regenerative and self-sustaining systems for water, food, and energy production is essential for supporting human colonies on other planets. A functional nutrition system must ensure sensory appeal, nutritional effectiveness, and safety for three to five years. NASA's goal is to create nutrient-rich, eco-friendly foods suitable for cooking in microgravity conditions. Despite progress like artificial gravity, adapting Earth-based nutritional systems for space necessitates substantial decreases in external resource use and waste generation. [44]

The space environment significantly differs from the terrestrial conditions on Earth, causing numerous challenges to space nutrition. Adverse conditions, such as microgravity, radiation, confined spaces, motion sickness, circadian rhythm disruptions, and a low-pressure atmosphere high in carbon dioxide and low in oxygen, affect astronauts. Nearly all physiological systems exhibit adaptive responses to microgravity and the rapid alternation of light and dark cycles, leading to loss of body mass, fluid transfer, electrolyte imbalance, dehydration, constipation, loss of potassium and calcium, reduction in red blood cell mass, and disruption in the intestinal microecology, leading to increase in harmful bacteria. One notable issue is less energy intake and the resultant weight loss; astronauts typically lose 2–5% of their pre-flight weight, with some experiencing over 10% loss despite adequate food supply. A negative energy balance may result in a weight loss of up to 5 kg each month. On a Mars mission, an initial weight reduction of 15% or more could have severe health consequences, worsening detrimental physiological changes, such as reduced muscle performance, cardiovascular issues, muscle fatigue, and heightened vulnerability to infections. **Table 17.2** lists the various challenges in the development of nutraceuticals for space and some potential measures to curb them. [40–41]

17.7 REGULATORY REQUIREMENTS

To mandate the standardized nutrition labeling of packaged foods to provide consumers with accurate information about the nutritional content of the products, the United States passed the Nutrition Labeling and Education Act (NLEA) in 1990. As per the Act, packaged foods must have nutrition labels on their packaging with information on a list of all ingredients, calorie content, and amount

TABLE 17.2
Various Areas of Interest in Developing Nutraceuticals for Space

Sr. No.	Area of Interest for Product Development	Measures to Tackle
1	Issues of palatability	Astronauts often experience reduced intake of space food due to altered sensory responses in space, leading to less-attractive taste, smell, and texture of food. Artificial fortification can destroy nutrients and create unpleasant tastes. Processed foods have a limited shelf life, losing their original flavor. There is a need for fresh food with distinctive flavors for long-term missions. Future research should focus on maximizing fresh food production within the constraints of spaceflight.
2	Meeting energy intake	Astronauts experience altered appetite and energy consumption during spaceflight. Enhancing food variety and quality and emphasizing dietary culture, may boost energy intake. Participating in food production and cooking relieves stress and induces happiness, increasing a feeling of being valued and mental health, thus boosting food intake and maintaining physical health.
3	Enhancing caloric intake	The type and nutrient content of food affect caloric intake. Energy-packed and nutrient-dense carbohydrates are preferred but limited by space constraints. Nuts are effective due to high energy and nutrients but have shelf life issues. Research should focus on lightweight, nutrient-rich foods and, techniques for fresh food cultivation to fulfill the requirements for carbohydrates, vegetables, and beverages.
4	Counterbalance the Effects of Space Environment	The space environment, including high CO_2 levels and varying light cycles, disrupts circadian rhythm and hormonal balance, affecting food intake and body weight. Continuous light exposure raises leptin levels, reducing food consumption. Research should explore reducing inflammatory foods, increasing anti-inflammatory foods, and designing space habitats to align circadian rhythms and optimize food production.
5	Increasing protein intake to counteract muscle atrophy	Microgravity reduces muscle mass and performance. Adequate protein intake is crucial for maintaining muscle health. The branched-chain amino acids aid in preventing muscle degradation. Foods abundant in these amino acids include fish, shrimp, milk, soy, corn, and cauliflower.
6	Combating intestinal microecology disorder	Microgravity-associated deterioration of gut flora can lead to infections. Calcium- and probiotics-rich foods like yogurt can help maintain intestinal health.
7	Nutritional strategies to mitigate radiation effects	Space radiation leads to both acute and chronic health problems. Consuming antioxidant-rich foods can aid in tolerating low levels of radiation exposure. Examples of such foods include tomatoes, garlic, nuts, oats, blueberries, broccoli, salmon, and green tea.
8	Mitigating bone loss with nutritional measures	Bone loss in space is multifaceted and can be mitigated with a diet rich in calcium, vitamin D, and vitamin K. Nutritional parallels between ageing orthopedic patients and astronauts provide insights. To prevent hip fractures, it is essential to maintain a balance of fiber, hydration, and bioactive compounds such as those found in coffee.

of various nutrients (such as fat, carbohydrates, fiber, cholesterol, sodium, sugars, protein, vitamins, and minerals), serving size, and the number of servings per container present in the product. It also aided in regulating the use of nutrient content claims on food labels, such as "low fat," "high fiber," or "reduced sodium," which, as per the NLEA Act, must now meet specific criteria set by the Food and Drug Administration (FDA). For any food product claiming for the mitigation, prevention, or treatment of any disease or health-related condition, The NLEA established rules on providing scientific evidence that needs to be approved by the FDA. The FDA is responsible for enforcing the NLEA and ensuring that food manufacturers comply with its labeling requirements. Violations of the NLEA can result in penalties, including fines and product recalls. [45]

17.8 THE FOOD SAFETY AND STANDARDS ACT

The Food Safety and Standards Authority of India (FSSAI) has established regulatory guidelines for the approval of nutraceuticals in the Indian market. The Food Safety and Standards Authority of India (FSSAI) is an autonomous statutory body that enacted the Food Safety and Standards Act (FSS) in 2006 to regulate food ingredients or products. There are eight categories clubbed under functional foods, namely, health supplements, nutraceuticals, food for special dietary use, foods containing probiotics, foods containing prebiotics, food for special Medical Purpose, Food containing Plant or Botanicals, and novel foods.

FSSAI not only establishes regulations and guidelines specific to nutraceuticals (Health Supplements, Nutraceuticals, Food for Special Dietary Use, Food for Special Medical Purposes, Functional Food, and Novel Food) Regulations also provide a product approval process for novel ingredients or formulations used in nutraceutical products. It also helps to conduct monitoring and surveillance activities to ensure compliance with nutraceutical regulations and standards, such as sampling and testing of the products to verify their safety, quality, and labeling accuracy. The Nutraceutical Advisory Group (NAG) is an expert panel established by the FSSAI to provide guidance and advice on matters related to the regulation, safety, and promotion of nutraceuticals in India. It comprises experts from various fields relevant to nutraceuticals, including nutrition, medicine, food technology, regulatory affairs, pharmacology, toxicology, and consumer advocacy who provide recommendations on the development of standards, guidelines, and policies related to nutraceuticals, including safety standards, labeling requirements, and quality assurance measures. [46]

FSSAI has established a specialized program Food Safety Training and Certification (FoSTaC) for Nutraceuticals which aims to provide comprehensive training covering topics such as Good Manufacturing Practices (GMPs), quality control, regulatory compliance, safety, and hygiene practices via the experts in food safety, nutraceuticals, and regulatory affairs. [47]

The Food Safety and Standards Act comprises 21 chapters in total, out of which the fourth article, that is, the 22 of the act talks various aspects of nutraceuticals, dietary supplements, functional foods, and related products, such the manufacturing, sale, distribution, labeling requirements, packaging, licensing and registration of food businesses, etc. Information such as the chemical composition of the engineered food, particle size, solubility, digestibility, surface chemistry, amount of nanomaterial, along with specific claims must be included for 'Novel food'. The Food Safety and Standards have undergone various amendments, with the latest update in 2021.

The amendment included regulations covering various forms of administration, including tablets, capsules, and syrups, for products containing combinations of vitamins and minerals. This includes products with a single vitamin and mineral as well, as long as they are formulated at levels not exceeding one or downward implying the regulations are specifying the permissible dosage levels for these products, ensuring that they meet certain safety and efficacy standards. It also restricts nutraceutical consumption in infants up to 24 months by separating the regulation as "Food for infant nutrition 2020'. he Act mandates prior approval from the Food Authority for specific health claims. Additionally, except for botanical extracts, any single purified chemical entity must have prior approval.

TABLE 17.3
Ingredients Specified in the Form of Schedules by the Food Authority of India 2026

Schedule	Ingredients
Schedule I	Vitamins and minerals
Schedule II	Essential amino acids and other nutrients
Schedule IV	List of plants and botanical ingredients
Schedule VI	List of ingredients as nutraceuticals
Schedule VII	List of strains as probiotics
Schedule VIII	List of prebiotic compounds

"Food for Special Dietary Use" (FSDU) refers to food products that are formulated to meet the dietary needs of specific groups of people with particular nutritional requirements or medical conditions such as infants and young children, individuals with food allergies, patients with medical conditions, and athletes or sportspeople. The Food Safety and Standards Act mandates the use of FSDU for sportspersons only when they are medically advised by a physician, certified dietician, or nutritionist, is meant for oral consumption, and is not meant for replacing all daily meals intended for slimming, weight management, and weight control. The Act also requires food businesses to ensure that nutrient quantities do not exceed the Recommended Dietary Allowances (RDA) or below. The businesses shall also notify the Food Authority in writing and submit additional data on safety on the use of vitamin-based esters, derivatives, and salts, alongside mineral salts and chelates **(Table 17.3)**.

Amendments also permitted the new forms of vitamins and minerals to be included in Schedule I of the regulation, such as Vitamin D (cholecalciferol), sources from lichen/algae, vitamin E (tocotrienols), selenious acid (source of selenium), copper oxide (copper (II) oxide, cupric oxide, and black copper oxide (sources of copper), natural sources of calcium sourced from shells, pearls, conch, corals, oysters, milk, etc. Amendments in Schedule IV included the addition of botanicals, the removal of botanicals such as *Areca catechu* L., *Carissa spinarum* L., and gave a revised range of usage for adults per day. [47–48]

The monitoring the safety of dietary supplements or functional foods by the collection, analysis, and evaluation of data related to adverse events or side effects of the products is known as nutrivigilance. On the other hand, phytovigilance deals with the safety of herbal products or phytomedicines to ensure the safe use of herbal products and identify any potential risks or concerns associated with their consumption. The various steps of obtaining registration of nutraceuticals in India are illustrated in **Figure 17.3.** [49]

17.9 REQUIREMENTS FOR LABELING AND ADVERTISEMENT OF SPECIFIC CLAIMS OF NUTRACEUTICALS AND ITS VARIANTS IN INDIA

The labeling or presentation of any nutraceutical product must not claim to have properties of preventing, mitigating, or treating a human disease but can contain statements regarding the form or function, or the overall well-being of the body, which is supported by scientific evidence. The label must contain the word *nutraceutical*, along with the quantity of each nutraceutical component in the product having nutritional value. It must also contain an advisory warning ("N for medicinal use), other possible risks from excessive consumption, and known side effects, contraindications, and interactions. It must also comply with the Legal Metrology Packaged Commodities Rules 2011, which lays out guidelines for labeling for packaged products. [50]

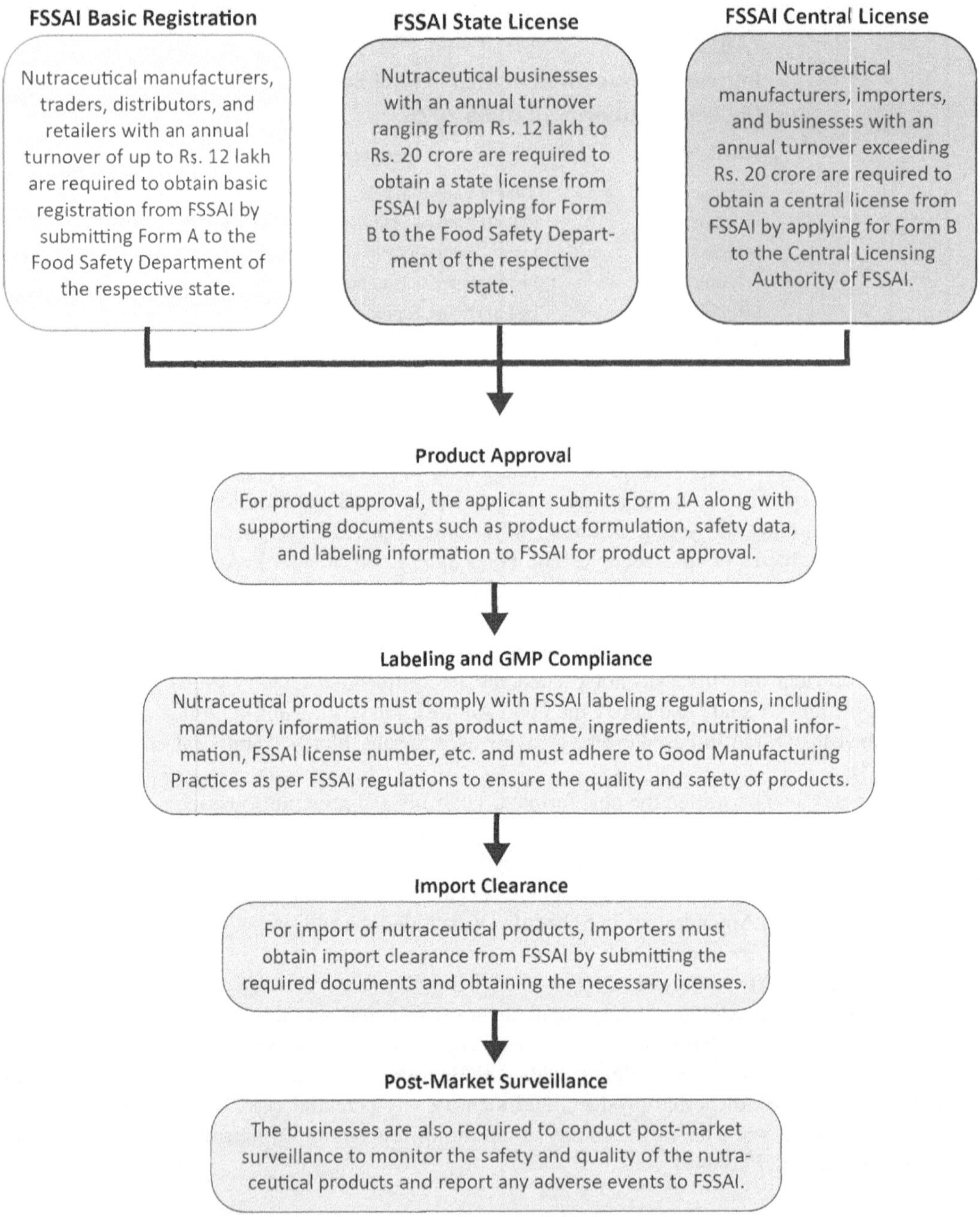

FIGURE 17.3 Process of registration of nutraceuticals in India.

The claims for advertisement of nutraceuticals must adhere to the guidelines. The claims must not be ambiguous, misleading, or false. They must convey truthful information to the consumers. The claims must not encourage excess consumption of any packaged food and in no way shall project the food product as a single means of nutrients or a source to replace meals under Food Safety and Standards Act 2006. Any package product claiming certain nutritional benefit or prevention of chronic health condition must have scientific basis to it by validation and characterization of the product ingredients. It is also mandatory to specify the number of servings of the food per day pertaining to a claim in any nutraceutical product in its label. Use of adjectives such as "natural," "fresh," "pure," "original," " 'traditional," "'authentic," "genuine", "real", etc., appearing in the label or advertisement of the food with potential to mislead consumers must contain a statement stating "This is only a brand name or trade mark and does not represent its true nature". No advertisements shall undermine the importance of healthy lifestyles and any declaration must be conspicuous and legible. [51]

17.10 CONCLUSION

The examination of global trends in the nutraceutical industry reveals a dynamic and rapidly evolving landscape shaped by diverse factors, including consumer preferences, technological advancements, and regulatory frameworks. The shift toward plant-based meat and milk products, driven by health, environmental, and ethical considerations, underscores a significant transformation in the food industry. These products not only offer nutritional benefits but also align with the increasing consumer demand for environmentally sustainable and ethically produced food options. Technological innovations have played a pivotal role in enhancing the quality and appeal of plant-based alternatives, making them more accessible and acceptable to a broader audience. Firms such as Beyond Meat and Impossible Foods have set new benchmarks in the industry by developing products that closely mimic the sensory attributes of traditional meat, thus appealing to both vegetarians and flexitarians alike.

The rising popularity of clean label nutrition reflects a broader trend towards transparency and simplicity in food production. Consumers are increasingly seeking products with recognizable, natural ingredients free from artificial additives and preservatives. This demand for clean label products is not just a fleeting trend but a fundamental shift in consumer behavior, emphasizing the importance of health, wellness, and sustainability. Furthermore, the nutraceutical market is experiencing significant growth, driven by the increasing health consciousness among consumers and the rising prevalence of lifestyle-related diseases. This growth is accompanied by a surge in product innovation, with companies developing new formats and delivery systems to enhance convenience and efficacy. Regulatory bodies are also adapting to ensure the safety and quality of these products, thereby fostering consumer trust and confidence.

REFERENCES

[1] Kalra EK. Nutraceutical-definition and introduction. *AAPS PharmSci.* 2003 Sep 10;5(3):27–8.

[2] Zheng S, Edney SM, Goh CH, Tai BC, Mair JL, Castro O, et al. Effectiveness of holistic mobile health interventions on diet, and physical, and mental health outcomes: a systematic review and meta-analysis. *eClinicalMedicine.* 2023 Dec 1;66:102309.

[3] Nutraceuticals market in India. 2018.

[4] Zheng S, Edney SM, Goh CH, Tai BC, Mair JL, Castro O, et al. Effectiveness of holistic mobile health interventions on diet, and physical, and mental health outcomes: a systematic review and meta-analysis. *eClinicalMedicine.* 2023 Dec 1;66:102309.

[5] Dinan TG, Cryan JF. Gut instincts: microbiota as a key regulator of brain development, ageing and neurodegeneration. *J Physiol.* 2017 Jan 15;595(2):489–503.

[6] Mayer EA, Tillisch K. The brain-gut axis in abdominal pain syndromes. *Annu Rev Med.* 2011 Feb 18;62(1):381–96.

[7] Flach J, van der Waal MB, Kardinaal AFM, Schloesser J, Ruijschop RMAJ, Claassen E. Probiotic research priorities for the healthy adult population: A review on the health benefits of Lactobacillus rhamnosus GG and Bifidobacterium animalis subspecies lactis BB–12. *Cogent Food Agric* [Internet]. 2018 [cited 2024 Jun 6];4(1). Available from: https://www.healthline.com/nutrition/lactobacillus-rhamnosus

[8] Gwak MG, Chang SY. Gut-brain connection: Microbiome, gut barrier, and environmental sensors. *Immune Netw.* 2021;21(3).

[9] Schell MA, Karmirantzou M, Snel B, Vilanova D, Berger B, Pessi G, et al. The genome sequence of *Bifidobacterium longum* reflects its adaptation to the human gastrointestinal tract. *Proc Natl Acad Sci.* 2002 Oct 29;99(22):14422–7.

[10] O'Callaghan A, van Sinderen D. Bifidobacteria and their role as members of the human gut microbiota. *Front Microbiol.* 2016;7(Jun).

[11] Ait-Belgnaoui A, Payard I, Rolland C, Harkat C, Braniste V, Théodorou V, et al. Bifidobacterium longum and lactobacillus helveticus synergistically suppress stress-related visceral hypersensitivity through hypothalamic-pituitary-adrenal axis modulation. *J Neurogastroenterol Motil.* 2018 Jan 30;24(1):138–46.

[12] Messaoudi M, Lalonde R, Violle N, Javelot H, Desor D, Nejdi A, et al. Assessment of psychotropic-like properties of a probiotic formulation (*Lactobacillus helveticus* R0052 and *Bifidobacterium longum* R0175) in rats and human subjects. *Br J Nutr.* 2011 Mar 14;105(5):755–64.

[13] Inulin (Prebiotic Fiber): Health Benefits and Risks [Internet]. [cited 2024 Jun 6]. Available from: https://www.verywellhealth.com/the-health-benefits-of-inulin-4587258

[14] Fructooligosaccharides: Benefits and Risks [Internet]. [cited 2024 Jun 6]. Available from: https://www.healthline.com/health/fructooligosaccharides#forms-and-where-to-find-them

[15] Martin-Gallausiaux C, Marinelli L, Blottière HM, Larraufie P, Lapaque N. Conference on diet and digestive disease symposium 2: Sensing and signalling of the gut environment: Scfa: Mechanisms and functional importance in the gut. *Proc Nutr Soc.* 2021;80(1):37–49.

[16] Souza AFC., Gabardo S, Coelho R de JS. Galactooligosaccharides: Physiological benefits, production strategies, and industrial application. *J Biotechnol.* 2022 Nov 20;359:116–29.

[17] Wilkins T, Sequoia J. Probiotics for gastrointestinal conditions: A summary of the evidence. *Am Fam Physician.* 2017 Aug 1;96(3):170–8.

[18] Kelly JR, Borre Y, O' Brien C, Patterson E, El Aidy S, Deane J, et al. Transferring the blues: Depression-associated gut microbiota induces neurobehavioural changes in the rat. *J Psychiatr Res.* 2016 Nov;82:109–18.

[19] Bibiloni R, Fedorak RN, Tannock GW, Madsen KL, Gionchetti P, Campieri M, et al. VSL#3 Probiotic-mixture induces remission in patients with active ulcerative colitis. *Am J Gastroenterol.* 2005 Jul;100(7):1539–46.

[20] Mimura T. Once daily high dose probiotic therapy (VSL#3) for maintaining remission in recurrent or refractory pouchitis. *Gut.* 2004 Jan 1;53(1):108–14.

[21] Toscano M, De Grandi R, Grossi E, Drago L. Role of the human breast milk-associated microbiota on the newborns' immune system: A mini review. *Front Microbiol.* 2017 Oct 25;8.

[22] Carr A, Maggini S. Vitamin C and immune function. *Nutrients.* 2017 Nov 3;9(11):1211.

[23] Read SA, Obeid S, Ahlenstiel C, Ahlenstiel G. The role of zinc in antiviral immunity. *Adv Nutr.* 2019 Jul;10(4):696–710.

[24] Markowiak P, Śliżewska K. Effects of probiotics, prebiotics, and Synbiotics on human health. *Nutrients.* 2017 Sep 15;9(9):1021.

[25] Lau KM, Lee KM, Koon CM, Cheung CSF, Lau CP, Ho HM, et al. Immunomodulatory and anti-SARS activities of Houttuynia cordata. *J Ethnopharmacol.* 2008 Jun;118(1):79–85.

[26] Calder PC. Omega-3 polyunsaturated fatty acids and inflammatory processes: nutrition or pharmacology? *Br J Clin Pharmacol.* 2013 Mar 5;75(3):645–62.

[27] Liu P, Chen G, Zhang J. A review of liposomes as a drug delivery system: current status of approved products, regulatory environments, and future perspectives. *Molecules.* 2022 Feb 17;27(4):1372.

[28] Jaiswal M, Dudhe R, Sharma PK. Nanoemulsion: An advanced mode of drug delivery system. 3 *Biotech.* 2015 Apr 8;5(2):123–7.

[29] Gupta T, Nema P, Soni S, Yadav V, Jain S, Soni V, et al. Preparation and in vitro evaluation of BBG-250 loaded liposomal formulation for anticancer potential. *Future J Pharm Sci.* 2024 Feb 12;10(1):1–16 [cited 2024 Jun 6]. Available from: https://fjps.springeropen.com/articles/10.1186/s43094-024-00581-w

[30] Zhou F, Yu T, Du R, Fan G, Liu Y, Liu Z, et al. Clinical course and risk factors for mortality of adult inpatients with COVID-19 in Wuhan, China: a retrospective cohort study. *The Lancet.* 2020 Mar;395(10229):1054–62.

[31] Martineau AR, Jolliffe DA, Hooper RL, Greenberg L, Aloia JF, Bergman P, et al. Vitamin D supplementation to prevent acute respiratory tract infections: systematic review and meta-analysis of individual participant data. *BMJ.* 2017 Feb 15;i6583.

[32] Calder P, Carr A, Gombart A, Eggersdorfer M. Optimal nutritional status for a well-functioning immune system is an important factor to protect against viral infections. *Nutrients.* 2020 Apr 23;12(4):1181.

[33] Jäger R, Kerksick CM, Campbell BI, Cribb PJ, Wells SD, Skwiat TM, et al. International society of sports nutrition position stand: protein and exercise. *J Int Soc Sports Nutr.* 2017 Jan 3;14(1).

[34] Kreider RB, Kalman DS, Antonio J, Ziegenfuss TN, Wildman R, Collins R, et al. International Society of Sports Nutrition position stand: safety and efficacy of creatine supplementation in exercise, sport, and medicine. *J Int Soc Sports Nutr.* 2017 Jan 3;14(1).

[35] Ashtary-Larky D, Bagheri R, Ghanavati M, Asbaghi O, Wong A, Stout JR, et al. Effects of beta-alanine supplementation on body composition: a GRADE-assessed systematic review and meta-analysis. *J Int Soc Sports Nutr.* 2022 Dec 31;19(1):196–218.
[36] Gleeson M. Immune function in sport and exercise. *J Appl Physiol.* 2007 Aug;103(2):693–9.
[37] Grgic J, Grgic I, Pickering C, Schoenfeld BJ, Bishop DJ, Pedisic Z. Wake up and smell the coffee: caffeine supplementation and exercise performance—an umbrella review of 21 published meta-analyses. *Br J Sports Med.* 2020 Jun;54(11):681–8.
[38] Shirreffs SM, Sawka MN. Fluid and electrolyte needs for training, competition, and recovery. J Sports Sci. 2011 Jan 9;29(Sup 1):S39–46.
[39] Kreider RB, Kalman DS, Antonio J, Ziegenfuss TN, Wildman R, Collins R, et al. International Society of Sports Nutrition position stand: safety and efficacy of creatine supplementation in exercise, sport, and medicine. *J Int Soc Sports Nutr.* 2017 Jan 3;14(1).
[40] Santini A. Nutraceuticals and functional foods: is it possible and sustainable for bridging health and food? *Foods.* 2022 May 30;11(11):1608.
[41] Hawley JA, Hargreaves M, Joyner MJ, Zierath JR. Integrative biology of exercise. *Cell.* 2014 Nov;159(4):738–49.
[42] Cucinotta FA. Review of NASA Approach to space radiation risk assessments for mars exploration. *Health Phys.* 2015 Feb;108(2):131–42.
[43] Tang H, Rising HH, Majji M, Brown RD. Long-term space nutrition: A scoping review. *Nutrients.* 2021 Dec 31;14(1):194.
[44] Grimm D, Grosse J, Wehland M, Mann V, Reseland JE, Sundaresan A, et al. The impact of microgravity on bone in humans. *Bone.* 2016 Jun;87:44–56.
[45] STATUTE-104-Pg2353.
[46] FSSAI guidance and notification on nutraceuticals-An insight.
[47] FSSAI [Internet]. [cited 2024 Jun 6]. Available from: https://www.fssai.gov.in/cms/food-safety-and-standards-regulations.php
[48] Food Safety Training and Certification - Home-[Internet]. [cited 2024 Jun 6]. Available from: https://fostac.fssai.gov.in/index
[49] Malve H, Bhalerao P. Past, Present, and likely future of nutraceuticals in India: Evolving role of pharmaceutical physicians. *J Pharm Bioallied Sci.* 2023;15(2):68–74.
[50] Frequently Asked Questions (FAQs) on the Food Safety and Standards (Health Supplements, Nutraceuticals, Food for Special Dietary Use, Food for Special Medical Purpose, Functional Food and Novel Food) Regulations [Internet]. Available from: www.fssai.gov.in.
[51] Food Safety and Standards (Advertising and Claims) Regulations, 2018.

18 Nutraceuticals in Czech Republic

Ales Franc and Vladimír Bíba

18.1 INTRODUCTION

The term *nutraceuticals* comes from the combination of the words *nutrition* and *pharmaceutical.* In the Czech Republic, this term usually refers to a "dietary supplement" (DS), which is presented as a food or its extract with a usually high content of vitamins, minerals, herbal and animal ingredients, trace elements, and other substances that promise a positive effect on human health and exceed their nutritional value. DSs are intended for direct consumption in small, measured quantities. It differs from "enriched food" (EF), which is food supplemented with specific nutritional components that are not contained in them under normal conditions or are found in them in a lower concentration [1]. The goal is to improve the nutritional profile of foods and help people achieve recommended nutrient intakes. A special group then consists of "foods for special medicinal purposes" (FFSMP), which are processed in a special way or have a special composition and are intended for the controlled dietary nutrition of patients, including infants. They are used under medical supervision as the sole or partial nutrition of patients with limited, reduced, or impaired ability to receive, digest, absorb, metabolize, or excrete usual foods or certain nutrients or their metabolites [2]. DS, EF, and FFSMP are not intended to treat or prevent disease.

The application forms of DS generally coincide with dosage forms known in the field of pharmaceuticals. Their history in the Czech Republic follows on from years of development of the food industry and healthy nutrition, which resulted in a separate category of nutraceuticals, including their legislation, registration, and production processes. This development has led to the emergence of a separate market for DS, in which both traditional food and pharmaceutical companies participate, with some emerging that exclusively produce nutraceuticals. This naturally led to the emergence of specific approaches, including registration and commercialization strategies, including regulated advertising and the creation of overviews of approved indications and health claims to reduce the possible risks associated with their use [3].

18.2 DEFINITION OF NUTRACEUTICALS IN CZECH REPUBLIC

DS within the Czech Republic forms a separate category of food. It is food whose purpose is to supplement the normal diet and which is a concentrated source of vitamins and minerals or other substances with a nutritional or physiological effect, contained in the food separately or in combination, intended for direct consumption in small, measured quantities. Narcotic, psychotropic, steroid, anabolic or hormonal substances, as well as substances that have been proven to have a toxic, genotoxic, teratogenic, hallucinogenic, narcotic, or other adverse effect on the human organism, cannot be added. It must also not contain ingredients of a new type that were not used for human consumption on the territory of the European communities to a significant extent before May 15, 1997 [4]. Tables 18.1 and 18.2 list plants and substances that must not be part of DS.

DSs are used prepared in the form of capsules, lozenges, tablets, pills, and in other similar forms, as well as liquid in ampoules, dropper bottles, and other similar forms of liquid or loose products intended for intake in small, measured quantities. DS is put into circulation only packaged [5].

DOI: 10.1201/9781003465041-18

TABLE 18.1
Plants and Their Parts Prohibited in Dietary Supplements

Aconitum spp., *Adonanthe* spp. (*Adonis* spp.), *Alkanna* spp., *Alstonia* spp., *Ammi visnaga* (L.) LAM., *Andira* spp., *Anchusa* spp., *Apocynum* spp., *Areca catechu* L., *Aristolochiaceae*, *Artemisia cina*, *Asarum* spp., *Aspidosperma quebracho-blanco* Schlecht., *Atropa* spp., *Azadirachta indica*, *Baccharis coridifolia*, *Baccharis megapotamica*, *Borago officinalis*, *Bragantia wallichii*, *Brachyglottis* spp., *Caltha palustris*, *Cassia acutifolia* (*Cassia senna*), *Cassia angustifolia* VAHL., *Catharanthus roseus* (L). G. DON., *Cephaelis* spp., *Cineraria* spp., *Citrullus colocynthis*, *Claviceps* spp., *Clematis* spp., *Cocculus* spp. (*Anamirta cocculus*), *Colchicum* spp., *Conium maculatum*, *Convallaria majalis L.*, *Corydalis* spp., *Croton tiglium*, *Cynoglossum* spp., *Datura* spp., *Digitalis* spp., *Diploclisia affinis.*, *Diploclisia chinensis*, *Dryopteris filix-mas*, *Duboisia* spp., *Ecballium elaterium* A. RICH., *Ephedra* spp., *Erechthites* spp., *Erythroxylon sp.*, *Eschscholtzia califomica* CHAM., *Eupatorium* spp. (except for *Eupatorium perfoliatum*), *Euphorbia* spp., *Exogonium purga*, *Frangula alnus* MILL., *Gelsemium sempervirens* (L.) AIT., *Genista tinctoria*, *Gossypium* spp., *Heliotropium* spp., *Helleborus* spp., *Hydrastis canadensis* L., *Hyoscyamus* spp., *Chelidonium majus* L., *Chenopodium ambrosioides*, *Ipomoea* spp., *Ledum palustre*, *Lithospermum officinale*, *Lobelia* spp., *Mallotus philippinensis*, *Mandragora officinarum* L., *Menispermum canadense*, *Menispermum dahuricum*, *Mucuna pruriens*, *Nerium oleander* L., *Nicotiana* spp., *Papaver somniferum* L., *Pausinystalia yohimbe* (K. SCHUM.) PIERRE., *Petasites* spp., *Physostigma venenosum* BALF., *Pilocarpus* spp., *Pinellia temata*, *Piper methysticum* FORSTER, *Podophyllum peltatum* L., *Polygonum sachalinensis*, *Polygonum cuspidatum*, *Pulsatilla vulgaris*, *Rauwolfia serpentina* (L.) BENTH., *Rhamnus cathartica* L., *Rhamnus purshiana* DC., *Rheum officinale*, *Rheum palmatum*, *Rhododendron ferrugineum* L., *Ricinus communis* L., *Ruta graveolens*, *Sarothamnus scoparius* (L.) WIMM. ex KOCH., *Sassafras albidum*, *Scopolia camiolica* JACQ., *Senecio* spp., *Schoenocaulon officinale* (SCHLECHT. et CHAM.) GRAY., *Sida cordifolia*, *Sinomenium acutum*, *Spartium junceum* L., *Spigelia* spp., *Stephania* spp., *Strophanthus* spp., *Strychnos* spp., *Symphytum* spp., *Tabebuia impetiginosa*, syn. *Handroanthus impetiginosus* (Mart. ex DC.) Matto, *Taxus* spp., *Thevetia peruviana* (PERS.) SCHUMANN., *Tussilago farfara*, *Urginea maritima* (L.) J. B. BAK., *Veratrum* spp., *Vladimiria souliei*, *Xysmalobium undulatum* R. BROWN.

TABLE 18.2
Active Ingredients Prohibited in Food Supplements

Aconitum spp., *Adonanthe* spp. (*Adonis* spp.), *Alkanna* spp., *Alstonia* spp., *Ammi visnaga* (L.) LAM., *Andira* spp., *Anchusa* spp., *Apocynum* spp., *Areca catechu* L., *Aristolochiaceae*, *Artemisia cina*, *Asarum* spp., *Aspidosperma quebracho-blanco* Schlecht., *Atropa* spp., *Azadirachta indica*, *Baccharis coridifolia*, *Baccharis megapotamica*, *Borago officinalis*, *Bragantia wallichii*, *Brachyglottis* spp., *Caltha palustris*, *Cassia acutifolia* (*Cassia senna*), *Cassia angustifolia* VAHL, *Catharanthus roseus* (L). G. DON., *Cephaelis* spp., *Cineraria* spp., *Citrullus colocynthis*, *Claviceps* spp., *Clematis* spp., *Cocculus* spp. (*Anamirta cocculus*), *Colchicum* spp., *Conium maculatum*, *Convallaria majalis* L., *Corydalis* spp., *Croton tiglium*, *Cynoglossum* spp., *Datura* spp., *Digitalis* spp., *Diploclisia affinis.*, *Diploclisia chinensis*, *Dryopteris filix-mas*, *Duboisia* spp., *Ecballium elaterium* A. RICH., *Ephedra* spp., *Erechthites* spp., *Erythroxylon* sp., *Eschscholtzia califomica* CHAM., *Eupatorium* spp., (except for *Eupatorium perfoliatum*), *Euphorbia* spp., *Exogonium purga*, *Frangula alnus* MILL., *Gelsemium sempervirens* (L.) AIT., *Genista tinctoria*, *Gossypium* spp., *Heliotropium* spp., *Helleborus* spp., *Hydrastis canadensis* L., *Hyoscyamus* spp., *Chelidonium majus* L., *Chenopodium ambrosioides*, *Ipomoea* spp., *Ledum palustre*, *Lithospermum officinale*, *Lobelia* spp., *Mallotus philippinensis*, *Mandragora officinarum* L., *Menispermum canadense*, *Menispermum dahuricum*, *Mucuna pruriens*, *Nerium oleander* L., *Nicotiana* spp., *Papaver somniferum* L., *Pausinystalia yohimbe* (K. SCHUM.) PIERRE., *Petasites* spp., *Physostigma venenosum* BALF., *Pilocarpus* spp., *Pinellia temata*, *Piper methysticum* FORSTER, *Podophyllum peltatum* L., *Polygonum sachalinensis*, *Polygonum cuspidatum*, *Pulsatilla vulgaris*, *Rauwolfia serpentina* (L.) BENTH., *Rhamnus cathartica* L., *Rhamnus purshiana* DC., *Rheum officinale*, *Rheum palmatum*, *Rhododendron ferrugineum* L., *Ricinus communis* L., *Ruta graveolens*, *Sarothamnus scoparius* (L.) WIMM. ex KOCH., *Sassafras albidum*, *Scopolia camiolica* JACQ., *Senecio* spp., *Schoenocaulon officinale* (SCHLECHT. et CHAM.) GRAY., *Sida cordifolia*, *Sinomenium acutum*, *Spartium junceum* L., *Spigelia* spp., *Stephania* spp., *Strophanthus* spp., *Strychnos* spp., *Symphytum* spp., *Tabebuia impetiginosa*, syn. *Handroanthus impetiginosus* (Mart. ex DC.) Matto, *Taxus* spp., *Thevetia peruviana* (PERS.) SCHUMANN., *Tussilago farfara*, *Urginea maritima* (L.) J. B. BAK., *Veratrum* spp., *Vladimiria souliei*, *Xysmalobium undulatum* R. BROWN., *Claviceps* spp. (stickwort—sclerotium), acetyltryptophan, dehydroepiandrosterone (DHEA), 5-hydroxytryptophan (5-HTP), colloidal silver, lactulose, melatonin, N-acetylcysteine

DSs are not intended to prevent or treat disease, nor may they refer to these properties. Nutritional and health claims can only be made under certain conditions. In practice, this means that the entity that manufactures, imports, or puts the product on the market is considered a food business operator, with all the ensuing consequences [6].

The boundary between DS, EF, FFSP, or even medicines, is very narrow, and the categorization of a specific preparation must be carried out on a case-by-case basis. The issue of borderline products in relation to medicinal products is dealt with by the State Institute for Drug Control, and nutritional consultations are provided by the State Health Institute [7].

18.3 HISTORY OF NUTRACEUTICALS IN CZECH REPUBLIC

The history of nutraceuticals in Czech Republic, although relatively short, is closely related to the development of dietetics and medicine. The first mentions of dietetics, which is the basis for nutraceuticals, can be found in literature from 1522, when a Karlovy Vary spa doctor, Václav Payer, devoted himself to treatment procedures in connection with healthy nutrition. Famous authors Magdaléna Dobromila Rettigová (1785–1845) and Marie Janků-Sandtnerová (1885–1946) also devoted space to diet, health, and fasting dishes in their cookbooks. In the 20th century, the development of medicine and natural sciences took place, which made it possible to lay the foundations of the modern concept of nutrition and dietetics. In the interwar period, there was an increased interest in healthy nutrition, and the first vitamin preparations appeared [8]. In 1927, the longtime spa chef, Jan Kettner, with the cooperation of Dr. Max Popper, published the book *Hygienic and Dietary Cuisine* [9]. The first Czech manual of dietetics, by the associate professor of internal medicine Otakar Kose, was published in 1928 [10]. The Prague school for dietary nurses, founded in 1934 on the initiative of the Women's Industrial Association, with the support of Dr. Bohumil Prusík, also has a relatively old tradition. This school was the first in Central Europe and one of the first in Europe at all. After the Second World War, in 1945, the SN (Society for Nutrition) was established, which from 1951 organized regular meetings of diet workers and, in 1952–1954, created the *Czechoslovak Diet System*, which was published in 1955 in Prague [11]. Since 1951, the IRHN (Institute for Research in Human Nutrition) has also been operating in Prague, whose activities were integrated into the newly established ICEM (Institute of Clinical and Experimental Medicine) in 1971 [12]. In 1960, due to the resolution of the College of the Minister of Health, dietetics became an integral part of complex treatment [13]. After 1989, thanks to the fall of communism in Eastern and Central Europe, there was a growing awareness of the importance of a healthy lifestyle and disease prevention. There were also the rapid development of the market for nutraceuticals and the creation of many Czech companies focused on the production and sale of DS. In 1997, the Food and Tobacco Products Act was adopted, which defined *nutraceuticals* as foods intended to supplement the normal diet with nutritional or other physiologically active substances. The law also set out the requirements for composition, labeling, and advertising [14]. In 2000, CASF (Czech Association for Special Foods) was registered, which brought together manufacturers and distributors of DS and actively participated in shaping the legislative framework in this area. CASF represents 24 companies and is an active member of international organizations, such as IADSA (International Alliance of Dietary/Food Supplement Associations) and FSE (Foods Supplements Europe) [15]. The list of members can be seen in Table 18.3. CANT (Czech Association of Nutritional Therapists) was also established in 2018, the aim of which is to improve nutritional care in Czech Republic [16]. Also, in 2018, a decree on dietary supplements and food composition was issued, which further regulates the market for nutraceuticals. The emphasis is on the quality and safety of products [17].

TABLE 18.3
List and Characteristics of Member Companies Associated with CASF

Alfa Medical Team s.r.o.

- Residence: Priemyselna, Modranka, Trnava, Slovak Republic
- Year of foundation: 2015
- Focus: sales promotion of pharmaceutical products

AMWAY, s.r.o.

- Residence: Jankovcova 1596/14a, Holesovice, Praha, Czech Republic
- Year of foundation: 1959
- Focus: dietary supplements and targeted dietary supplements

BENU Ceska republika, a.s.

- Residence: K Perovne 945/7, Praha, Czech Republic
- Year of foundation: 1993
- Focus: securing a full range of pharmaceutical products for the BENU chain of pharmacies.

Biovit Impex CO. Czech Republic. s.r.o.

- Residence: Michelska 1425/84, Praha, Czech Republic
- Year of foundation: 1999
- Focus: dietary supplements

Dr. Max Pharma s.r.o.

- Residence: Na Florenci 2116/15, Praha, Czech Republic
- Year of foundation: 2006
- Focus: securing a full range of pharmaceutical products for the chain of pharmacies Dr. Max

EGYM Bohemia s.r.o.

- Residence: Na Pankraci 1062/58, Praha, Czech Republic
- Year of foundation: 1995
- Focus: natural products containing active substances of plant origin

FAVEA Plus a.s

- Residence: Safrankova 1238/1, Praha, Czech Republic
- Year of foundation: 2005
- Focus: dietary supplements produced by innovative technologies

K2 Pharm s.r.o.

- Residence: Ratiborska 1651/177a, Opava - Katerinky, Czech Republic
- Year of foundation: 2013
- Focus: innovations and news with dietary supplements and cosmetics

KUK Bohemia spol. s r.o.

- Residence: Vinohradska 89, Praha, Czech Republic
- Year of foundation: 1994

Focus: special food and pharmaceutical additives

NOVENTIS s.r.o.

- Residence: Filmova 174, Zlin, Czech Republic
- Year of foundation: 1993
- Focus: medicines and dietary supplements medicines

NUTREND D.S., a.s.

- Residence: Chvalkovice 604, Olomouc, Czech Republic
- Year of foundation: 1993
- Focus: sports nutrition, dietary supplements, functional foods

Pragomed s.r.o.

- Residence: Na Hrebenkach 1755/18, Praha, Czech Republic.
- Year of foundation: 2008
- Focus: original and innovative food supplements

(*Continued*)

TABLE 18.3 *(Continued)*
List and Characteristics of Member Companies Associated with CASF

Rosenpharma a.s.
- Residence: Veverske Kninice 265, Czech Republic
- Year of foundation: 2003
- Focus: medicines, food supplements, cosmetics

Simply You Pharmaceuticals a.s.
- Residence: Rohacova 188/37, Praha, Czech Republic
- Year of foundation: 1997
- Focus: medicines and food supplements

Vegall Pharma, s.r.o.
- Residence: Zelezna 63, Vrbno pod Pradedem, Czech Republic
- Year of foundation: 2005
- Focus: natural dietary supplements, herbal products

Walmark a.s.
- Residence: Oldrichovice 44, Trinec, Czech Republic
- Year of foundation: 1990
- Focus: medicines, dietary supplements

Woykoff a.s.
- Residence: Hauptova 594, Praha, Czech Republic
- Year of foundation: 2006
- Focus: dietary supplements, veterinary medicine

Zentiva a.s.
- Residence: U Kabelovny 529/16, Praha, Czech Republic
- Year of foundation: 2003
- Focus: medicines, dietary supplements

18.4 LEGISLATION

DSs are regulated in the EU by a comprehensive set of legal regulations that ensure their safety and quality. This section summarizes the key documents that define the rules for the marketing and labeling of DS. These regulations are also reflected in the national legislation in Czech Republic.

18.4.1 European Legislation

Since the Czech Republic has been a member of the EU since 2004, its legislation, which is listed here chronologically, also applies to it. The basic document is Directive 2002/46/EC of the EP and of the Council, which regulates DS [18]. It introduces their definition and categorizes the nutrients that can be added to them, including the requirements for their labeling. It also establishes rules for their marketing. This directive is amended by several EU regulations, among others by EC Regulation No. 1170/2009 [19], which declares a list of vitamins, minerals, and their forms that can be added to food, including DS, and declares their permitted amounts. EC Regulation No. 178/2002 sets out the general principles and requirements of food law in the EU. It also establishes the EFSA (European Food Safety Authority) and establishes procedures for safety assessment [20]. EC Regulation No. 1924/2006 defines the rules for the use and labeling of food, defines the terms "nutritional claim" and "health claim," including their list and conditions for their use [21]. EU Regulation No. 1169/2011 defines the rules for providing food information to consumers. This information includes the label of the food, list of ingredients, nutritional information, information on allergens, origin of the food, instructions for use, and expiration date [22]. EU Commission Regulation No. 432/2012 establishes a list of approved health claims for food labeling, which does not include reducing the risk of illness and health of children. The list is continuously updated by a lot of EU regulations [23]. EU Regulation No. 2015/2283 lays down the rules for placing new foods on the market that were not used in the EU before May 15, 1997, and for placing on the market, an EC permit is therefore required [24].

18.4.2 National Legislation

Act No. 110/1997 Coll. is a legal regulation on food and tobacco products and on the amendment and addition of some related laws. It establishes the obligations of entrepreneurs in the production of food and tobacco products and their introduction into circulation and regulates state supervision of compliance with the obligations arising from this law. The law applies to all stages of food production, processing, labeling, and distribution [14]. Decree No. 58/2018 Coll. is about dietary supplements and food composition. This decree reflects the regulations of the EU and, following them, regulates the requirements for the composition of DS and food, their labeling, and the method of use. The decree exhaustively determines which vitamins and minerals, and their forms, can be used to produce DS [15]. Decree No. 417/2016 Coll., on certain methods of food labeling, incorporates the relevant EU regulation and regulates some methods of food labeling and the method of batch labeling. Together, these pieces of legislation form a comprehensive framework for DS control and ensure that these products are safe for consumers and properly labeled. Conditions of use of some substances in dietary supplements are shown in Tables 18.4 and 18.5.

TABLE 18.4
Conditions for Using Certain Plants in Dietary Supplements

Name of Plant	Part of Plant Used	Highest Permissible Amount in Daily Dose
Cimicifuga racemosa	Dried root and rhizome	40 mg
Citrus aurantium	Extract (calculated as synephrine)	20 mg synefrine
Crataegus spp.	Dried leaf, flower, fruit	1.000 mg
Dioscorea sp.	Dried tuber	100 mg
Echinacea spp.	Dried root and herb	500 mg
Garcinia cambogia	Peel (calculated as hydroxycitric acid)	2.000 mg
Ginkgo biloba	Standardized leaf extract	60 mg
Ginkgo biloba	Dried leaves	4.000 mg
Hypericum perforatum	Dried herb	900 mg
Panax ginseng	Dried root	2.000 mg
Ptychopetalum olacoides	Dried wood	500 mg
Rhodiola rosea	Standardized root extract (4% rosavin)	100 mg
Schisandra chinensis	Dried berries	600 mg
Tribulus terrestris	Dried herb and fruit	2.000 mg
Turnéra diffusa	Dried leaves	1.000 mg
Uncaria tomentosa	Dried root	1.000 mg
Valeriana officinalis	Dried root	1.000 mg

TABLE 18.5
Conditions of Using Some Substances in Dietary Supplements

Active Ingredient	Highest Permissible Amount in Daily Dose
Acetylkarnitin	500 mg
Monakolin K from *Monascus purpureus*	10 mg
DMAE (dimethylaminoethanol)	20 mg
Coenzym Q10 (ubichinon a ubichinol)	200 mg
Ortic acid	50 mg
Pycnogenol	100 mg
Taurin	2.000 mg

18.5 NUTRITIONAL HEALTH AND "ON HOLD" CLAIMS

A *claim* is any communication about a food that is not required by law, including a pictorial, graphic, or symbolic representation, which states, suggests, or implies that a food has particular characteristics. Claims are divided into *nutritional* and *health claims*. A *nutritional claim* focuses on the characteristics of a food, such as nutrient content, energy value, or the absence of certain substances. A *health claim*, on the other hand, emphasizes the connection between food and human health. The use of any food claims in the EU is strictly regulated. Only the claims listed in the appendix of the relevant regulation and meeting the specified conditions can be used. These claims must be true, supported by scientific evidence, and in accordance with applicable EU regulations [21]. Table 18.6 provides the list of active ingredients with authorized health claims, and Table 18.7 provides example of some nutritional, health, and "on hold" claims.

TABLE 18.6
Active Ingredients with Authorized Health Claims

Activated carbon, ALFA-cyclodextrin, arabinoxylan, beta-glucan, betaine, proteins, biotin, chicory inulin, potassium, plant stanol esters, cocoa flavanols, fluoride, folate, free sterol form, phosphorus, fructose, glucomannan, guar gum, magnesium, hydroxypropyl)methylcellulose (HPMC), chitosan, chlorine, chromium, ionic drinks, iodine, creatine, docosahexaenoic acid (DHA), docosahexaenoic acid and eicosapentaenoic acid (DHA/EPA), linoleic acid, folic acid, oleic acid, pantothenic, α-linolenic acid and linoleic acid, lactitol, lactase, lactulose, manganese, meat or fish, copper, melatonin, molybdenum, *Monascus purpureus*, monounsaturated and/or polyunsaturated fatty acids, non-fermentable carbohydrates, indigestible carbohydrates, niacin, pectins, polyphenols from olive oil, slowly digestible starch, foods with a low or reduced content of saturated fatty acids, foods with a low or reduced content of sodium, resistant starch, plant sterols and plant stanols, carbohydrates, selenium, prunes of cultivars "domestic plums" (Prunus domestica L.), calcium, water-soluble tomato concentrate (WSTC I and II), vitamin A, vitamin B1, vitamin B12, vitamin B2, vitamin B6, vitamin C, vitamin D, vitamin E, vitamin K, sugar beet fiber, fiber from wheat bran, fiber from barley grains, fiber from oat grains, walnuts, zinc, iron, rye fiber, live yogurt cultures, sugar-free chewing gum.

TABLE 18.7
Examples of Approved Nutritional, Health, and "On Hold" Claims

Nutritional Claims	Health Claims	On Hold
Fat-free	Activated charcoal contributes to the reduction of excessive flatulence after meals	Antioxidant
Low calorie	Beta-glucans contribute to maintaining normal blood cholesterol levels	Appetite
Low fat	Betaine contributes to normal homocysteine metabolism	Blood sugar level, sugar metabolism
Low saturated fat	Consuming alpha-cyclodextrin as part of a meal containing starch contributes to limiting the rise in blood glucose levels after this meal	Blood sugar level, sugar metabolism
Low sodium/salt	Consuming alpha-cyclodextrin as part of a meal containing starch contributes to limiting the rise in blood glucose levels after this meal	Bone health, teeth mineralization
Low sugar	Consuming arabinoxylan as part of a meal contributes to limiting the rise in blood glucose levels after this meal	Digestive health
No added sodium/salt	Consuming beta-glucans from oats or barley as part of a meal contributes to limiting the rise in blood glucose levels after this meal	Digestive health, weight management and loss
No added sugar	Proteins contribute to the growth of muscle mass	Gallbladder health
No calories	Magnesium is involved in the process of cell division	Hair and nail health
Reduced calorie	Taking (hydroxypropyl)methylcellulose with a meal helps limit the rise in blood glucose levels after that meal	Invigoration of the body
Saturated fat–free		Mental well-being, stress
Sodium-free/salt-free		Metabolism of proteins, carbohydrates, and lipids
Sugar-free		Natural defenses, resistance
Very low sodium/salt		Respiratory health

18.6 REGISTRATION

The operator of a food business must fulfill the notification obligation before placing a DS or OP on the market for the first time, by logging into his account via a link on the website of the Ministry of Agriculture, where he fills in the composition and other properties of the product and, if necessary, inserts a label. *Label* means any mark, data, pictorial, or other description that is written, printed, marked, or affixed to a package or container. It does so through the Food Notification System (FNS), which is an electronic system for fulfilling the information obligation for DS and EF and is filled out via the Ministry of Agriculture portal, which is available from November 11, 2021. The operator of the food business provides the Czech text, including all mandatory data, which is listed on the product packaging. The effects declared by the manufacturer are not verified. Even so, it is allowed to state the effects on packaging and in promotional materials, if they do not conflict with the applicable legal regulations for food labeling and DS. However, it should be emphasized that the declaration of effects does not guarantee their truth [25]. Just as there is pharmacovigilance for drugs, there is also nutravigilance. If the user experiences problems of a non-infectious nature after the application of nutraceuticals, this fact can be reported through the online form on the website of the State Health Institute [26].

18.7 LABELING

In addition to the labeling requirements for packaged foods, packaging for consumers shall state "dietary supplement" and the name with numerical data on the quantity of vitamins, minerals, or other substances characterizing the product based on the recommended daily dose. The recommended daily dosage is also indicated with a warning about exceeding it and the necessity to store it out of reach of children. There is also a warning that dietary supplements are not a substitute for a varied diet. Here, the risks and restrictions associated with the intake of certain substances, possibly in defined concentrations, should be mentioned: Already, 800 μg of vitamin A and more in a daily dose is unsuitable for pregnant women; *Ginkgo biloba* may reduce blood clotting; the product containing *Cimicifuga racemosa* must be discontinued, and a doctor consulted, in case of any suspicion of liver disease; Monacolin K is unsuitable for children, young people, pregnant and lactating women, people taking hypolipidemic drugs; and people with kidney, liver, and muscle disorders [15]. In addition, the label must not attribute or refer to properties related to the prevention, treatment, or cure of human disease to the DS. There must be no statement that states or implies that a balanced and varied diet cannot provide sufficient nutrients. Nutritional and health claims about food supplements can be made under the terms of the directly applicable regulation of the EU on the requirements for making nutritional and health claims in food labeling [22]. Substances whose placing on the market is subject to a directly applicable regulation of the EU on novel foods may only be present in food supplements according to this regulation [25]. Table 18.8 gives a list of words that can already be considered medical claims and which should be avoided.

TABLE 18.8
List of Words That Can Be Considered as Health Claims

abscess, acne, acute, allergy, alopecia, Alzheimer's disease, anemia, anorexia, apathy, arrhythmia, arthritis, atherosclerosis, atherosclerosis, atopic, autism, bacillus, bacteria, benign, bleeding, blood clot, bronchitis, bulimia, burn, burning, cancerous, chickenpox, cirrhosis, clotting, common cold, cough, Crohn's disease, cyst, damage, deafness, diarrhea, diarrheal, disease, edema, eczema, encephalitis, endometriosis, expectoration, fainting, fever, fistula, foam, frigidity, fungal, fungus, gout, goiter, healing, hemorrhoids, hernia, hives, hot flashes, hypertension, hypotension, impotence, incontinence, infarction, inflammation, laxative, leg ulcers, leukemia, lips, loss of appetite, Lyme, lymphadenopathy, malignant, malnutrition, mastitis, medication, medicine, migraine, mononucleosis, motion sickness, multiple sclerosis, muscular dystrophy, myocardium, myopathy, myositis, nausea, necrosis, neurasthenia, neurological, neuropathic, nervous breakdown, neurasthenia, nursing home care, nutrition, obesity, obese, osteoarthritis, osteoporosis, otitis, oxygen, pain, painful, Parkinson's disease, pathogenic, periodontal disease, pneumonia, poisoning, post-traumatic stress disorder, pregnancy, premature birth, prolapse, prostate cancer, psoriasis, purulent, radiation, rash, relapse, renal failure, respiratory failure, retardation, rheumatic fever, rheumatism, rhinitis, salmonellosis, scarlet fever, schizophrenia, scratch, secretion, sepsis, shortness of breath, sick, stroke, stye, swelling, swollen, symptom, symptoms, syndrome, tachycardia, temperature, thrush, tonsillitis, tumor, varicose veins, vomiting, wart, weight loss, wheelchair.

18.8 ADVERTISEMENT

Advertising means an announcement, demonstration, or other presentation, disseminated mainly through communication media, aimed at promoting business activity, and the consumption or sale of goods. It must be true and based on verified information and must not exaggerate the properties of the product. It must be stated on the product that it is a "dietary supplement" [27]. The use of health claims on food packaging, their presentation, and their advertising is regulated [22]. Furthermore, advertising must not be deceptive and mislead the consumer regarding the properties of the product, its effects, its price, etc. It must be ethical and must not abuse the consumer's trust and be discriminatory or offensive. Even though DS cannot declare or refer to preventive and curative properties, in practice, manufacturers and sellers often make claims that are, to say the least, misleading. They give the impression that the product has preventive or curative properties, even if they do not explicitly state this. Most often, these are claims about the medicinal properties of the individual components of the product in connection with the demonstration of its effects in clinical studies. These claims should be treated with caution. The manufacturer often refers to studies performed with the active substance outside the living organism or to studies performed on patients, but these studies are difficult to evaluate objectively because they do not apply the criteria for clinical evaluation of medicinal products. If a DS contains the same active ingredients as a registered medicinal product, the manufacturer often refers to clinical studies conducted with the medicinal product in question, thereby unfairly "promoting" the use of the DS for its intended use [28]. Advertising ethics is supervised by the Advertising Council. Compliance with the Act on Food and Tobacco Products is overseen by the State Agricultural and Food Inspection [14].

18.9 CONCLUSION

Currently, the DS market in Czech Republic is developing dynamically and responding to the growing consumer demand for products with a positive effect on health. Emphasis is placed on research and innovation, which ensures high quality and product safety. DSs, or nutraceuticals, are becoming an important part of the contemporary healthy lifestyle [29].

REFERENCES

[1] Kudlová E. (2022) Živiny potenciálně nedostatkové v rostlinné stravě a jejich možné rostlinné zdroje. *Vyziva a Potraviny* 2:25–28.
[2] Karásková E. (2017). Umělá mléčná kojenecká výživa–současná doporučení. *Pediatr praxi* 18:186–189.
[3] Mach I. (2012) Doplňky stravy. Grada Publishing 35–36.
[4] Djaoudene O, Romano A, Bradai YD, Zebiri F, Ouchene A, Yousfi Y, Madani K (2023) A global overview of dietary supplements: regulation, market trends, usage during the COVID-19 pandemic, and health effects. *Nutrients* 15:3320.
[5] Chaloupkova P, Petrty M, Verner V, Kokoska L. (2020). Dietary supplements versus functional foods: consumers' attitudes to their consumption. Br Food J 122:3853–3868.
[6] Verhagen H, Vos E, Francl S, Heinonen M, van Loveren H. (2010) Status of nutrition and health claims in Europe. *Arch Biochem Biophys* 501:6–15.
[7] Baudischova L, Straznicka J, Pokladnikova J, Jahodar L. (2018). The quality of information on the internet relating to top-selling dietary supplements in the Czech republic. *Int J Clinl Pharm* :40183–189.
[8] Krobot M, Hawk V, Spáčilová V, Kapounová, Z (2021). A historical perspective and opportunities for dietetics education, practice and professional organizations in the Czech Republic. *Profese Online* 14:9–8.
[9] Kettner J. (1927) Hygienická a dietní kuchyně; Praktický rádce pro živnosti hostinské a domácnosti. A. Svěcený. Praha 1–116.
[10] Kose O. (1928) Příručka dietetiky. *Prakt lek* 17:1–63.
[11] Doberský P (1955) Nový dietní systém pro nemocnice. Státní zdravotnické nakladatelství, *Edice zdravotnické aktuality* 1–314.
[12] Soubor prací Ústavu pro výzkum a výživu lidu v Praze 1956–1957 (1958) *Acta Inst Aliment Hum Pragae* 2:1–279.
[13] Společnost pro výživu, od historie k současným aktivitám (2018). Společnost pro výživu—https://www.vyzivaspol.cz/wp-content/uploads/2018/06/Aktivity-SPV.pdf
[14] Zákon o potravinách a tabákových výrobcích a o změně a doplnění některých souvisejících zákonů č. 110/1997 Sb.
[15] Weerasena B, Ferlito C, Bellavite P. (2021) Dietary supplements in the Malaysian context: Policy recommendations for enhancing freedom of choice and healthcare savings. *CME Policy Paper* 1:1–50.
[16] Vágnerová T, Klímová E (2019) Standard nutriční péče v geriatrii: Obecné principy nutriční péče. Česká asociace nutričních terapeutů, 1–11.
[17] Vyhláška o doplňcích stravy a složení potravin č. 58/2018 Sb.
[18] Directive (EC) No. 2002/46 of the European Parliament and of the Council of 10 June 2002 on the approximation of the laws of the Member States relating to food supplements (Text with EEA relevance).
[19] Commission Regulation (EC) No. 1170/2009 amending Directive 2002/46/EC of the European Parliament and of Council and Regulation (EC) No. 1925/2006 of the European Parliament and of the Council as regards the lists of vitamin and minerals and their forms that can be added to foods, including food supplements.

[20] Regulation (EC) No. 178/2002 of the European Parliament and of the Council laying down the general principles and requirements of food law, establishing the European Food Safety Authority and laying down procedures in matters of food safety.
[21] Regulation (EC) No. 1924/2006 of the European Parliament and of the Council of 20 December 2006 on nutrition and health claims made on foods.
[22] Regulation (EU) No. 1169/2011 of the European Parliament and of the Council of 25 October 2011 on the provision of food information to consumers, amending Regulations (EC) No 1924/2006 and (EC) No. 1925/2006 of the European Parliament and of the Council, and repealing Commission Directive 87/250/EEC, Council Directive 90/496/EEC, Commission Directive 1999/10/EC, Directive 2000/13/EC of the European Parliament and of the Council, Commission Directives 2002/67/EC and 2008/5/EC and Commission Regulation (EC) No 608/2004 Text with EEA relevance.
[23] Commission Regulation (EU) No. 432/2012 of 16 May 2012 establishing a list of permitted health claims made on foods, other than those referring to the reduction of disease risk and to children's development and health Text with EEA relevance.
[24] Regulation (EU) No. 2015/2283 of the European Parliament and of the Council of 25 November 2015 on novel foods, amending Regulation (EU) No. 1169/2011 of the European Parliament and of the Council and repealing Regulation (EC) No. 258/97 of the European Parliament and of the Council and Commission Regulation (EC) No. 1852/2001 (Text with EEA relevance).
[25] Food Notification System. Ministry of Agriculture, 11 November 2021—https://eagri.cz/public/portal/mze/potraviny/doplnky-stravy.
[26] Nutravigilance. Centrum zdravím výživy a potravin—https://nutrivigilance.szu.cz/.
[27] Act amending Act No. 40/1995 Coll., on the regulation of advertising and amending Act No. 468/1991 Coll., on the operation of radio and television broadcasting, as amended, and Act No. 132/2003 Coll., amending Act No. 40/1995 Coll., on the regulation of advertising and amending Act No. 468/1991 Coll., on the operation of radio and television broadcasting, as amended.
[28] Muela-Molina C, Salvador PO, Ana GA. (2021). False and misleading health-related claims in food supplements on Spanish radio: an analysis from a European regulatory framework. *Public Healt Nutr* 24:5156–5165.
[29] Veselá K, Navrátilová M, Beranová M. (2018) Evaluation of consumer behaviour on the food suplements market. Digital Mirrors – part I. Marketing identity. Trnava 559–570.

Index

Note: Page numbers in *italics* indicate a figure and page numbers in **bold** indicate a table on the corresponding page.

A

advertisement regulation
  Australia, 235–236
  CARICOM, 51
  Czech Republic, 339
  India, 5, 326–328
  Indonesia, 225
  South Africa, 172
  US, 92, 92–93
Africa, *see also* South Africa
  nutraceutical market, 278
  product classification and health authorities, 277
Africa and Latin America (LATAM), nutraceutical market, 278
allergen labeling
  EU 150, 157
  South Korea 258
Argentina, regulatory approval procedures, 105–107
  National Administration of Medicines, Food, and Medical Technology (ANMAT), 105–106
ASEAN Alliance of Health Supplement Associations (AAHSA), 32, **33**
Asia-Pacific (APAC) nutraceuticals market, 278
Australia
  Australian Regulatory Guidelines for Complementary Medicines, 18, 19
  complementary medicines, 232–233
    consumer confidence, 240–241
    consumer purchase choice, 239
    conventional medicine, integration with, 238–239
    designated active ingredients, 233
    growth of manufacturing and exports, 238
    increased demand, 237–238
    increased export potential, 238
    manufacturing costs and supply chain issues, 241
    Modern Manufacturing Strategy (MMS), 238
    product export challenges, 240
    regulation of practitioner, 240
    regulatory changes, 238
    regulatory compliance, 239
    workforce, 241
  complementary medicines regulation
    advertising, 235–236
    assessed listed medicines, 234
    Australian Register of Therapeutic Goods (ARTG), 233
    good manufacturing practices, 236
    labels, 235
    listed medicines, 233–234
    presentation, 235
    registered medicines, 234
    scheduling of registered, 234
  Food Standards Australia New Zealand (FSANZ), 55
  Health Product Regulation Group (HPRG), 17, 22
  post-market regulatory activities of complementary medicines, 232–233
    adverse events reporting, 236
    compliance reviews, 237
    inspection of manufacturers, 236
    product recalls, 237
    sampling of medicines, 236–237
  regulatory frameworks, 275
  Therapeutic Goods Administration, 17–18
  warning letters on drug–nutraceutical interactions, 61
Australian Register of Therapeutic Goods (ARTG), 233, 234
Australian Regulatory Guidelines for Complementary Medicines (ARGCM), 18, 19, 232
Australia New Zealand Food Regulation Ministerial Council (ANZFRMC), 21
Australia New Zealand Therapeutic Products Authority (ANZTPA), 18
Australian Therapeutic Goods Administration (TGA), 17–18, 22, 29, 29–30, 61, 232, 235, 236, 275
  post-market regulatory activities, 236–237
  Therapeutic Goods Act of 1989, 232
  Therapeutic Goods Advertising Code Council (TGACC), 235
  Therapeutic Goods Regulations of 1990, 232
Australian Self-Medication Industry (ASMI), 235
Ayurveda, 309

B

Bahrain
  GCC member states, 282–283
  nutraceutical market, 287
  regulatory frameworks, 275
Bahrain Standards and Metrology Directorate (BSMD), 282–283
biologically active supplements (BASs), Russia, 32–34
Botanical-Be company, 68
Brazil
  (ANVISA), 37, 97, 99, 102–103, 275
  approval requirements, 102
  food supplements in, 37
  labeling regulations, 103
  regulation of nutraceuticals, 102–103
Brazilian Health Regulatory Agency (ANVISA),), 37, 97, 99, 102–103, 275
Bureau of Indian Standards (BIS), 2, 5

C

Canada
  Health Canada, 13–15, **30–31**, 30–32
  Natural Health Product Act, 2004, 13, **13**, 14
  novel food and regulations in, 14–15
  regulatory frameworks, 274

warning letters on drug–nutraceutical interactions, 59–60
Caribbean Public Health Agency (CARPHA), 43
CARICOM (Caribbean Community), 43–44, *44*
annual growth rate, 44–45
drug interaction, 45–46
medicines registration, 50
medicines regulation, 50
National Regulatory Authority (NRA), 50
pharmaceutical registration in, **48–49**
quality assurance, 46
registration requirements
future directions and recommendations for, 51–52
harmonization challenges in, 50–51
traditional herbal medicine system, 46, **47**
China, 18–20, 192
Blue hat certification, 19, 202–203
central regulatory authorities, 194–195
China Food and Drug Administration (CFDA), 18–20, *19*, 20, 194, 196, 199–200, 202–203, 210, 274
dietary supplements, 203, **204**
Food Hygiene Law in, 1995, 198
Food Safety Law of China, 19, 194, 198, 199, 203
Food Sanitation Law, 197
functional health foods, 199–203
health food category, 18–20, *19*
legislative measures, 194–195
nutraceuticals in, 195–197
Regulations on Health Food Registration, 195
regulatory frameworks, 195–198, 274
State Administration for Market Regulation (SAMR), 18, 194–195, 198, 199
State Food and Drug Administration (SFDA), *see* China Food and Drug Administration (CFDA)
TCM, *see* traditional Chinese medicine (TCM)
China's National Medical Products Administration (NMPA), 32
clean label nutrition
benefits of, 308
challenges, 308
consumer awareness and education, 307–308
demand for, 307
future prospective, 308–309
health and wellness trends, 308
regulatory and certification bodies, 308
transparency and trust, 308
clean nutrition and mental wellness, 310–311
essential nutrients, 310
practical strategies, 310–311
Codex Alimentarius, US, 142–144, **143**
Codex Committee on Nutrition and Foods for Special Dietary Uses (CCNFSDU), 142
Colombia
dietary supplement, 40
(INVIMA) in, 40, 97, 104–105
labeling regulations, 105
legal documentation, 104–105
regulation of nutraceuticals, 104–105
complementary medicines in South Africa, 163–165
advertising, 172–173
classification, 163–164
discipline-specific traditional claims, 165
import/export, requirements to, 165, 168–171
labeling, 173, *174*, *175*
purpose of regulation of, 173
regulatory authorities, *see* regulatory authorities, in South Africa
Complementary Medicines Australia (CMA), 17, 232, 235
compound annual growth rate of nutraceutical (CAGR), 44–45
convenient nutraceuticals, 311
effervescent powders, 311
gummies, 311
marketing and branding, 312
Millennial influence, 312
product innovation, 312
protein bars, 311
ready-to-eat products, 311–312
regulatory considerations, 312
COVID-19 pandemic, 45, 215, 278, 306, 307, 309, 320
current Good Manufacturing Practices (cGMP), 9, 84, 85
product recall procedure, 65–67
violations, 64–65
customs clearance safety control measures, Korea, 262–263
import declaration and inspection, 262–263
safety control, 263
Czech Association for Special Foods (CASF), 333
Czechoslovak Diet System, 333
Czech Republic, 331, 333
active ingredients prohibited, **332**
advertisement, 339
conditions of use of substances, **336**
dietary supplements, 331, 338, 339
European legislation, 335–336
Directive 2002/46/EC, 335
EC Regulation No. 178/2002, 335
EC Regulation No. 432/2012, 336
EC Regulation No. 1170/2009, 335
EC Regulation No. 1924/2006, 335
EU Regulation No. 2015/2283, 336
Food and Tobacco Products Act, 333
foods for special medicinal purposes (FFSMP), 331
history, 333
labeling, 338–339
list of members with CASF, **334–335**
national legislation, 336–337
Act No. 110/1997 Coll., 336
Decree No. 58/2018 Coll., 336
Decree No. 417/2016 Coll., 337
nutritional health, 337, **337–338**
on hold claims, 337, **337**
plants and parts prohibited, **332**
registration, 338

D

dietary reference values (DRV), 12
dietary supplements, *see* nutraceuticals
drug–nutrient interactions, 45–46

E

Eagle Energy, 62
echinacea, 83
Edible Oils Packaging (Regulation) Order, India, 1998, 3

effervescent powders, 311
Emirates Authority for Standardization and Metrology (ESMA), 275
enriched foods, 110, 331
Essential Commodities Act of 1955, India, 2, 3, 4
established nutraceuticals, 184
European Food Safety Agency (EFSA), 25–27, **26**, 55, 116, 141, 149
  warning letters on drug–nutraceutical interactions, 59
European Medicines Agency (EMA), 12, 117
European Parliament and Council implemented Regulation (EC) No. 178/2002, 55
European Rapid Alert System, 127, 129, 130, 131
European Union (EU)
  Directive 2001/82/EC, 55
  future trends and developments, 159
  Food Labelling Regulations of 1996, 117–118
  Food Notification System (FNS), 338
  Good Agricultural Practice (GAP), 151, 194, 211
  Good Manufacturing Practices (GMPs), 9, 14, 85, 112, 158
    clearance, 236
    for complementary medicines, 236
    definition of, 268
    for health functional foods, 268
    key aspects, 236
    objectives of, 268
    regulatory framework, 268
  Germany, *see* Germany
  health claims, 156
    criteria for, 155–156
    examples for, 156
  historical perspective (Directive 2002/46/EC), 10–12
  labeling and packaging requirements, 156–157
    allergen information, 157
    language and presentation, 157–158
    nutrient content and functionality claims, 157
  market trends of, 120–121, *121*
  novel food supplements legislation in, 12–13
    Regulation (EU) 2017/2470, 12
    Regulation (EC) No. 258/1997, 12
    Regulation (EC) No. 2283/2015, 12
  pre-market approval process, 153
    application procedure, 154
    data requirements, 155
    safety and efficacy assessments, 154
  Rapid Alert System for Food and Feed (RASFF), 10, 58–59
  regulations and restrictions, 151
    functional ingredients, 153
    herbal extracts, 152–153
    vitamins and minerals, 152
  regulatory compliance, 158–159
    consumer safety, 147
    industry integrity, 147
    market access and global trading, 147–148
    research objectives, 148
  regulatory framework, 148–149
    contaminants, 151
    European Food Safety Authority, 149
    health claims regulation, 149–150
    labeling requirements, 150–151
    novel foods, 149
  scientific substantiation process, 156

F

facility registration, FDA, 87–88
Family Dollar Inc., 71
FD&C Act, US, 7, 62, 63
Federal Office of Consumer Protection and Food Safety (BVL), Germany
  Codex Alimentarius, 144
  federal registration, 125–130
  food categories/product registrations and health claims, 136–142
  genetically modified food/feed, authorization procedures, 128–130, *129*
  guidance documents, 130
  monitoring and compliance mechanisms, 130
  organizational structure, 128–130, *129*
  procedural guidance, 130
  tablets and smartphones, 142
Federal Trade Commission (FTC), USA
  advertisements, dietary supplement, **92**, 92–93
folic acid, 83
food additives
  China, 19–20
  EU, 11
  halal, 300, **303**
  Korea, 250, 261
    designation or revision, 251
    types and functions, 250, **250–251**
  US, 83
    Food Additives Amendment of 1958, 7
Food and Drug Administration (FDA), *see* United States Food and Drug Administration (USFDA)
Food and Drug Supervisory Agency of Indonesia, 225
food business operators (FBOs)
  in EU, 10–11, 148
  India's licensing system for 185–186
food equipment labeling, 258
Food for Special Dietary Use (FSDU), India, 5, 55, 326
Food Hygiene Law, 1995, China, 198
food labeling regulations, Korea
  functions, 258
  requirements, 257
  types of, 258
Food Notification System (FNS) in EU, 338
Food Product Order (FPO), India, 2
food recall plan in India, 187
Food Safety and Standards Act (FSSA), India, 2, 5, 178, 179–180, 185, 189, 325–326, **326**
Food Safety and Standards Authority of India (FSSAI), 2–6, **3–4**, 55, 178–180, 185, 274, 325–327
  amendment (2016), 5–6
  Edible Oils Packaging (Regulation) Order, 1998, 3
  Food Product Order, 1955, 2
  Food Safety Training and Certification (FoSTaC), 325
  food recall plan, 187
  Food Safety and Standards Act, 2006, 5
  labeling requirements, 186–187, **187**
  license number, **188**
  licensing system
    basic registration, 186
    central license, 186
    required documents for, 186
    state license, 186
    validity of, 186

logo, **188**
Meat Food Products Order, 1973, 2–3
Milk and Milk Products Order, 1992, 3–4
Prevention of Food Adulteration Act, 1954, 2
registration process of nutraceuticals, 185
Solvent Extracted Oil, De-oiled Meal and Edible Flour (Control) Order, 1967, **4**, 4–5
terminology, nutraceuticals, 27–28
toxic contaminants, 75
Vegetable Oil Products (Control) Order, 1947, 3
Food Safety Law of China, 19, 194, 198, 199, 203
Food Safety Modernization Act, 2011, US, 9–10
food supplement claims, Korea, 249–250
disease risk reduction claims, 250
general health claims, 250
nutrient claims, 250
food supplement market, Germany, 124–125
fortified foods, 110, 184, **315**
fructooligosaccharides (FOS), 319
Functional Food Science in Europe (FUFOSE), 117
functional health claim (FHC), 16
functional health foods, 196

G

galactooligosaccharides (GOS), 319
genetically modified foods (GMFs), 300
regulatory standards in Germany, 131–136, *135*
genetically modified organisms (GMOs)
halal, 302
Germany, 131
administrative offenses, 136
criminal offenses, 136
labeling requirements, 133
packaging requirements, 132–133
safety assessments, 134
scope of regulation, 132
traceability, 133–134
without genetic engineering, 134–136, *135*
South Korea, 258
Germany
Codex Alimentarius, 142–144, **143**
consumer health market in, 123–124
e-commerce growth, 123–124
Federal Ministry of Food and Agriculture (BMEL), 125–126, 131
Federal Institute for Risk Assessment (BfR), 126–127, *127*
federal registration, 125–130
food categories/product registrations and health claims in, 136–142
Food Information Implementation Ordinance (LMIDV), 132
future opportunities in, 144
genetically modified foods, regulatory standards, 131–136, *135*
guidance documents, 130
market diversity, 124–125
market domination, 122–123
monitoring and compliance mechanisms, 130
nutraceutical market in, 121, *122*
procedural guidance, 130
tablets and smartphones, user-friendly apps, 142
Ginkgo biloba, 338
ginseng *(Panax ginseng)*, 82–83, 192
good importers, Korea, *see* importing goods, Korea
Greater Arab Free Trade Area (GAFTA), 280
green tea extract (GTE), 74–75, 83
Gulf Cooperation Council (GCC), 34, **34**, 278
establishment and membership, 278
geographical coverage, 278
integration, 278
member states, 281–282
Bahrain, 282–283
Kuwait, 284–285
Oman, 284
Qatar, 284
Saudi Arabia, 283–284
United Arab Emirates, 282
Yemen, 285
nutraceuticals market, 287–289
objectives, 278–280
organizational structure, 280–281
Ministerial Council, 280
Secretariat General, 280–281
Supreme Council, 280
regulatory requirements, 289–290
Gulf Cooperation Council (GCC) Standardization Organization (GSO), 281–282
core values, 281–282
food supplement guideline (GSO 55:2020)
collagen, 295
enzymes, 295
general requirements, 290, 294–295
ginseng, 295
halal compliance, 290
ingredients, 295
labeling, 297
nutritional facts, 297–298
packaging, 296
permitted ingredients, 290, **291–294**, 295
Royal Jelly, Propolis, and Pollen, 295
scope, 290
transport and storage, 296
variance limits, **299**
warnings and cautions, **299**
international cooperation, 286
mission, 281
objective, 281
president, 285–286
standards development, 286
vision, 281

H

HACCP (hazard analysis and critical control points), 248
halal products, 300
accreditation authorities, 301
certification bodies, 300–301
classification of non-halal animals, **303**
general requirements for, 301
labeling, 302–304
packaging and wrapping materials, 302
regulation in Indonesia, 226–228
certification, 227
gelatin capsules, 228
Law No. 33 of 2014, 227
POM agency, 227

vitamins A, E, and C, 228
retail, handling, and service outlets, 302
scope, 300
storage, display, and transportation, 302
harmonization challenges, 50–51
Health Canada, 30–32, *see also* Canada
interaction of nutraceuticals, 59–60
organizational hierarchy of, **30–31**
health claims, 8, 11, 13, 18, 59, 61, *see also specific countries*
health functional foods (HFFs), Korea, 245, 249, *see also* Good Manufacturing Practices (GMPs)
adverse events management, 269–270
consumption, 249
distribution stag, 249
functional ingredient certification application procedure, 251–254, *253*
application process, 253
content and requirements of submission materials, 253
fees, 254
processing time, 254
required documents, 252, *252*
generic, 247
individually recognized ingredients, 247
manufacturing stage, 248
notified functional ingredients, 247
labeling, 258
procedure for reporting, 255–257, *256*
categories, 256
conditions for denial, 257
notification of changes, **257**
required documentation, 256
specialized distribution, 256–257
product-specific, 247
recognition process, 248–249
regulatory process, 259
safety evaluation, 268–269
safety management, 269
unauthorized claims risks, 269
health supplements, *see* nutraceuticals
herbals, 184
holistic nutrition, 307

I

immune system boosters, 320, *321*
analyses and reviews, 321
delivery system innovations, 320
ingredients and mechanisms, 320
liposomal encapsulation, 320
mechanistic studies, 321
nanoemulsions, 320
randomized controlled trials, 321
sustained-release formulations, 320
importing goods, Korea, 260–264
India
advertisement, 327–328
history of food regulation, 2–6, 178–179, **179**
labeling, 326
novel food regulations in, 6
registration of nutraceuticals in, 327
regulatory frameworks, 274
food authority, *181*
Food for Special Dietary Use (FSDU), 5, 55, 326
Food Safety and Standards Act of 2006, 178, 179–180, **180**
Food Safety Knowledge Assimilation Network (FSKAN), 6, *6*
FSSAI, *see* Food Safety and Standards Authority of India (FSSAI)
Prevention of Food Adulteration Act of 1954, 2, 179
required forms, 180–181, **181**
Indonesia
disease distribution, 215–216
halal products, 226–228
health supplements containing non-halal ingredients, 228–230
health supplements registration, 217, 218–225
malnutrition, 215, 216
metabolic disorders, 215, *216*
nutraceuticals use, 215–216, 217–218
nutritional status, 215–216
obesity, 216
smoking and alcohol consumption, 215–216
traditional medicine and health supplements advertising registration, 218
traditional medicine registration, 218–225
Indonesian National Agency for Drug and Food Control (BPOM), 35–36
Indonesian traditional medicines, 217–218
animals prohibited, 223
assessment graph, *226*
classification, 218
Directorate of Traditional Medicines, Health Supplements, and Cosmetics, 217
functional food, 218
health supplements registration, 217, 218
ingredients permitted for supplements, **220**
ingredients prohibited in health supplements, **223**
jamu, 218
phytomedicines, 218
quality requirements, 218–219
quality testing parameters, 220, **221**
registered herbal medicines (OHT), 218
registration, 218–225
advertising and claim provisions, **225**
BPOM, 225
CPPOB, 221
export registration, requirements, 222–223
labeling criteria, 220, 224, **224**
Marketing Permit Agency, 224
pre-clinical/clinical test documents, 222
pre-registration stage, 219
priority services, 222
safety monitoring, 222
stage, 219
tax documentation, 221
solvent residue limits, **219**
stability tests, 220
Sub-Directorate for the Evaluation of Pre-Clinical/Clinical Tests, 217
Sub-Directorate of Cosmetic Registration, 217
Sub-Directorate of Registration of Products and Advertising, 217
supervision of, 225–226
Innovacion Naturals, 71

inulin (prebiotics), 318–319
interactions of drugs and nutraceuticals, 59–65, **60**
Islamic Sharia, 300

J

Jamaica's nutraceutical industry, 45
Japan
 health food regulations in, **15**, 15–17, *16*, **17**
  foods for specified health uses (FOSHU), 16, 16–17, 22, 27, 275
  food with functional claims (FFC), 27
  food with nutritional function claims (FNFC), 16, 17, 27
 Japan Pharmaceuticals and Medical Devices Agency (PMDA), 27, **27**
 regulatory frameworks, 275
 warning letters on drug–nutraceutical interactions, 61

K

kava-kava, 116
Kuwait, *see also* Gulf Cooperation Council (GCC)
 GCC member states, 279, 281, 284–285
 Public Authority for Food and Nutrition (PAFN), 275
 regulatory frameworks, 275
Kuwait Standards and Industrial Services Affairs (KOWSMD), 284–285

L

labeling, 338
 complementary medicines, 235
  in South Africa, 173, *174*, *175*
 EU, 156–157
  Czech Republic, 338–339
  Germany, 133
 Gulf countries, 297
 South Korea, 257–258
 India, 186–187, **187**, 326
 halal products, 302–304
 Indonesian traditional medicines, 220, 224, **224**
 Latin America, 99
  Brazil, 103
  Colombia, 105
 UK, 117–118
 US, 84–85
Latin America, 96–97
 categorisation of nutraceuticals, 99
 consumer awareness, 96
 health authorities/regulatory agencies, 97–98
 labeling, 99
 regulatory approval procedures, 96–97, 100
  Argentina, 105–107
  Brazil, 102–103
  Colombia, 104–105
  Mexico, 100–102
 scientific evidence, 98–99
liposomal encapsulation, 320

M

Maldives Food and Drug Authority (MFDA), 36
market, *see* nutraceutical market
medicinal foods, UK, 110–111
Meta Herbal, 67
Mexico
 Federal Commission for the Protection against Sanitary Risks (COFEPRIS), 36–37
 regulatory approval procedures, 100–102
Middle East
 health authorities, 276, **277**
 nutraceutical market, 278
 product classification, 276, **277**
 regulatory frameworks, 275, 278
Middle East/North Africa (MENA) nutraceutical market, 278
moxibustion (heat therapy), 205

N

nanoemulsions, 320
natural/traditional nutraceuticals, 184
New Zealand
 Medsafe, 36
 Dietary Supplements Regulations, 1985, 20–21
 Food Standards Australia New Zealand (FSANZ), 55
 New Zealand Food Safety Authority (NZFSA), 20–21
 regulatory frameworks, 275
N-methyltyramine, 76
non-traditional nutraceuticals, 184
novel foods
 Canada, 14–15
 EU, 12–13, 139–142
 India, 6
 Korea, 259–260
Nutraceutical Advisory Group (NAG), 325
nutraceutical market
 Africa and Latin America (LATAM), 278
 Asia-Pacific (APAC), 278
 Middle East/North Africa (MENA), 278
nutraceuticals, historical perspectives, 1, 306, *see also* safety and toxicity
 in Australia, 17–18
 in Canada, **13**, 13–15
 in China, 18–20, *19*
 definition, 180, 195
 in Europe, 10–13
 in India, 2–6, **3**, **4**, *6*,
 intended use and trade name, **315–316**
 in Japan, **15**, 15–17, *16*, **17**
 in New Zealand, 20–21, **21**
 in USA, 7–10
nutraceuticals, future trends, 306, 328
 Ayurveda and immunity, 309
 clean label nutrition, 307–309
 clean nutrition and mental wellness, 310–311
 consumer trends, 314, **315–316**
 environmental impact, 313
 ethical considerations, 313–314
 Food Safety and Standards Authority of India (FSSAI), 325–326
 holistic nutrition, 307
 immune system boosters, 320–321
 labelling, India, 326–328
 nutraceutical market, 311–313
 plant-based meat and milk, 313
 prebiotics-probiotics, 317–320
 regulatory requirements, 323, 324, 325
 space nutrition, 323

sports nutrition, 321–323
technological advancements, 314
nutraceuticals' role in human health, 182
categories of, 182, *183*
drawback, 182–183
health benefits, 182
Nutrition Labeling and Education Act (NLEA) of 1990, 8, 323, 325
nutrivigilance, 58

O

Oman, *see also* Gulf Cooperation Council (GCC)
Directorate General for Standards and Metrology (DGSM), 284
GCC member state, 279, 281, 284
Public Authority for Consumer Protection (PACP), 275
regulatory frameworks, 275
OptiBiotix Health Plc, 120

P

phytochemicals, 184
plant-based meat and milk, 313
consumer trends, 314
environmental benefits of, 313–314
ethical considerations, 313–314
market growth, 314
technological advancements, 314
prebiotic foods, 110
prebiotics, 184
for CNS health, 318, *318*
fructooligosaccharides, 319
galactooligosaccharides, 319
inulin, 318–319
UK and MHRA, 111
probiotics, 110, 120, 184
for CNS health, 317, *318*
Bifidobacterium longum, 317
Bifidobacterium longum R0175, 317
Lactobacillus helveticus R0052, 317
Lactobacillus rhamnosus GG, 317
UK and MHRA, 111
probiotics and prebiotics combination, 319
Symprove, 319
VSL#3, 320
product recall, 237
protein bars, 311

Q

Qatar, *see also* Gulf Cooperation Council (GCC)
GCC member state, 279, 281, 284
nutraceutical market, 287
Qatar General Organization for Standardization (QS), 275, 284
regulatory frameworks, 275
Qigong, 206
quality assurance, 65–66

R

ready-to-eat products, 311–312
recombinant nutraceuticals, 184
recommended dietary allowance (RDA), 77
regulation, 272
accurate labeling and transparency, 273
adverse events, 273
batch-to-batch consistency, 273
consumer confidence and trust, 273
consumer safety assurance, 272–273
dosage and formulation standards, 273
international trade and compliance, 273
product integrity and quality control, 273
substantiating efficacy claims, 273
regulatory frameworks, 274–275
Russia
(ROSPOTREBNADZOR), 32–34
biologically active supplements, 32–34

S

safety and toxicity, 55
ADR reporting systems, 57–59, *58*
current Good Manufacturing Practices (cGMP), 84, 85
product recall procedure, by regulatory agencies, 65–67
violations, 64–65
drug-nutraceutical interactions, 59–65, **60**
future aspects of, 77
historical review of, 55–57, *56*, **57**
National Center for Toxicological Research (NCTR), 76
National Toxicology Program (NTP), 76
recall dietary supplements (case study), 67–71, **69–70**
recall procedure, 65–67
toxic contaminants, 75–77
toxicological aspects
Citrus aurantium L. (bitter orange), 73–74
ephedra (Chinese Ma Huang), 71–73
green tea extract, 74–75
vitamin D, 74
Saudi Arabia, *see also* Gulf Cooperation Council (GCC)
GCC member state, 279, 281, 283–284
regulatory frameworks, 275
Saudi Food and Drug Authority (SFDA), 275
Saudi Standards, Metrology, and Quality Organization (SASO), 283–284
South Africa
complementary medicines import/export, 168–171
National Department of Health requirements, 169
South African Health Products Regulatory Authority requirements, 169–170
South African Pharmacy Council requirements, 170–171
Customs and Excise Act 91 of 1964, 169
Foodstuffs, Cosmetics, and Disinfectants Act 54 of 1972, 169
Good Pharmacy Practice in, 171
regulatory authorities in, 165–168
importer of complementary medicines, requirements for, 165
Medicines Act, 163–165
Minister of Health, 165
National Department of Health license, 166, *166*
South African Health Products Regulatory Authority (SAHPRA), 166, *167*, 169–170
South African Pharmacy Council, 166, *168*
South African Pharmacy Council (SAPC)
certificate, 166, *168*

import/export complementary medicines, requirements to, 170–171
pharmacy grading certificate, 165, *168*
registration, 165, *168*
Southeast Asia, nutraceutical classification and health authorities, 276, **276–277**
South Korea's food regulation
1962 Food Sanitation Act, 61, 244
customs clearance safety control measures, 262–263
food additives, 250, 261
designation or revision, 251
types and functions, 250, **250–251**
food labeling regulations, 257–258
food supplement claims, 249–250
disease risk reduction claims, 250
general health claims, 250
nutrient claims, 250
functional benefits, 246
functional ingredients, 246
functional labeling, 246
functional marking, 246
general food, 245
health food, 245
health functional foods, 245–246, 259
consumption stage, 249
distribution stage, 249
general food, health food *vs.*, 245–246
generic, 247
individually recognized ingredients, 247
manufacturing stage, 248
vs. medicines, 246
notified functional ingredients, 247
product-specific, 247
recognition process, 248–249
regulation and labeling, 246
regulatory process, 259
reporting sales of, 255–257
healthy diets and safe food consumption, **265–266**
imported food safety management, 260–263
Korean Food and Drug Administration (KFDA), 249, 268
licensing and sale, 254–257
manufacturing standards, 246
medicines, 246
novel foods, 259–260
proactive safety control measures, 261–262
reference amounts, 266–268
regulatory body scope, 244–245
safety, 261–262
Food Safety Framework Act of 2008, 244
Food Safety Management System, 260
Ministry of Food and Drug Safety (MFDS), 248, 250, 257, 262, 264
Special Act on Imported Food Safety Control, 260, 261
space nutraceuticals, 323, **324**
sports nutrition, 321–322
future directions, 322–323
mechanistic studies, 322
meta-analyses, 322
nutraceutical ingredients and benefits, 322
randomized controlled trials, 322
systematic reviews, 322
Sri Lanka (NMP), 36
structure/function claims, 8
substances generally recognized as safe (GRAS), 7, 54
Sugar MDs LLC, 68
sustained-release formulations, 320
synbiotics, *see* probiotics and prebiotics combination

T

terminology, nutraceuticals
ASEAN Alliance of Health Supplement Associations, 32, **33**
Australia Therapeutic Goods Administration, **29**, 29–30
Brazilian (ANVISA), 37
Canada (Health Canada), **30–31**, 30–32
China National Medical Products Administration, 32
Colombia (INVIMA), 40
European Food Safety Authority, 25–27, **26**
Food Safety and Standards Authority of India, 27–28
Gulf Cooperation Council, 34, **34**
Indonesian National Agency for Drug and Food Control (BPOM), 35–36
Japan Pharmaceuticals and Medical Devices Agency, 27, **27**
Maldives Food and Drug Authority, 36
Mexico (COFEPRIS), 36–37
New Zealand (Medsafe), 36
Russia (Rospotrebnadzor), 32
Sri Lanka National Medicines Policy, 36
Thailand Food and Drug Administration, 35
United Kingdom Medicines and Healthcare products Regulatory Agency, 24–25, **25**
United States Food and Drug Administration, 24–25, **25**
Thailand Food and Drug Administration (TFDA), 35
toxicity, *see* safety and toxicity
traditional Chinese medicine (TCM), 192–193, 204–205
central regulatory authorities, 194–195
components of, 204
dietary supplements, 203, *203*, **204**
disharmony patterns, 207
five elements hypothesis, 204
ginseng *(Panax ginseng)*, 192
Jun-Chen-Zuo-Shi formulation strategy (君臣佐使), 208
legislative measures, 194–195
licensing requirements, 210–211
National Administration of Traditional Chinese Medicine, 194
nomenclature of, 211–212
origins of, 204–205
product dossier requirements, 200–201
product registration, 210–211
qi (vital life force), 207
regulatory compliance, 193
role of herbal medicine in, 207–210
State Administration of, 210
treatments using, 205–207
yin and yang, 204, *205*
traditional herbal medicine system, in CARICOM, 46, **47**
traditional Indonesian medicines, 217–218
Traditional Medicines in the Islands (TRAMIL), 46–47
trans-Tasman regulatory system, 55
tryptophan, 116

U

United Arab Emirates (UAE)
GCC member state, 279, 281, 282
Ministry of Industry and Advanced Technology (MoIAT), 282
nutraceutical market, 287
regulatory frameworks, 275
United Kingdom (UK)
and Medicines and Healthcare Products Regulatory Agency, 35
categories of, 109–111
food supplements, labeling and packaging of, 117–118
Food Supplements Regulations of 2003, 112
health benefits of, 118
medical claims, 116–117
registering and licensing, 112–115
regulations of, 117
vitamins and minerals, sources of, 116–117
United States (US)
adults dietary supplements usage in, 82
advertisement regulation, **92**, 92–93
Dietary Supplement Health and Education Act, 8–9, 54
dietary supplements manufacturing regulations, 86–88, **87**
echinacea, 83
folic acid, 83
Food Safety Modernization Act, 2011, 9–10
ginseng, 82–83
green tea, 83
historical perspective, 7–9
Dietary Supplement Health and Education Act of 1994, 8–9
Food Additives Amendment, 1958, 7
FD&C Act, 1938, 7
Food, Drug, and Cosmetic Act, 1938, 7
Nutrition Labeling and Educational Act of 1990, 8
Proxmire Amendments, 8
Pure Food and Drug Act of 1906, 7
hygienic practices, 85, **86**
import regulations, 88, **89**
labeling, 84–85
manufacturing, regulations for, 86–88, **87**
nutraceutical guidelines, 90–92, **91**
Nutrition Labeling and Education Act, 1990, 8
office of dietary supplements (ODS), 9, 72
Proxmire Amendments, 1976, 7
Public Health Security and Bioterrorism Preparedness and Response Act of 2002, 88–90, **89–90**
regulatory frameworks, 274
safety, risk, and quality regulations, 85, **86**
trends, 83, 93
United States Food and Drug Administration (USFDA), 24–25, *25*, 54
coordinated outbreak response and evaluation (CORE), 67
facility registration, 87–88
Modernization Act, 8
Proxmire Amendments, 8
sanitation programs, 7
toxicity, *see* safety and toxicity
warning letters on drug–nutraceutical interactions, 59
Eagle Energy (RE: 617712), 62
Harold J. Leighton, 63–65
Mancuso, 62–63
US Center for Food Safety and Applied Nutrition (CFSAN), 24
adverse event reporting system, 58, 71, 73
US, Dietary Supplement Health and Education Act (DSHEA), 8–9, 21–22, 50, 54, 83–84, 93–94, 274
current good manufacturing practices, 9
history, 7
office of dietary supplements, 9
US, Dietary Supplement Ingredient Directory, 8, 90
US, Food, Drug, and Cosmetic Act (FDCA)
of 1938, 7
Federal, 62–63
US, Foreign Supplier Verification Program for Importers of Food for Humans and Animals, 88–89, **89**

W

warning letters on drug–nutraceutical interactions
in Australia, 61
in Canada, 59–60
in European Union, 59
Food and Drug Administration (FDA)
Eagle Energy (RE: 617712), 62
Harold J. Leighton, 63–65
Mancuso, 62–63
implications and consumer awareness, 62
issued reasons, 61–62
in Japan, 61
new dietary ingredient, 76–77
in United States, 59
WEFUN capsules recall, 68

Y

Yellow Card Scheme, 58
Yemen Standardization, Metrology, and Quality Control Organization (YSMO), 285